Studies in Mechanobiology, Tissue Engineering and Biomaterials

Volume 14

Series Editor

Amit Gefen, Ramat Aviv, Israel

For further volumes:
http://www.springer.com/series/8415

Amit Gefen
Editor

Multiscale Computer Modeling in Biomechanics and Biomedical Engineering

Springer

Editor
Amit Gefen
Faculty of Engineering
Department of Biomedical Engineering
Tel Aviv University
Tel Aviv
Israel

ISSN 1868-2006 ISSN 1868-2014 (electronic)
ISBN 978-3-642-36481-5 ISBN 978-3-642-36482-2 (eBook)
DOI 10.1007/978-3-642-36482-2
Springer Heidelberg New York Dordrecht London

Library of Congress Control Number: 2013933586

Printed on acid-free paper

Springer is part of Springer Science+Business Media (www.springer.com)

Preface

Multiscale modeling (MSM) has been a fundamental theoretical modeling approach in several engineering disciplines including for example materials engineering and fluid mechanics for many years, but the capacities of MSM have increased tremendously over the past several years with the exponential growth in computer power. In biomechanics and biomedical engineering, a multiscale model typically represents several hierarchies in the structures of organs and tissues, possibly down to cellular or even molecular scales, and there are links for exchange of information between these different hierarchical scales. For example, in a problem involving tissue loading, e.g., in musculoskeletal biomechanics there will be different model scales to describe how loads are transferred between organs, and then at the levels of tissue structures, and eventually at the level of cellular structures such that it would be possible to connect between continuum-scale mechanical loads and loads at the scale of individual cells. Moreover, with the recent progress that was made in computational and systems biology it is now possible to, e.g., assess the effects of such potential cell-level loads on their biological function, such as viability, synthesis of biomolecules, and events in a cell's life cycle. One particularly interesting problem that is often being addressed by means of MSM is how mechanotransduction eventually shapes tissue function, since it is now possible to connect computationally between phenomena that take place at a cell-scale and those that affect whole-tissue behavior, and incorporate in the modeling not only mechanics but also transport, thermodynamics, and biochemistry.

The frontier of MSM methods and techniques in bioengineering research today is described in this volume through contributions of internationally leading groups in this field, from the UK, France, The Netherlands, Italy, Ireland, New Zealand, and the United States. Biomedical engineers, medical physicists, applied mathematicians, and computer scientists who are interested in the state-of-science and current challenges in MSM should find this book very useful. Likewise, medical researchers in fields such as orthopedics, cardiology and vascular surgery, oncology and cancer research, respiration and pulmonary medicine, infections and wound healing, and others who wish to be updated about the technologies and latest achievements in this exciting arena of research are also potential readers. Faculty and graduate students as well as medical practitioners will be able to use this volume for learning about the latest achievements and great promises that MSM brings in to biomedical research.

Amit Gefen
Editor
Multiscale Computer Modeling in Biomechanics
and Biomedical Engineering
Series Editor
Studies in Mechanobiology, Tissue Engineering
and Biomaterials

Contents

Part I
Connective Tissue

Application of Neural Network and Finite Element Method for Multiscale Prediction of Bone Fatigue Crack Growth in Cancellous Bone

Ridha Hambli and Nour Hattab

Abstract Fatigue damage in bone in the form of microcracks results from the repetitive loading of daily activities. It is well known that the resistance of bone at the organ level to fatigue fracture is a function of its resistance to the initiation and propagation of local microcracks at a mesoscopic scale which can lead to macrocrack growth at the organ level. Multiscale investigation of the relationship between the effect of the fatigue microcrack growth at microscopic scales and the whole bone behaviour is a subject of great interest in the research field of the biomechanics of human bone. Several finite element models (FE) have been developed in recent years in order to provide better insight and description regarding bone fatigue microcrack growth. Despite the progress in this field, there is still a lack of models integrating multiscale approaches to assess the accumulation of apparent fatigue microcracks in relation with trabecular architecture into practical FE simulations. In this chapter, a trabecular bone multiscale model based on FE simulation and neural network (NN) computation is presented to simulate the accumulation of trabecular bone crack density and crack length at a given trabecular bone site during cyclic loading. The FE calculation is performed at macroscopic level and a trained NN incorporated into a FE code is employed as a numerical device to perform the local mesoscopic computation (the behaviour law needed to compute the outputs at mesoscale is substituted by the trained NN). The input data for the NN are some trabecular morphological and material factors, the

R. Hambli (✉)
Prisme Institute – MMH, 8, Rue Léonard de Vinci,
45072 Orleans Cedex 2, France
e-mail: ridha.hambli@univ-orleans.fr

N. Hattab
ISTO, UMR 7327 - CNRS/Université d'Orléans, Campus Géosciences,
1A, rue de la Férollerie, 45071 Orleans Cedex 2, France
e-mail: nour.hattab@univ-orleans.fr

Stud Mechanobiol Tissue Eng Biomater (2013) 14: 3–30
DOI: 10.1007/8415_2012_146

Published Online: 19 September 2012

applied stress and cycle frequency. The output data are the average crack density and length computed at a given trabecular bone site.

1 Introduction

Bone is a hierarchically organized material at different length scales. At the macroscopic scale, it is composed of cortical or compact bone and trabecular or cancellous bone (Fig. 1).

Bone strength strongly depends on the trabecular structure of the bone, which can be assessed by changes in its morphological and mechanical properties over time [15, 23, 36, 49]. In general, modelling the trabecular bone behaviour must address changes in its structure, at multiple levels, allowing for a more accurate description of the bone tissue. This process occurs hierarchically at different spatio-temporal scales and involves interacting phenomena (deformation, damage, adaptation, etc.) [15, 23, 36, 49]. In particular, the adaptation of trabecular bone to cyclic fatigue loads involves a complex physiological response that is targeted to local sites of damage. Fatigue damage in bone results from the repetitive loading of daily activities in the form of microcracks with an average crack length of about 100 μm [46] and diffusely damaged areas. Such alterations are relevant for cancellous bone with high metabolic activity and numerous bone quality changes [16, 30]. Physically, fatigue microcracks at the mesoscale are assessed by crack density (Cr.Dn) and crack length (Cr.Le) [14]. Assessment of the hierarchical effect of the Cr.Dn and Cr.Le accumulation within trabecular bone on the whole bone (organ) quality is of major biological and clinical importance for the investigation of bone diseases, fractures and their treatment. The effect of damage microcracks on the mechanical properties of bone is complex since a crack can affect the mechanical properties of the surrounding matrix, it can act as a local stress riser, and it can further exacerbate the heterogeneous and anisotropic character of bone [40]. Moreover, evidence that fatigue damage decreases bone organ quality, increases fracture susceptibility, and serves as a remodeling stimulus motivates the development of numerical multiscale modeling approaches. Most theoretical and numerical studies of bone damage evolution at the macroscopic level have focused on the development of continuum approaches in which the total damage is a scalar quantity defined either as the normalized number of cycles ($D = f(N/Nf)$) or in terms of the changes in elastic modulus and residual strain during life [4, 22, 23, 32, 43, 47]. However, these approaches ignore the fact that the physical damage in bone at the trabecular level takes the form of Cr.Dn and Cr.Le non-linear accumulation.

Fracture mechanics laws have also been applied to investigate crack growth in bone. Taylor and Lee [46] developed a theoretical model to predict fatigue damage and failure in bone based on simulation of the growth of every crack in a piece of bone material. The limitation of fracture mechanics based approaches is that it is

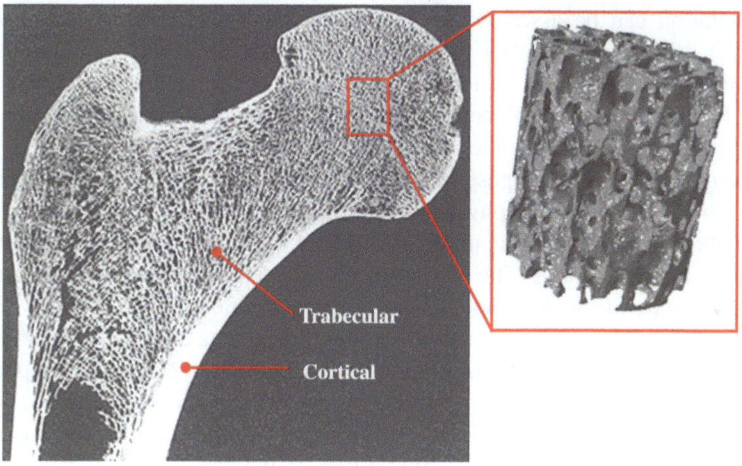

Fig. 1 Cross-section of human femur showing trabecular and cortical bone from http://www.theodora.com/anatomy/the_femur.html

very difficult to predict the behaviour of every crack explicitly in a large population of cracks, such as will occur in a typical piece of bone. In such cases the model must consider the difference in crack locations, sizes, lengths, distribution and loading conditions.

Existing homogenization theories can be applied in multiscale analysis to assess the effective properties of a hierarchical material. However, in the non-linear case it is generally necessary to perform numerical calculation at each iteration for each structural level, due to the fact that at the micro-level, the homogeneous components may change their mechanical behaviour, depending on the stress level (fracturing, yielding, damage, etc.). This approach becomes expensive in order to obtain the effective outputs for a global FE model. A more realistic modelling of physical Cr.Dn and Cr.Le growth within bone must include a multiscale approach to describe microcrack accumulation from the trabecular level (mesoscopic) to the whole femur (macroscopic).

Despite progress in the field of bone fatigue modelling, there is still a lack of models integrating Cr.Dn and Cr.Le accumulation into practical numerical simulation. In the last few years, artificial neural networks (NN) have been used in many engineering applications as a tool for multiscale analysis to couple models on different spatial scales, to identify model parameters or to simulate the material itself. In this paper, a rapid multiscale approach for the simulation of trabecular bone Cr.Dn and Cr.Le accumulation using hybrid FE analysis and a NN (FENN) method is developed. The input data for the NN are the applied apparent stress, the number of cycles, the bone volume fraction (BV/TV), the ash density and the apparent elastic modulus. The output data are the averaged Cr.Dn and Cr.Le at a specific bone site. First, we show that with a given number of numerical experiments on a set of different trabecular bone samples, the Cr.Dn and Cr.Le accumulation can be assessed. Second, a trained NN is able to generalize the acquired data. Third, the NN model can be incorporated into an FE multiscale simulation procedure as a material formulation on the integration point level at every FE iteration—e.g. the behaviour law needed to compute the outputs at the mesoscale is substituted by the trained neural network.

The potential of the proposed FENN method is its ability to incorporate local trabecular information with physical meaning at the continuum whole femur level. This is beneficial for investigating for example the role of damage accumulation on femoral neck fractures. The NN approach is beneficial if the numerical analysis of the complex model is time-consuming or even unfeasible [3, 17, 21, 33, 44]. A further advantage of the method is that the training data can be directly extracted from experimental data. The development of a multiscale procedure coupling FE and NN computation is also motivated by the need in the field of bone biomechanics for an efficient coupling between scales in a multiscale approach for bone analysis. Such detailed numerical simulations on the mesoscale can capture many of the relevant bone features. Performing FE analysis at the entire femur level with its trabecular architecture generates a complete mesh composed of some millions of FE which requires a huge computational time.

2 Hybrid Finite Element and Neural Network Multiscale Concept

The proposed hybrid FENN method is a simulation procedure to link multiscale simulations. From a numerical point of view, the macro scale passes information to the mesoscale in the form of macroscopic variables and boundary conditions obtained at each FE of the mesh after solving the macroscopic analysis using FE simulation. At the mesoscopic scale, the local boundary conditions (derived from the macro FE results) are applied to the mesoscopic bone model and the trained NN allows for the rapid computation of the meso model responses. And finally, the mesoscale passes information to the macroscale in the form of averaged updated outputs (Cr.Dn and Cr.Le) (Fig. 2). Changes in the material distribution of the continuum model will have an effect on the stress/strain field, thus affecting the mechanical state of the bone in the subsequent iteration. At the completion of every iteration, a new FE analysis is performed to update the Cr.Dn and Cr.Le distribution in the continuum model at the macro level. The proposed methodology follows this iterative procedure until convergence is achieved.

From a practical point of view, the following five steps summarize the application of the proposed FENN approach (Fig. 3):

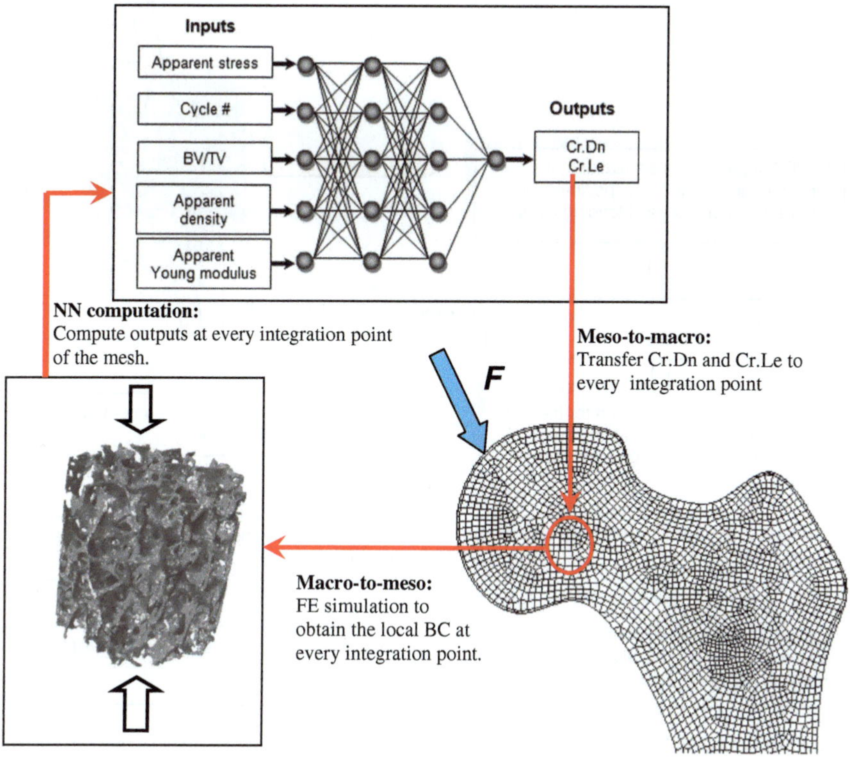

Fig. 2 Multiscale hierarchical FENN approach for bone and Meso-to-macro transition: the NN is incorporated into the FE code Abaqus via the routine UMAT. During the FE calculation at the macro level, the NN is called at every integration point to compute the averaged Cr.Dn and Cr.Le outputs representing the local trabecular bone architecture

(i) Performing suitable numerical experiments to simulate fatigue damage accumulation of different 3D micro-CT trabecular bone specimens taken from different human proximal femurs for different combinations of loading conditions.
(ii) Averaging the sample outputs in terms of Cr.Dn and Cr.Le accumulation.
(iii) Steps (i) and (ii) supply training data in the form of an inputs–outputs patterns table for NN training.
(iv) Training the neural network based on the results of step (iii).
(v) Incorporation of the NN into the macro FE model to link meso-to-macro scales.

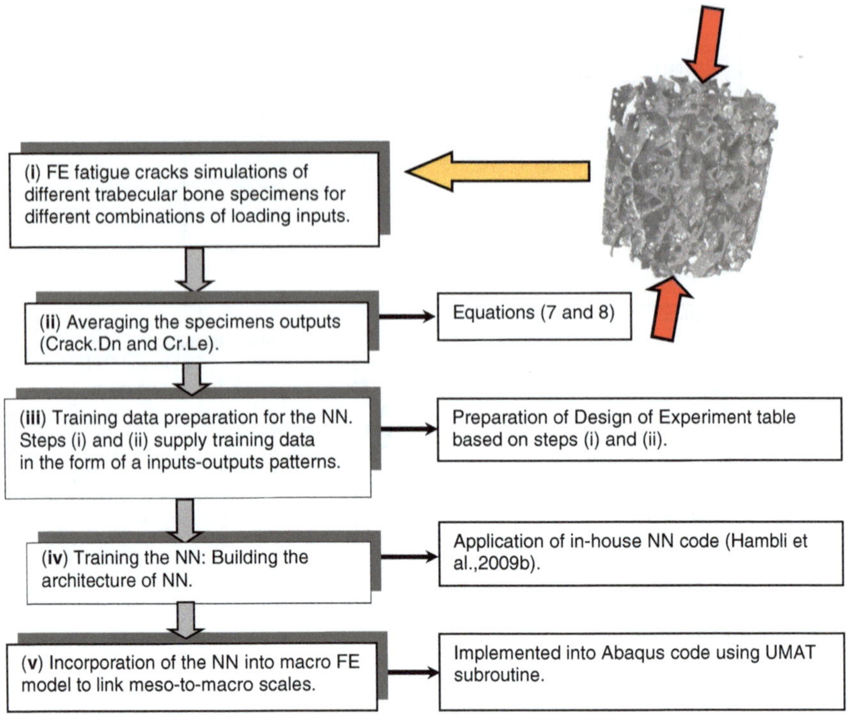

Fig. 3 Building and incorporation of the trained NN in FE code in five steps (i)–(v). The interdependencies of steps (i)–(v) is expressed in terms of cascades of step execution sequences

3 Neural Network for Approximation and Interpolation

NNs have recently been widely used for the analysis of an increasing number of problems in science and technology [17–21, 23, 28, 33, 48, 51, 53]. NNs can be used for the mapping of input to output data without knowing "a priori" the relationship between the data. One of the distinct characteristics of the NN is its ability to learn and generalize from experience and examples and to adapt to changing situations. Once the NN has been satisfactorily trained and tested, it is able to generalize rules and will be able to respond very rapidly (a few seconds) to input data to predict the required output, within the domain covered by the training examples [27, 29, 44]. NN architecture is composed of an input layer, a certain number of hidden layers and an output layer in forward connections (Fig. 4). Each neuron in the input layer represents a single input parameter. These values are directly transmitted to the subsequent neurons of the hidden layers. The neurons of the last layer represent the NN outputs [18, 27, 29].

The output y_i^m of a neuron i in a layer m is calculated as [18, 29, 44]:

$$y_i^m = f\left(v_i^m\right) \tag{1}$$

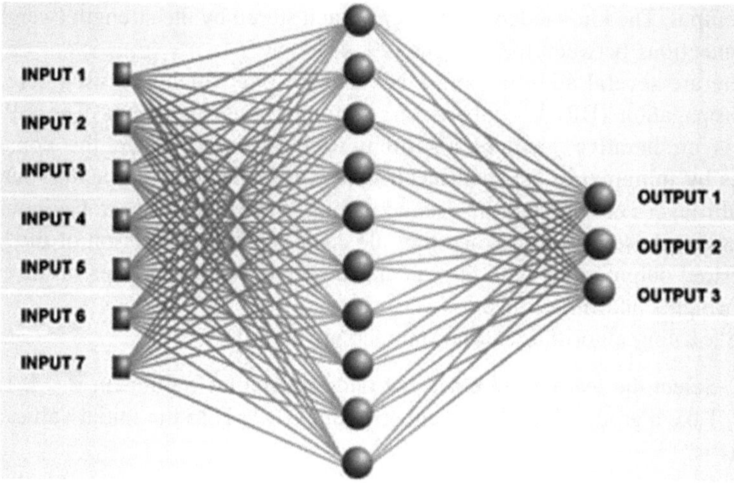

Fig. 4 Neural network architecture composed of input, hidden and output layers

$$v_i^m = \sum_{j=1}^{L} w_{ji}^{m-1} y_j^{m-1} + b_i^m \tag{2}$$

where f is the activation function, L is the number of connections to the previous layer, w_{ji}^{m-1} corresponds to the weights of each connection and b_i^m is the bias, which represents the constant part in the activation function.

From among activation functions the sigmoid (logistic) function is the one most usually employed in NN applications. It is given by:

$$f\left(v_i^m\right) = \frac{1}{1 + \exp\left(-\theta\, v_i^m\right)} \tag{3}$$

where θ is a parameter defining the slope of the function.

It has been reported that the θ parameter influences the speed of NN learning and that the optimal value of the *slope parameter* is problem-dependent [29, 44]. Nevertheless, a constant value of the θ parameter of 1 ($\theta = 1$) is generally applied [17, 29, 44].

3.1 Training Algorithm

The training process in the NN involves presenting a set of examples (input patterns) with known outputs (target output). The system adjusts the weights w_{ji}^{m-1} of the internal connections to minimize errors between the network output and target output. The knowledge is represented and stored by the strength (weights) of the connections between the neurons [29, 44].

There are several algorithms in a NN and the one which was used here is the back-propagation (BP) Levenberg–Marquardt training algorithm. The BP algorithm is an iterative gradient algorithm designed to compute the connection weights by minimizing the total mean-square error between the actual output of the multi-layer network and the desired output. In particular, at the beginning, the weights are chosen randomly and the rule consists of a comparison of the known and desired output value with the calculated output value by means of the current set of weights and thresholds.

The learning algorithm can be summarized as follows:

Step 1. Select the learning rate $\eta = 0.1$ and momentum coefficient $\alpha = 0.1$.
Step 2. Take a group of random numbers within $(-1, 1)$ as the initial values of the weights w_{ji}^{m-1}.
Step 3. Compute the outputs of all neurons layer by layer, starting with the input layer using Eqs. (1)–(3).
Step 4. Compute the system mean square error by:

$$E = \frac{1}{2}\sum_{i=1}^{P} (D_i - y_i)^2 \tag{4}$$

where y_i is the actual output of the ith output node while D_i is the corresponding desired output. P denotes the number of output nodes.
Step 5. If E is small enough or the learning iteration is too high, stop learning.
Step 6. Compute the learning errors for every neuron layer by layer:

$$\delta_m = (D_m - y_m)\, v_m \tag{5}$$

Step 7. Update the weights along the negative gradient of the error E:

$$w_{ji}(t+1) = w_{ji}(t) + \eta\, \delta_i\, y_j + \alpha\left[w_{ji}(t) - w_{ji}(t-1)\right] \tag{6}$$

Step 8. Repeat by going to Step 3.

In the present work, an in-house NN program called Neuromod written in Fortran [17, 21] was developed to perform the training and the prediction. Neuromod includes a module which allows for the automatic selection of the best architecture of the network based on the following steps:

- Select an initial configuration (typically, one hidden layer with the number of hidden units set to half the sum of the number of input and output factors.
- Iteratively, conduct a number of calculations with each configuration, retaining the best network (in terms of verification error) found.
- On each calculation, if under-learning occurs (the network does not achieve an acceptable performance level) try adding more neurons to the hidden layer(s). If

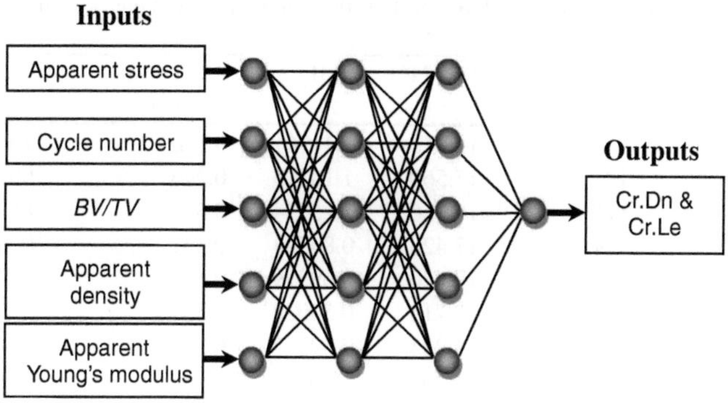

Fig. 5 Neural network architecture composed of input, hidden and output layers

this does not help, try adding an extra hidden layer. If over-learning occurs (verification error starts to rise) try removing hidden units (and possibly layers).

In the present work, the selected architecture is based on double hidden layers with five neurons of each layer with a learning rate factor $\eta = 0.01$ and momentum coefficient $\alpha = 0.01$ (Fig. 5).

3.2 *Trabecular Bone Specimens for Training*

Fracture in trabecular bone is a complex process that depends strongly on the volume fraction (the relative fraction of bone tissue vs. void space), the architecture (the geometrical arrangement of the bone tissue, connectivity, and mean trabeculae thickness), the mechanical properties of the bone tissue, and the applied loads. For a realistic application of the multiscale approach proposed, different trabecular specimens must be used and trained to cover the morphological ranges of the whole trabecular zone of the proximal femur (lowest and highest porosities and densities). From a biological perspective every bone site has a different morphology. Therefore, different samples must be used to capture a realistic human proximal femur morphology. In the current study 23 cylindrical cores (7.04 mm in diameter and 5.5–10 mm long) from fresh-frozen elderly human proximal femur ($n = 12$) and greater trochanter ($n = 11$) trabecular bone were harvested from eleven males (50–94 years old) ($n = 12$). All the specimens were machined such that the main trabecular orientation was aligned with the axis of the core [37]. None of the donors had a history of metabolic bone disease or cancer. Fifteen samples were used to generate training data for the NN and eight samples were kept (not used in the training phase) to validate the previously trained NN (comparison between FE and NN predicted results). The 23 specimens were scanned using micro-computed tomography using the Skyscan 1072 system at an approximate spatial resolution of

Table 1 Distribution and microarchitectural information of specimens by anatomic site, *FN* femoral neck, *GT* greater trochanter

Training/ validation	Age	Anatomic site	*BV/TV* (%)	SMI	Ash density (g/cm^3)	Apparent Young's modulus (MPa)
T	64	GT	12.612	1.41361	0.178	404
V	88	FN	15.560	1.18315	0.296	694
T	92	GT	16.956	1.25774	0.322	867
T	74	FN	17.420	0.89504	0.331	929
T	84	GT	17.620	0.69379	0.335	957
V	89	GT	17.715	0.54028	0.337	970
T	92	FN	18.225	0.89345	0.346	1044
T	91	FN	18.532	1.03561	0.352	1090
T	86	FN	18.562	1.0013	0.353	1095
V	94	GT	19.330	1.05269	0.367	1215
T	95	GT	20.348	0.90966	0.387	1387
T	84	FN	23.048	0.8242	0.438	1913
T	98	GT	23.682	0.42063	0.450	2052
V	82	FN	25.035	0.95246	0.476	2368
T	90	GT	26.200	0.78813	0.498	2663
T	68	GT	27.266	0.90167	0.518	2952
V	84	GT	28.988	0.97644	0.551	3457
T	50	GT	30.410	0.56246	0.578	3912
T	51	FN	31.019	0.44852	0.589	4117
V	61	FN	33.330	0.32045	0.633	4956
T	86	FN	34.232	0.22346	0.650	5309
V	–	FN	39.482	0.15987	0.750	7672
T	88	FN	56.347	0.09023	1.071	13206

Fifteen specimens were used for training (T) and eight specimens were kept for validation (V). *SMI* structure model index

20 μm (microarchitectural information is given in Table 1). Some representative images from different anatomic sites are shown in Fig. 6.

The digital image-based modeling technique using micro-CT (Skyscan 1072 system) and voxel FE at 20 μm voxel sizes were used to simulate the fatigue behaviour of the trabecular samples under varying cyclic compressive stresses. Despite the small size of the bone samples, the model satisfied the continuum assumption for trabecular bone (i.e. at least five intertrabecular lengths in size) [24].

3.3 Factors Affecting the Fatigue Damage Accumulation in Bone

Several factors can affect the evolution of Cr.Dn and Cr.Le within trabecular bone. These factors include the trabecular bone architectural parameters (*BV/TV*, SMI, porosity, Tb.N, Tb.Sp, Tb.Th, ...), bone material parameters (elastic modulus,

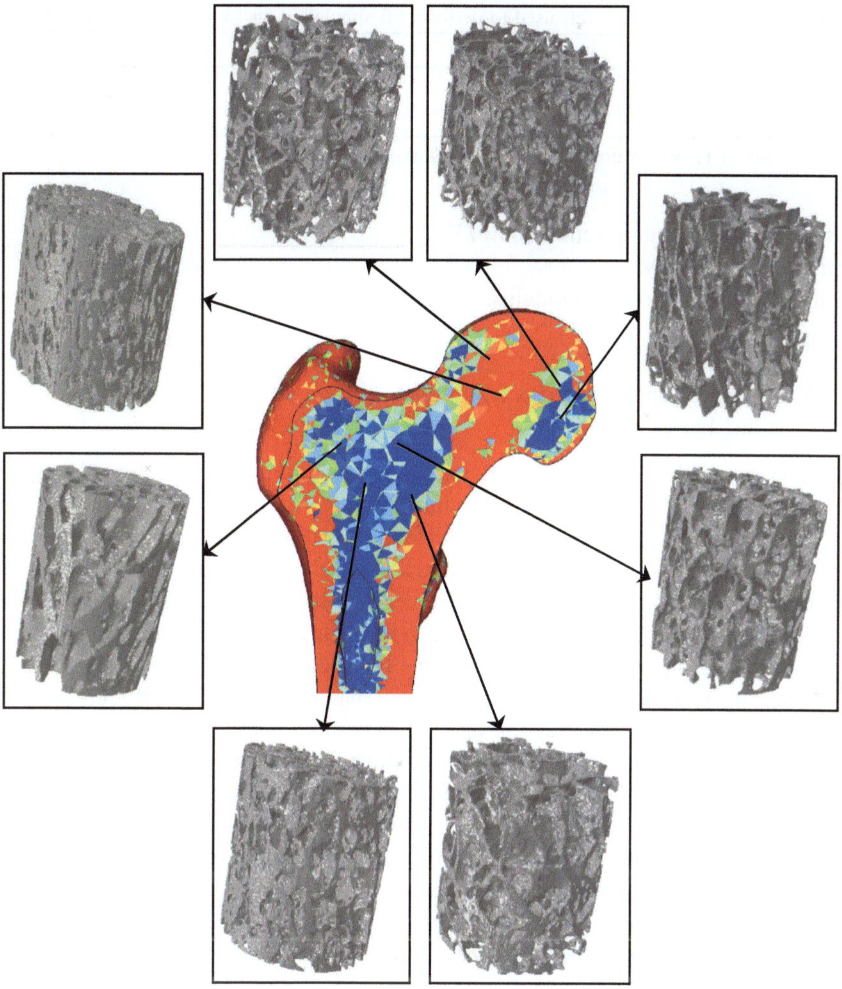

Fig. 6 Example of harvested representative trabecular cores from different anatomic sites of a human femur

mineralization, collagen cross links, …), loading parameters (stress, frequency), biological parameters (Ca, hormones, …) and the environment (age, gender, risk factors, drugs, …). For a given trabecular site (a given morphology), it is more important and more useful to consider the stress applied in the long term because the processes by which living bones can repair cracks, and adapt their shape respond generally to altered stress levels [4, 52]. In the current investigation, five factors were selected to develop the NN model.

Three bone relevant morphological and material factors (Table 1):

Table 2 Loading factors levels used to perform FE simulation on 15 specimens (Table 1)

Loading inputs	Level	Values
Number of cycles	5	$1.e^{1}$, $1.e^{3}$, $1.e^{4}$, $1.e^{5}$, $1.e^{7}$
Applied apparent stress (MPa)	5	25, 50, 70, 90, 110

The results of FE simulation were used as training data for the NN

(i) the bone volume fraction (BV/TV),
(ii) the ash density,
(iii) the apparent elastic modulus.

and two loading factors:

(iv) the applied apparent stress amplitude,
(v) the number of cycles.

The outputs are Cr.Dn (#/mm^3) and Cr.Le (μm/mm^3). This set of outputs represents one of the main fatigue damage parameters observed [4, 36, 50].

Five values for each factor were selected, generating full factorial combinations of the inputs (5 × 5) applied to the 15 training specimens. Hence, 375 (5 × 5 × 15) micro-CT FE simulations were performed on the 15 specimens with a total computation time of about 350 h on a 64 GB dual-core computer to study the effect of each input combination on every trabecular bone specimen.

The stress training values applied were based on a previously established relation between volume fraction and the maximum stress measured in a monotonic strength experiment [45]. The lower stress amplitude was chosen to be 10 % of the ultimate stress (σ_u) while the upper stresses were assigned to four different amplitudes, varying from 20 to 90 % of the ultimate stress. In addition, reported *S–N* curves for human cancellous and cortical bone matrix show that the cycles to failure vary from $1.e^{1}$ to $1.e^{7}$ [7, 45]. Based on these results, five cycle values were used here to train the NN, ranging from $1.e^{1}$ to $1.e^{7}$ (Table 2).

The training data for the NN were extracted by homogenization from results of FE simulations performed on 15 specimens. Eight specimens were not used to prepare the training data but were kept to check the validity of the previously trained NN (NN prediction first on the 15 training specimens followed by FE predictions (using the 8 remaining specimens).

4 Finite Element Modeling Approaches

Two FE models were applied in the current investigation:

(i) Meso FE model needed to generate virtual data for the NN training based on different trabecular specimen simulations coupled to fatigue damage.

(ii) Macro FE coupled with the NN (FENN) model used to perform the multiscale prediction incorporating the NN to describe the whole femur behaviour. Damage effects were not explicitly considered in the macro FE analysis. At this level, damage is described implicitly during the local NN computation and fatigue damage outputs are passed back to the macro model.

4.1 Virtual Testing: Micro-CT Based Finite Element Simulation of Trabecular Bone Fatigue

Available experimental and/or numerical data are necessary to train the NN. This section presents the numerical tests applied to investigate the accumulation of fatigue cracks in trabecular bone samples. To prepare the training data for the NN, the analysis was divided into two separate problems: (i) macroscale FE analyses were performed on a human proximal femur to determine the boundary conditions (stress amplitude) applied at every FE of the mesh during one-legged stance for different frequencies; (ii) the results of the macroscale FE analysis were used to define the local applied boundary conditions on the micro-CT voxel FE models of trabecular bone specimens.

The average Cr.Dn and Cr.Le accumulation of every trabecular bone specimen at every cycle were computed using:

$$Cr.Dn = \frac{1}{V_S} \int_{V_S} n_{be} dV \tag{7}$$

$$Cr.Le = \frac{1}{V_S} \int_{V_S} L_{be} dV \tag{8}$$

where n_{be}, L_{be} and V_S denote respectively the number of broken elements, the length of broken elements, and the apparent sample volume.

Several micro-CT based models of trabecular bone have been developed based on Continuum Damage Mechanics (CDM) [18, 22, 32, 47]. CDM can be used to monitor the damaging process, as a result of cyclic loading up to the time of the appearance of cracks [6, 34]. In this case, the trabecular bone tissue was modelled as a non-linear elastic isotropic behaviour law coupled to fatigue damage given by:

$$\sigma_{ij} = (1 - D)E^{l}_{ijkl}\varepsilon_{kl} \tag{9}$$

where D is the isotropic damage variable at tissue level and E^{l}_{ijkl} is the local (meso) isotropic elasticity stiffness tensor.

When dealing with loading histories composed of well-defined discrete cycles, an evolution law in terms of the number of cycles and their amplitudes is often considered more practical in the literature. Such a cycle-based formulation can be obtained in the form of [6]:

$$\frac{\partial D}{\partial N} = f\left(\frac{N}{N_f}\right) \quad (10)$$

where N_f denotes the cycle at failure.

For high cycle fatigue under purely elastic strain it is possible to apply Chaboche's model expressed by [6]:

$$\partial D = \left[1 - (1-D)^{\beta+1}\right]^{\lambda} \left(\frac{\sigma_{eqM} - \bar{\sigma}}{M\,(1-D)}\right)^{\beta} \partial N \quad (11)$$

where λ is a parameter depending mainly on the loading conditions given by:

$$\lambda = 1 - \eta \left\langle \frac{\sigma_{eqM} + \sigma_{eqm} - 2\sigma_D}{2\sigma_u - \sigma_{eqM} + \sigma_{eqm}} \right\rangle \quad (12)$$

η, β and M are material parameters which can be obtained from experimental fatigue tests from *S–N* curves [6].

σ_{eqM}, σ_{eqm}, σ_D and σ_u are respectively the equivalent maximum peak stress, the equivalent minimum peak stress, the fatigue limit and the ultimate stress.

Operator $\langle \ldots \rangle$ denotes $\langle x \rangle = x$ if $x \geq 0$ and $\langle x \rangle = 0$ if $x < 0$.

Integration of Eq. (11) leads to the Chaboche model expressed by [6]:

$$D = 1 - \left[1 - \left(\frac{N}{N_f}\right)^{\frac{1}{1-\alpha}}\right]^{\frac{1}{1+\beta}} \quad (13)$$

Furthermore, to distinguish between the differences in fatigue accumulation in tension and compression in bone [36, 42], the cycle at failure N_f can be computed using the experimental results of Martin et al. [36]:

$$N_f^c = 1.479 \times 10^{-21}\,\Delta\varepsilon^{-10.3} \text{ for compressive loads} \quad (14)$$

$$N_f^t = 3.630 \times 10^{-32}\,\Delta\varepsilon^{-14.1} \text{ for tensile loads} \quad (15)$$

where $\Delta\varepsilon$ is the amplitude of applied microstrain.

The material properties for bone used for the simulation are given in Table 4.

A practical and sufficiently accurate way to represent fracture propagation at continuum level is the so-called "kill element" technique. When the damage parameter reaches its critical value D_c inside an element, its stiffness matrix is set to zero, leading to the redistribution of the stress state in the vicinity of the fractured zone (crack initiation). Once a crack is initiated, the propagation direction is simulated by the propagation of the broken elements of the mesh. At continuum level, the local critical damage value in tension is generally equal to 1 ($D_c^T \approx 1.0$) [42, 54]. To avoid numerical convergence problems, the critical damage value at fracture was set to $D_c^T = 0.95$. When the kill element method is used in a compressive region, it is necessary to model the self-contact in the gap created the element removal. The alternative is to keep the elements, but to reduce their stiffness to a low, but not null value. In compression, the critical damage value at fracture was set to $D_c^C = 0.5$ [19, 20]. The kill element technique was implemented into Abaqus/Standard via the user subroutine UMAT.

With the concept of CDM, there is no difference between crack initiation and propagation. Both result from the failure of an element with a characteristic dimension (typical size of a crack). Thus, crack initiation and propagation are studied in a unified approach [6, 34]. Moreover, the CDM has no intrinsic material characteristic length in the constitutive law which leads to mesh size dependence results i.e. the crack growth depends on the FE mesh size. The average crack lengths found in bones are typically 100 μm [4, 46]. In the present work, this size corresponds to the mesh characteristic length at trabeculae level of about five finite elements (20 μm/element). Hence, to prevent mesh dependence that generally affects the damage propagation rate, numerical fatigue fracture occurs when the damage value reaches a critical value at a set of five serial broken elements (a crack length of about 100 μm). Concerning the whole specimen in the case of compressive cyclic tests, the definition of the apparent critical damage value at failure (D_c^a) is rather arbitrary, varying between 0.1 and 0.5 [2, 45]. In this study, the apparent failure criterion was set to $D_c^a = 0.4$ [45].

A major drawback of cumulative damage models is the computational cost associated with modeling every loading cycle. In order to reduce the computation time, the integration of the damage growth rate was based on the cycle blocks approach. In this case, the real cycle number is reduced (divided) into equivalent cycle blocks. Damage accumulation is computed over the cycle blocks and extrapolated over the real corresponding cycles.

Within the framework of the cycle blocks approach, fatigue damage evolution can be obtained by:

$$D_{n+p} = D_n + \Delta D \tag{16}$$

where D_{n+p} is the damage at iteration $n + p$ where p denotes the number of cycles in one block set, D_n is the damage at iteration n and ΔD is the damage increment computed for one jump of p cycles. Further, to compute the damage at every cycle, it is possible to extrapolate the damage state using:

$$D_{n+1} = D_n + \frac{\Delta D}{p} \tag{17}$$

To ensure convergence of the Newton–Raphson iterations of the numerical calculation and to avoid discontinuities in the responses (NN computation), the compressive stress applied was decomposed into several increments. In this way, for all these increments an equilibrium solution of the system was found.

4.2 Finite Element Simulation of Proximal Femur Elastic Behaviour

In the current study, the application of the multiscale FENN method was restricted to the trabecular bone only. Nevertheless, the FENN method can be extended to describe the multiscale fatigue crack accumulation process of cortical bone.

Trabeculae are composed of lamellae, lacunae, canaliculi and cement lines. The lamellae are arranged longitudinally along the trabeculae within trabecular packets. Hence, trabecular bone exhibits anisotropic behaviour and its average (trabeculae level) properties depend on its microstructure. Since the investigation scale of the present study corresponds to one or some trabeculae, it can be considered that the bone behaviour is purely elastic coupled to damage with isotropic average properties from the nanoscale level.

At the macroscopic level model (whole proximal femur), fatigue damage was not accounted for. Its coupling effect was considered in the mesoscopic formulation at the trabecular level.

The behaviour law for cortical and cancellous bone at the macro (apparent) level is expressed by:

$$\sigma_{ij} = E^a_{ijkl}\varepsilon_{kl} \tag{18}$$

where σ_{ij} is the stress, ε_{kl} is the strain and E^a_{ijkl} is the apparent (macro) isotropic elasticity stiffness tensor.

The transition from the mesoscale to the macroscale is accomplished by employing the trained NN. The mesoscale model (NN) able to predict detailed responses was incorporated into the macroscale model as a material formulation on the integration point level at every FE iteration—e.g. the behaviour law needed to compute the outputs at the mesoscale was substituted by the trained NN (Fig. 2). The NN model was incorporated into the Abaqus FE code via the user routine UMAT to link the meso and the macro scales.

5 Multiscale Simulation of 3D Proximal Femur Fatigue Crack Accumulation

To apply the proposed hybrid FENN method, a proximal femur (male, age = 77) was imaged using the quantitative computed tomography (QCT) technique. The 3D FE mesh was generated from the reconstructed QCT images using 66680 tetrahedral elements (Fig. 7). Individual conversion of Hounsfield units (HU) to equivalent ash density ($\rho_{ash}(\mathrm{g/cm^3})$) was performed based on the relation [11]:

$$\rho_{ash} = -9 \cdot 10^{-3} + 7 \cdot 10^{-4} HU \tag{19}$$

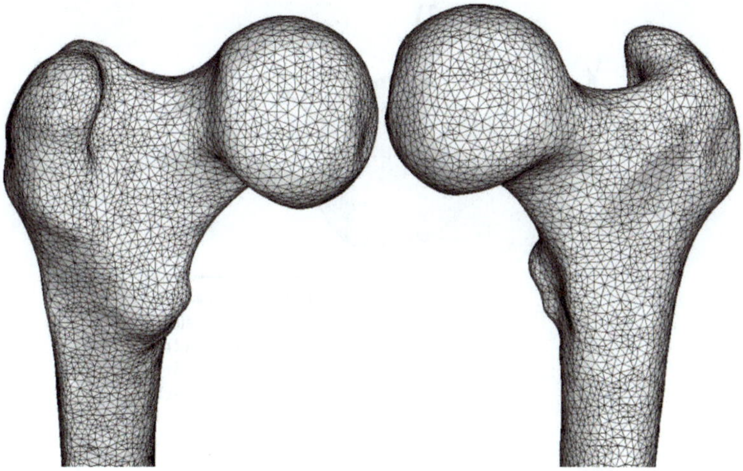

Fig. 7 Three-dimensional mesh of the proximal femur constructed using 66680 tetrahedral elements

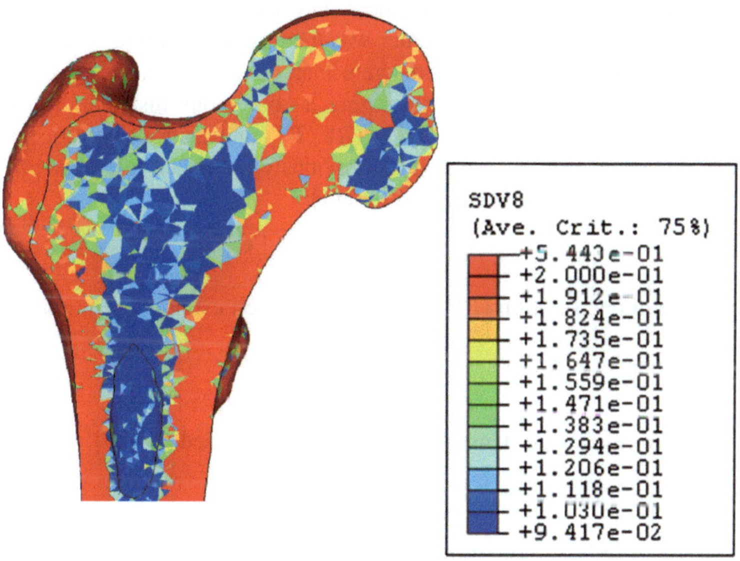

Fig. 8 Partitioned proximal femur with 23 discrete groups of bone volume fraction (BV/TV). The result was obtained using Eqs. (19) and (21) with $\rho_t = 2.31\,\text{g/cm}^3$

In the current work, 23 trabecular specimens with varying characteristics were selected to be assigned to different locations of the trabecular region of the proximal femur based on a partitioning procedure defined by the interval of the apparent density.

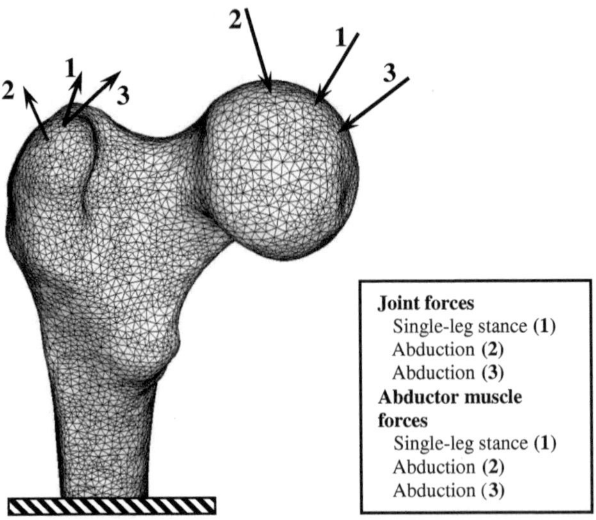

Fig. 9 Three-dimensional FE model of the femur and boundary conditions

The mean *HU* number of each element was averaged from the values of the voxels it contained. The proximal femur was then partitioned into 23 regions with different values of ash density (23 discrete material groups with an iso-value of density for each group, thus approximating a continuous distribution) (Fig. 8). Subsequently, the bone volume fraction (BV/TV) and isotropic elastic modulus (E) were calculated from ρ_{ash} based on previously published relations.

The Young's modulus is expressed by [11]:

$$E\,(\mathrm{MPa}) = 14664\ {\rho_{ash}}^{1.49} \tag{20}$$

The bone volume fraction (BV/TV) was calculated using the relationship [25]:

$$BV/TV = \frac{\rho_{ash}}{\rho_t} \tag{21}$$

where ρ_t is the true tissue density which can be expressed by [25]:

$$\rho_t \approx 1.41 + 1.29\,\alpha \tag{22}$$

α is the ash function ($\alpha = 0$ for osteoid and $\alpha = 0.7$ for fully mineralized bone). The dry tissue densities corresponding to these ash fraction values are 1.41 and 2.31 $\mathrm{g/cm^3}$ respectively [35].

To illustrate the capabilities of the FENN multiscale method, the fatigue process of the 3D proximal femur was performed. The daily loading history was simulated consisting of joint reaction and abductor muscle forces similar to those proposed by Carter et al. [5] for normal activity (Fig. 9).

The model was run in alternating applied loads and unload ($F = 0\ N$) during 1.E5 cycles with a number of cycles in one block set ($P = 500$) (200 computation

Table 3 Selected load conditions that simulate high loads during walking

Cycles/day	Joint forces (N)	Orientation (°)		Abductor muscle forces (N)	Orientation (°)	
		FP	SP		FP	SP
12000	(1) 3244	24	6	(1) 984	28	15
4000	(2) 1621	−15	35	(2) 491	−8	9
4000	(3) 2167	56	−20	(3) 655	35	16

Orientation refers to the frontal (FP) and sagittal (SP) planes [5]

Table 4 Material properties for bone for the FEM simulation from Hambli [20]

Parameters	Notation	Trabecular bone (meso level)	Trabecular bone (macro level)	Cortical bone
General parameters				
Elastic modulus	E (MPa)	10000	Eq. (20)	Eq. (20)
Poisson ratio	ν	0.3	0.3	0.3
Bone density	ρ (g/cm^3)	1.2	Eq. (19)	Eq. (19)
Damage law parameters				
Fatigue parameter	η	0.7	–	–
Fatigue exponent	β	0.4	–	–
Ultimate stress	σ_u (MPa)	120	–	–
Fatigue limit	σ_D (MPa)	50	–	–
Critical damage value at fracture	D_c	0.95	–	–
Cycles	N	1.E5	1.E5	1.E5
Number of cycles in one block set	p	500	500	

loops), a fixed number of cycles per day and a fixed orientation (Table 3). The transition from the mesoscale to the macroscale was accomplished by employing the trained NN. The total computation time using the FENN model was about 40 min on 64 GB dual-core computer.

The material properties for bone used for the simulation are given in Table 4.

6 Results

6.1 Fatigue Cracks Within Trabecular Bone

Figure 10 shows an example of the fatigue damage contour after 1.E5 loading cycles of the specimen ($BV/TV = 26.2\ \%$) under excessive cyclic compressive load ($\sigma = 110\,\text{MPa}$). Fatigue damage distribution was found to be highly inhomogeneous across the specimen and only a small percentage of trabeculae are cracked. The "kill element" technique was applied to predict the initiation and growth of the induced fatigue cracks.

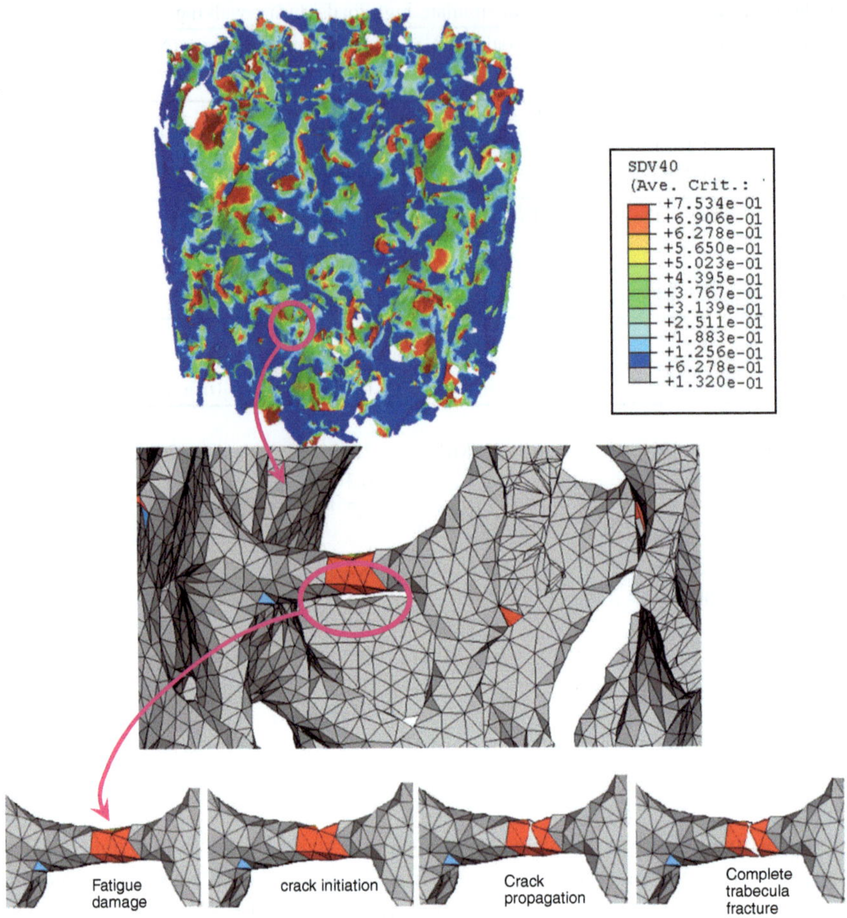

Fig. 10 Contour of fatigue damage within a trabecular bone specimen (cycle = 1.E5, $\sigma = 110\,\text{MPa}$, $BV/TV = 26.2\,\%$). The enlargement shows the location of a fatigue crack and its corresponding initiation and propagation process

In Fig. 11, an enlargement of the predicted fatigue damage contour is given and compared with reported experimental results. Micro-CT images of trabecular bone specimens were obtained at 10 μm resolution by Wang et al. [50] to measure the fraction of damaged bone tissue and to capture localized regions of damage in trabecular bone specimens (Fig. 11b).

The predicted damage was found to be located in a limited number of trabeculae at low apparent stress levels during the first cycle stage. It can be seen that the predicted damage contour of Fig. 11a is qualitatively similar to the experimental ones of Wang et al. [50]. Both results show that the damage is localized in small areas of the bone. Such local damage generates local non-linearities. This is in agreement with experiments by Dendorfer et al. [9]. The authors showed that

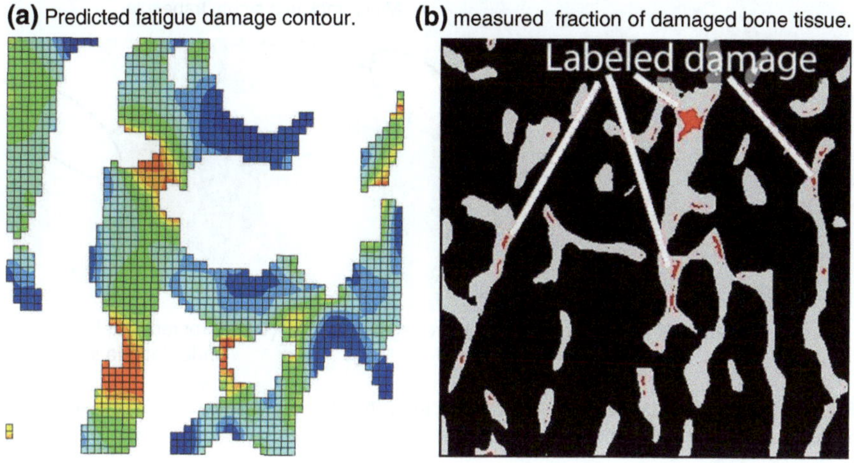

Fig. 11 **a** Enlargement showing the predicted contour of the fatigue damage. **b** Fraction of damaged bone tissue and localization in trabecular specimens from Wang et al. [50]

accumulated residual and maximum strains within a trabecular bone specimen generate strain concentrations followed by the formation of a fracture line through the specimen. Martin et al. [36] reported that relatively small amounts of micro-damage can cause significant reductions in the bone's mechanical properties even before the appearance of microcracks. Similar results were obtained with numerical studies, which demonstrated that damage is visible below apparent compressive yield strain and local tissue yielding initiates at low apparent stress levels [38, 39]. The predicted fatigue damage of single trabeculae (Fig. 12a) lead to a crack initiation and propagation (Fig. 12b) which appears to be local, in agreement with experimental observations (Fig. 12c) [10].

After the training phase, the eight remaining trabecular bone samples were used in order to check the validity of the NN prediction. First Cr.Dn and Cr.Le were predicted with the NN and second, micro-CT FE fatigue simulations were performed for verification. Figure 13a, b shows an example of Cr.Dn and Cr.Le evolution versus cycles predicted by both the NN and the FE methods (specimen: $BV/TV = 25.035\ \%$). One can observe that the accumulation of cracks increases rapidly during the first cycles, followed by a reduced rate of accumulation and an accelerated growth rate prior to the specimen failure. Similar experimental results on cortical bone were obtained by O'Brien et al. [41]. The shape of the Cr.Dn accumulation can be explained by the following three stages: (i) strain localization due to trabeculae shapes and interconnections during the early cycles; (ii) accumulation of local residual strain leading to damage growth and (iii) final catastrophic rupture of the localized bone tissue which triggers the growth of one macroscopic crack through the trabecular structure.

It can be seen that very good agreement is obtained between the two predictions. An average coefficient of correlation of 0.9984 was found between the NN

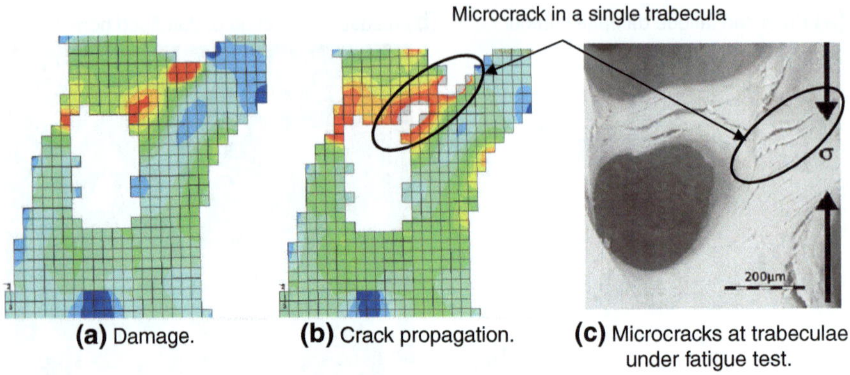

Fig. 12 Predicted fatigue damage (**a**) and cracking (**b**) of a single trabecula of a transversally loaded trabecula; **c** SEM image of a fractured trabecula [10]

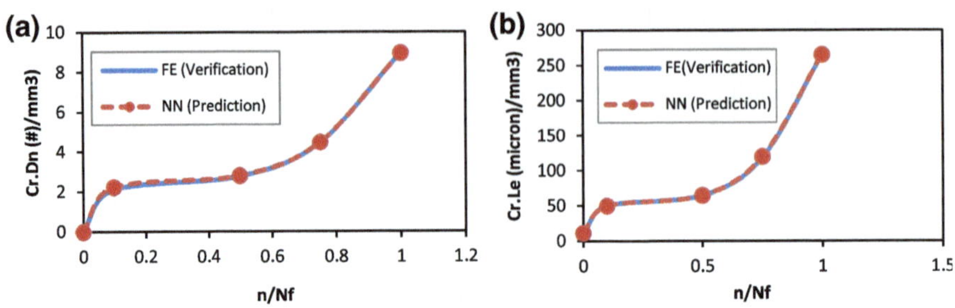

Fig. 13 Comparison between NN Prediction and FE verification of Cr.Dn and Cr.Le accumulation versus cycles. Example of sample ($BV/TV = 25.035\ \%$) (Table 1). First, NN prediction was performed. Second, FE verification was applied to check the validity of the NN. **a** Crack density, **b** crack length

prediction and the FE one. The NN computation time was about 1 s. The FE computation time was about 1 h (64 GB dual-core computer). The results indicate that the NN model was able to predict rapidly and accurately the average Cr.Dn and Cr.Le accumulation within trabecular bone under varying applied compressive stress.

6.2 *Fatigue Cracks Within a Proximal Femur Organ: Multiscale Results*

The previously trained NN was exploited as a tool to rapidly estimate the Cr.Dn and Cr.Le accumulation at a given bone site for a human proximal femur. Since the proposed NN model incorporates the modeling of the real scanned 3D trabecular networks, stress concentration due to the trabecular shapes and connections can be

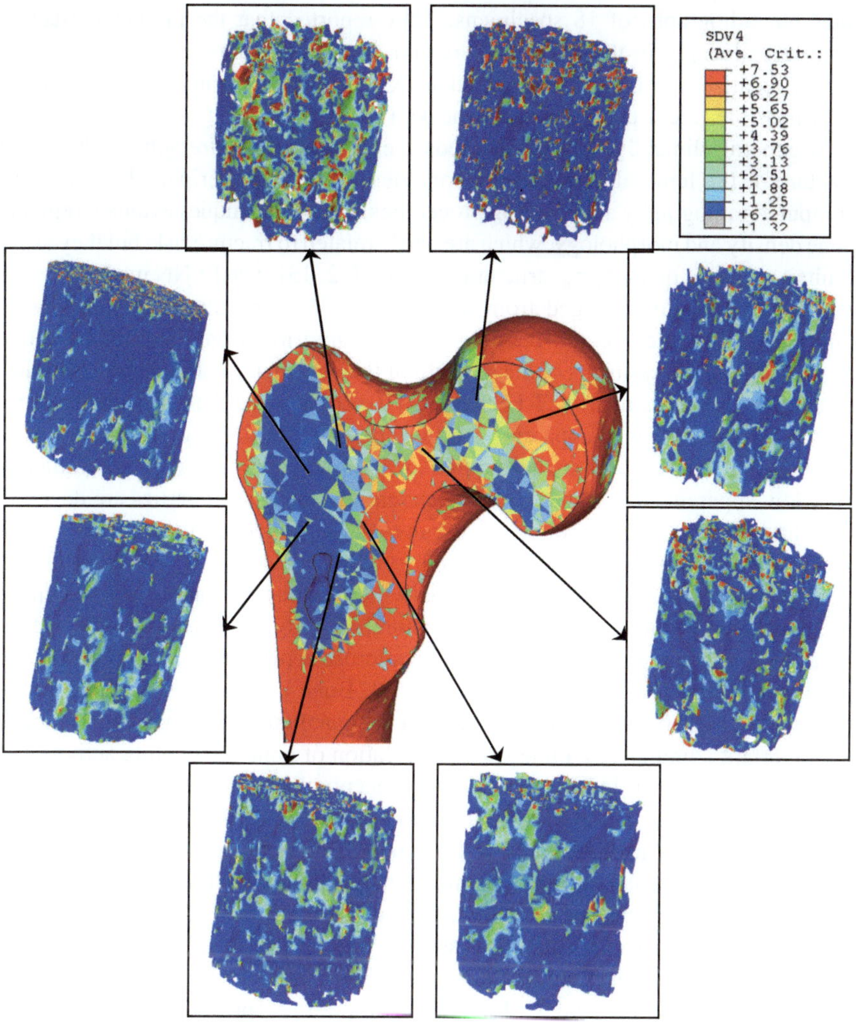

Fig. 14 Averaged contour of Cr.Dn after 1.E5 loading cycles obtained using the proposed multiscale FENN approach and corresponding Cr.Dn distribution on specific trabecular bone specimens obtained by FE

expected to lead to more accurate prediction of damage accumulation, as suggested by Cowin [8].

Figure 14 shows the contour of Cr.Dn after 1.E5 loading cycles obtained using the proposed multiscale FENN approach. The trabecular bone specimen results are averaged and passed back to the entire femur in the form of averaged Cr.Dn and Cr.Le using the validated NN.

Observations of the Cr.Dn contour of the proximal femur reveal a very similar distribution of BV/TV (Fig. 8). Keyak et al. [31] compared experimental fracture sites on radiographs of 18 specimens. They reported that the predicted fracture sites were located in the sub-capital region in all specimens. FENN results predicted localized Cr.Dn in the femoral neck which is qualitatively in agreement with possible fracture sites in femur necks [1, 12, 13].

In general, clinically available methods of estimating bone strength and the risk of fracture of the femoral neck include bone densitometry or peripheral quantitative computed tomography and imaging procedures. These techniques evaluate regional bone density and morphology, which are partly related to fracture risk, but they are of limited value for quantifying structural strength [12, 13]. The FENN method, which incorporates results averaged from real scanned 3D trabecular bone cores using micro-CT FE models, could possibly achieve rapid multiscale non-invasive and precise clinical assessments of the strength of the proximal femur and possible sites of fracture. Previous micro-CT FE models [10, 26] showed that microdamage initiation occurred prior to apparent yield at relatively low local principal strains in compression. The authors suggested that local tissue yielding can in fact initiate at very low apparent strains and that the apparent mechanical properties are degenerated through these localized effects. Morgan et al. [38] suggested that relatively small amounts of microdamage have a major effect on the mechanical properties of bone.

Repetitive loading from everyday activities results in fatigue microdamage accumulation leading to overall bone structural fragility. The histological evidence indicates that fatigue damage occurs at the microstructural level (microcracks) and at ultrastructural levels (diffuse damage) [50]. Bone microcracks resulting in the disruption of osteocytic communication via the canalicular network may in fact be an important stimulus providing spatial regulation of bone remodeling activity [8]. The balance between local remodeling and accumulation of trabecular bone fatigue microcracks is believed to play an important role in the maintenance of skeletal integrity. However, the local mechanical parameters associated with the initiation and propagation of fatigue microcracks are not well understood. Despite the high clinical relevance of trabecular bone damage, the relationship between local multiple microcracks and local stresses and strains at trabecular level is not well understood. A quantitative assessment of trabecular level stresses and strains associated with Cr.Dn and Cr.Le accumulation may provide insight into the improvement of fracture risk assessment methods and bone repair simulations.

7 Summary and Conclusions

In this chapter, the focus was on the development and the implementation of a novel multiscale approach referred to as the FENN method for crack density and crack length accumulation in trabecular bone using finite element simulation and neural network computation. The transition from mesoscale to macroscale is done by means of the trained NN. The mesoscale model (NN) able to predict detailed responses was incorporated into the macroscale model as a material formulation on the integration point level at every FE iteration—e.g. the behaviour law needed to compute the outputs at mesoscale is substituted by the trained NN. The hierarchical combination of trained NN (mesocale) and FE (macroscale) models for the prediction of bone fatigue microcrack growth can be applied using other factors (age, gender, drugs, …) integrating information extracted from available data and experiments. The aim here was to illustrate the potential of the FENN method in the field of bone biomechanics to incorporate local responses at continuum macroscopic level both accurately and rapidly. Due to the restricted amount of clinical data, it was very difficult to directly compare the proposed FENN numerical results to in vivo experimental ones. Nevertheless, the results provide evidence for the capability of the proposed FENN method to supply outputs of the fatigue microcrack behaviour of trabecular bone. The FENN model advances current FE models by explicitly describing the Cr.Dn and Cr.Le evolution under cyclic compressive loads and by the rapid computation time (the NN approach was about *4.32e5* times faster than the FE computation). Such a method can contribute toward the development of more sophisticated multiscale FE models to predict the behaviour of living bone by explicitly including the effects of fatigue crack influence.

The application of NN models is beneficial since a multilevel numerical analysis of bone behaviour simulation is time-consuming. The complexity of the NN is determined during training such that the minimal network is used that can accurately represent the training data. However, despite the success of the proposed hybrid FENN approach, certain limitations apply. First, for the sake of simplification, the behaviour of the trabecular bone was considered to be isotropic and fatigue data were obtained from published experimental results on human vertebral trabecular bone. Since the scale of investigation of the present work corresponds to one or some trabeculae, one can assume isotropic averaged properties from the nanoscale level. Alternatively, an anisotropic behaviour description of trabecular bone tissue can be used to simulate the anisotropic fatigue damage behaviour of bone. Furthermore, material anisotropy and fatigue data extracted from the proximal femur can be assigned to the FE model. However, limited experimental data are available to characterize such micromechanical properties of trabecular bone. Third, the bone repair process resulting from bone remodeling was not considered in the present study. Nevertheless, the overall structure of the proposed multiscale FENN approach will remain unchanged. There will still be a need to perform multiscale simulations to predict the accumulation of local fatigue microcracks within the trabecular bone of human proximal femurs.

Acknowledgments This work was supported by the French National Research Agency (ANR) through TecSan program (Project MoDos, no. ANR-09-TECS-018). The authors are grateful to Dr Loulou H. for providing the micro-CT trabecular bone meshes.

References

1. Bessho, M., Ohnishi, I., Matsuyama, J., Matsumoto, T., Imai, K., Nakamura, K.: Prediction of strength and strain of the proximal femur by a CT-based finite element method. J. Biomech. **40**, 1745–1753 (2007)
2. Bowman, S.M., Guo, X.E., Cheng, D.W., Keaveny, T.M., Gibson, L.J., Hayes, W.C., McMahon, T.A.: Creep contributes to the fatigue behaviour of bovine trabecular bone. J. Biomech. Eng. **120**, 647–654 (1998)
3. Bugmann, G.: Normalized radial basis function networks. Neurocomput, Special Issue Radial Basis Func. Netw. **20**, 97–110 (1998)
4. Burr, D.B.: Remodeling and the repair of fatigue damage, Calcified Tissue Int **53**(Suppl 1), S75–S81 (1993)
5. Carter, D.R., Fyhrie, D.P., Whalen, R.T.: Trabecular bone density and loading history: regulation of tissue biology by mechanical energy. J. Biomech. **20**, 785–795 (1987)
6. Chaboche, J.L.: Continuum damage mechanics a tool to describe phenomena before crack initiation. Nucl. Eng. Des. **64**, 233–247 (1981)
7. Choi, K., Goldstein, S.A.: A comparison of the fatigue behaviour of human trabecular and cortical bone tissue. J. Biomech. **25**(12), 1371–1381 (1992)
8. Cowin, S.C.: Mechanosensation and fluid transport in living bone. J. Musculoskel. Neuron. Interact. **2**(3), 256–260 (2002)
9. Dendorfer, S., Maier, H.J., Hammer, J.: Anisotropy of the fatigue behaviour of cancellous bone. J. Biomech. **41**(3), 636–641 (2008)
10. Dendorfer, S., Maier, H.J., Hammer, J.: Fatigue damage in cancellous bone: an experimenytal approach from continuum to micro scale. J. Mech. Behav. Biomed. Mater. **2**, 113–119 (2009)
11. Dragomir-Daescu, D., Op Den Buijs, J., McEeligot, S., Dai, Y., Entwistle, R.C., Salas, C., Melton III, J., Bennet, E., Khosla, S., Amin, S.: Robust QCT/FEA models of proximal femur stiffness and fracture load during a sideways fall on the hip. Ann. Biomed. Eng. **39**(2), 742–755 (2011)
12. Faulkner, K.G., Cummings, S.R., Black, D., Palermo, L., Gluer, C.C., Genant, H.K.: Simple measurement of femoral geometry predicts hip fracture: the study of osteoporotic fractures. J. Bone Miner. Res. **8**, 1211–1217 (1993)
13. Faulkner, K.G., Cummings, S.R., Black, D., Palermo, L., Gluer, C.C., Genant, H.K.: Simple measurement of femoral geometry predicts hip fracture: the study of osteoporotic fractures. J. Bone Miner. Res. **21**, 101–108 (1993)
14. Fazzalari, N.L., Kuliwaba, J.S., Forwood, M.R.: Cancellous bone microdamage in the proximal femur: influence of age and osteoarthritis on damage morphology and regional distribution. Bone **31**(6), 697–702 (2002)
15. Ghanbari, J., Naghdabadi, R.: Nonlinear hierarchical multiscale modeling of cortical bone considering its nanoscale microstructure. J. Biomech. **42**(10), 1560–1565 (2009)
16. Haddock, S.M., Yeh, O.C., Mummaneni, P.V., Rosenberg, W.S., Keaveny, T.M.: Similarity in the fatigue behaviour of trabecular bone across site and species. J. Biomech. **37**, 181–187 (2004)
17. Hambli, R.: Statistical damage analysis of extrusion processes using finite element method and neural networks simulation. Finite Elem. Anal. Des. **45**(10), 640–649 (2009)
18. Hambli, R.: Application of neural networks and finite element computation for multiscale simulation of bone remodeling. J. Biomech. Eng. **132**(11), 114502 (2010)
19. Hambli, R.: Numerical procedure for multiscale bone adaptation prediction based on neural networks and finite element simulation. Finite Elem. Anal. Design **47**(7), 835–842 (2011)
20. Hambli, R.: Apparent damage accumulation in cancellous bone using neural networks. J. Mech. Behav. Biomed. Mater. **4**(6), 868–878 (2011)
21. Hambli, R., Chamekh, A., Bel Hadj Salah, H.: Real-time deformation of structure using finite element and neural networks in virtual reality applications. Finite Elem. Anal. Des. **42**(11), 985–991 (2006)

22. Hambli, R., Soulat, D., Gasser, A., Benhamou, C.L.: Strain–damage coupled algorithm for cancellous bone mechano-regulation with spatial function influence, Vol. 198. Comput. Methods Appl. Mech. Eng. **33–36**(1), 2673–2682 (2009)
23. Hambli, R., Katerchi, K., Benhamou, C.L.: Multiscale methodology for bone remodelling simulation using coupled finite element and neural network computation. Biomech. Model. Mechanobiol. **10**(1), 133–145 (2011)
24. Harrigan, T.P., Jasty, M., Mann, R.W., Harris, W.H.: Limitations of the continuum assumption in cancellous bone. J. Biomech. **21**, 269–275 (1988)
25. Hernandez, C.J., Beaupre, G.S., Keller, T.S., Carter, D.R.: The influence of bone volume fraction and ash fraction on bone strength and modulus. Bone **29**, 74–78 (2001)
26. Hernandez, C.J., Gupta, A., Keaveny, T.M.: A biomechanical analysis of the effects of resorption cavities on cancellous bone strength. J. Bone Miner. Res. **21**(8), 1248–1255 (2006)
27. Hornik, K.: Approximation capabilities of multilayer feedforward networks. Neural Netw. **4**, 251–257 (1991)
28. Hurtado, J.E.: Analysis of one-dimensional stochastic finite elements using neural networks. Probab. Eng. Mech. **17**(1), 35–44 (2002)
29. Jenkins, W.M.: An introduction to neural computing for the structural engineer. Struct. Eng. **75**(3), 38–41 (1997)
30. Karim, L., Vashishth, D.: Role of trabecular microarchitecture in the formation, accumulation, and morphology of microdamage in human cancellous bone. J. Orthop. Res. **29**(11), 1739–1744 (2011)
31. Keyak, J.H., Rossi, S.A., Jones, K.A., Les, C.M., Skinner, H.B.: Prediction of fracture location in the proximal femur using finite element models. Med. Eng. Phys. **23**, 657–664 (2001)
32. Kosmopoulos, V., Schizas, C., Keller, T.S.: Modeling the onset and propagation of trabecular bone microdamage during low-cycle fatigue. J. Biomech. **41**, 515–522 (2008)
33. Lefik, M., Boso, D.P., Schrefler, B.A.: Artificial neural networks in numerical modelling of composites. Comput. Methods Appl. Mech. Eng. **198**, 1785–1804 (2009)
34. Lemaitre, J.: A continuous damage mechanics model for ductile fracture. J. Eng. Mater. Technol. **107**, 83–89 (1985)
35. Martin, R.B.: Porosity and specific surface of bone. Crit. Rev. Biomed. Eng. **10**, 179–222 (1984)
36. Martin, R.B., Burr, D.R., Sharkey, N.A.: Skeletal tissue mechanics. Springer, New York (1998)
37. Morgan, E.F., Keaveny, T.M.: Dependence of yield strain of human trabecular bone on anatomic site. J. Biomech. **34**, 569–577 (2001)
38. Morgan, E.F., Yeh, O.C., Keaveny, T.M.: Damage in trabecular bone at small strains. Eur. J. Morphol. **42**(1–2), 13–21 (2005)
39. Nagaraja, S., Couse, T.L., Guldberg, R.E.: Trabecular bone microdamage and microstructural stresses under uniaxial compression. J. Biomech. **38**(4), 707–716 (2005)
40. Nicolella, D.P., Nicholls, A.E., Lankford, J.: Micromechanics of creep in cortical bone. Trans. Orthop. Res. Soc. **23**, 137 (1998)
41. O'Brien, F.J., Taylor, D., Clive Lee, T.: Microcrack accumulation at different intervals during fatigue testing of compact bone. J. Biomech. **36**, 973–980 (2003)
42. Pattin, C.A., Caler, W.E., Carter, D.R.: Cyclic mechanical fatigue property degradation during fatigue loading of cortical bone. J. Biomech. **29**(1), 69–79 (1996)
43. Prendergast, P.J., Taylor, D.: Prediction of bone adaptation using damage accumulation. J. Biomech. **27**(8), 1067–1076 (1994)
44. Rafiq, M.Y., Bugmann, G., Easterbrook, D.J.: Neural network design for engineering applications. Comput. Struct. **79**(17), 1541–1552 (2001)
45. Rapillard, L., Charlebois, M., Zysset, P.K.: Compressive fatigue behaviour of human vertebral trabecular bone. J. Biomech. **39**, 2133–2139 (2006)
46. Taylor, D., Lee, T.C.: A crack growth model for the simulation of fatigue in bone. Int. J. Fatigue **2**, 387–395 (2003)
47. Taylor, M., Cotton, J., Zioupos, P.: Finite element simulation of the fatigue behaviour of cancellous bone. Meccanica **37**, 419–429 (2002)

48. Unger, J.F., Konke, C.: Coupling of scales in multiscale simulation using neural networks. Comput. Struct. **86**(21–22), 1994–2003 (2008)
49. Viceconti, M., Taddei, F., Jan, S.V.S., Leardini, A., Cristofolini, A., Stea, S., Baruffaldi, F., Baleani, M.: Multiscale modelling of the skeleton for the prediction of the risk of fracture. Clin. Biomech. **23**, 845–852 (2008)
50. Wang, X., Masse, D.B., Huijie, L., Hess, K.P., Ross, R.D., Roeder, R.K., Niebur, G.L.: Detection of trabecular bone microdamage by micro-computed tomography. J. Biomech. **40**(15), 3397–3403 (2007)
51. Waszczyszyn, Z., Ziemianski, L.: Neural network in mechanics of structures and materials—new results and prospects of applications. Comput. Struct. **79**, 2261–2267 (2001)
52. Weiner, S., Traub, W.: Bone structure: from ångstroms to microns. FASEB J. **6**, 879–885 (1992)
53. Yagawa, G., Okuda, H.: Neural networks in computational mechanics. Arch. Comput. Methods Eng. **3–4**, 435–512 (1996)
54. Zioupos, P., Tong Wang, X., Currey, J.D.: Experimental and theoretical quantification of the development of damage in fatigue tests of bone and antler. J. Biomech. **29**(8), 989–1002 (1996)

Multiscale Approach to Understand the Multiphysics Phenomena in Bone Adaptation

Thibault Lemaire and Salah Naili

Abstract The ability of bone tissue to adapt itself to its physical environment is the research focus of several teams all over the world. If the physical *stimuli* playing a role in bone remodelling are often identified, how they act and are converted into a cellular response is still an open question. The aim of this paper is, in a first part, to propose an overview on the physical factors participating in the bone remodelling process. In a second part, we present some recent developments concerning the implications of hydro–electro-chemical couplings that could modify the bone adaptation process. Since the phenomena that are involved in this mechanism are related both to the mechanical solicitations of the tissue and the physical phenomena in the vicinity of bone cells, different scales, from the organ to the cell, should be considered to go deeper in its understanding. That is why a multiscale strategy based on periodic homogenization has been carried out to propagate the multiphysics description at the cellular scale toward the macroscopic scale of the tissue. This multi-level approach is so adapted to connect macroscopic physical information to microscopic phenomena, *et vice versa*. Thus, using convenient simulations, we have brought a new light on classical interrogations dealing with bone adaptation. These five questions are: i. Can the sole hydro-mechanical coupling describe the poro-mechanical behaviour of bone or should we consider a modified Biot model including electro-chemical effects?; ii. Similarly, is the classical Darcy law sufficient to describe the bone interstitial fluid flow?; iii. What is the nature of the stress-induced electric potentials that can be measured in vivo?; iv. What are the consequences of the electro-chemical couplings on the

T. Lemaire (✉) · S. Naili
Université Paris-Est Laboratoire Modélisation et Simulation Multi Echelle,
MSME UMR 8208 CNRS 61 Avenue du Général de Gaulle,
94010 Créteil cedex, France
e-mail: thibault.lemaire@univ-paris-est.fr

Stud Mechanobiol Tissue Eng Biomater (2013) 14: 31–72
DOI: 10.1007/8415_2012_149

Published Online: 29 September 2012

shear sensitivity of the osteocytes?; v. What are the consequences of the microscopic physico-chemical properties of the bone microstructure on the mass transport within the lacuno-canalicular system? Finally, from these simple model-driven observations, we propose a new perspective to alter the current bone adaptation paradigm.

1 Introduction

The mechano-transduction process consists of a chain of three main functions: mechano-reception, mechano-transmission and mechano-activation. Whether they are biological or more generally industrial, these functions make that some systems act as mechano-sensors and/or senso-actors. Thus, these systems are able to gather and transmit information on their neighborhood and on themselves, modifying so their properties in response to various physico-chemical solicitations.

The living materials, which are intrinsically active systems, are typical adaptive systems that may change their response in function of the context. Their activity generates changes in mass, composition and shape. This phenomenon is called remodelling.

As such bone tissue can sense, react and adapt itself to its environmental vicinity. For instance, when submitted to a cycling loading generated by walking activity, bone is able to change its morphology and its mechanical properties by forming, respectively resorbing, bone matter in the high, respectively low, stress zones. For instance, it is well known that the bone quality decreases during space flight [22, 101].

Thus bone formation and resorption is the result of a series of events transforming a physical information into a biological response. This process including all the phenomena characterizing the bone cell ability to sense mechanical *stimuli* and possibly to reply is called the mechano-transduction of bone remodelling. As sketched in Fig. 1, the cycle of the mechano-transduction of bone remodelling can be coarsely summarized as: a macroscopic external physical *stimulus* (i) is propagated within the bone tissue (ii), and then sensed at the microscale by sensitive cells (typically the osteocytes, OCY) (iii), that induce signals emission (iv) to activate effector cells that will resorb (osteoclasts, OCL) old tissue and create (osteoblasts, OBL) new one (v), thus modifying the macroscopic properties of the organ (vi).

This adaptation results in the optimization of bone morphology to obtain the best mechanical resistance using the minimum mineral quantity. This was already postulated in the nineteenth century by the surgeon Julius Wolff [162]. The Wolff's law is part of the general trend in the biological community of the late nineteenth century. This trend can be crystallized by the statement of Spencer: "Life is definable as the continuous adjustment of internal relations to external relations" [145]. If this theory has been roughly checked thanks to experiments and clinical observations, many avenues of research are still open [36, 137].

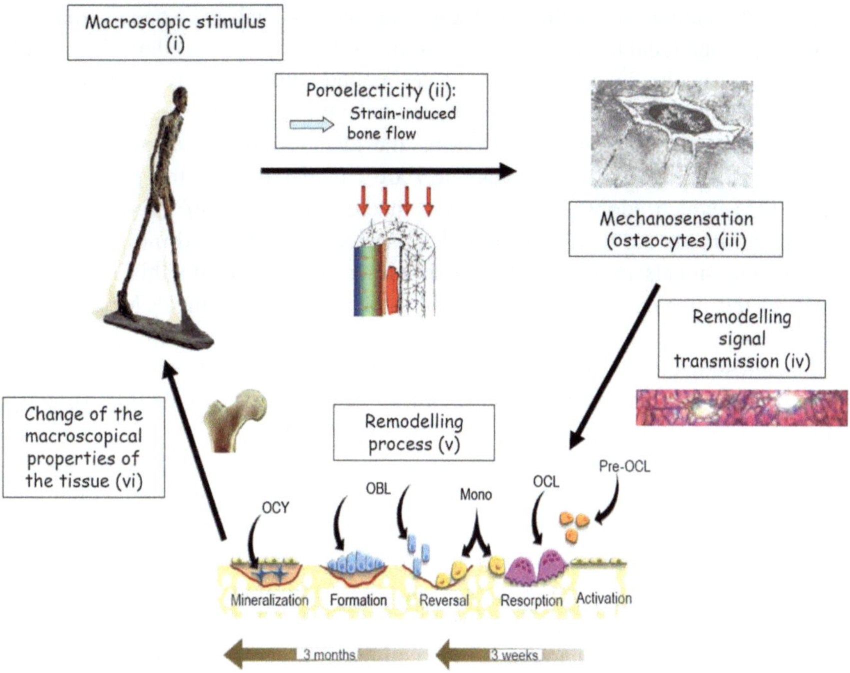

Fig. 1 Simplified chain of the mechano-transduction of bone remodelling (reproduced with permission from [75])

In particular, the understanding of the transduction of physical signal is still in debate [19]. The current trend consists in considering that the central role in the mechano-reception is played by the osteocytes [18, 64, 66, 159].

These are bone cells embedded in the bone matrix volume. Their bodies are located in ellipsoidal cavities of 10–30 μm called the *lacunae*. These *lacunae* are connected thanks to *canaliculi* whose diameters are on the order of magnitude if 0.1 μm. Cytoplasmic osteocyte cell processes that develop from the cell body occupy the central zone of the *canaliculi* so that the interstitial fluid pathway roughly corresponds to an annular geometry. Thus, the stellar osteocytes form a network in the bone volume and communicate one to the other through the gap junctions. This particular configuration of the couple osteocytes/lacuno-canalicular porosity (OCY/PLC) explains that it is often considered as the mechano-sensitive organ of bone.

In summary, if it is commonly accepted that the effector cells are the osteoclasts and osteoblasts and the mechano-sensitive cells are the osteocytes [92], the understanding of the nature of the triggering signals for the different cellular activities is much debated. The following section proposes a brief review of the classical hypotheses related to the mechanism by which bone cells sense their mechanical environment and initiate the deposition or resorption of bone tissue.

Notwithstanding the nature of these triggering signals, it appears that it is correlated to the local hydro-mechanical state of the tissue and/or the cells vicinity. The great difficulty in carrying out experimental investigations of in situ bone poro-mechanical behaviour renders theoretical investigations crucial. In most of the representations of bone remodelling, the mechanical *stimuli* acting on cells (pressure, shear stress, drag forces, etc.) are calculated from the poro-elasticity theory [15], which adopts intrinsically a macroscopic point of view [24]. These mechanical inputs are then somehow downscaled and converted into biochemical microscopic signals regulating the remodelling activity [1]. In this manner, the nature of the incoming signals is thought to be purely mechanical. Moreover, the microscopic phenomena are not directly involved since the fluid flow is quantified by a macroscopic textural parameter, the hydraulic permeability. In other words, even if involving microscopic biochemical signals, these modelling strategies remain purely macroscopic.

In this contribution, we would like to emphasize recent developments that may strongly modify the current bone mechano-sensation paradigm. Using a multiscale strategy, we propose to investigate the multiphysics effects due to the physico-chemical phenomena that occur at the microstructural scale of bone tissue. In particular, we trace how the fluid-flow and mass transport models for mechano-transduction should be changed by considering additional effects related to electro-chemical couplings that characterize the cellular vicinity. Our strategy consists in discussing, at both the macroscale and the microscale, the importance of the multiphysical phenomena featuring in bone behaviour using physiologically-based simulations. That is why, after having described the bone structure and the different physical stimuli that may affect its behaviour, the ingredients of our model are introduced. Grounded on several former studies [57, 74–76, 84–86] invoking the periodic homogenization method, a multiscale description linking the macroscopic and microscopic features is proposed. Then, some numerical examples are presented to investigate the real implications of underlying electro-chemical couplings in the bone remodelling signals expression. Even if their consequences are not visible at the macroscale, these multiphysical effects could be significant at the cellular scale and thus should be taken into account in new scenarios of bone adaptation. To illustrate this, the discussion provides five model-driven examples proving the necessity of creating a new paradigm in bone remodelling including multiphysics considerations.

2 Some Basic Aspects in Bone Physiology

When aiming at understanding the complex mechanisms governing biological systems, the in vivo insuperable problems quickly arise and make it very difficult to progress without tremendous efforts. In the bone biology field, you have to be prepared to move mountains when carrying out experiments intending at analyzing the remodelling process. An alternative avenue of research could be the in silico

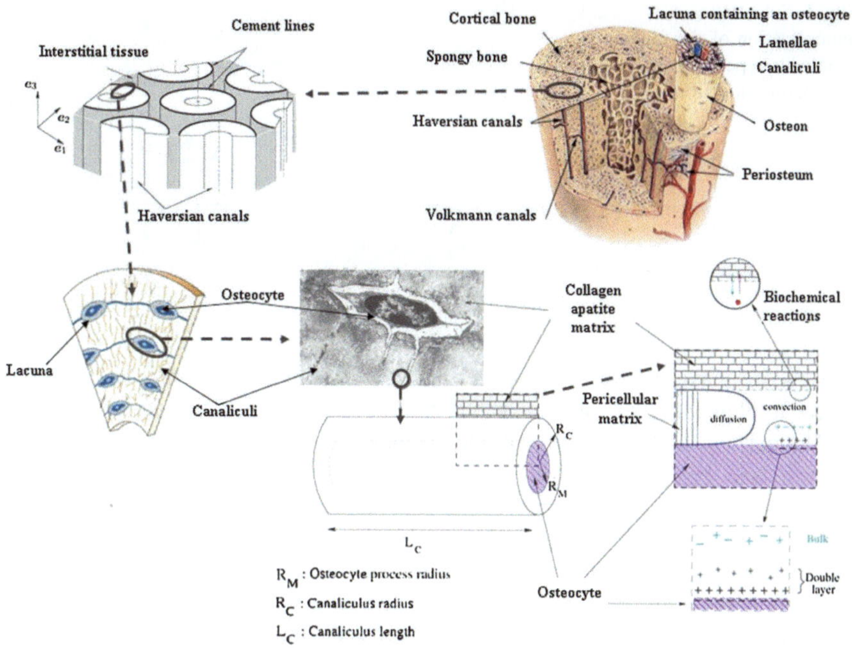

Fig. 2 Multiscale bone structure (reproduced with permission from [75])

method, which consists in mimicking bone dynamic behaviour at different scales. Two main approaches develop in parallel: the first one treats bone adaptation through its tissue evolution, whereas the second one is more focussed on describing the complex cellular interactions operating in the remodelling process. A review of all these computer-driven investigations of bone remodelling can be found in Webster and Müller [157]. Concerning the remodelling models at the organ or tissue scales, they are often continuum or micromechanical based and describe the variation of bone density as a function of both biological and mechanical *stimuli*. In this section, we propose on overview of these different *stimuli*.

2.1 Structure of Bone Tissue

Bone is a multiscale complex structure [16, 17] presenting two types of tissue (see Fig. 2a): i. the trabecular bone (or spongy bone), a very porous tissue (porosity $\sim 85\,\%$), located in the interior region of bone and containing bone marrow where hematopoiesis takes place; ii. the cortical bone (or compact bone), less porous (porosity $\sim 3\,\%$), located at the periphery of long bones. Representing 90 % of the

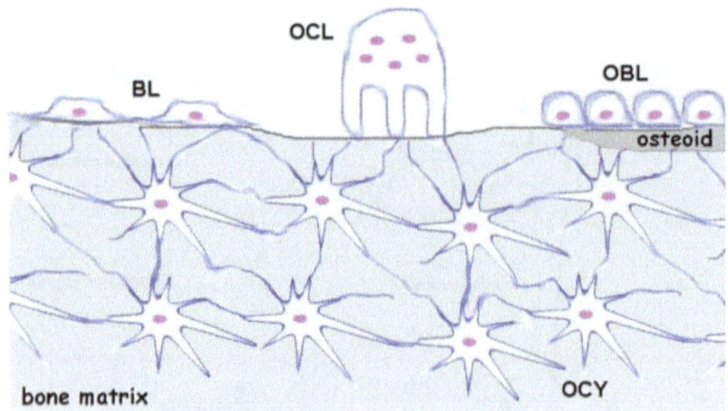

Fig. 3 Schematic representation of the four bone cells types (reproduced with permission from [82])

total bone mass, cortical tissue permits the locomotion, stores and releases chemical elements like calcium or phosphorous and protects the organs.

The cortical unit structure, called osteon, is a cylinder whose radius is about 10^{-4} m (see Fig. 2b). An osteon is constituted by the collagen-apatite matrix containing vascular porosity (Haversian and Volkmann's canals) and elliptic holes named *lacunae*. Each *lacuna* holds one mechanical sensor cell (osteocyte, see Fig. 2d) swimming in fluid environments. These osteocytes develop within little channels (*canaliculi*) connecting them together and so forming a stellar network within bone volume (see Fig. 2c). At the microscale, the representative volume is a fraction of the lacuno-canalicular system (see Fig. 2e). Canaliculi are described by two concentric straight cylinders whose the cross section is circular with radii R_C and R_M such as $R_C > R_M$. The interstitial fluid occupies the annular space between the canalicular wall and the osteocyte process membrane. The canaliculus length is noted L_C. Since the radii of the osteocyte and of the canaliculi vary with age, species, bone location, osteocyte age, etc. [26], a major difficulty in modelling consists in the allocation of these values. Thus if $R_C = 130 \pm 65\,\text{nm}$ for mice [170], for sheep and dogs R_C ranges from 100 to 500 nm [66, 129]. As we are interested in human bone, in this study, we will typically consider the values of You et al. [165], namely $R_C = 100\,\text{nm}$. Since the ratio R_C/R_M has given to be around 2 [154], the value of R_M will consequently be 50 nm. According to [165], the osteocyte process length is three hundred times its radius, so we typically have $L_C \equiv 15 \times 10^{-6}$ m.

2.2 Bone Cells

In bone tissues, four types of cells can be distinguished. Figure 3 gives a schematic representation of the four bone cells types. Multinucleated cells are the osteoclasts (OCL) which can remove bone. New bone is built by the osteoblasts (OBL) which can synthesize the osteoid matrix (white zone) which will be mineralized (grey zone). Some osteoblasts embedded in the bone volume can evolute in osteocytes (OCY) that are connected with bone lining cells (BL).

2.2.1 Lining Cells

The bone lining cells (BL) are located on the walls of the Volkmann and Haversian canals as well as on endosteal and periosteal zones of bone. They are undifferentiated original cells from the mesenchyme that can differentiate into osteoblastic cells. These stem cells are increasingly used in bone tissue engineering to generate tissue to be grafted [50].

2.2.2 Osteoblastic Cells

Osteoblastic cells (OBL) are mono-nucleate cells that are responsible for bone formation. Osteoblasts produce osteoid, which is composed mainly of Type I collagen and are also responsible for mineralization of the osteoid matrix. During the remodelling process, some osteoblasts are trapped in the bone matrix and thus form a lacuna where they become osteocytes (OCY). Furthermore, some osteoblasts at the remodelling surface become bone lining cells.

2.2.3 Osteocytes

Osteocytes (OCY) are the result of the evolution of osteoblast embedded in *lacunae*. Scattered in the bone volume, they are networked to each other via long cytoplasmic extensions (osteocyte process) that occupy tiny canals called canaliculi. This endows osteocytes with a star-shaped configuration. In the canaliculi which are used for paracrinal communication, exchange of nutrients and dispose of waste products, osteocytes can also communicate through gap junctions by the transmission of chemical agents.

2.2.4 Osteoclasts

Osteoclasts (OCL) are multinucleated cells that destroy bone tissue by removing its mineralized matrix and breaking up the organic bone (collagen). This process is known as bone resorption and is regulated by several hormones, including calcitonin from the thyroid gland, parathyroid hormone from the parathyroid gland, and growth factor interleukin 6.

2.3 Stages of Bone Remodelling Process

The processes of bone resorption and bone growth are taking place continuously throughout the skeletal system to protect it from daily wear. Bone remodelling is a cyclic progress that lasts approximatively 4 months when considering mature tissue. By observing bone tissues, resorption and formation stages of a basic multicellular unit (BMU) have been described [9]. Note that some authors prefer calling it bone modelling unit. The work of this BMU consists in resorbing and synthesizing bone material to generate a secondary osteon. Bone is so constituted by billions of modelling units likely to create new osteons. Since each BMU is chronologically and geographically separated from the others, it is possible to adapt the quantity and the architecture of bone. This tacitly suggests that the adaptation process depends on systemic or local factors. Indeed, the cellular events responsible for remodelling are locally controlled, certainly by autocrine or paracrine factors generated in the bone micro-environment [64, 87, 159]. Classically, the bone remodelling process is decomposed into 6 phases. Basically, this phenomenon sequentially occurs as summed up hereafter [51, 92].

2.3.1 Activation

In response to biochemical messages due to external physical *stimuli*, a BMU is activated. The *stimuli* that initiates bone remodelling have not been yet identified and a possible scenario is explored in the following of this article. The current concept of bone remodelling is based on the hypothesis that osteoclastic precursors become activated and differentiate into osteoclast. Thus, the lining cells are joined by the mono-nucleate precursors of osteoclasts or pre-osteoclasts.

2.3.2 Resorption

The bone resorption cascade involves a series of steps to obtain the removal of both the mineral and organic constituents of bone material. Roughly, the osteoclastic differentiated cells resorb old bone material. These cells ultimately undergo apostosis.

2.3.3 Reversal Phase

After the maximum eroded depth has been achieved by the osteoclasts, there is a reversal stage that lasts around 9 days. During this stage, osteoblasts counteract the osteoclats activity; resorption is stopped at a surface corresponding to the cement line that surrounds the future osteon.

2.3.4 Formation

This stage consists in the bone matrix formation in response to a complex cascade of events. The osteoblasts synthesize the organic extracellular matrix called osteoid and composed by collagen fibers arranged in concentric *lamellae*. This deposit is made from the cement line to the central zone of the future secondary osteon until the Haversian canal is formed.

2.3.5 Mineralization

Bone crystals composed of hydroxyapatite are transported within the collagen matrix. The mineralization phenomenon starts after 13 days at an initial rate of 1 μm/day.

2.3.6 Quiescence

When the secondary osteon is achieved, osteoblasts gradually flatten and become quiescent lining cells covering the Haversian surface. Some of the osteoblasts are simply destroyed or differentiate into osteocytes when embedded in the bone matrix.

The last three stages last longer than the first three ones. However, many questions remain on hold concerning this remodelling process. What is the physical signal pathway? If the effector cells are the lining cells that are activated to initiate the remodelling process, what is the nature of the signal felt by the mechano-sensitive cells?

2.4 *Candidates for the Mechano-Sensitive Role*

There are experimental evidences that most of the bone cells can sense a mechanical solicitation (membrane deformation, fluid shear effects, etc.) and consequently change their metabolic activity [31]. Anyway, two main scenarios, that may occur in parallel, have been proposed to explain the mechanically induced bone formation or resorption [30].

2.4.1 The Bone Surface Cells

On the one hand, the cells located at the bone surface (osteoblasts, lining cells and osteoclasts) may simultaneously be mechano-sensors and senso-actors. Thus, these cells would be able to sense the strain of their pericellular matrix in their immediate environment.

2.4.2 The Osteocyte: A Good Candidate for Mechano-Sensation

On the other hand, many authors stipulate that osteocytes are the bone mechano-sensing and mechano-transducing cells. Indeed, osteocytes are the most abundant cells in bone tissue, since there are 10 times more osteocytes than osteoblasts, which are themselves more numerous than osteoclasts. Furthermore, the special spatial organization of the osteocytes within the bone cortex forms a tridimensional network connected to the remodelling actors located at the bone surface.

Cell-to-cell communication of electrical signals and small molecules through gap junctions has been demonstrated in osteoblasts. There is evidence for similar gap junctions in osteocytes, and it seems likely that they participate in such communication with osteoblasts, and bone-lining cells as well.

2.5 Bone Remodelling Signals

There has been considerable speculation that osteocytes produce a signal proportional to mechanical loading by sensing different remodelling signals within bone tissue through stretch-activated ion channels, interstitial fluid flow, electrical potentials, or some other phenomenon. In this subsection, the main *stimuli* that can induce bone remodelling are presented.

2.5.1 *Stimuli* Originating in the Solid Matrix of Bone

According to the Wolff's law [162] that roughly states that bone is preferentially deposited in the area characterized by a high mechanical solicitation and removed where it is not mechanically needed, the first research of the bone remodelling were linked to the skeleton deformations.

- *Bone matrix micro-strains* Considering different physical activities, micro-strains of the human skeleton have been measured thanks to micro-gauges exhibiting values ranging between 0.04 and 0.3 % [21, 71]. The key parameters that directly influence the biological response of bone tissue have been shown to be the strain amplitude [136] and the strain rate [13, 49, 72, 134]. In particular, it is known that bone adaptation is proportionally governed by the strain rates [116, 149].
- *Hiatus between in vivo and in vitro micro-strains* In physiological conditions typically corresponding to normal locomotion activities, the bone tissue strains that can be measured in vivo remain rather small since quantitative data obtained for running horses and men and fast flying birds present maximal values around 0.2–0.3 % [21, 133]. These measurements are paradoxical when compared to the *in vitro* necessary strains that induce a cellular response which are one or two order higher, from 1 to 10 % [34, 135]. In vivo, such a huge strain

level would cause bone fracture. This paradox reinforces the idea that the direct mechano-sensation of the mechanical loading by the bone surface cells (OBL, BL and OCL) is certainly not the main sensing pathway.

- *Micro-cracks* Notwithstanding the reversible micro-strains, physiological observations of bone tissue exhibit that normal bone presents micro-cracks [2, 114]. These cracks originate within the bone cortex and tend to merge and propagate along the cement lines that form the outer layer of the osteons [113, 140]. This indicates that one role of the cement line could be to deflect the bone micro-cracks propagation and thus limit the fracture risk. In the wake of these imaging results, new scenarios of the bone remodelling initiation have been proposed. For instance, the stress concentration phenomenon that is inherent to these micro-cracks has been proposed to be the key textural phenomenon inducing the remodelling process [62]. Sites of remodelling in cortical bone have been indeed shown to occur in conjunction with micro-cracks [20]. In particular, it has been observed experimentally that a strong association between microdamage, osteocyte viability and modulation of remodelling activity does exist [151]. This supports the idea that osteocyte apoptosis may play a role in the signalling mechanisms by which bone is remodelled after microcrack formation [111]. It has also been suggested that a micro-damage occurring inside the osteonal volume may generate a cell transducing mechanism based on ruptured osteocyte processes [48]. Concomitantly, micro-cracks are likely to alter the fluid flow and convective transport through the bone tissue and thus modify the hydraulic behaviour of the fluid in the vicinity of the sensitive cells [40, 73, 103, 108, 110]. As shown hereafter, the fluid environment of osteocytes plays also a crucial role of bone mechano-sensation.
- *Bone piezo-electricity: a 60 years old idea* From the birth of electrophysiology in the wake of Galvani's work in the late eighteenth century [120] to the contemporary electromagnetic medicine, the action of electricity on living tissue fascinates. Focussing on bone electricity, the year 2012 is the sixtieth anniversary of the discovery of the piezo-electricity of bone which was reported by Dr. Yasuda from Kyoto, Japan [70].

The piezo-electric properties of dry bone are not due to the apatite crystals, which are centrosymmetric and thus non-piezo-electric, but to collagen molecules [37]. Collagen exhibits the polar uniaxial orientation of molecular dipoles in its structure and can so be considered as a sort of dielectric material. In the 1960s, electric measurements in bone tissue [164] motivated the hypothesis that bone adaptation could be explained thanks to collagen piezo-electricity. Historically, it was argued that a mechanically loaded bone induces compression on its concave side and tension on its convex side [10]. Due to the piezo-electricity of bone collagen, negative charges are visible on its compressive side and positive charges on its tensile parts [95]. Thus, it is stipulated that, in the electro-negative zone, osteoblasts would be stimulated, increasing bone formation, whereas osteoclasts activity, and therefore bone resorption, would be improved in the electro-positive zone. If the polarization implications seem to be quite easy to understand when focussing on the solely mineralization process [112, 163], the mechanisms by which charges and piezo-electric properties affect osteoblasts and osteoclasts responses are more difficult to be identified [11, 107]. Moreover, this piezo-electric craze faded away in the 1980s when more compelling mechanisms related to the interstitial fluid movement began being studied [55]. Only recent studies proposed that piezo-electricity could induce an increase in the electric charge of the bone tissue and thus engenders an opposite electro-osmotic flow limiting the total interstitial flow, and thus increasing the apparent stiffness and the mass transport properties of bone tissue [3, 78]. Moreover, it has been recently proposed that the mineral crystals of the bone matrix could act as an electric storage system [106]. Notwithstanding this renewed interest in bone piezo-electricity, important components of this hypothesis have to be demonstrated and several questions remain. In particular, due to the poro-mechanical coupling, the stress (or strain) field of the solid phase of bone cannot be separated of concomitant interstitial fluid movement. Spurred on by the pioneering paper of [121] in the late 1970s, the mainstream idea in the bone community became that the mechano-transduction of bone remodelling was a flow-induced phenomenon [35].

2.5.2 Bone Fluid Flow Signals

It had been for a long time believed that the sole function of bone interstitial fluid movement in the lacuno-canalicular pores was to provide nutrients and remove wastes. The strain induced micro-flows were first proposed by Piekarski and Munro [121]. However, these lacuno-canalicular micro-flows have only been experimentally observed 20 years later by tracer studies [67–69, 155]. This difficulty to carry out convenient in vivo experiments to measure hydraulic fluid velocities and interstitial fluid pressure within bone tissue motivated the model-driven investigations of the bone behaviour.

- *Evidence of the fluid flow and stretch in bone cell activity* Since it forms the immediate environment of bone cells and the pathway for nutrient supply and waste removal, the role of bone interstitial fluid in bone activity is evident [66]. In vivo, when comparing the effect of static [72] and cyclic mechanical loading [135], only the later mechanical condition positively influences bone formation for peak strains of 0.1 %. The difference between these two loading types is that the dynamical loading induces interstitial fluid movement. In parallel, several *in vitro* studies proved that bone cells were sensitive to neighboring fluid flow [54, 96, 161]. Thus, several studies demonstrated that bone cells are more responsive to fluid flow than to mechanical strain. For instance, the strain-induced osteoblastic response measured by Binderman et al. [14] is 6 times lower than the flow-induced response observed by Reich and Frangos [127, 128]. When focussing on the osteocytes, if their in vivo sensitivity to the mechanical loading was known in the late 1980s [143], the strong influence of the fluid shear stress induced by neighboring flow on the cellular activity was proven *in vitro* the decade after [4, 18, 65, 160].

- *Pressure effects* By comparing biochemical responses of osteoblast and osteocytes when submitted to a fluid flow, it was showed that the shear stress induced by the flow was the predominant mechanical effect felt by the cells [65]. Notwithstanding this strong evidence, some studies still impute an important role of the pressure in the bone cells behaviour [104, 105].
- *Flow shear stress* In the footsteps of Piekarski and Munro [121], Weinbaum et al. [158] proposed that the fluid flow due to physiological loading was the primary *stimulus* that enabled osteocytes to sense and respond to their mechanical environment. Their theoretical investigation predicted shear effects induced by the fluid flow on the osteocyte process membrane of a few Pascals. This value is roughly of the same level as for the endothelium in vascular capillaries, and typically corresponds to *in vitro* measurements of bone fluid shear stress [65]. The success of the "flow shear stress" as a valuable candidate for the mechano-sensation role in bone tissue is still visible through the recent works on bone remodelling based on this signal [1, 56, 58, 60, 110].
- *Worship what you have burned, and burn what you have worshiped: shear force versus drag force* Even if the shear stress effects predicted in the lacuno-canalicular and vascular systems are comparable, the morphological differences between dendritic cell and endothelial cell should induce different mechano-sensation scenarios. Indeed, due to the densely packed central actin filament bundle in the osteocyte process, this portion of the osteocyte should be much more rigid than the endothelial cell body. Thus, the same group that made the success of the shear effects within bone tissue proposed another point of view to investigate the interactions between the interstitial fluid flow and the ultra-structure of the osteocyte. Indeed, concomitantly to the progress in the imaging of the canalicular structure [97, 166], forgetting the shear effects, the Weinbaum's group proposed that the stimulus was the drag force exerted by the fluid flow on the pericellular matrix surrounding the cell processes. These forces would be transmitted by tethering filaments and canalicular projections which connect the membrane of the cell process to the canalicular wall, generating a strain amplification of the cell membrane in the hoop direction [46, 156, 165]. Other recent studies provide evidence that primary cilia projecting from the surface of cultured bone cells can translate fluid flow into cellular responses [90]. Note that cilia are present only in lacuna in close proximity (25 μm) to the periosteal surface, that these cilia are parallel and not perpendicular to the cell body surface and, thus, could not bend in response to fluid flow under these confined conditions [159].
- *Cell-to-cell communication* The transport of biochemical species within bone is not only essential for the survival of osteocytes which are not directly in contact with the vascular supply [100], but it is central in the paracrinal cell-to-cell communication as well. Indeed, another stage in the pathway of bone mechanotransduction is the production of signalling molecules by the osteocyte, which can alter the bone remodelling activity of osteoclasts and osteoblasts. Even

when loads are normally transferred into a fluid flow, which is subsequently adequately sensed by the osteocyte, it is still possible that the transduction of the mechanical signal to a chemical signal that takes place within the osteocyte is affected. As a consequence, the production of signalling molecules by mechanically stimulated osteocytes will be altered. In vivo, this biochemical cell-to-cell communication becomes more and more understood [28]. Diffusive mechanism alone gives a prohibitively long transit time generating very different chemical environments of osteocytes according to their position [68, 124]. Based on in vivo tracer perfusion studies showing the link between mechanical loading and mass transport within bone tissue [67, 89, 155], other transport mechanisms have been proposed: load-induced fluid flow and concomitant convective mixing [68, 121, 152] or Taylor-Aris dispersion [141]. However, based on a theoretical study, it was predicted that neither diffusion nor stress induced fluid flow is capable of sustaining osteocyte viability, and thus that cyclic stress could stimulate an active mass transport in bone [119].

- *The electro-chemical avenue of research* When studying fluid transfer in nanoporous materials such as the lacuno-canalicular system, other concurrent phenomena to pressure-driven flow, such as electro-osmosis or osmosis, may exist. These coupled effects could generate an opposite flow, resulting in a decrease in the apparent permeability [3, 78], ionic permselectivity effects [59, 76], and a modification of the Biot model and the Darcy law [74, 86, 102]. This phenomenon is due to an important property inherent to most of the biological porous media: they present a surface charge on their pore surface. Indeed, within cortical bone, the lacuno-canalicular pores tend to present a negative surface charge, due to the presence of fatty acids on the bone matrix [94] and of phospholipids on the cell membrane [98]. This negative charge is compensated by the adsorption of cations on the surface forming the inner compact layer commonly referred to as the immobile Stern layer. However, the majority of the excess of positively charged counter-ions are located in the electrolyte aqueous solution in the vicinity of the solid phase, forming an outer diffuse layer composed of mobile charges. Together with the fixed charged groups of the solid matrix, these ions form the so-called electrical double layer, as shown in Fig. 4 [52, 75]. When advected by the interstitial fluid, the mobile charge population of the double-layer generates the macroscopically observed streaming currents and the concomitant streaming potential [74]. The gradient of this potential engenders the electrophoretic movement of the mobile charges opposing to the streaming current. Due to the viscous drag interaction, the ions pull the liquid with them resulting in an electro-osmotic seepage flow opposing the pressure-gradient driven flow, and thus limiting the apparent permeability. Moreover, due to the possible overlap of similarly charged ionic layers, the Donnan pressure may cause swelling effects changing the macroscopic hydro-mechanical behaviour of the porous materials, as, for instance, visible in clayey media [79, 88].

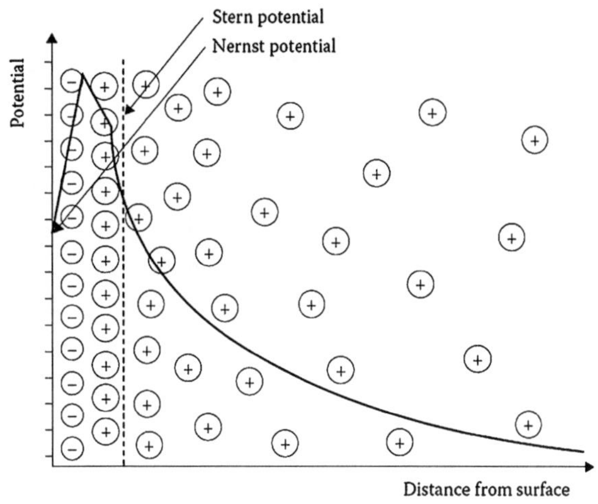

Fig. 4 Equilibrium electrostatic potential in an electrolyte solution bordered by a plane negative surface (reproduced with permission from [115]). The microstructure of the ionic distribution at the interface induces the phenomenon of Debye shielding by the ion cloud of the opposite sign (double layer phenomenon)

Even if highly studied in the 1980s to prove the lacuno-canalicular fluid movement thanks to the stress-generated streaming potentials [123, 138, 139], these hydro–electro-chemical couplings have been poorly studied in the context of the flow induced mechano-transduction.

Additionally, some ionic exchanges appear between the bone matrix and the bone fluid [33, 91] as well as between bone cells and the interstitial fluid. Indeed, being living entities, bone cells receive nutrients, throw their waste and communicate with one another through the interstitial fluid [100]. This biochemical activity and the surface chemical reactions lead to a change of the physico-chemical properties of the fluid/solid interface. In particular, the negative surface charge density is modified when exchanging ionic species and thus depends on the ionic concentration [146]. Because of the electro-viscous couplings, these exchanges influence the interstitial fluid transport [76, 80] and, *in fine*, the fluid mechano-transduction signals.

2.5.3 Multiphysics Mechano-Transduction of Bone Remodelling: A Multiscale Strategy

Here, we propose to adopt a fresh new angle to elucidate the multiphysics consequences on the mechano-transduction of bone remodelling. Indeed, combining a description of electrokinetics in bone tissue [84–86] with the description of piezo-biomaterials [99], we intend to derive a fully coupled electro–hydro-mechanical model of bone tissue [74]. Since the anatomical origin of bone electricity is located at the microscopical scale, our electro–chemo-mechanical description combines the collagen fibers deformation effects and the advection of ionic double layers through canalicular network. Then, adopting the periodic asymptotic homogenization procedure proposed by Auriault and Sanchez-Palencia [8] based on the

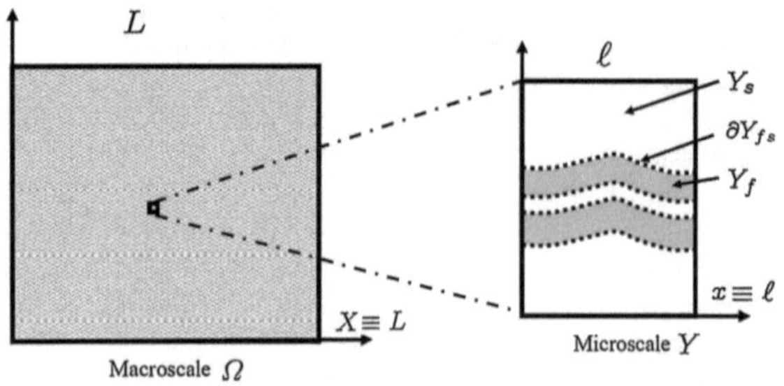

Fig. 5 Representation of the macroscopic medium Ω (reference length L) and the periodic representative cell Y (reference scale ℓ). In the representative periodic cell, the fluid and solid domains Y_f (*grey*) and Y_s (*white*) and the solid-fluid interface ∂Y_{fs} (*dashed boundary*) are shown (reproduced with permission from [77])

determination of scaling laws, the consequences at the macroscale of these microscopic phenomena are obtained.

Our strategy consists of discussing, at both the macroscale and the microscale, the importance of the multiphysical phenomena featuring in bone behaviour using physiologically-based simulations. That is why, in the first section, the ingredients of our model are introduced. Then, some numerical examples are presented to illustrate the interest of our multiscale approach. Through these simulations, we are able to distinguish the phenomena that are purely microscopic from those which can be macroscopically observed. Finally, from these multiscale considerations, we propose some avenues which it is worth exploring when aiming at in silico representing the bone remodelling process.

3 Multiphysics Model of the Bone Behaviour

The first step in our investigation consists in giving an accurate description of the phenomena. The solid, the fluid and the fluid-solid interface are treated separately. The representative periodic cell is built from anatomical considerations. Cortical bone is seen as a fully saturated porous medium Ω. The relevant porosity level Ω_f for bone fluid electro-chemical phenomena corresponds to the lacuno-canalicular porosity [84] whereas the origin of piezo-electricity is located in the collagen-apatite matrix Ω_s. Thus the representative elementary volume Y corresponds to a periodic portion of the lacuno-canalicular network as shown in Fig. 5. In this representative periodic cell, the fluid and solid domains are noted Y_f (grey) and Y_s (white) respectively and the solid-fluid interface ∂Y_{fs}. In cortical bone, the canalicular pores are occupied by a dendrite of the osteocyte cells (seen here as a part of the solid phase) giving at the fluid domain an annular geometry. Note that we neglect the influence of the larger porosity level of cortical bone corresponding to vasculature. This assumption is grounded on the hierarchical multi-porous structure of bone described as a set of nested porosities like a set of Russian nested dolls. This idea is developed using a phenomenological approach in Cowin et al. [27], Gailani and Cowin [39] and a homogenization procedure in Rohan et al. [132]. As a result, the interstitial fluid transport outcomes are governed by the phenomena at the smallest porous level.

3.1 Equations at the Microscale of the Lacuno-Canalicular Structure

Neglecting the magnetic effects, at the microscale, the multiphysics description integrate the solid matrix electro-mechanics, the interstitial fluid electrokinetics and the solid-fluid interfacial phenomena.

3.1.1 Piezo-Electricity in the Solid Matrix of Bone Tissue

Due to the piezo-electric property of the collagen matrix of bone, this phase can be considered as a sort of dielectric material characterized by a permittivity tensor ε_s and exhibiting a quasi-permanent space charge q_s. The electric potential in the solid ϕ_s is thought to occur when a number of collagen molecules are stressed in the same way in response to a gradient of the displacement vector field $\mathbf{u}$, moving charge carriers from the inside to the surface of the specimen. As a result, the constitutive laws for the stress tensor $\mathbf{S}_s$ and the electrical displacement vector field $\mathbf{D}_s$ in the solid involve both the displacement and the electrical potential effects, the piezo-electric coupling being quantified by the the piezo-electric third-order tensor $\mathbf{P}_Z$:

$$\mathbf{S}_s = \mathbb{C} : \varepsilon(\mathbf{u}) + \mathbf{P}_Z^T \cdot \nabla\phi_s, \tag{1}$$

$$\mathbf{D}_s = \mathbf{P}_Z : \nabla\mathbf{u} - \varepsilon_s \cdot \nabla\phi_s, \tag{2}$$

where $\mathbb{C}$ represents the fourth-order elasticity tensor of the solid, ∇ is the gradient operator, $\star^T$ stands for the transpose operator, $\varepsilon(\star) = 1/2(\nabla\star + \nabla\star^T)$ is the operator that gives the symmetric part of the gradient of the quantity $\star$. Here $\varepsilon(\mathbf{u})$ represents the symmetric part of the gradient of the displacement of the solid phase (i.e. the infinitesimal strain second-order tensor). By neglecting the body forces, the mechanical equilibrium and the Maxwell-Gauss equations in the solid phase are:

$$\nabla \cdot \mathbf{S}_s = \mathbf{0}, \tag{3}$$

$$\nabla \cdot \mathbf{D}_s = q_s, \tag{4}$$

where $\nabla\cdot$ designates the divergence operator,

3.1.2 Electrokinetics in the Fluid Phase of Bone Tissue

The multiphysics description of the transport phenomena in the bone interstitial fluid requires to combine electrostatics, fluid movement and ionic transport.

- *Double-layer and streaming potentials* The interstitial fluid, which is composed by a Newtonian and incompressible water-like solvent containing biochemical agents, is a dielectric material characterized by a spherical permittivity tensor $\varepsilon_f = \varepsilon_f \mathbf{I}$, $\mathbf{I}$ being the unit second-order tensor. For simplification purpose, if we represent all the charged species in the fluid by monovalent ions (seen as point charges) characterized by their molar concentration n^+ and n^-, the charge density in the fluid q_f is then expressed by:

$$q_f = F(n^+ - n^-), \tag{5}$$

where F is the Faraday constant. When advected by the strain-induced interstitial fluid velocity, the mobile charge population of the double-layer generates the streaming currents. In parallel, to conserve charge, the movement of the net charge generates an electric potential, often referred to as streaming potential Ψ_b.
As shown in Fig. 4, the surface charge of the pores induces asymmetric Boltzmann distributions of the cationic and anionic concentrations which are governed by the reduced double-layer potential $\bar{\varphi}$:

$$n^\pm = n_b \exp(\mp\bar{\varphi}). \tag{6}$$

Note that the reduction of electric potentials $\star$ involves the Faraday constant F, the ideal gas constant R and the absolute temperature T, so that $\bar{\star} = F\star/RT$. In summary, the electric potential in the fluid ϕ_f is decomposed into the sum of the double layer potential φ and the streaming potential Ψ_b [86]:

$$\phi_f = \varphi + \Psi_b. \tag{7}$$

Furthermore, the double-layer potential φ obeys the Poisson-Boltzmann equation [52, 74]:

$$\nabla \cdot \nabla(\bar{\Psi}_b + \bar{\varphi}) = \frac{1}{L_D^2} \sinh \bar{\varphi}. \tag{8}$$

The Debye length $L_D = \sqrt{\varepsilon_f RT/(2F^2 n)}$ characterises the thickness of the diffuse ionic layer. When the pore size is large when compared to the Debye length, the hyperbolic sine of Eq. (8) can be linearized ($\sinh \bar{\varphi} \simeq \bar{\varphi}$) to obtain the Debye-Hueckel approximation [84].

- *Fluid movement* If the interstitial fluid is assumed Newtonian and incompressible, its constitutive law is [74]:

$$\mathbf{S}_f = -(p_b + \pi)\mathbf{I} + 2\mu_f \varepsilon(\mathbf{v}), \tag{9}$$

where $\mathbf{S}_f$ is the stress tensor in the fluid, $\mathbf{v}$ is the fluid velocity, p_b is the hydraulic pressure, π is the Donnan pressure and μ_f is the fluid dynamic viscosity. Note that the osmotic Donnan swelling pressure $\pi = 2RTn_b(\cosh\bar{\varphi} - 1)$ due to the double layer electrostatics repulsion [79] is here introduced.

Moreover, the canalicular space is partially occupied by pericellular fibers generating a sub-microscopic friction effect and thus slowing down the fluid movement [85, 166]. To take into account this effect occurring in the fluid volume, a viscous force $\mathbf{F}_b = -(\mu_f/k_f)\,\mathbf{v}$ depending on the fluid velocity $\mathbf{v}$ is introduced. Here, k_f represents the isotropic pericellular fibers permeability. Note that for few fibers, the pore space is very permeable (i.e. k_f is high) and $\mathbf{F}_b$ vanishes. In parallel, noting $\mathbf{E}_f$ the electric field in the fluid, the electric body force $\mathbf{F}_e = q_f\,\mathbf{E}_f$ has to be taken into account. Thus the equilibrium equation for the fluid reads:

$$\nabla \cdot \mathbf{S}_f + \mathbf{F}_b + \mathbf{F}_e = \mathbf{0}. \tag{10}$$

In addition, assuming the electrolyte to be incompressible, the mass conservation equation for the fluid reads:

$$\nabla \cdot \mathbf{v} = 0. \tag{11}$$

- *Ionic transport* Regarding the ionic transport, the total cationic and anionic flux vectors are the sum of the convection flux vector, the diffusion flux vector and the electric current vector. Thus, the Nernst-Planck convection-diffusion-electromigration equations that govern the ionic transport are [76]:

$$\frac{\partial n^\pm}{\partial t} + \nabla \cdot (n^\pm \mathbf{v}) = \nabla \cdot \left[\mathbf{D}_\pm \cdot (\nabla n^\pm \pm n^\pm \nabla \bar{\phi}_f)\right], \tag{12}$$

where t is the time, $\mathbf{D}_\pm$ are the water-ions diffusion tensors for cations and anions, respectively. Using the Boltzmann distributions of the ionic species of Eq. (6) and the electric potential decomposition of Eq. (7), this equation becomes:

$$\frac{\partial(n_b \exp(\mp\bar{\varphi})}{\partial t} + \nabla \cdot (n_b \exp(\mp\varphi)\mathbf{v}) = \nabla \cdot [\mathbf{D}_\pm \cdot (\exp(\mp\varphi)(\nabla n_b \pm n_b \nabla \bar{\Psi}_b))], \tag{13}$$

In this equation, the terms in the divergence corresponds to the electro-diffusive ionic flux vectors $\mathbf{J}_\pm = -\mathbf{D}_\pm \cdot (\exp(\mp\bar{\varphi})(\nabla n_b \pm n_b \nabla \bar{\Psi}_b))$.

3.1.3 Solid-Fluid Interface Conditions

The solid-fluid interface ∂Y_{fs} is characterized by its normal unit vector $\mathbf{n}$ directed toward the solid domain. Concerning the surface charge density, the presence of the Stern layer and the ionic exchanges quantified by the coefficients $\alpha^{\pm}$ (homogeneous to lengths [76]) may generate an electrical jump σ_{fs} by crossing the interface from the solid part to the fluid phase. As a result, two different electric surface charges are introduced, σ_f and $\sigma_s = \sigma_f + \sigma_{fs}$, referring respectively to the surface charge density seen from the fluid or the solid. The set of interface conditions is then given by:

$$\phi_f = \phi_s, \tag{14}$$

$$\left(-\varepsilon_f \cdot \nabla\phi_f\right) \cdot \mathbf{n} = \sigma_f, \tag{15}$$

$$-\left(\mathbf{P_Z} : \nabla\mathbf{u} - \varepsilon_s \cdot \nabla\phi_s\right) \cdot \mathbf{n} = \sigma_s, \tag{16}$$

$$\mathbf{v} = \frac{d\mathbf{u}}{dt}, \tag{17}$$

$$-\mathbf{J}_{\pm} \cdot \mathbf{n} = \alpha^{\pm}\frac{\partial n_b}{\partial t}, \tag{18}$$

$$\mathbf{S_s} \cdot \mathbf{n} = \mathbf{S_f} \cdot \mathbf{n}. \tag{19}$$

They correspond, respectively, to the electric potentials continuity and the electric flux conditions at the interface, to the no-slip condition, to the ionic exchange property of this interface and to the continuity of the normal stress.

3.2 *Upscaling Procedure*

An asymptotic periodic homogenization process [7, 102] is carried out to propagate our microscopic description of the phenomena at the upper scale. The principle of this procedure is to consider a periodic representative cell Y or representative volume element (RVE), whose size must be large enough to contain all relevant microscopic heterogeneities and small enough to ensure the macroscopic homogeneity. Let ℓ and L be characteristic lengths of the micro- and macroscales respectively, and x and X the associated coordinate systems. In order to keep the independence between these two scales, the ratio $\eta = \ell/L$ must be small in comparison with 1. In our problem, the length ℓ associated with the microscale is typically the pore size length and the length L is typically of the order of the cortical tissue size.

The general procedure of the periodic homogenization consists in: i. writing the equations at the microscale in a non-dimensional fashion and giving scaling laws for each non-dimensional number; ii. expanding all the fields in terms of the small parameter η and collecting the terms corresponding to each power of η (cascade of equations); iii. identifying the slow variables that do not vary at the microscopic scale and proposing closure problems for the supplementary fluctuating terms; iv. deriving the modelling at the macroscale by averaging the remaining quantities at the microscale over the representative cell.

Remark The obtention of the macroscopic equations requires to average quantities of the microscale over the representative cell Y. Thus, we define $\langle * \rangle = \frac{1}{|Y|}\int_Y * \, dV$ the averaging operator on the representative cell Y. When focussing on the fluid phase occupying the representative cell, the average over the fluid domain Y_f is: $\langle * \rangle_f = \frac{1}{|Y_f|}\int_{Y_f} * \, dV_f$. Finally, $\langle * \rangle_{int} = \frac{1}{|Y|}\int_{\partial Y_{int}} * \, dS$ represents the solid-fluid interface averaging operator.

When focussing on the multiphysics behaviour of bone tissue, the homogenization procedure has been carried out for the Poisson-Boltzmann (8) and the fluid flow (10) equations in a previous study [86], whereas the upscaling of the Nernst-Planck equations (13) and its associated ionic exchanges has been summarized in Kaiser et al. [57] and Lemaire et al. [76]. Furthermore, when neglecting the pericellular matrix and the ionic exchanges at the solid-fluid interface, the homogenization of the piezo-poro-mechanics model of bone tissue is extensively presented in [74].

3.3 Consequences at the Scale of Bone Tissue

Thanks to the upscaling procedure, it is possible to obtain the model at the macroscale that will mimic the macroscopic behaviour of bone tissue.

3.3.1 Macroscopic Unknowns

First, the homogenization procedure allows to identify the variables purely macroscopic (slow variables). So, the solid displacement $\mathbf{u}_{[0]}$, the bulk concentration $n_{b_{[0]}}$, the streaming potential $\Psi_{b_{[0]}}$ and the fluid pressure $p_{b_{[0]}}$ are slow variables that do not vary at the microscopic scale [74].

Remark All these quantities are indexed [0] since they correspond to the first term in the expansion of the corresponding physical quantity into a sequence of the small parameter η.

As summarized in Table 1, the macroscopic unknowns are the Darcy velocity $\mathbf{V}$, the solid displacement $\mathbf{u}_{[0]}$, the total stress tensor $\mathbf{S_{tot}} = \left(< \mathbf{S}_{s[0]} > + < \mathbf{S}_{f[0]} > \right)$, the ionic concentration $n_{b_{[0]}}$, the streaming potential $\Psi_{b_{[0]}}$, the fluid pressure $p_{b_{[0]}}$ and the porosity η_f. This corresponds to 16 scalar unknowns.

3.3.2 What Stays Beyond Macroscopic Observation

Due to the scaling laws inherent to cortical bone, both the double layer and the piezo-electric effects vanish through the homogenization process. Notwithstanding the fact that the double layer electric potential φ and the piezo-electric potential ϕ_s do not filter through the upscaling process, their consequences at the macroscopic scale can still exist. In particular, as indicated by Ahn and Grodzinsky [3], the variations of the piezo-electric potential ϕ_s directly modify the pore surface charge, and thus the double layer potential. As shown hereafter, even if purely microscopic, this double layer potential can induce important consequences at the macroscale.

3.3.3 Ionic Electro-Diffusive Transport

In the remodelling process, the paracrine communication between the mechano-sensors (osteocytes) and the effector cells (osteoclasts and osteoblasts) requires to develop specific transport processes. Due to the narrow space of the *canaliculi*, the convective effect vanishes through the upscaling process. Taking into account the possible ionic exchanges between the cell and its fluid environment and the electromigration effects, two macroscopic electro-diffusive Nernst-Planck equations can be obtained for monovalent ions (see Remark) [76]:

$$\begin{aligned} &\frac{\partial}{\partial t}\Big([n_{b_{[0]}}\Big(\eta_f\langle\exp(\mp\bar{\varphi}_{[0]})\rangle_f\Big)\Big)] + \frac{\partial n_{b_{[0]}}}{\partial t}\langle\alpha_{\pm[0]}\rangle_{int} \\ &\quad = \nabla_X \cdot \Big[\mathbf{D}_{\pm}{}^{*}(\nabla_X n_{b_{[0]}} \pm n_{b_{[0]}}\nabla_X\bar{\Psi}_{b_{[0]}})\Big]. \end{aligned} \tag{20}$$

Remark The nabla operator is here indexed X since it corresponds to the macroscopic spatial derivative operator, that is to say with respect to the macroscopic coordinate X.

This equation exhibits three contributions to the ionic transport: i. the temporal term involving the influence of the porosity η_f, the averaged double-layer effects and the surface exchange term $\alpha_{\mp}$; ii. a Brownian diffusion term in response to the salinity gradient; iii. an electromigration term in response to the gradient of the streaming potential.

These two last terms are quantified using effective diffusion tensors $\mathbf{D}^*_{\pm}$ involving, in addition to the diffusion coefficients of the ions $D_{\pm}$, the porosity η_f and the electro-tortuosity tensors $\boldsymbol{\vartheta}_{\pm}$:

$$\mathbf{D}^*_\pm = \eta_f D_\pm \vartheta^{-1}_\pm. \tag{21}$$

The explicit definition of $\vartheta_\pm$ is obtained during the homogenization process and is detailed in Kaiser et al. [59].

3.3.4 Modified Darcy Law

As obtained in clayey materials [81, 102], the fluid flow, in addition to the hydraulic pressure gradient, may be governed by supplementary driving phenomena: the electro-osmotic seepage flow induced by the streaming potential fluctuations and the osmotic flow in response to the chemical gradient. As a result, the macroscopic fluid flow is described through a modified Darcy law including, in addition to the pressure gradient induced flow (indexed by P), chemo-osmosis (indexed by C) and electro-osmosis (indexed by E):

$$\mathbf{V} = \mathbf{V}_P + \mathbf{V}_C + \mathbf{V}_E = -\mathbf{K}_P \nabla_X p_{b_{[0]}} - \mathbf{K}_C \nabla_X n_{b_{[0]}} - \mathbf{K}_E \nabla_X \Psi_{b_{[0]}}. \tag{22}$$

The macroscopic permeability tensors $\mathbf{K}_k$ $(k = P, C, E)$ are obtained through the homogenization process as detailed in Lemaire et al. [74, 85, 86].

3.3.5 Coupled Biot Problem

Bone fluid flow is generated by the strain of the solid matrix. Classically, in bone biomechanics, the calculation of the hydraulic velocities caused by the mechanical loading are based on the poro-elasticity theory [15]. Here, a Biot-like constitutive relation derived from our microscale analysis is obtained [74]:

$$\mathbf{S}_{tot} = \mathbb{C}^* : \varepsilon_X(\mathbf{u}_{[0]}) - \boldsymbol{\alpha}^* p_{b_{[0]}} + \boldsymbol{\tau}^*. \tag{23}$$

In this equation, ε_X corresponds to the macroscopic part of the operator ε, that is to say built from ∇_X. Thus, $\varepsilon_X(\mathbf{u}_{[0]})$ corresponds to the macroscopic strain tensor. Furthermore, the homogenized fourth-order elasticity tensor $\mathbb{C}^*$ and the homogenized Biot second-order tensor $\boldsymbol{\alpha}^*$ are obtained following the classical treatment of poro-elasticity as proposed by Auriault and Sanchez-Palencia [8]. Moreover, the macroscopic electro-chemical tensor $\boldsymbol{\tau}^*$ representing the macroscopic effects of the fluid electro-chemical phenomena is similar to the one previously obtained for the multiphysical description of clayey materials [81, 102]. This tensor accounts for: i. the spherical Donnan pressure effect; ii. the action of the Maxwell tensor; iii. the electro-chemical effects occurring at the solid-fluid interface [74].

As a result, by neglecting the body forces, the macroscopic equilibrium equation simply reads:

$$\nabla_X \cdot \mathbf{S}_{tot} = \mathbf{0}. \tag{24}$$

Table 1 Inventory of the unknowns and equations present in the multiphysics poro-elastic model at the macroscopic scale

Unknowns		Equations	
Physical quantity	Number	Macroscopic equations	Number
Solid displacement $\mathbf{u}_{[0]}$	3	Macroscopic balance equation (24)	3
Total stress tensor $\mathbf{S}_{tot}$	6	Modified Biot constitutive law equation (23)	6
Darcy velocity $\mathbf{V}$	3	Modified Darcy law equation (22)	3
Ionic concentration $n_{b_{[0]}}$	1	Electro-diffusive Nernst-Planck equation (20)	2
Fluid pressure $p_{b_{[0]}}$	1	Mass conservation equation (25)	1
Streaming potential $\Psi_{b_{[0]}}$	1	Fluid mass conservation equation (26)	1
Porosity η_f	1		
Total of scalar unknowns	16	Total of scalar equations	16

The coupled Biot equation is derived from the mass conservation and the Darcy law:

$$\nabla_X \cdot \mathbf{V} = -\boldsymbol{\alpha}^* : \frac{\partial}{\partial t} \boldsymbol{\varepsilon}_X \left(\mathbf{u}_{[0]}\right) + \boldsymbol{\Upsilon} \frac{\partial}{\partial t} p_{b[0]} + \boldsymbol{\varpi}. \tag{25}$$

Again, in addition to the classical Biot formulation involving the Biot tensor $\boldsymbol{\alpha}^*$ [8] and the number Υ which are explicitly derived through the upscaling procedure, another electro-chemical effect ϖ has to be also taken into account. This quantity, which is due to the electrically induced pores deformation is explicitly given in Lemaire et al. [74].

Finally, it can be shown that the fluid mass conservation could be expressed in terms of the porosity, the Darcian velocity and the displacement [102]:

$$\frac{\partial \eta_f}{\partial t} + \nabla_X \cdot \mathbf{V} + \eta_f \nabla_X \cdot \frac{\partial}{\partial t} \mathbf{u}_{[0]} = 0. \tag{26}$$

3.4 Summary of the Macroscopic Description of Bone Tissue

A problem dealing with studying the multiphysics evolution of a poroelastic material is defined by a macroscopic set of time-space partial differential equations over the macroscopic domain involving different physical unknowns. In addition, convenient macroscopic boundary conditions have to be provided. To check that our multiscale strategy results in a macroscopic problem that is convenient with the number of the unknowns, we recapitulate in Table 1 the final macroscopic description of bone tissue.

It is important to indicate that, in addition to these macroscopic laws, the purely microscopic cellular problems have to be solved on the periodic cell to express all the homogenized parameters. Moreover, the double layer potential φ, solution of the Poisson-Boltzmann problem, has also to be found since it appears in some of these effective parameters.

Now that a multiphysics representation of bone tissue has been built, we are able to simulate its poro-elastic behaviour, and thus to improve our understanding of the mechano-transduction of the bone remodelling signals.

4 Some Illustrative Results Through 5 Questions Dealing with Bone Remodelling Signals

The great difficulty of carrying out experimental investigations of in situ bone poro-mechanical behaviour renders theoretical investigations crucial. Modelling and computational approaches become thus increasingly common tools for testing hypothesis for the regulation of bone remodelling [1, 150].

In this section, we propose to illustrate the great interest of adopting a coupled viewpoint to treat the in silico issues of bone remodelling. The classical continuum or micro-mechanically based models of bone behaviour use the sole poro-mechanical description [24, 132]. The mechanical *stimuli* acting at the cell scale (pressure, shear stress, drag forces, etc.) are downscaled from this poro-mechanical description and somehow converted into biochemical signals regulating the remodelling activity [1]. In this manner, the nature of the incoming signals is thought to be purely mechanical.

Hereafter, we propose to illustrate the necessity to strongly modify the paradigm of in silico bone adaptation by answering to 5 questions related to the mechano-sensation and mechano-transduction of bone remodelling signals.

4.1 First Question: Is the Classical Poro-Elasticity Sufficient to Describe Bone Behaviour?

To elucidate the cellular and molecular mechanisms of bone adaptation or bone pathologies, the in vivo fluid environment of bone cells has to be thoroughly described. Indeed, it is commonly accepted that the *stimuli* initiating the bone remodelling process are correlated to the hydro-mechanical state of bone tissue (see Sect. 2).

However, due to the heterogeneity of bone material and the multiscale structure of the bone porosity, from the vascularity (on the order of magnitude of $10\,\mu m$) to the lacuno-canalicular network (canalicular space of $0.01-0.1\,\mu m$ and lacunar space of $1-10\,\mu m$) [25], a precise determination of the fluid velocity and pressure fields surrounding the osteocytes remains a tremendous feat.

Since the interstitial bone fluid flow is induced by the skeleton strains, the Biot poro-elasticity theory [15] has been used to mimic bone behaviour. It is thus

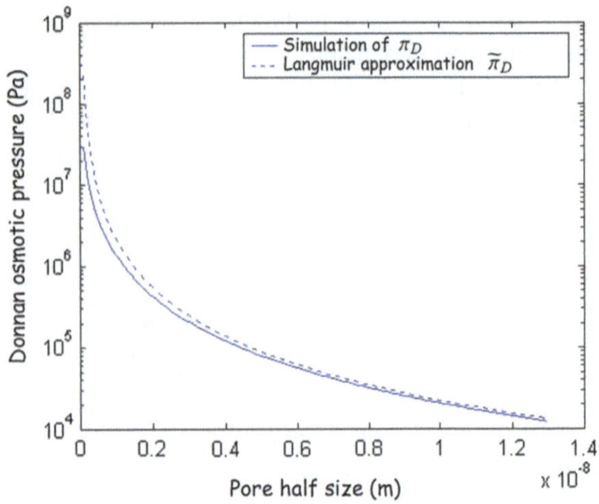

Fig. 6 Donnan pressure versus the pore size (reproduced with permission from Lemaire et al. [75])

justified to wonder if the new form of the Biot equation involving an electrical coupling effect through the Maxwell tensor $\boldsymbol{\tau}^*$ [see Eq. (23)] should not be preferred.

According both to experimental studies [88] and theoretical works [29], the most important contribution to the electro-chemical tensor $\boldsymbol{\tau}^*$ is the Donnan osmotic pressure term $-\pi_D\mathbf{I}$. An approximation of this parameter $\widetilde{\pi}_D$ exists and is proposed for a platelet geometry of the pores through the Langmuir formula which only depends on the half inter-platelets distance d [53].

This approximation is acceptable for a pore size d larger than a few nanometers (what is true in bone) and for high values of the surface charge density of the interface. This surface charge density being unknown, its order of magnitude can be estimated around $-0.2\,\text{C.m}^{-2}$ considering either the glyco-amino-glycan surface charge [93] or the cationic absorption capacity on the hydroxyapatite surface [47].

Moreover, the lower the salinity, the thicker the double-layer and so the stronger the electro-chemical effects are going to be. As a result, we consider a low salinity $n_b = 10^{-4}\,\text{mol.l}^{-1}$ to plot in Fig. 6 the evolution of the Donnan pressure π_D versus the pore half size d. The geometry of the pore consists in face-to-face platelets presenting the negative surface charge and separated by a distance $2d$. The calculation of the Donnan pressure results from evaluation of the double-layer potential thanks to a custom code first developed to study clayey materials [80].

The Langmuir approximation becomes more pertinent for large pores. Nevertheless, this approximation tends to slightly overestimate the Donnan pressure. Moreover, for typical pore sizes of the bone *canaliculi* (10^{-8}–10^{-7}m) [166], the Donnan pressure is one order of magnitude lower than the atmospheric pressure. Since the typical stresses within bone lay around 0.1–1 MPa [121, 167], the electro-chemical effects included in the modified Biot constitutive can be neglected and classical poro-elasticity seems appropriate to describe the averaged hydro-mechanical phenomena of cortical tissue.

A limitation of this optimistic answer to the first question appears under the light of the more recent descriptions of the lacuno-canalicular structure. Indeed, when considering the typical fiber-to-fiber distance in the pericellular matrix that partially occupies the canalicular space [166] or the canalicular structures described by Anderson and Knothe Tate [6] or McNamara et al. [97] the pore size may be nanometric, inducing a Donnan pressure of the same order of magnitude as the hydraulic pressure.

4.2 Second Question: Is the Classical Darcy Law Sufficient to Describe Interstitial Fluid Flow?

In parallel to the Biot law that is involved in the strain-induced interstitial fluid flow estimation, the possible implications of multiphysics phenomena in the fluid transport have to be investigated. In the classical Biot theory, the ability of bone to transmit fluid is quantified through the Darcy law by the intrinsic permeability (m^2). In anisotropic media, the permeability is characterized by a second-order tensor. This tensor is both symmetric and positive definite. For isotropic media, this tensor is characterized by an unique scalar κ. This scalar is a textural parameter only depending on the porous network geometry [78]. The more precise the quantification of this parameter at the osteocyte scale (lacuno-canalicular permeability), the better the understanding of the mechano-sensation process becomes. However, as visible in Table 2, the determination of this parameter is quite controversial.

To understand the uncertainties on the values of this parameter, according to our multiphysics description of the Darcy law, some coupled effects such as electro-osmosis could generate an opposite flow, resulting in a decrease in the apparent permeability [3, 78]. Using the coupled Darcy law given by (22), it has been shown that, for physiological conditions, the osmotic and electro-osmotic parts of the macroscopic Darcian velocity introduced represent less than 7 % of the whole interstitial macroscopic flow (see Fig. 7), depending on the value of the pericellular fibers permeability k_f quantifying the volume viscous force $\mathbf{F}_b$ induced by the fibrous matrix that partially occupies the canalicular space [85].

Thus, a classical purely hydraulic Darcy law is sufficient to roughly describe the bone fluid macroscopic movement. A strong limitation of this result is that the pericellular matrix is made of glycans proteins that do also present a charge density. This indicates that the meaningful pore size that should be considered for the electro-chemical transport could be the typical pericellular fiber-to-fiber distance instead of the canalicular radius. In particular, it was shown that for so thin pores of a few nanometers, the electrically induced flow could compensate the hydraulic flow, resulting in an apparent permeability decrease [78].

Table 2 Estimations of the lacuno-canalicular intrinsic permeability, state of the art (reproduced with permission from Lemaire et al. [78])

Reference	Permeability (m^2)	Remarks
Theoretical estimations		
[158]	1.02×10^{-20}	Isotropic pericellular matrix
[168]	1.47×10^{-20}	Adapted from [158]
[153]	$\{0.1\text{–}13\} \times 10^{-20}$	Adapted from [158]
[144]	2.2×10^{-22}	FEM and fitting of the electrical data of [117]
[44]	$\{0.67\text{–}7.5\} \times 10^{-20}$	Anisotropic pericellular matrix, adapted from [158]
[12]	$\{1\text{–}1500\} \times 10^{-22}$	Data of [130, 32], adapted from [158]
[169]	3.79×10^{-21}	Adapted from [158]
[85]	$\{1 - 10\} \times 10^{-20}$	Data of [166], adapted from [158]
[5]	$\{2.65\text{–}8.73\} \times 10^{-18}$	Personal data, scaled-up computational model
[43]	$\{1.05\text{–}105\} \times 10^{-20}$	Poiseuille modified model
[61]	$\{3\text{–}10\} \times 10^{-19}$	Personal data, adapted from [158]
[78]	$\{1\text{–}10\} \times 10^{-19}$	FEM, anatomical data of [166]
Experimental estimations (coupled with poroelasticity)		
[118]	4.1×10^{-24}	Nanoindentation
[41]	6.5×10^{-23}	Nanoindentation
[38]	$\{1\text{–}10\} \times 10^{-25}$	Stress relaxation of single osteons
[42]	2.8×10^{-23}	Step loading of intact bone

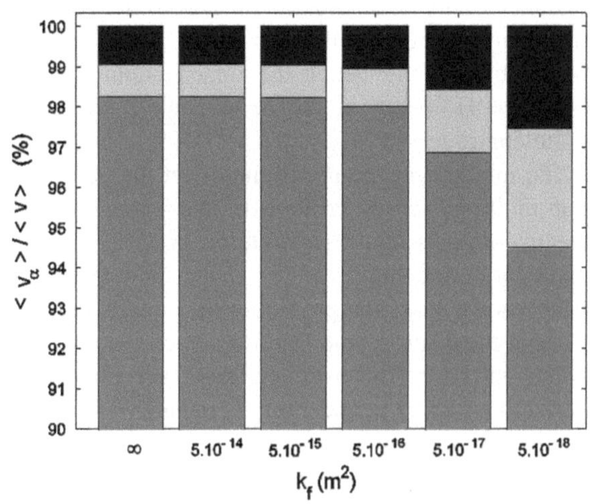

Fig. 7 Comparison between the three driving parts of the fluid flow for various values of the pericellular fibers parameter k_f: Poiseuille effect (*dark grey*), Osmosis (*grey*), electro-osmosis (*black*) (reproduced with permission from Lemaire et al. [85])

4.3 Third Question: What is the Nature of Bone In Vivo Electricity?

As explained in the Sect. 2, the nature of bone electricity is a recurrent question in bone biomechanics. The in vivo stress induced electric potentials that have been reported by Lanyon et al. [71] could indeed be explained either by the piezo-electric behaviour of the collagen-apatite matrix that constitutes the solid part of bone, or by the stress-induced streaming potentials. The use of our model to interpret available experimental data could provide a definitive answer to this question. Here we refer to the in vivo recording of stress-generated potentials during walking measured by Cochran et al. [23] thanks to electrodes implanted on a canine radius. In this study, a simultaneous recording of bone strain is provided. Despite the parasite effects generated by locomotion (interactions with other legs action), the loading can be approximatively represented by vertical cyclic loading conditions.

According to the results of our homogenization procedure, piezo-electricity and double layer potential vanish at the macroscale. Consequently, the macroscopic electric potential should be identified as streaming potentials.

To check the validity of this statement, we propose to recover the electric potentials measured on a walking dog by Cochran et al. [23]. Following the idea of Salzstein and Pollack [138], an estimation of the streaming potentials induced by the stress-generated fluid flow can be made. Indeed, invoking the Onsager reciprocity, the streaming potential field are shown to be proportional to the macroscopic pressure field [74]. Thus, mimicking the walking induced strain of Cochran et al. [23] by harmonic axial loading, the macroscopic pressure field can be computed [109]. In this calculation, the electrical parameters of the cortical microstructure are those used by Lemaire et al. [84], whereas the poro-elastic parameters are those of Rémond et al. [131]. In addition, the electric conductivity is taken as the one of a 0.01 M chloride potassium solution at 37 °C (0.1735 S.m−1).

The comparison in Fig. 8 between experimental and numerical results suggests that the solution of the poro-elastic model can satisfactory represent electric phenomena induced by walking activity. This is an indirect validation of the poro-elastic models of cortical bone and it suggests that the streaming potentials developing in response to the strain-induced bone fluid movements are likely to be the physiologically observed electric potentials.

4.4 Fourth Question: What are the Consequences of the Electro-Chemical Couplings on the Shear Sensitivity of the Osteocytes?

If osteocytes have been proposed to play a major role in sensing mechanical signals and regulating bone remodelling [63, 142, 148], they were also shown to directly participate in the calcium homeostasis by regulating the dissolution and deposition of calcium in the lacuno-canalicular space [125, 126]. The presence of such chemical gradients (such as directional calcium fluxes between the bone matrix and the interstitial fluid [91]) can result in an osmotic fluid flow. This directional movement of calcium from or into the bone can affect the fluid flow within the canaliculus and, consequently, the shear stress felt by the osteocytes, as illustrated by Fig. 9.

Shear stresses are linked to the velocity gradient. Thus it is necessary to investigate how the velocity gradients are affected near the cell process membrane,

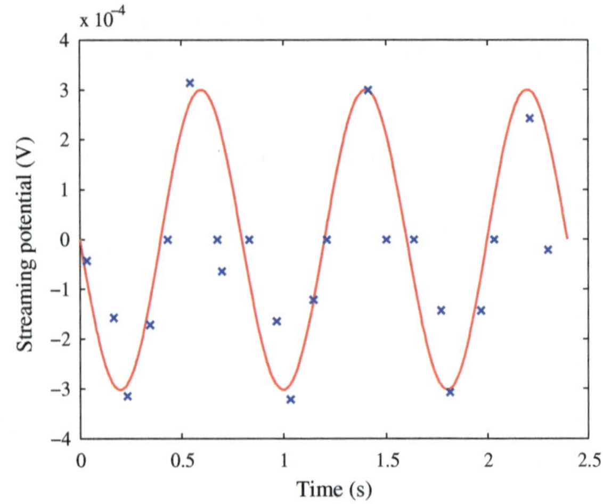

Fig. 8 Recovery of the in vivo electric potential data of Cochran et al. [23] (*crosses*) by our streaming potential model (reproduced with permission from Lemaire et al. [77])

where the electric double layer develops. Interstitial fluid velocity profiles and shear stress values at the cell membrane have been computed for physiological parameters values [58]. The total average fluid velocity in the canaliculus and its hydraulic, osmotic and electro-osmotic components as expressed by the modified Darcy law (22) are provided in Table 3. The hydraulic component governs the total fluid flow within the bulk of the canalicular pore space accounting for more than 95 % of the transport, while the electro-osmotic and osmotic phenomena are negligible. However, due to the electro-chemical over-velocities that develops in the double layer in the close proximity of the osteocyte membrane [85, 86], the contribution of the osmotic and electro-osmotic phenomena cannot be neglected. This is apparent looking at the shear stress on the cell membrane (see Table 3). The major contribution is still due to the hydraulic gradient (52 %), but neither the electro-osmotic ($\sim$43 %) nor the osmotic contributions ($\sim$5 %) are negligible.

When transferring this results in the context of the cellular mechano-sensitivity to shear effects, the osteocytes responding to fluid shear stress [65], the results reported in Table 3 indicate that the hydraulic and the electro-chemical parts of the shear stress may cause an annihilation of the total shear stress stimulation felt by the cell. Thus, chemical fluxes due to calcium deposition and dissolution from the canalicular walls can be expected to modulate osteocyte mechano-sensitivity both in the presence of unidirectional flow due to the pressure differential within the blood circulatory system and oscillatory flow due to physical activity [58].

4.5 *Fifth Question: What is the Ionic Transport Main Mechanism Within Bone Tissue?*

The mass transport mechanism within the lacuno-canalicular network is another open question in the domain of bone biomechanics. In the viewpoint of bone remodelling, it is necessary to properly identify how the chemical species can be

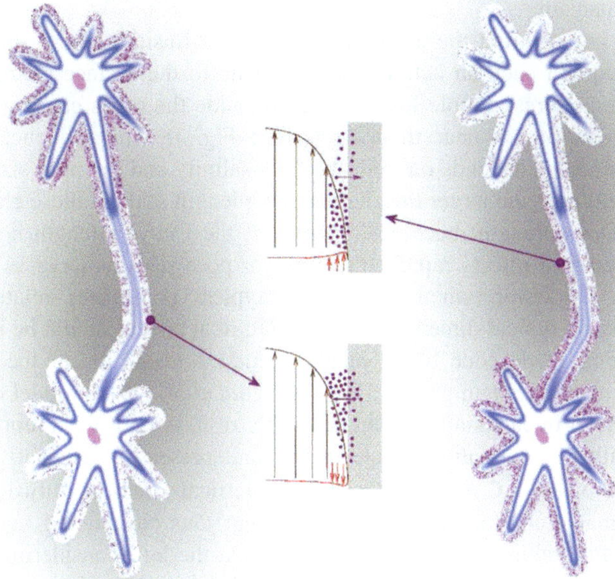

Fig. 9 Example of the consequences of physiological calcium fluxes inside the lacuno-canalicular system (adapted with permission from Kaiser et al. [58]). The electro-chemical (in *red*) and hydraulic (in *brown*) velocities may act in the same direction (profile on the *top*) or in a concurrent way (profile on the *bottom*)

Table 3 Average coupled parts of the fluid velocity and shear stress on the cell membrane computed using the physiological values of Kaiser et al. [58] (reproduced with permission from Kaiser et al. [58])

	Fluid velocity V_α (m.s^{-1})	Shear stress τ_α (Pa)
P	1.88×10^{-6} (97.4 %)	0.22 (52.0 %)
C	2.80×10^{-9} (0.1 %)	0.02 (4.7 %)
E	4.80×10^{-8} (2.5 %)	0.19 (43.3 %)
Total	1.93×10^{-6}	0.4295

moved from the osteocytes that are imbedded in the bone tissue toward the lining cells.

As indicated by [119], neither diffusion nor stress induced fluid flow could capable of sustaining an efficient mass transport inside the lacuno-canalicular system. Based on the estimation of the osteocytic glucose consumption rate, these authors highlighted the need for a new interpretation of experimental results where it was found that cyclic loading increases marker penetration within the lacuno-canalicular system [68], and proposed that cyclic stress stimulates an active transport mechanism.

If we take into account the electrical effects on diffusion, a new angle can be proposed to explain such an active transport. Due to the surface charge characterizing the bone pores, double layers develop inside the pores causing an asymmetric ionic distribution near their walls (see Fig. 4). The influence of these electrical phenomena depends on both the fluid salinity and the pore size. Indeed, the lower the salinity, the larger the characteristic length scale of these electrostatic phenomena (Debye length) becomes. Moreover, the Debye's length being nanometric, the electrical effects rapidly fade out for pore sizes as large as the canalicular radius. However, when considering typical pore sizes related to the pericellular matrix typical dimensions, these electrical effects should be important.

Using a home-made code based on the recursive resolution of the Cartesian Poisson-Boltzmann problem proposed in Derjaguin et al. [29], a model of straight channel described in cartesian coordinate (half-size h) is built for a representative pore filled with a water-saturated electrolyte and presenting a negative surface charge density of -0.2 C.m^{-2}. The reduced electrical effective diffusion parameters (see Remark) are calculated for monovalent ions considering various pore sizes and salinity values, as shown by Fig. 10. On the left, considering a typical fiber-to-fiber distance of the pericellular matrix elements $2h = 10$ nm [158, 166], we present the evolution of the reduced effective diffusivity parameters with the salinity. On the right, the pore size dependance of these parameters is shown considering a salinity of 0.01 M.

Remark The one-dimensional diffusivity parameters derived from Eq. (21) are reduced thanks to the diffusion coefficients of the ions $D_{\pm}$ and the porosity η_f.

These two graphs show that the electro-chemical effects may strongly affect the effective diffusion process. The evolutions of the cationic, respectively anionic, effective diffusion parameters exhibit the permselectivity property of the lacuno-canalicular system which acts here as a negatively charged nanoporous medium which tends to enhance the cationic transport and limit the anionic one [45, 122]. In particular, for pore sizes corresponding to the pericellular fiber-to-fiber distance, the electrostatic exclusion-enrichment effect becomes the dominating mass transport mechanism. Thus, depending on their charge, the chemical species transport may be enhanced by one or two orders of magnitude. This process could be a possible explanation of the active transport mechanism in bone claimed by Petrov and Pollack [119].

Such a charge effect within the lacuno-canalicular space has been observed in the in vivo tracer experiments of Tami et al. [147]. Indeed these authors showed that the transport mechanisms of negatively charged and neutral dextran particles were not the same within the lacuno-canalicular pores. If the neutral species remained confined in the vasculature, the anions did penetrate inside the

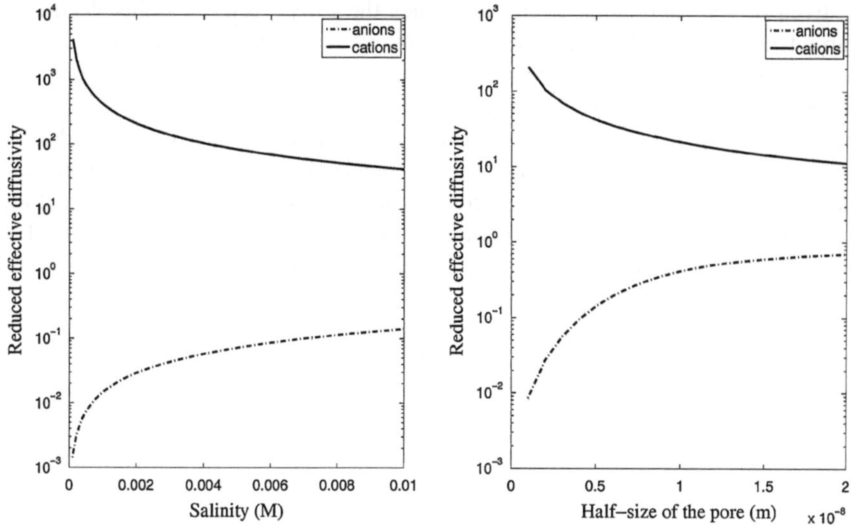

Fig. 10 Reduced effective electro-diffusity of cations (*solid line*) and anions (*dashed line*) as a function of the salinity (*left*, pore size of 10 nm) and the half-size of the pore (*right*, salinity of 0.01 M) (reproduced with permission from Kaiser et al. [59])

extravascular pores. Note that this experimental study brings up some fundamental questions on the sign of the surface charge of the canalicular pores. Indeed, if this charge is classically thought to be negative [3], such an anionic enrichment effect would rather indicate a positively charged pore surface. In the concluding remarks of this work, a possible explanation to this paradoxical observation is proposed.

5 Conclusion: Toward a New Paradigm of Bone Remodelling

The aim of this study was to offer new perspectives for in silico bone remodelling representations. Based on a multiphysical hierarchical treatment of the phenomena governing the tissue behaviour, this work quantitatively proved the accuracy of common macroscopic bone tissue models. Nevertheless, qualitatively, it also puts into relief the weakness of purely hydro-mechanical approaches when studying remodelling signals at the cellular scale. Indeed, even if washed out at the tissue level, the microscopic electro-chemical phenomena are visible in the neighbourhood of the cell.

Through this study, it was shown that the consequences of the microscopic effects in the classical remodelling transduction scenarios may be considerable. Since the key phenomena are located at the scale of the cell, it would be vain to carry on working only at the organ scale. For instance, we stressed on the major consequences on the shear stress of chemical fluxes occurring at the cellular scale.

Even if huge steps have been made since the early works of Julius Wolff, the current bone remodelling paradigm still draws from the vision of the German surgeon: the general point of view consists in analyzing the different biophysical phenomena in the light of the mechanical state of the skeleton. Although in vivo chemical transport seem to be crucial in the behaviour of bone tissue and despite the efforts done to properly describe the biochemical autocrine and paracrine cascade leading to the removal and formation of tissue [28], mass transport at the organ scale remains the poor relation of the bone remodelling paradigm. The peculiar role of electrically charged species in this biochemical remodelling pathway is obvious, with the calcium ions at the top of the list.

If we follow up the possibilities emerging from this work, a step toward a new paradigm of bone adaptation can be made. Indeed, adopting a multiscale vision, we are able to include significant innovative elements in the representation of the remodelling process linked to transport phenomena within bone volume. As illustrated by Fig. 11, we could propose that the formation and resorption of bone are mainly controlled by the ability to provide in situ calcium ions from the blood supply to the osteocytes inside the bone system. This mechanism could be regulated by ionic permselectivity induced by the electric surface charge of the lacuno-canalicular pores. The cationic/anionic permselectivity is the ratio of cations/anions based on the total number of ions that pass through the selective nanoporous lacuno-canalicular material. In response to a mechanical solicitation of bone and the concomitant piezo-electric effects of the collagen-apatite matrix, the tensile part of the tissue, respectively the compressive one, would generate a positively charged environment that decreases the cationic flux, respectively a negatively charged one increasing this flux [95]. As a result, the calcium transport from the blood supply toward the osteocytes environment would depend on the local loading conditions. This asymmetry would engender different extracellular calcium concentrations, modifying the osteocytic signaling pathway and so the bone adaptation [83]. Thus, the in vivo tracer experiments of Tami et al. [147] could be a clue of this permselectivity effect in the lacuno-canalicular transport. In this study, it was shown that the load-induced transport mechanisms of negatively charged and neutral dextran particles are different. If the neutral species remained confined in the vasculature, the anions did penetrate inside the extravascular pores. If these authors also indicate that this trend is more pronounced in the tension area of bone, no information is given concerning the compressive zone of bone. This has to be done now to reinforce our proposal.

By giving some answer to recurrent questions in bone biology, a reexamination of the common point of view of bone remodelling models can be proposed, raising up new questions. In particular, the focus should be made on the phenomena occurring at the cellular scale. We hope that the concomitant advances in bone modelling at the micro- and nanoscale (homogenization, molecular dynamics, etc.) and in microscopic experiments (imaging process, micro-sensors, etc.) will result in an inflexion of the current paradigm of bone adaptation replacing stress-controlled transport phenomena in the heart of the problem. Thus, through such an interscale approach,

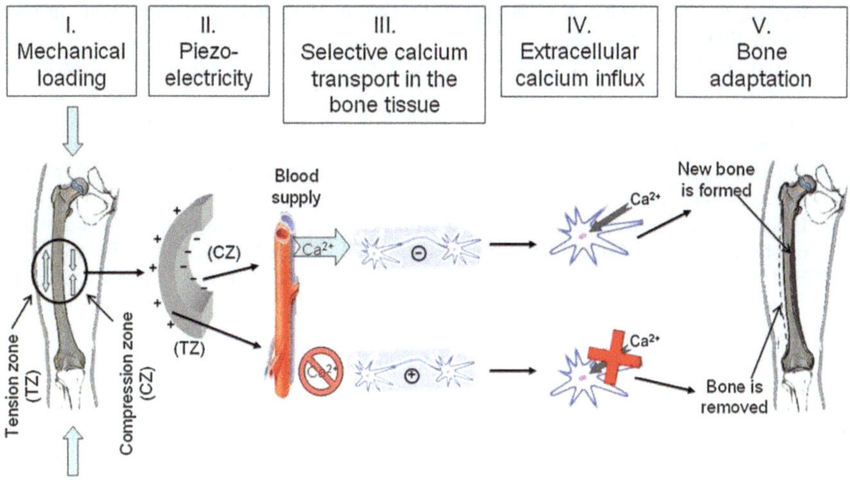

Fig. 11 Toward a new paradigm for bone adaptatione centered on the charged species transport phenomena. Piezo-electricity and calcium permselectivity may interact in the mechano-adaptation chain of bone tissue (reproduced with permission from Lemaire and Naili [83])

the unification of bone biomechanics at the tissue scale and bone mechanobiology describing the complex cellular interactions could become possible.

References

1. Adachi, T., Kameo, Y., Hojo, M.: Trabecular bone remodelling simulation considering osteocytic response to fluid-induced shear stress. Philos. T. R. Soc. A **368**, 2669–2682 (2010)
2. Ager, J.W. III, Balooch, G., Ritchie, R.O: Fracture, aging, and disease in bone. J. Mater. Res. **21**, 1878–1892 (2006)
3. Ahn, A.C., Grodzinsky, A.J: Relevance of collagen piezoelectricity to wolff's law: a critical review. Med. Eng. Phys. **31**, 733–741 (2009)
4. Ajubi, N.E., Klein-Nulend, J., Nijweide, P.J., Vrijheid-Lammers, T., Alblas, M.J., Burger, E.H: Pulsating fluid flow increases prostaglandin production by cultured chicken osteocytes-a cytoskeleton-dependent process. Biochem. Biophys. Res. Commun. **225**, 62–68 (1996)
5. Anderson, E., Kreuzer, S., Small, O., Knothe Tate, M.: Pairing computational and scaled physical models to determine permeability as a measure of cellular communication in micro- and nano-scale pericellular spaces. Microfluid. Nanofluidics **4**, 193–204 (2008)
6. Anderson, E.J., Knothe Tate, M.L: Idealization of pericellular fluid space geometry and dimension results in a profound underprediction of nano-microscale stresses imparted by fluid drag on osteocytes. J. Biomech. **41**, 1736–1746 (2008)
7. Auriault, J.-L., Adler, P.M.: Taylor dispersion in porous media: analysis by multiple scale expansions. Adv. Water Res. **18**, 217–226 (1995)
8. Auriault, J.-L., Sanchez-Palencia, E.: Etude du comportment macroscopique d'un milieu poreux saturé déformable. J. Mécanique **16**, 575–603 (1977)
9. Baron, R., Tross, R., Vignery, A.: Evidence of sequential remodeling in rat trabecular bone—morphology, dynamic histomorphometry, and changes during skeletal maturation. Anat. Rec. **208**, 137–145 (1984)

10. Bassett, C.A.L., Pawluk, R.J., Becker, R.O: Effects of electric currents on bone in vivo. Nature **204**, 652–654 (1964)
11. Baxter, F., Bowen, C., Turner, I., Dent, A.: Electrically active bioceramics: a review of interfacial responses. Ann. Biomed. Eng. **38**, 2079–2092 (2010)
12. Beno, T., Yoon, Y.-J., Cowin, S.C., Fritton, S.P: Estimation of bone permeability using accurate microsctructural measurements. J. Biomech. **39**, 2378–2387 (2006)
13. Biewener, A.A., Taylor, C.R: Bone strain: a determinant of gait and speed. J. Exp. Biol. **123**, 383–400 (1986)
14. Binderman, I., Shimshoni, Z., Somjen, D.: Biochemical pathways involved in the translation of physical stimulus into biological message. Calcif. Tissue Int. **36**, S82–S85 (1984)
15. Biot, M.A: General theory of three-dimensional consolidation. J. Appl. Phys. **12**, 155–164 (1941)
16. Buckwalter, J.A., Glimcher, M.J., Cooper, R.R., Recker, R.: Bone biology. Part I: Structure, blood supply, cells, matrix, and mineralization. J. Bone Joint Surg. Am. **77**, 1256–1275 (1995)
17. Buckwalter, J.A., Glimcher, M.J., Cooper, R.R., Recker, R.: Bone biology. Part II: Formation, form, modeling, remodeling, and regulation of cell function. J. Bone Joint Surg. Am. **77**, 1276–1289 (1995)
18. Burger, E.H., Klein-Nulend, J.: Mechanotransduction in bone: role of the lacuno-canalicular network. Faseb. J. **13**(Suppl), S101–S112 (1999)
19. Burger, E.H., Klein-Nulend, J., Smit, T.H: Strain-derived canalicular fluid flow regulates osteoclast activity in a remodelling osteon—a proposal. J. Biomech. **36**((10), 1453–1459 (2003)
20. Burr, D.B., Martin, R.B: Calculating the probability that micro-cracks initiate resorption spaces. J. Biomech. **26**, 613–616 (1993)
21. Burr, D.B., Milgrom, C., Fyhrie, D., Forwood, M., Nyska, M., Finestone, A., Hoshaw, S., Saiag, E., Simkin, A.: In vivo measurement of human tibial strains during vigorous activity. Bone **18**, 405–410 (1996)
22. Cann, C.E., Adachi, R.R: Bone resorption and mineral excretion in rats during spaceflight. Am. J. Physiol. **224**, R327–331 (1983)
23. Cochran, G.V.B., Dell, D.G., Palmieri, V.R., Johnson, M.W., Otter, M.W., Kadaba, M.P: An improved design of electrodes for measurement of streaming potentials on wet bone in vitro and in vivo. J. Biomech. **22**, 745–750 (1989)
24. Cowin, S.C: Bone poroelasticity. J. Biomech. **32**, 217–238 (1999)
25. Cowin, S.C: Bone mechanics handbook, 2nd edn. CRC Press, Boca Raton, FL (2001)
26. Cowin, S.C: Mechanosensation and fluid transport in living bone. J. Musculoskel. Neuron. Interact. **2**, 256–260 (2002)
27. Cowin, S.C., Gailani, G., Benalla, M.: Hierarchical poroelasticity: movement of interstitial fluid between porosity levels in bones. Philos. T. R. Soc. A **367**, 3401–3444 (2009)
28. Crockett, J.C., Rogers, M.J., Coxon, F.P., Hocking, L.J., Helfrich, M.H: Bone remodelling at a glance. J. Cell Sci. **124**, 991–998 (2011)
29. Derjaguin, B.V., Churaev, N., Muller, V.: Surface Forces. Plenum Press, New York (1987)
30. Duncan, R.L., Turner, C.H: Mechanotransduction and the functional response of bone to mechanical strain. Calcif Tissue Int. **57**((5), 344–358 (1995)
31. Ehrlich, P.J., Lanyon, L.E: Mechanical strain and bone cell function: a review. Osteoporosis Int. **13**((9), 688–700 (2002)
32. Ferretti, M., Muglia, M.A., Remaggi, F., Cane, V., Palumbo, C.: Histomorphometric study on the osteocyte lacuno-canalicular network in animals of different species. II. Parallel-fibered and lamellar bones. Ital. J. Anat. Embryol. **104**, 121–131 (1999)
33. Freemont, A.J: The tissues we deal with (i) bone. Curr. Orthopaed. **12**, 181–192 (1998)
34. Fritton, S.P., Mcleod, K.J., Rubin, C.T: Quantifying the strain history of bone: spatial uniformity and self-similarity of low-magnitude strains. J. Biomech. **33**, 317–325 (2000)
35. Fritton, S.P, Weinbaum, S.: Fluid and solute transport in bone: flow-induced mechanotransduction. Annu. Rev. Fluid Mech. **41**, 347–374 (2009)

36. Frost, H.M: From wolff's law to the utah paradigm: insights about bone physiology and its clinical applications. Anat. Rec. **262**, 398–419 (2001)
37. Fukada, E., Yasuda, I.: On the piezoelectric effect of bone. J. Phys. Soc. Jpn **12**, 1158–1162 (1957)
38. Gailani, G., Benalla, M., Mahamud, R., Cowin, S.C., Cardoso, L.: Experimental determination of the permeability in the lacunar-canalicular porosity of bone. J. Biomech. Eng. **131**((10), 101007 (2009)
39. Gailani, G., Cowin, S.: Ramp loading in Russian doll poroelasticity. J. Mech. Phys. Sol. **59**, 103–120 (2011)
40. Galley, S.A., Michalek, D.J., Donahue, S.W: A fatigue microcrack alters fluid velocities in a computational model of interstitial flow in cortical bone. J. Biomech. **39**, 2026–2033 (2006)
41. Galli, M., Oyen, M.L: Fast identification of poroelastic parameters from indentation tests. Comput. Model. Eng. Sci. **48**, 241–269 (2009)
42. Gardinier, J.D., Townend, C.W., Jen, K.-P., Wu, Q., Duncan, R.L., Wang, L.: In situ permeability measurement of the mammalian lacunar-canalicular system. Bone **46**, 1075–1081 (2010)
43. Goulet, G., Coombe, D., Martinuzzi, R., Zernicke, R.: Poroelastic evaluation of fluid movement through the lacunocanalicular system. Ann. biomedical Engineering **37**, 1390–1402 (2009)
44. Gururaja, S., Kim, H., Swan, C., Brand, R., Lakes, R.: Modeling deformation-induced fluid flow in cortical bone's canalicular-lacunar system. Ann. Biomed. Eng. **33**, 7–25 (2005)
45. Han, J., Fu, J., Schoch, R.B: Molecular sieving using nanofilters: past, present and future. Lab Chip **8**, 23–33 (2008)
46. Han, Y., Cowin, S.C., Schaffler, M.B., Weinbaum, S.: Mechanotransduction and strain amplification in osteocyte cell processes. Proc. Natl. Acad. Sci. U.S.A. **101**, 16689–16694 (2004)
47. Harding, I.S., Rashid, N., Hing, K.A: Surface charge and the effect of excess calcium ions on the hydroxyapatite surface. Biomaterials **26**, 6818–6826 (2005)
48. Hazenberg, J.G., Freeley, M., Foran, E., Lee, T.C., Taylor, D.: Microdamage: a cell transducing mechanism based on ruptured osteocyte processes. J. Biomech. **39**, 2096–2103 (2006)
49. Hert, J., Liskova, M., Landa, J.: Reaction of bone to mechanical stimuli. 1. Continuous and intermittent loading of tibia in rabbit. Folia Morphol. **19**, 290–300 (1971)
50. Hilfiker, A., Kasper, C., Hass, R., Haverich, A.: Mesenchymal stem cells and progenitor cells in connective tissue engineering and regenerative medicine: is there a future for transplantation?. Langenbecks Arch. Surg. **396**, 489–497 (2011)
51. Hill, P.A: Bone remodelling. Brit. J. Orthodont **25**, 101–107 (1998)
52. Hunter, R.J: Zeta Potential in Colloid Science: Principles and Applications. Academic Press, London (1981)
53. Israelachvili, J.N: Intermolecular and Surface Forces, 3rd edn. Academic Press, New York (2011)
54. Johnson, D.L., McAllister, T.N., Frangos, J.A: Fluid flow stimulates rapid and continuous release of nitric oxide in osteoblasts. Am. J. Physiol. Endoc. Metab. **271**, E205–E208 (1996)
55. Johnson, M.W., Chakkalakal, D.A., Harper, R.A., Katz, J.L., Rouhana, S.W: Fluid flow in bone in vitro. J. Biomech. **15**, 881–885 (1982)
56. Kaiser, J., Lemaire, T., Naili, S., Komarova, S.V., Sansalone, V.: Calcium fluxes within cortical bone fluid may affect osteocyte mechanosensitivity. Comput. Meth. Biomech. Biomed. Eng. **14**, S141–S142 (2011)
57. Kaiser, J., Lemaire, T., Naili, S., Sansalone, V.: Multiscale modelling of fluid flow in charged porous media including cationic exchanges: application to bone tissues. C.R. Mecanique **337**, 768–775 (2009)
58. Kaiser, J., Lemaire, T., Naili, S., Sansalone, V., Komarova, S.V.: Do calcium fluxes within cortical bone affect osteocyte mechanosensitivity? J. Theoret. Biol. **303**, 75–86 (2012a)

59. Kaiser, J., Lemaire, T., Naili, S., Sansalone, Lemaire, T.: Effective chemical transport within cortical bone: a comparison between textural and electrochemical effects (submitted, 2012b)
60. Kameo, Y., Adachi, T., Hojo, M.: Transient response of fluid pressure in a poroelastic material under uniaxial cyclic loading. J. Mech. Phys. Sol. **5**, 1794–1805 (2008)
61. Kameo, Y., Adachi, T., Sato, N., Hojo, M.: Estimation of bone permeability considering the morphology of lacuno-canalicular porosity. J. Mech. Behav. Biomed. Mat. **3**, 240–248 (2010)
62. Kasiri, S., Taylor, D.: A critical distance study of stress concentrations in bone. J. Biomech. **41**, 603–609 (2008)
63. Klein-Nulend, J., Bacabac, R.G., Mullender, M.G: Mechanobiology of bone tissue. Path. Bio. **53**, 576–580 (2005)
64. Klein-Nulend, J., Bakker, A.: Osteocytes: mechanosensors of bone and orchestrators of mechanical adaptation. Clin. Rev. Bone Miner. Metabol. **5**, 195–209 (2007)
65. Klein-Nulend, J., van der Plas, A., Semeins, C.M., Ajubi, N.E., Frangos, J.A., Nijweide, P.J., Burger, E.H: Sensitivity of osteocytes to biomechanical stress in vitro. Faseb J. **9**, 441–445 (1995)
66. Knothe Tate, M.L: Whither flows the fluid in bone? An osteocyte's perspective. J. Biomech. **36**(10), 1409–1424 (2003)
67. Knothe Tate, M.L., Knothe, U., Niederer, P.: Experimental elucidation of mechanical load-induced fluid flow and its potential role in bone metabolism and functional adaptation. Am. J. Med. Sci. **316**, 189–195 (1998a)
68. Knothe Tate, M.L., Niederer, P., Knothe, U.: In vivo tracer transport through the lacunocanalicular system of rat bone in an environment devoid of mechanical loading. Bone **22**, 107–117 (1998b)
69. Knothe Tate, M.L., Steck, R., Forwood, M.R., Niederer, P.: In vivo demonstration of load-induced fluid flow in the rat tibia and its potential implications for processes associated with functional adaptation. J. Exp. Biol. **203**, 2737–2745 (1998c)
70. Kubo, T.: Piezoelectricity of bone and electrical callus. J. Orthop. Sci. **17**, 105–106 (2012)
71. Lanyon, L.E., Hampson, W.G., Goodship, A.E., Shah, J.S: Bone deformation recorded in vivo from strain gauges attached to the human tibial shaft. Acta Orthop. Scand. **46**, 256–268 (1975)
72. Lanyon, L.E., Rubin, C.T: Static vs dynamic loads as an influence on bone remodelling. J. Biomech. **17**(12), 897–905 (1984)
73. Lemaire, T., Borocin, F., Naili, S.: Mechanotransduction of bone remodelling: role of micro-cracks at the periphery of osteons. C.R. Mecanique **336**, 354–362 (2008)
74. Lemaire, T., Capiez-Lernout, E., Kaiser, J., Naili, S., Rohan, E., Sansalone, V.: A multiscale theoretical investigation of electric measurements in living bone. Piezo-electricity and electrokinetics. Bull. Math. Biol. **73**, 2649–2677 (2011)
75. Lemaire, T., Capiez-Lernout, E., Kaiser, J., Naili, S., Sansalone, V.: What is the importance of multiphysical phenomena in bone remodelling signals expression? A multiscale perspective. J. Mech. Behav. Biomed. Mat. **4**, 909–920 (2011b)
76. Lemaire, T., Kaiser, J., Naili, S., Sansalone, V.: Modelling of the transport in charged porous media including ionic exchanges. Mech. Res. Commun. **37**, 495–499 (2010a)
77. Lemaire, T., Kaiser, J., Sansalone, V., Rohan, E., Naili, S.: What is the nature of bone in vivo electricity? Comput. Meth. Biomech. Biomed. Eng. **14**, S143–S144 (2011c)
78. Lemaire, T., Lemonnier, S., Naili, S.: On the paradoxical determinations of the lacuno-canalicular permeability of bone. Biomech. Model. Mechanobiol. **11**, 933–946 (2012)
79. Lemaire, T., Moyne, C., Stemmelen, D.: Imbibition test in a clay powder (mx-80 bentonite). Appl. Clay Sci. **26**, 235–248 (2004)
80. Lemaire, T., Moyne, C., Stemmelen, D.: Modelling of electro-osmosis in clayey materials including ph effects. Phys. Chem. Earth **32**, 441–452 (2007)
81. Lemaire, T., Moyne, C., Stemmelen, D., Murad, M.A.: Electro-chemo-mechanical couplings in swelling clays derived by homogenization: electroviscous effects and

onsager's relations. In: Poromechanics II, proceedings of the second Biot conference on poromechanics, Grenoble, France, pp. 489–500. Balkema Publishers, Lisse (2002)
82. Lemaire, T., Naili, S.: Stimuli physiques du remodelage osseux. In: Reconstruction osseuse et cutanée: Biomécanique et Techniques de l'Ingénieur, pp. 57–71. Sauramps Medical, Montpellier (2008)
83. Lemaire, T., Naili, S.: Possible role of calcium permselectivity in bone adaptation. Med. Hyp. **78**, 367–369 (2012)
84. Lemaire, T., Naili, S., Rémond, A.: Multiscale analysis of the coupled effects governing the movement of interstitial fluid in cortical bone. Biomech. Model. Mechanobiol. **5**(1), 39–52 (2006)
85. Lemaire, T., Naili, S., Rémond, A.: Study of the influence of fibrous pericellular matrix in the cortical interstitial fluid movement with hydroelectrochemical effects. J. Biomech. Eng. **130**(1), 011001 (2008)
86. Lemaire, T., Sansalone, V., Naili, S.: Multiphysical modelling of fluid transport through osteo-articular media. An. Acad. Bras. Cienc. **82**, 127–144 (2010)
87. Lemaire, V., Tobin, F.L., Greller, L.D., Cho, C.R., Suva, L.J.: Modeling the interactions between osteoblast and osteoclast activities in bone remodeling. J. Theoret. Biol. **229**, 293–309 (2004b)
88. Low, P.F: The clay/water interface and its role in the environment. Prog. Colloid Polym. Sci. **95**, 98–107 (1994)
89. Mak, A.F.T., Qin, L., Hung, L.K., Cheng, C.W., Tin, C.F: A histomorphometric observation of flows in cortical bone under dynamic loading. Microvasc. Res. **59**, 290–300 (2000)
90. Malone, A.M.D., Anderson, C.T., Tummala, P., Kwon, R.Y., Johnston, T.R., Stearns, T., Jacobs, C.R.: Primary cilia mediate mechanosensing in bone cells by a calcium-independent mechanism. Proc. Natl. Acad. Sci. U.S.A. **104**, 13325–13330 (2007)
91. Marenzana, M., Shipley, A.M., Squitiero, P., Kunkel, J.G., Rubinacci, A.: Bone as an ion exchange organ: evidence for instantaneous cell-dependent calcium efflux from bone not due to resorption. Bone **37**, 545–554 (2005)
92. Martin, R.B., Burr, D.B., Sharkey, N.A: Skeletal Tissue Mechanics, 1st edn. Springer, New York (1998)
93. Mattern, K.J., Nakornchai, C., Deen, W.M: Darcy permeability of agarose-glycosaminoglycan gels analyzed using fiber-mixture and donnan models. Biophys. J. **95**, 648–656 (2008)
94. Mbuyi-Muamba, J.-M., Dequeker, J., Gevers, G.: Biochemistry of bone. Baillière Clin. Rheum. **2**, 63–101 (1988)
95. Mc Elhaney, J.H: The charge distribution on the human femur due to load. J. Bone Joint Surg. **49**, 1561–1571 (1967)
96. McAllister, T.N., Frangos, J.A: Steady and transient fluid shear stress stimulate no release in osteoblasts through distinct biochemical pathways. J. Bone Miner. Res. **14**, 930–936 (1999)
97. McNamara, L.M., Majeska, R.J., Weinbaum, S., Friedrich, V., Schaffler, M.B: Attachment of osteocyte cell processes to the bone matrix. Anat. Rec. **292**, 355–363 (2009)
98. Messer, H.H: Bone cell membranes. Clin. Orthop. Relat. Res. **166**, 256–276 (1982)
99. Miara, B., Rohan, E., Zidi, M., Labat, B.: Piezomaterials for bone regeneration design—homogenization approach. J. Mech. Phys. Sol. **53**, 2529–2556 (2005)
100. Mishra, S., Knothe Tate, M.L: Effect of lacunocanalicular architecture on hydraulic conductance in bone tissue: implications for bone health and evolution. Anat. Rec. A **273**, 752–762 (2003)
101. Morey, E.R., Baylink, D.J: Inhibition of bone formation during space flight. Science **201**, 1138–1141 (1978)
102. Moyne, C., Murad, M.A: Electro-chemo-mechanical couplings in swelling clays derived from a micro/macro-homogenization procedure. Int. J. Sol. Struct. **39**, 6159–6190 (2002)
103. Muir, P., Sample, S.J, Barrett, J.G., McCarthy, J., Vanderby, R. Jr., Markel, M.D., Prokuski, L.J., Kalscheur, V.L: Effect of fatigue loading and associated matrix microdamage on bone blood flow and interstitial fluid flow. Bone **40**, 948–956 (2007)

104. Nagatomi, J., Arulanandam, B.P., Metzger, D.W., Meunier, A., Bizios, R.: Frequency- and duration-dependent effects of cyclic pressure on select bone cell functions. Tissue Eng. **7**, 717–728 (2001)
105. Nagatomi, J., Arulanandam, B.P., Metzger, D.W., Meunier, A., Bizios, R.: Effects of cyclic pressure on bone marrow cell cultures. J. Biomech. Eng. **124**, 308–314 (2002)
106. M. Nakamura, R. Hiratai, and K. Yamashita. Bone mineral as an electrical energy reservoir. J. Biomed. Mater. Res. A, 100A: 1368–1374, 2012.
107. Nakamura, S., Kobayashi, T., Nakamura, M., Yamashita, K.: Enhanced in vivo responses of osteoblasts in electrostatically activated zones by hydroxyapatite electrets. J. Mater. Sci. Mater. M. **20**, 99–103 (2009)
108. Nguyen, V.-H., Lemaire, T., Naili, S.: Anisotropic poroelastic hollow cylinders with damaged periphery under harminically axial loading: relevance to bone remodelling. Multidisc. Model. Mater. Struct. **5**, 205–222 (2009)
109. Nguyen, V.-H., Lemaire, T., Naili, S.: Poroelastic behaviour of cortical bone under harmonic axial loading: theoretical study at the osteonal tissue scale. Med. Eng. Phys. **32**, 384–390 (2010)
110. Nguyen, V.-H., Lemaire, T., Naili, S.: Influence of interstitial bone micro-cracks on strain-induced fluid flow. Biomech. Model. Mechanobiol. **10**, 963–972 (2011)
111. Noble, B.: Bone microdamage and cell apoptotsis. Europ. Cells Mater. **6**, 46–56 (2003)
112. Noris-Suárez, K., Lira-Olivares, J., Ferreira, A.M., Feijoo, J.L., Suárez, N., Hernández, M.C., Barrios, E.: In vitro deposition of hydroxyapatite on cortical bone collagen stimulated by deformation-induced piezoelectricity. Biomacromolecules **8**, 941–948 (2007)
113. Norman, T.L., Wang, Z.: Microdamage of human cortical bone: incidence and morphology in long bones. Bone **20**, 375–379 (1997)
114. O'Brien, F.J., Taylor, D., Dickson, G.R., Lee, T.C.: Visualisation of three-dimensional micro-cracks in compact bone. J. Ana. **197**, 413–420 (2000)
115. C. Oddou, T. Lemaire, J. Pierre, and B. David. Hydrodynamics in porous media with applications to tissue engineering. In K. Vafai, editor, Porous media: applications in biological systems and biotechnology, pages 75–119. CRC Press, 2011.
116. Otter, M., Shoenung, J., Williams, W.S: Evidence for different sources of stress-generated potentials in wet and dry bone. J. orthop. Res. **3**, 321–324 (1985)
117. Otter, M.W., Palmieri, V.R., Wu, D.D., Seiz, K.G., MacGinitie, L.A., Cochran, G.V.B: A comparative analysis of streaming potentials in vivo and in vitro. J. Orthop. Res. **10**, 710–719 (1992)
118. Oyen, M.L: Poroelastic nanoindentation responses of hydrated bone. J. Mater. Res. **23**, 1307–1314 (2008)
119. Petrov, N., Pollack, S.R: Comparative analysis of diffusive and stress induced nutrient transport efficiency in the lacunar-canalicular system of osteons. Biorheology **40**, 347–353 (2003)
120. Piccolino, M.: Animal electricity and the birth of electrophysiology: the legacy of luigi galvani. Brain Res. Bull. **46**, 381–407 (1998)
121. Piekarski, K., Munro, M.: Transport mechanism operating between blood supply and osteocytes in long bones. Nature **269**, 80–82 (1977)
122. Plecis, A., Schoch, R.B., Renaud, P.: Ionic transport phenomena in nanofluidics: experimental and theoretical study of the exclusion-enrichment effect on a chip. Nano Lett. **5**, 1147–1155 (2005)
123. Pollack, S.R., Petrov, N., Salzstein, R., Brankov, G., Blagoeva, R.: An anatomical model for streaming potentials in osteons. J. Biomech. **17**, 627–636 (1984)
124. Price, C., Zhou, X., Li, W., Wang, L.: Real-time measurement of solute transport within the lacunar-canalicular system of mechanically loaded bone: direct evidence for load-induced fluid flow. J. Bone Miner. Res. **26**, 277–285 (2011)
125. Qing, H., Ardeshirpour, L., Divieti Pajevic, P., Dusevich, V., Jähn, K., Kato, S., Wysolmerski, J., Bonewald, L.F: Demonstration of osteocytic perilacunar/canalicular remodeling in mice during lactation. J. Bone Miner. Res. **27**, 1018–1029 (2012)

126. Qing, H., Bonewald, L.F: Osteocyte remodeling of the perilacunar/canalicular matrix. Int. J. Oral Sci. **1**, 59–65 (2009)
127. K.M. Reich and J.A. Frangos. Effect of flow on prostaglandin e2 and inositol triphosphate levels in osteoblasts. Am. J. Physiol., 261: C428–432, 1991.
128. Reich, K.M., Frangos, J.A.: Protein kinase c mediates flow-induced prostaglandin e2 production in osteoblasts. Calcif. Tissue Int. **52**, 62–66 (1993)
129. Reilly, G.C., Knapp, H.F., Stemmer, A., Niederer, P., Knothe Tate, M.L: Investigation of the morphology of the lacunocanalicular system of cortical bone using atomic force microscopy. Ann. Biomed. Eng. **29**, 1074–1081 (2001)
130. Remaggi, F., Cane, V., Palumbo, C., Ferretti, M.: Histomorphometric study on the osteocyte lacuno-canalicular network in animals of different species. I. Woven-fibered and parallel-fibered bones. Ital. J. Anat. Embryol. **103**, 145–155 (1998)
131. Rémond, A., Naili, S., Lemaire, T.: Interstitial fluid flow in the osteon with spatial gradients of mechanical properties: a finite element study. Biomech. Model. Mechanobiol. **7**, 487–495 (2008)
132. Rohan, E., Naili, S., Cimrman, R., Lemaire, T.: Multiscale modelling of a uid saturated medium with double porosity: relevance to the compact bone. J. Mech. Phys. Sol. **60**, 857–881 (2012)
133. Rubin, C.T: Skeletal strain and the functional significance of bone architecture. Calcif. Tissue Int. **36**, S11–S18 (1984)
134. Rubin, C.T., Lanyon, L.E: Limb mechanics as a function of speed and gait: a study of functional strains in the radius and tibia of horse and dog. J. Exp. Biol. **101**, 187–211 (1982)
135. Rubin, C.T., Lanyon, L.E: Regulation of bone formation by applied dynamic loads. J. Bone Joint Surg. Am. **66**, 397–402 (1984)
136. Rubin, C.T., Lanyon, L.E: Regulation of bone mass by mechanical strain magnitude. Calcif. Tissue Int. **37**, 411–417 (1985)
137. Ruff, C., Holt, B., Trinkaus, E.: Who's afraid of the big bad wolff?: "wolff's law" and bone functional adaptation. Am. J. Phys. Anthropol. **129**, 484–498 (2006)
138. Salzstein, R.A., Pollack, S.R: Electromechanical potentials in cortical bone-ii: experimental analysis. J. Biomech. **20**, 271–280 (1987)
139. Salzstein, R.A., Pollack, S.R., Mak, A.F.T., Petrov, N.: Electromechanical potentials in cortical bone. I: a continuum approach. J. Biomech. **20**, 261–270 (1987)
140. Schaffer, M.B., Choi, K., Milgrom, C.: Aging and matrix microdamage accumulation in human compact bone. Bone **17**, 521–525 (1995)
141. Schimdt, S.M., McCready, M.J., Ostafin, A.E: Effect of oscillationg fluid shear on solute transport in cortical bone. J. Biomech. **28**, 2337–2343 (2005)
142. Sikavitsas, V.I., Temenoff, J.S., Mikos, A.G: Biomaterials and bone mechanotransduction. Biomaterials **22**, 2581–2593 (2001)
143. Skerry, T.M., Bitenski, L., Chayen, J., Lanyon, L.E: Early strain-related changes in enzyme activity in osteocytes following bone loading in vivo. J. Bone Miner. Res. **4**, 783–788 (1989)
144. Smit, T.H., Huyghe, J.M., Cowin, S.C: Estimation of the poroelastic parameters of cortical bone. J. Biomech. **35**, 829–835 (2002)
145. Spencer, H.: First Principles. Williams and Norgate, 2nd edn, London (1867)
146. Sposito, G.: The Surface Chemistry of Soils. Oxford University Press, Oxford (1981)
147. Tami, A.E., Schaffler, M.B., Knothe Tate, M.L: Probing the tissue to subcellular level structure underlying bone's molecular sieving function. Biorheology **40**, 577–590 (2003)
148. Tatsumi, S., Ishii, K., Amizuka, N., Li, M., Kobayashi, T., Kohno, K., Ito, M., Takeshita, S., Ikeda, K.: Targeted ablation of osteocytes induces osteoporosis with defective mechanotransduction. Cell Metab. **5**, 464–475 (2007)
149. Turner, C.H., Yoshikawa, T., Forwood, M.R., Sun, T.C., Burr, D.B: High frequency components of bone strain in dogs measured during various activities. J. Biomech. **28**(1), 39–44 (1995)

150. van Oers, R.F.M., Ruimerman, R., Tanck, E., Hilbers, P.A.J., Huiskes, R.: A unified theory for osteonal and hemi-osteonal remodeling. Bone **42**, 250–259 (2008)
151. Verborgt, O., Tatton, N.A., Majeska, R.J., Schaffler, M.B: Spatial distribution of bax and bcl-2 in osteocytes after bone fatigue: complementary roles in bone remodeling regulation?. J. Bone Miner. Res. **17**, 907–914 (2002)
152. Wang, L., Cowin, S.C., Weinbaum, S., Fritton, S.P: Modeling tracer transport in an osteon under cyclic loading. Ann. Biomed. Eng. **28**, 1200–1209 (2000)
153. Wang, L., Fritton, S.P., Cowin, S.C., Weinbaum, S.: Fluid pressure relaxation depends upon osteonal microstructure: modeling an oscillatory bending experiment. J. Biomech. **32**, 663–672 (1999)
154. Wang, L., Fritton, S.P., Weinbaum, S., Cowin, S.C: On bone adaptation due to venous stasis. J. Biomech. **36**, 1439–1451 (2003)
155. Wang, L., Wang, Y., Han, H., Henderson, S.C., Majeska, R.J., Weinbaum, S., Scaffler, M.B: In situ measurement of solute transport in the bone lacunar-canalicular system. Proc. Natl Acad. Sci. U.S.A. **102**, 11911–11916 (2005)
156. Wang, Y., McNamara, L.M., Schaffler, M.B., Weinbaum, S.: A model for the role of integrins in flow induced mechanotransduction in osteocytes. Proc. Natl Acad. Sci. U.S.A. **104**, 15941–15946 (2007)
157. Webster, D., Müller, R.: In silico models of bone remodeling from macro to nano-from organ to cell. WIREs Syst. Biol. Med. **3**, 241–251 (2011)
158. Weinbaum, S., Cowin, S.C., Zeng, Y.: A model for the excitation of osteocytes by mechanical loading-induced bone fluid shear stresses. J. Biomech. **27**, 339–360 (1994)
159. Weinbaum, S., Duan, Y., Thi, M., You, L.: An integrative review of mechanotransduction in endothelial, epithelial (renal) and dendritic cells (osteocytes). Cell Molec. Bioeng. **4**, 510–537 (2011)
160. Westbroek, I., Ajubi, N.E., Ablas, M.J., Semeins, C.M., Klein-Nulend, J., Burger, E.H., Nijweide, P.J: Differential stimulation of prostaglandin g/h synthase-2 in osteocytes and other osteogenic cells by pulsating fluid flow. Biochem. Biophys. Res. Commun. **268**, 414–419 (2000)
161. Williams, J.L., Iannotti, J.P., Ham, A., Bleuit, J., Chen, J.H: Effects of fluid shear stress on bone cells. Biorheology **31**, 163–170 (1994)
162. Wolff, J.: Das Gesetz der Transformation der Knochen. Hirschwald, Berlin (1892)
163. Yamashita, K., Oikawa, N., Umegaki, T.: Acceleration and deceleration of bone-like crystal growth on ceramic hydroxyapatite by electric poling. Chem. Mater. **8**, 2697–2700 (1996)
164. Yasuda, I.: Piezoelectricity of living bone. J. Kyoto Pref. Univ. Med. **53**, 2019–2024 (1964)
165. You, L., Cowin, S.C., Schaffler, M.B., Weinbaum, S.: A model for strain amplification in the actin cytoskeleton of osteocytes due to fluid drag on pericellular matrix. J. Biomech. **34**, 1375–1386 (2001)
166. You, L.D., Weinbaum, S., Cowin, S.C., Schaffler, M.B: Ultrastructure of the osteocyte process and its pericellular matrix. Anat. Rec. **278**, 505–513 (2004)
167. Zhang, D., Weinbaum, S., Cowin S., C.: On the calculation of bone pore water pressure due to mechanical loading. Int. J. Sol. Struct. **35**, 4981–4997 (1998)
168. Zhang, D., Weinbaum, S., Cowin S., C.: Estimates of the peak pressures in bone pore water. J. Biomech. Eng. **120**, 697–703 (1998)
169. Zhou, X., Novotny, J., Wang, L.: Modeling fluorescence recovery after photobleaching in loaded bone: potential applications in measuring fluid and solute transport in the osteocytic lacunar-canalicular system. Ann. Biomed. Eng. **36**, 1961–1977 (2008)
170. Zhou, X., Novotny, J.E., Wang, L.: Anatomic variations of the lacunar-canalicular system influence solute transport in bone. Bone **45**, 704–710 (2009)

Multiscale Elastic Models of Collagen Bio-structures: From Cross-Linked Molecules to Soft Tissues

Michele Marino and Giuseppe Vairo

Abstract Mechanics of collagen bio-structures at different scales (nano, micro, and macro) is addressed, aiming to describe multiscale mechanisms affecting the constitutive response of soft collagen-rich tissues. Single-scale elastic models of collagen molecules, fibrils, and crimped fibers are presented and integrated by means of consistent inter-scale relationships and homogenization arguments. In this way, a unique modeling framework based on a structural multiscale approach is obtained, which allows to analyze the macroscale mechanical behavior of soft collagenous tissues. It accounts for the dominant mechanisms at lower scales without introducing phenomenological descriptions. Comparisons between numerical results obtained via present model and the available experimental data in the case of tendons and aortic walls prove present multiscale approach to be effective in capturing the deep link between histology and mechanics, opening to the possibility of developing patient-specific diagnostic and clinical tools.

1 Introduction

Soft tissues are throughout the whole human body and they include tendons, ligaments, skin, fibrous tissues, muscles and blood vessels. They link, support, and are part of other bio-structures and organs, playing a key role in the biomechanics

M. Marino (✉) · G. Vairo
Department of Civil Engineering and Computer Science, University of Rome "Tor Vergata", via del Politecnico 1, 00133 Rome, Italy
e-mail: m.marino@ing.uniroma2.it

G. Vairo
e-mail: vairo@ing.uniroma2.it

Stud Mechanobiol Tissue Eng Biomater (2013) 14: 73–102
DOI: 10.1007/8415_2012_154

Published Online: 5 October 2012

of many body systems (e.g., musculo-skeletal, cardiovascular) [1]. Collagen, elastin and ground substance are the main constituents of the extracellular matrix in soft tissues, and their arrangement significantly affect the tissue mechanical response. For instance, stiffness and strength features in soft tissues mainly depend on the arrangement and the amount of collagen, which is organized in agreement with a precise hierarchical multiscale scheme [2]. Since the fundamental mechanical role of collagen, soft tissues are usually referred to as collagenous tissues.

The structured organization of soft tissues, and thereby their mechanical behavior, is highly related to the biochemical processes occurring within them [2]. In fact, altered tissue response in disease (e.g., aneurism, keratoconus, arthofibrosis) arises from pathological tissue remodeling, inducing unphysiological histology and biochemical composition. Typical disorders, such as tissue hyper-extensibility or weakness, can be associated with alterations at different scales [3–7]: in content of tissue constituents, in shape of collagen fibers, in collagen genetic pattern, in density of inter-molecular cross-links. Nevertheless, available non-invasive techniques do not allow to measure directly a number of important histological, mechanical, and biochemical properties of collagenous tissues such as, for instance but not exclusively, collagen content and fiber waviness, collagen cross-linking, elastin amount and stiffness of elastin networks.

In this context, the biomechanical analysis and modeling of collagen-rich tissues can be retained a frontier challenge aiming to understand many physio-pathological processes occurring at very different length scales, as well as to identify relationships among alterations and diseases. Accordingly, dominant mechanisms occurring at different scales should be accounted for and consistently coupled in a unique modeling approach. Moreover, in order to enhance model reliability for diagnostic and clinical practice, model parameters should be few and associated with clear physical properties of the tissue, avoiding to introduce phenomenological descriptions. These requirements can be satisfied if the tissue structured hierarchical arrangement is explicitly described, possibly reducing model complexity by means of multiscale homogenization techniques. Such an approach, employed for example in [8], will be referred to as a structural multi-scale method, and consists in developing mechanical models at very different length scales, which are coupled each other by means of consistent inter-scale relationships. In some way, the structural multiscale approach exploits the rationale followed by nature in "designing" tissues and "building" organs.

Structural multiscale models open to the possibility of developing virtual simulation tools, that are patient-specific not only for the geometric description of the tissue domains, but also for the accurate representation of the tissue mechanical properties. As a result, the effects of changes in histological arrangement or biochemical processes on the overall macroscopic functionality of tissues and organs could be predicted. Thereby, really customized pharmacological treatments and therapeutic strategies could be conveniently designed and applied. Finally, parametric biomechanical simulations of tissues and organs based on a multiscale structural framework might be coupled with non-invasive in vivo histological and

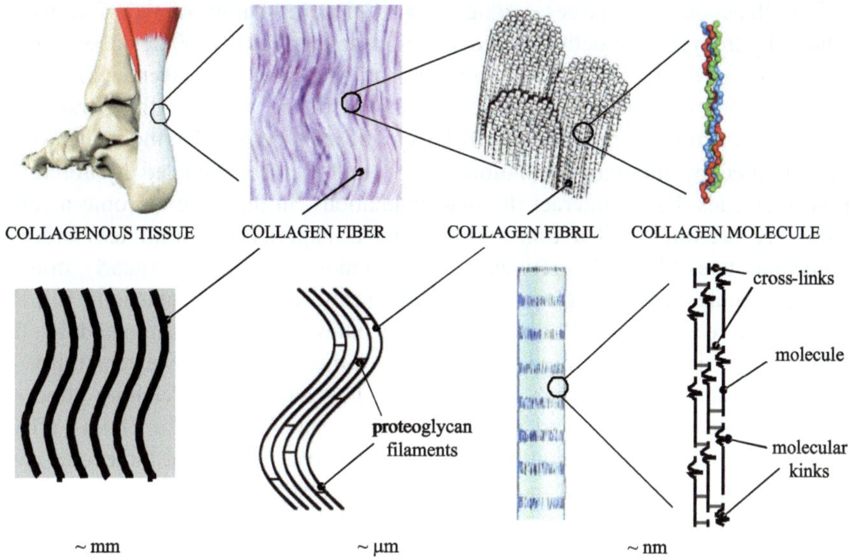

Fig. 1 Hierarchical arrangement of collagen in a regular soft tissue such as a tendon

functional measures. Accordingly, following an inverse-like scheme, indirect estimates of histo-mechano-chemical features, otherwise unknown, could be furnished aiming to improve diagnostic procedures.

2 Biochemical, Histological and Mechanical Properties

Collagen in soft collagenous tissues is arranged according to a precise hierarchical pattern, based on a number of different bio-structures characterized by very different length scales (Fig. 1): from the nanoscale (molecules), to the microscale (fibers), up to the macroscale (tissue). Biomechanics of a single-scale bio-structure and inter-scale coupling effects strictly depend on biochemical, histological and mechanical features at the different scales, and highly affect tissue mechanics at the macroscale.

2.1 Collagen Molecules and Fibrils

The collagen molecule subunit (tropocollagen) can be regarded as a one-dimensional structure about 300 nm long and 1–2 nm in diameter, made up of three polypeptide strands, each one being a left-handed helix. The three helices are twisted together into a triple helix (namely, a "super helix"), representing a cooperative quaternary structure stabilized by covalent cross-links [1].

In collagenous soft tissues, collagen molecules are mainly of type I and they exhibit hydroxyproline-deficient sequences characterized by 60 residues (about 20 nm long), referred to as labile domains and indicated as molecular kinks [2]. Their measure of length ℓ_{kinks} is comparable with the value of the persistence length[1] ℓ_p for collagen (about 14 nm). This evidence confirms that molecular kinks are activated by thermal fluctuations [2, 9] and can be extended by forces at molecular ends that counteract thermal undulations. In this case entropic mechanisms are activated and a transition regime is experienced, from less ordered molecular states (thermally-activated kinks) to more ordered ones (nearly straight macromolecule). In this regime, usually referred to as entropic elasticity [9], the mechanical response of collagen macromolecules is mainly dominated by the flexural behavior of the polypeptide helices rather than by the extensibility of intra-molecular covalent bonds. Accordingly, neglecting any stretching effect of the intra-molecular bonds and in agreement with the Worm-Like Chain (WLC) model [10, 11], the pair of equilibrated forces F^s to be applied at molecular ends for obtaining the end-to-end molecular length ℓ_m results in:

$$F^s(\ell_m) = \rho\left[\frac{1}{4(1-\ell_m/\ell_c)^2} - \frac{1}{4} + \frac{\ell_m}{\ell_c}\right], \tag{1}$$

where ℓ_c is the molecular contour length and $\rho = k_B T/\ell_p$, T being the absolute temperature and k_B the Boltzmann constant. Equation (1) exhibits a pole for $\ell_m = \ell_c$, highlighting that the WLC-model is not able to capture an extension of the end-to-end molecular length over ℓ_c involving entropic mechanisms only. Nevertheless, well-established evidences on collagen show a significant level of molecular extensibility beyond ℓ_c [12]. Accordingly, when ℓ_m approaches ℓ_c the applied force contributes to activate the stretch of molecular covalent bonds, inducing the onset of energetic mechanisms [9, 11].

Theoretical models accounting for the extensibility of biopolymer macromolecules over ℓ_c have been recently proposed in [13–15], and the transition regime from entropic towards energetic elasticity for collagen has been numerically investigated by using Molecular Dynamical Simulations (MDSs) [9]. MDS-based results proposed in [9] have been successfully recovered by the theoretical approach developed in [15], wherein the transition mechanisms were consistently described by a physically-based lumped-parameter equilibrium formulation, avoiding phenomenological transition parameters as made in [13, 14].

It is worth pointing out that, as experimental [12] and MDS-based [9, 16] results suggest, the mechanical response of collagen molecules due to energetic effects is highly non-linear in the first stage, following a pseudo-exponential law, and then tends asymptotically towards a linearly elastic behavior for $\ell_m \gg \ell_c$. Such an evidence can be justified by observing that the non-linearities are essentially due to

[1] The persistence length is the maximum contour length over which the corresponding molecular segment appears as straight under thermal fluctuations.

the unrolling mechanisms of the triple helical structure, that tends to disappear when $\ell_m \gg \ell_c$. In the literature, both linearly [8, 9] and non-linearly [15] elastic models aiming to describe collagen energetic elasticity can be found.

In biological soft tissues, a single tropocollagen subunit self-assembles in the extracellular matrix with four other collagen molecules, with regularly staggered ends, to form units that in turn assemble themselves into even larger arrays, called fibrils. A collagen fibril, characterized by a diameter between 50 and 500 nm, can be thought as a mesoscale structure between molecule at the nanoscale and fiber at the microscale. Within this organized bio-structure, molecules interact each other by means of both inter-molecular covalent cross-links (each of them connecting two molecules) and weak bonds (including hydrogen bonds and other electromagnetic weak interactions), the former being dominant with reference to the fibril's elastic behavior [17].

Recent experimental evidences [18] revealed a three-dimensional crystallographic patterns of collagen molecules within fibrils. Nevertheless, simple arrangement models were proved to be effective in capturing the mechanical key aspects of fibrils, related to molecules and to their mutual interactions. For instance, according to the Hodge–Petruska scheme [19], fibrils can be successfully modelled as staggered arrays of parallel macromolecules with an axial offset of about 67 nm and an equilibrium center-to-center distance of about 1.5 nm between two transversally adjacent molecules.

2.2 *Collagen Fibers and Soft Collagenous Tissues*

Collagen fibrils are densely packed in bundles called fibers. Adjacent fibrils within fibers are stabilized by lateral fibril-to-fibril proteoglycan filaments [20]. As confirmed by the specialized literature, a controversial matter is if proteoglycans play or not a significant role in loading transfer among adjacent fibrils. If fibers consisted in chains of short fibrils interconnected by proteoglycans, then the among-the-fibrils load-transfer mechanisms would be highly affected by inter-fibrils links [21]. Nevertheless, many recent experimental studies [22, 23] reveal extremely few fibril ends, not confirming the thesis of short fibrils. Moreover, other experimental/numerical results suggest that proteoglycan-based cross-links have a marginal and unlikely mechanical role in the elastic behavior of a collagen fiber [24], although they contribute to fundamental physiological processes. Accordingly, these evidences support the hypothesis that the load-transfer mechanism among fibrils within fibers is nearly proteoglycan-independent.

Soft collagenous tissues are generally fibrous connective tissues which can be either dense or loose, depending on the collagen amount; they consist primarily of elastin, amorphous ground substance, cells and collagen fibers [25]. Collagen fibers can be arranged in agreement with a regular (e.g., tendons) or an irregular (e.g., skin) pattern, and regular tissues (that is, with a regular fiber arrangement) can be conveniently classified in uni- (e.g., tendons and ligaments) or multi- (e.g., arterial

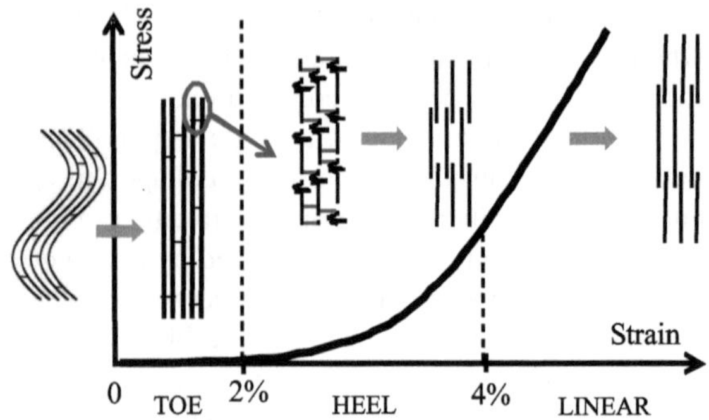

Fig. 2 Typical stress/strain curve for unidirectional regular soft collagenous tissues such as tendons and ligaments

walls) directional. A multi-directional tissue is intended to be made up of a number of stacked thin layers, each of them with a regular uni-directional fiber arrangement.

As confirmed by well-established studies [2], constitutive response of soft collagenous tissues depends primarily on the mechanical behavior of collagen fibers and molecules. For instance, in the case of a uni-directional tissue subjected to a uni-axial tensile test along the fiber direction, a progressive fiber straightening and the disappearance of nanoscale kinks within molecules are experienced, resulting in an increase of the overall tissue stiffness. Accordingly, the stress/strain curves are typically J-shaped and can be subdivided into three main regions (Fig. 2):

- Toe region: the region of small tissue strains (up to 2 %), related to the removal of the microscopic crimp in collagen fibers.
- Heel region: at strains 2–4 %, characterized by a significant stiffening response due to the entropic straightening of nanoscale molecular kinks [2, 9].
- Linear region: when the tissue is stretched beyond the heel region, most kinks are straightened and no further extension is possible by entropic mechanisms. Therefore, the mechanical response is primarily affected by the stretching of the collagen triple-helices and by molecular rearrangement (collagen sliding) [26].

It should be pointed out that, despite of similar features at the nanoscale, different types of collagenous tissues can be characterized by significantly different mechanical responses, mainly depending on histological composition and organization at lower scales. In the following, two different regular soft collagenous tissues, with different complexity in terms of collagen arrangement, geometry and loading conditions, will be addressed: tendons and the tunica media of the aortic walls.

2.2.1 Tendons

Tendons are regular dense connective tissues, transmitting muscular forces and playing a crucial role in the functioning of joints and musculo-skeletal system. The mean axis of collagen fibers in tendons is mainly aligned along the loading direction, and their geometric features depend on the animal species. For instance, addressing rat tail tendons, collagen fiber crimp period and amplitude are in the order of 200 and 10 μm, respectively [27, 28], and fiber radius varies between 2 and 15 μm [29] (while in humans it can be as high as 300 μm). Finally, collagen volume fraction has been reported to be about 50 % [30], with a little variability depending on tendon location and on physiological conditions.

Due to collagen content and organization, uni-axial tendon stress/strain curves are J-shaped, exhibiting the highly non-linear mechanical response that has been previously described (Fig. 2).

2.2.2 Aortic Wall

The aorta is the largest artery in the body, and its structure and mechanical response have a crucial role in the fluid-structure interaction mechanisms relevant to the cardiovascular system. The walls of the aorta are made up of three different tissue layers: a thin inner layer (intima), a thick elastic middle layer (media), and a thin outer layer (adventitia). Among these layers, the media is the most important from the mechanical point of view, as a result of its thickness and stiffness. Histologically, the tunica media is made up of concentric layers consisting in smooth muscle cells embedded in an organized network of loose connective tissue. Many authors (e.g., [31, 32]) describe the media tissue as a thick cylinder comprising different layers, usually denoted as medial lamellar units (MLUs), that have practically the same structural arrangement.

Geometrical features of the aortic media are extremely variable among different living species, depending on the animal size and weight, as well as on the location along the vessel length (e.g., in the thoracic—T—or abdominal—A—zones). Referring to humans, many authors (e.g., [31, 33–35]) indicate that the aortic radius at zero pressure is about 6–9 mm (T) or 5–8 mm (A), that the media thickness-to-radius ratio is about 0.1–0.2, and that the MLU number is about 60 (T) or 30 (A). The three-dimensional histological structure of a single MLU has been recently investigated [36] and described as a thick sub-layer of elastin sided by an interlamellar substance made up of water, elastin, smooth muscle cells and collagen. Collagen results in about 20–30 % of the aortic wall dry-weight [37] and is organized in crimped fibrils with radius varying from 25 to 50 nm [38]. Fibrils are in turn arranged in both thick and thin bundles (namely, fibers). Analysis by means of scanning electron microscopy reveals fiber period in the order of 5 μm and fiber amplitude-to-period ratio about 0.2–0.5 [36]. Fibers are organized in such a way that no complex mesh within each interlamellar layer appears, but they are disposed to realize sub-layers with a uni-directional regular character. In other words, fibers are arranged in circumferential laminae with the fiber axis helically wrapped around the vessel direction, with the wrapping angle (i.e., the angular deflection of the fiber axis with respect to the vessel axis) varying across the MLU thickness, in agreement with a multi-directional tissue structure. Histological evidences [36] confirmed that the wrapping angle exhibits a symmetric uni-modal distribution with a mean angle of about 90° (i.e., the mean fiber axis direction corresponds to the circumferential's). Regarding such a distribution as a Gaussian one, the full width at half maximum is about 40° and then, in agreement also with sequential confocal microscopic images [36], a variation of the wrapping angle (from the inner towards the outer side of a single lamellar unit) from about 90° up to 70° and from about 90° up to 110° is experienced.

From a mechanical point of view, this complex collagen organization results in a tissue constitutive response highly non-linear, characterized by a progressive stiffening for values of the circumferential strains that increase from the zero-load state. It should be remarked that aortic segments in living bodies are pre-stressed and pre-stretched under zero loads. In fact, when the vessel is excised, the aortic segment shortens, and when a ring cross-section of an artery free of external loads is cut radially, an open sector appears [39–41]. Longitudinal pre-stretch and/or pre-stress generally increases along the length of the aorta, whereas the circumferential one smoothly varies along the longitudinal direction of the vessel.

3 Modeling Approaches

At the macroscopic level, several constitutive models for collagen-rich tissues can be found in the specialized literature. Most of them are deduced from phenomenological evidences and generally employ exponential and power-law functions [42, 43], based on parameters having no direct physical or morphological meaning. Other approaches, namely structural approaches, aim to link model parameters with structural properties of the tissue, either by micro–macro homogenization techniques, describing effects related to collagen fibers as linearly elastic [44–46], or by assuming an orthotropic hyper-elastic macroscopic behavior, accounting for the main constituents of the tissue and for some microscale features [47–49]. In this case, non-linearities related to the mechanical response of tissue constituents (mainly geometric non-linearities related to the fiber crimp, and material non-linearities induced by nanoscale mechanisms within and among collagen molecules) are often taken into account by choosing a suitable representation of the fiber strain-energy density (as in a phenomenological approach). Therefore, any direct relationship with the molecular scale is usually neglected and the corresponding models are not able to give predictive indications on nanoscale effects, that probably play the most important role in many diseases (e.g., cross-linking variations are strictly related to aortic rupture and tendon hyper-extensibility [3–7]). Analogously, some microscale features (such as the crimp shape or the thickness of collagen fibers) are not explicitly modeled in many cases, despite of their physiopathological importance (e.g., spontaneously ruptured tendons show reduced thickness and crimp angle [7]).

On the other hand, in a patient-specific framework that aims to give predictive diagnostic indications as well as to describe the tissue response evolution, each model parameter should be directly associated with in vivo measurable features.

This can be obtained by means of a structural multiscale technique, coupling a structural rationale with multiscale homogenization approaches. Accordingly, the equivalent responses of tissue substructures at different scales are analytically derived and consistently integrated, allowing to include at the macroscale the dominant mechanisms occurring at smaller scales [8, 50].

In this context, the authors recently proposed elastic constitutive models for collagen-rich tissues, based on multiscale homogenization techniques that explicitly incorporate nanoscale and microscale mechanisms, as well as their coupling effects [8, 51, 52]. By simulating histological alterations at nano, micro and macro scales, present models have been proved to be able to highlight and to analyze the deep link between histology and mechanical response of both collagenous tissues and body structures [52, 53]. Such an approach allows also to include, through convex analysis arguments and in the same modeling framework, damage evolution at different scales, induced by both mechanical and non-mechanical sources [15].

In the following, a mechanical model of soft collagenous tissues is discussed, addressing the purely elastic response and neglecting any inelastic and damage mechanism. The model generalizes the one proposed and applied in [8, 51], and follows a structural multiscale rationale. The macroscale tissue response is recovered by integrating single-scale models of collagenous bio-structures at very different length scales: molecules (nanoscale), fibrils (mesoscale) and crimped fibers (microscale). Following a structural approach, the ordered histology of both uni-directional and multi-directional tissues is explicitly taken into account.

As a result of a multi-step homogenization procedure, the homogenized tissue at the macroscale is treated as a non-linearly elastic anisotropic continuum, passing from its reference configuration to the actual one via a quasi-static deformation path ϕ, governed by a time-like variable τ. In turn, bio-structures at lower scales undergo to ϕ-induced quasi-static transformations. At each scale this process is herein described following an incremental strategy. For the sake of notation, in what follows the symbol $\dot{x}$ denotes the partial derivative of x with respect to τ.

4 Nanoscale Mechanics: Molecules

A collagen molecule is modelled as an equivalent zero-dimensional nano structure, whose reference end-to-end length is $\ell_{m,o}$, and is lower than its contour length ℓ_c. Let A_m be a measure of the molecular cross-sectional area (assumed to be constant during the overall deformation process), and $\varepsilon_m = \ell_m/\ell_{m,o} - 1$ a measure of the molecular nominal strain, ℓ_m being the actual molecular end-to-end length.

Entropic and energetic mechanisms are assumed to act as in series and they contribute to the overall molecular stretch measure ε_m by ε_m^s and ε_m^h, respectively, so that by compatibility

$$\varepsilon_m = \varepsilon_m^s + \varepsilon_m^h. \tag{2}$$

Denoting with $r_\ell = \ell_{m,o}/\ell_c$, the molecular tangent modulus E_m^s that describes the entropic mechanisms can be introduced, in agreement with the WLC model [8], as

$$E_m^s(\varepsilon_m^s) = \frac{\rho}{A_m}\left\{\frac{r_\ell}{2[1 - r_\ell(1+\varepsilon_m^s)]^3} + r_\ell\right\}. \tag{3}$$

Moreover, the molecular tangent modulus E_m^h dealing with energetic mechanisms is represented as [15]:

$$E_m^h(\varepsilon_m^h) = \frac{\hat{E} r_\ell}{1 + e^{-\eta(r_\ell \varepsilon_m^h - \varepsilon_o^h)}} + \hat{E}_o r_\ell, \tag{4}$$

where $\hat{E}$, $\hat{E}_o$, η and ε_o^h are model parameters.

Expression (4) is able to grasp the experimental evidence that the molecular mechanical response is characterized in the first stage by a pseudo-exponential law, and then asymptotically tends to a linearly elastic behavior for high values of ℓ_m/ℓ_c. The introduced parameters have a clear physical meaning and, thereby, their values can be set by numerical atomistic computations or, when available, by experiments: $\hat{E}$ and $\hat{E}_o$ govern the asymptotic behavior of $E_m^h(\varepsilon_m^h)$ for $\varepsilon_m^h \to \pm\infty$, η describes the slope of the function $E_m^h(\varepsilon_m^h)$ for $\varepsilon_m^h = \varepsilon_o^h$, and ε_o^h is the molecular strain contribution within the energetic regime at which E_m^h practically attains its mean value.

Therefore, the molecular tangent modulus accounting for the series elasticity induced by entropic and energetic mechanisms results in:

$$E_m(\varepsilon_m) = \frac{E_m^s(\varepsilon_m^s) E_m^h(\varepsilon_m^h)}{E_m^s(\varepsilon_m^s) + E_m^h(\varepsilon_m^h)}, \tag{5}$$

and the strain contributions due to entropic and energetic effects, that is the functions $\varepsilon_m^s = \varepsilon_m^s(\varepsilon_m)$ and $\varepsilon_m^h = \varepsilon_m^h(\varepsilon_m)$, are obtained by solving the following differential problem:

$$\dot{\varepsilon}_m^s = \frac{E_m(\varepsilon_m)\dot{\varepsilon}_m}{E_m^s(\varepsilon_m^s)}, \qquad \dot{\varepsilon}_m^h = \frac{E_m(\varepsilon_m)\dot{\varepsilon}_m}{E_m^h(\varepsilon_m^h)}. \tag{6}$$

Accordingly, the description of the entropic/energetic transition directly derives from the mutual competition of the stiffnesses associated to E_m^s and E_m^h: when $E_m^s \ll E_m^h$ molecular mechanics is mainly governed by the entropic mechanisms ($E_m \approx E_m^s$), when $E_m^s \gg E_m^h$ by the energetic ones ($E_m \approx E_m^h$), whereas when $E_m^s \approx E_m^h$ both mechanisms significantly contribute. As the analysis of Eqs. (6) reveals, present model of the entropic/energetic transition is based on equilibrium and compatibility conditions, despite of the phenomenological approaches introduced in [13, 14].

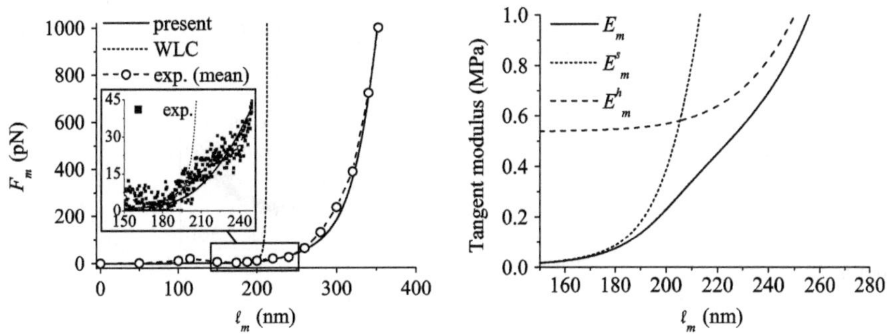

Fig. 3 *Left*: comparison among F_m-ℓ_m curves obtained for a collagen molecule by experimental tests [54], classical WLC, and present model. *Right*: molecular (E_m), entropic (E^s_m), and energetic (E^h_m) tangent moduli vs. ℓ_m for a collagen molecule. Parameters: $T = 310.15$ K, $\ell_p = 14.5$ nm, $\ell_c = 215$ nm, $\ell_{m,o} = 1$ nm, $\hat{E}_o = 0.1$ GPa, $\hat{E} = 10$ GPa, $\eta = 10$, $\varepsilon^h_o = 0.65$

For the uni-axial traction problem of a collagen molecule, the force pair F_m applied at the molecular ends can be computed in terms of the actual molecular length ℓ_m as $F_m = A_m\sigma_m$, where the molecular stress measure results from:

$$\sigma_m(\varepsilon_m) = \int_0^{\varepsilon_m} E_m(\xi)d\xi, \tag{7}$$

E_m being expressed by Eq. (5).

In Fig. 3, the F_m-ℓ_m curve obtained by means of the present approach is compared with the experimental data proposed in [54] and with results computed by the classical WLC model, that is by employing Eq. (1). As reported by many authors [9, 54], the WLC model fits well the molecular response experienced in a low-force regime, but predicts an unrealistic molecular inextensibility at high forces. On the contrary, the proposed approach exhibits an excellent agreement in the overall force range herein addressed, accounting for the molecular compliance during the entropic/energetic transition. In Fig. 3b the molecular tangent modulus is plotted versus the molecular extension, highlighting that $E_m \approx E^s_m$ within the entropic regime, and $E_m \approx E^h_m$ in the energetic one.

Another verification of consistency and soundness of the proposed model for the entropic/energetic transition follows from Fig. 4a: it shows the computed strain rates $\dot{\varepsilon}^s_m$ and $\dot{\varepsilon}^h_m$ (normalized with respect to $\dot{\varepsilon}_m$) plotted versus the molecular length ℓ_m. Reference is made to the numerical analyses proposed in [9], wherein for a collagen molecule with $\ell_c = 301.7$ nm, MDS-based results predict that the entropic regime dominates for $\ell_m < 280$ nm and the energetic one for $\ell_m > 317$ nm. This evidence is fully recovered by present results obtained by solving Eqs. (6): $\dot{\varepsilon}^h_m \to 0$ for $\ell_m < 280$ nm and $\dot{\varepsilon}^h_m/\dot{\varepsilon}_m \to 1$ for $\ell_m > 317$ nm. It is worth pointing out that present equilibrium and compatibility conditions ensure that the pole at $\varepsilon^s_m = 1/r_\ell - 1$ for the function $E^s_m(\varepsilon^s_m)$ is never reached ($\dot{\varepsilon}^s_m \to 0$ for $\ell_m > 317$ nm).

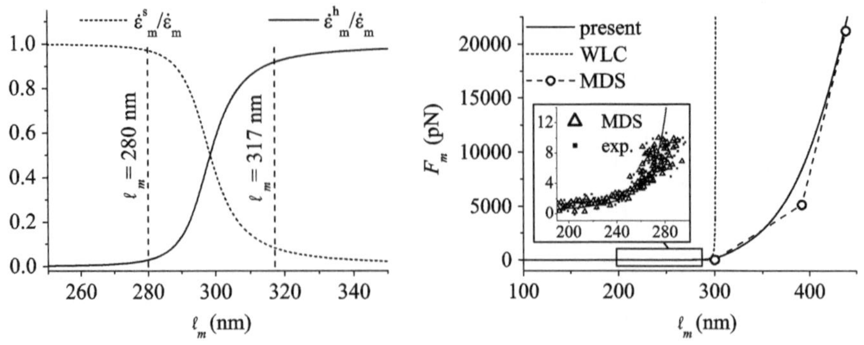

Fig. 4 *Left*: strain rates associated with entropic ($\dot{\varepsilon}_m^s$) and energetic ($\dot{\varepsilon}_m^h$) mechanisms (normalized with respect to $\dot{\varepsilon}_m$) computed by the present approach. *Right*: collagen F_m-ℓ_m curves obtained by experimental tests [12], MDS [9], classical WLC, and present model. Parameters: $T = 310.15$ K, $\ell_p = 16$ nm, $\ell_c = 301.7$ nm, $\ell_{m,o} = 1$ nm, $\hat{E}_o = 5$ GPa, $\hat{E} = 100$ GPa, $\eta = 10$, $\varepsilon_o^h = 0.35$

Moreover, in Fig. 4b, the F_m-ℓ_m curve obtained in this case is successfully compared with the experimental data proposed in [12] (available only for low molecular forces), as well as with the numerical results in [9].

5 Mesoscale Mechanics: Fibrils

Collagen fibrils are modelled as long right cylinders. Let **f** be the fibril axis direction, and A_f a measure of the fibril cross-sectional area. From a computational point of view, fibrils are thought as a collection of N_m identical one-dimensional molecules, with axis aligned along **f** and mutually interacting through N_{cl} identical cross-links, which connect pairs of adjacent molecules (see Fig. 5). Molecules in reference configuration are assumed to be long $\ell_{m,o} = \ell_c - \ell_{kinks}$, ℓ_{kinks} being the length measure of molecular kinks.

Let n_s be the average number of molecules along the fibril length and $\ell_{f,o}$ the reference fibril length. Disregarding axial offset among molecules, the following relationship holds: $\ell_{f,o} \approx n_s \ell_{m,o}$. Moreover, let λ be an average measure of cross-link occurrence for each molecule, such that $N_{cl} = \lambda N_m$. In the case of a homogeneous traction and defining δ as the sway of a cross-link along the fibril length, it is consistent to assume that both molecular strain ε_m and cross-link stretch δ are space-independent quantities within the fibril domain.

Furthermore, denoting with $\Delta\ell_f$ the fibril length variation and with ε_f the fibril nominal strain, the following kinematic assumption is enforced:

$$\Delta\ell_f = n_s(\ell_{m,o}\varepsilon_m + \delta) \rightarrow \varepsilon_f = \frac{\Delta\ell_f}{\ell_{f,o}} = \left(\varepsilon_m + \frac{\delta}{\ell_{m,o}}\right), \qquad (8)$$

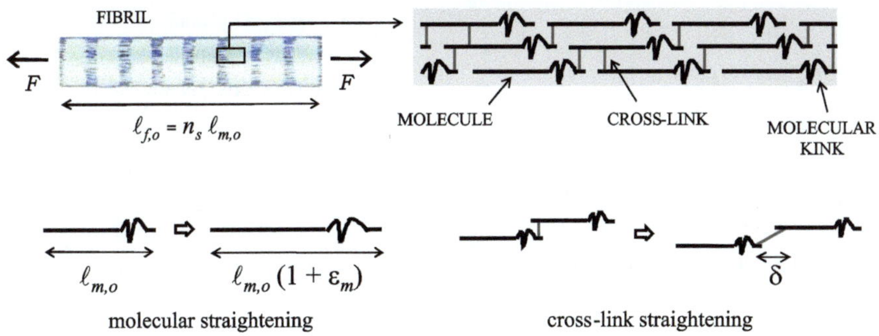

Fig. 5 Fibril model. Notation

that accounts for series mechanisms occurring between molecular and cross-link stretching, and that is consistent with experimental evidences [26]. It is worth pointing out that parameter λ does not intervene in the compatibility equation (8) because covalent cross-links occurring upon a given molecule are assumed to act in parallel.

Assuming a linearly elastic behavior with sway stiffness k_{cl} for each cross-link, a measure of the nominal fibril stress along **f** is

$$\sigma_f = \mu\sigma_m = \lambda k_{cl}\delta/A_f, \tag{9}$$

with σ_m expressed by Eq. (7) and where μ is the average measure of the ratio between solid (occupied by molecules) and total cross-section. Accordingly, by combining Eqs. (8) and (9), the fibril tangent elastic modulus along **f** results in:

$$E_f(\varepsilon_f) = \mu\left[\frac{1}{E_m(\varepsilon_m)} + \frac{A_m}{\lambda k_{cl}\ell_{m,o}}\right]^{-1}, \tag{10}$$

where the contribution to the molecular elongation due to the fibril strain, that is the function $\varepsilon_m = \varepsilon_m(\varepsilon_f)$, is obtained by solving the following inter-scale equilibrium differential problem:

$$\dot{\varepsilon}_m = \frac{E_f(\varepsilon_f)}{\mu E_m(\varepsilon_m)}\dot{\varepsilon}_f. \tag{11}$$

In the following, since fibrils within tissues are made up of densely packed molecules and in agreement with evidences proposed by [55], the model parameter μ is set equal to one.

In Fig. 6, numerical results obtained considering a uni-axial traction of a collagen fibril are shown, highlighting the capability of the proposed approach to predict the evolution of both molecular and fibril tangent moduli versus ε_f. In agreement with recent theoretical evidences at the atomistic scale [2, 56], lower values of the fibril stiffness with respect to the molecular ones are successfully reproduced, as induced by the stretching of inter-molecular cross-links. Moreover,

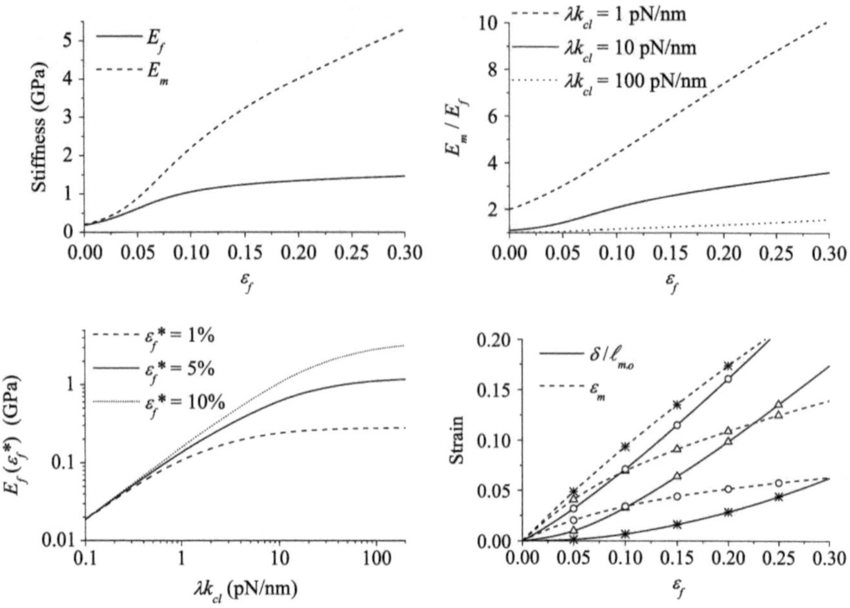

Fig. 6 *Top left*: fibril (E_f) and molecular (E_m) tangent moduli vs. fibril nominal strain ε_f (computed for $\lambda k_{cl} = 10$ pN/nm). *Top right*: ratio between molecular and fibril moduli E_m/E_f vs. fibril nominal strain ε_f for different values of λk_{cl}. *Bottom left*: fibril tangent modulus E_f (in $\log_{10}$ scale) vs. λk_{cl} (in $\log_{10}$ scale) for different values of fibril nominal strain ε_f^*. *Bottom right*: molecular nominal strain ε_m and cross-link normalized extension $\delta/\ell_{m,o}$ vs. fibril strain ε_f for different values of λk_{cl}: (○) $\lambda k_{cl} = 1$ pN/nm; (△) $\lambda k_{cl} = 10$ pN/nm; (∗) $\lambda k_{cl} = 100$ pN/nm. Constant parameters: $\ell_p = 14.5$ nm, $\ell_c = 287$ nm, $\ell_{kinks} = 22$ nm, $\hat{E}_o = 1$ GPa, $\hat{E} = 100$ GPa, $\eta = 10$, $\varepsilon_o^h = 0.35$, $A_m = 1.41$ nm^2, $T = 310.15$ K, $\mu = 1$

a quantitative estimate of the fibril stiffness variation due to changes in cross-link mechanics is provided, showing that fibril stiffness significantly varies both in terms of absolute values and with respect to the molecule's. An increase in cross-link occurrence produces an increase in fibril modulus up to a saturation level that corresponds to the molecular stiffness. Furthermore, numerical results clearly show the non-linear dependence of molecular and cross-link strain measures on both fibril strain and occurrence/stiffness of cross-links.

These results at the mesoscale recover and justify also other significant evidences. Since identical collagen molecules likely exhibit identical nano-mechanical responses, the wide range of values for fibril/fiber modulus, generally reported in the specialized literature as a result of microscale experimental investigations (0.2–12 GPa [2]), can be justified via the proposed results as a consequence of different occurrence and mechanical response of cross-links. Moreover, since after the removal of fibril (fiber) geometrical crimp the mechanical response of the fibrils' material (corresponding to cross-linked collagen molecules) can be considered as representative of the elastic behavior of the overall tissue [2, 8], present results recover the tissue stiffening experienced at the macroscale, confirming as this occurrence can be associated with the formation of mature cross-links [57, 58].

6 Microscale Mechanics: Fibers

A collagen fiber is modeled as a homogeneous beam with a circular cross-section of radius r_F and with a periodic planar centerline in both reference and actual configurations (Fig. 7). Let the along-the-chord nominal strain measure be defined as $\varepsilon_F = L/L_o - 1$, L (respectively L_o) being the actual (reference) fiber period. Moreover, let the Cartesian frame $(\mathbf{t}, \mathbf{n})$ be introduced, with the unit vector $\mathbf{t}$ aligned with the actual fiber chord direction, and let x be the coordinate along $\mathbf{t}$. Accordingly, the centerline position vector results in $\mathbf{r}(x, \varepsilon_F) = f(x, \varepsilon_F)\,\mathbf{n} + x\,\mathbf{t}$, where $f(x, \varepsilon_F)$ denotes the centerline curve, whose slope with respect to $\mathbf{t}$ is defined by the angle $\alpha(x, \varepsilon_F)$.

Following a constrained Hu-Washizu variational approach, and disregarding any shear and Poisson-related effect, the tangent equivalent along-the-chord fiber modulus E_{eq} results in [59]:

$$E_{eq}(\varepsilon_F) = E_c I_F \langle\cos\alpha\rangle \left[I_F\langle\cos^2\alpha\rangle + A_F\langle f^2\rangle\right]^{-1}, \tag{12}$$

where E_c is the tangent elastic modulus of the fiber material along the direction perpendicular to the fiber cross-section, $A_F = \pi r_F^2, I_F = \pi r_F^4/4$, and symbol $\langle\cdot\rangle$ denotes the curvilinear average operator defined along the curvilinear coordinate s following the fiber centerline, that is

$$\langle\cdot\rangle = \frac{1}{\mathcal{L}(\varepsilon_F)} \int_0^{\mathcal{L}(\varepsilon_F)} \cdot\, ds, \tag{13}$$

$\mathcal{L}(\varepsilon_F)$ being the length over a period of the actual fiber centerline. By assuming the fiber material behavior to be governed by the fibril non-linear constitutive response, that is enforcing the constitutive inter-scale condition $E_c = E_f(\varepsilon_f)$, nano- and mesoscale mechanical effects are straightforwardly included in the mechanical description of a collagen fiber. Nominal strain ε_f associated with ε_F, that is the function $\varepsilon_f = \varepsilon_f(\varepsilon_F)$, is computed from the following inter-scale compatibility relationship:

$$\varepsilon_f = \mathcal{L}(\varepsilon_F)/\mathcal{L}(0) - 1, \tag{14}$$

with

$$\mathcal{L}(\varepsilon_F) = \int_0^L \frac{d\zeta}{\cos\alpha}. \tag{15}$$

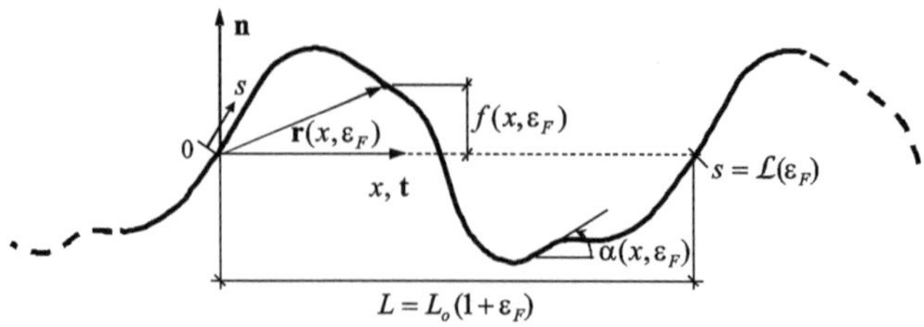

Fig. 7 Planar fiber model. Notation

The actual shape $f(x, \varepsilon_F) = \mathbf{r}(x, \varepsilon_F) \cdot \mathbf{n}$ at the along-the-chord strain level ε_F is determined by solving the following differential problem, resulting by applying the Principle of Virtual Works [59]:

$$\dot{\mathbf{r}}(x, \varepsilon_F) = \frac{E_{eq}(\varepsilon_F)}{E_f(\varepsilon_f)} [a_1(x, \varepsilon_F)\,\mathbf{n} + a_2(x, \varepsilon_F)\mathbf{t}]\dot{\varepsilon}_F, \tag{16}$$

with

$$\begin{aligned} a_1(x, \varepsilon_F) =& - x\bar{a}(\varepsilon_F) + \int_0^x \left[\sin\alpha - \frac{A_F}{I_F}\frac{\zeta f}{\cos\alpha}\right] d\zeta \\ & - x\frac{A_F}{I_F}\int_x^L \frac{f}{\cos\alpha}\, d\zeta, \end{aligned} \tag{17}$$

$$\begin{aligned} a_2(x, \varepsilon_F) =& f(x, \varepsilon_F)\bar{a}(\varepsilon_F) + \int_0^x \left[\cos\alpha + \frac{A_F}{I_F}\frac{f^2}{\cos\alpha}\right] d\zeta \\ & + f(x, \varepsilon_F)\frac{A_F}{I_F}\int_x^L \frac{f}{\cos\alpha}\, d\zeta, \end{aligned} \tag{18}$$

$$\bar{a}(\varepsilon_F) = \frac{1}{2\langle\cos\alpha\rangle}\left[\langle\sin 2\alpha\rangle - 2\frac{A_F}{I_F}\langle\zeta f\rangle\right]. \tag{19}$$

In order to show effectiveness and soundness of the proposed microscale description, numerical results obtained for an isolated collagen crimped fiber are discussed, addressing the fiber along-the-chord response. The reference centerline curve $f(x, 0)$ is defined as the sum of two sinusoidal waves depending on the parameters $\omega \in \mathbb{N}$ $(\omega \geq 0)$ and $\chi \in \mathbb{R}$:

$$f(x, 0) = H_o \sin(2\pi x/L_o) + \chi H_o \sin(2\omega\pi x/L_o) \tag{20}$$

and, as a notation rule, $H_{max,o} = \max\{f(x, 0)\}$. Moreover, nanoscale parameters defining the fiber's material behavior are assumed to be equal to $\ell_p = 14.5$ nm, $\ell_c = 287$ nm, $\hat{E}_o = 1$ GPa, $\hat{E} = 100$ GPa, $\eta = 10$, $\varepsilon_o^h = 0.35$, $\ell_{kinks} = 22$ nm, $A_m = 1.41$ nm^2, $T = 310.15$ K, $\mu = 1$ and $\lambda k_{cl} = 10$ pN/nm.

By integrating Eq. (16), reference and actual (at $\varepsilon_F = 0.1$) shapes of curvilinear fibers with different centerline curves are shown in Fig. 8. The typical stiffening

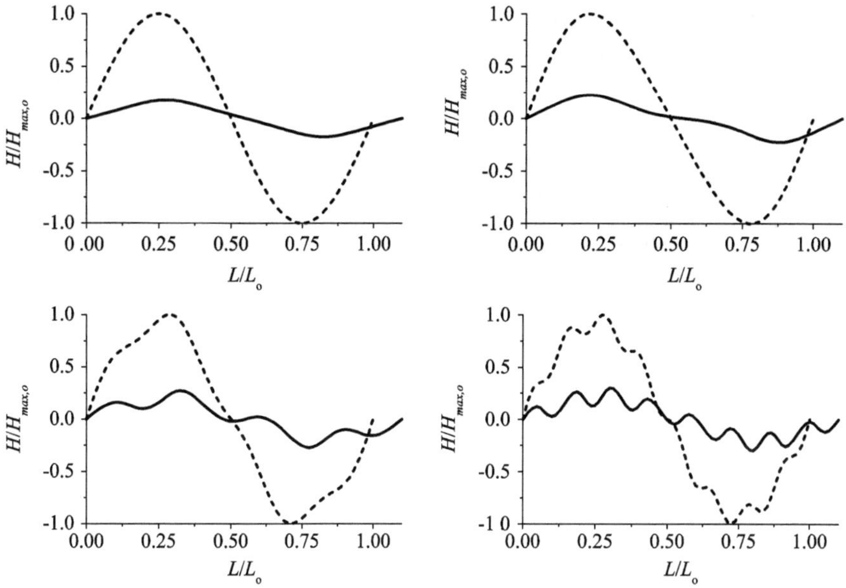

Fig. 8 Actual fiber configuration at $\varepsilon_F = 0.1$ (*continuous lines*) for different reference centerline shapes (*dotted lines*) corresponding to $\omega = 0$ (*top left*), $\omega = 2$ (*top right*), $\omega = 4$ (*bottom left*), and $\omega = 8$ (*bottom right*). Parameters: $H_o/L_o = 0.1$, $r_F/L_o = 0.025$, $L_o = 100\,\mu\text{m}$, $\chi = 0.1$

response related to the decrease of the fiber crimp is highlighted in Fig. 9, wherein the centerline shape is proved to affect the fiber mechanical response mostly for high values of the aspect ratio H_o/L_o, resulting in a significant dependence on the parameters χ and ω.

The fiber behavior is highly non-linear because of both material (at nanoscale and mesoscale) and geometric effects. Nevertheless, if the equivalent modulus E_{eq} is normalized with respect to the fibril's one E_f, the effects related only to geometric non-linearities can be highlighted. Accordingly, Fig. 10 shows the influence of the shape parameters (H_o/L_o, ω, and r_F/L_o) on the fiber mechanical response related to geometric non-linearities, in terms of the tangent modulus in the reference configuration (i.e., at $\varepsilon_F = 0$) as well as in terms of modulus variation versus the fiber strain level ε_F.

7 Macroscale Mechanics

Previous microscale approach is employed to describe the mechanical behavior of collagenous fibers within regular soft tissues. Fibers are assumed to be embedded into a linearly elastic isotropic matrix, whose Young modulus and Poisson ratio are E_M and ν_M, respectively. Neglecting any fiber–matrix interaction effect, crimped fibers are reduced to equivalent reinforcing straight fibers, exhibiting an elastic

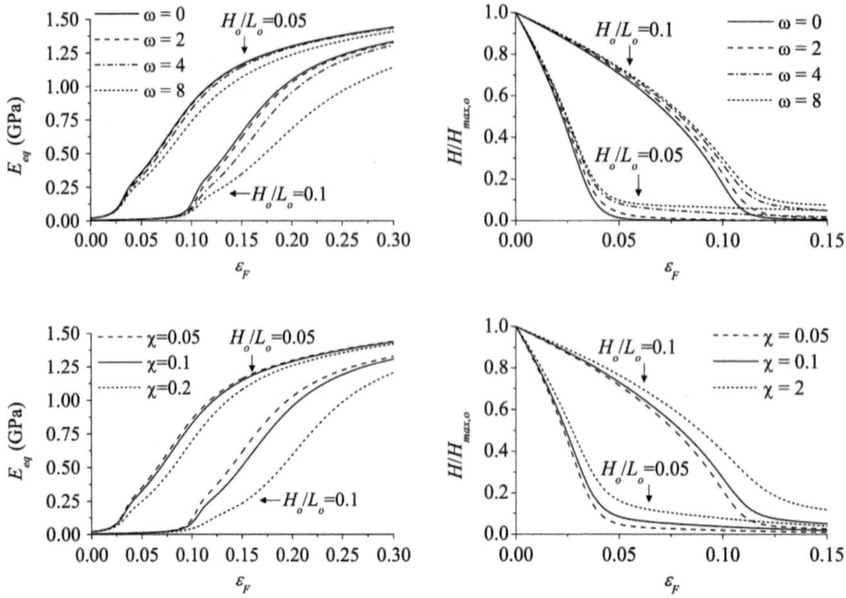

Fig. 9 Along-the-chord modulus E_{eq} (*left*) and dimensionless amplitude variation $H(\varepsilon_F)/H_{max,o}$ (*right*) vs. the fiber along-the-chord nominal strain ε_F for different reference centerline curves. Parameters: $r_F/L_o = 0.025$, $L_o = 100\,\mu\text{m}$, $\chi = 0.1$, $\omega = 4$

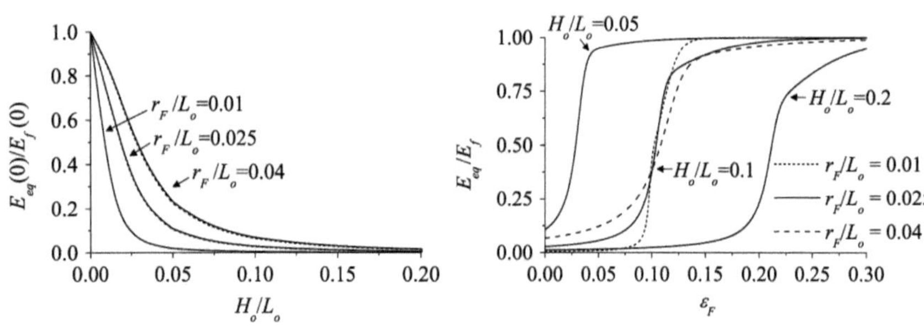

Fig. 10 Influence of the aspect ratios H_o/L_o and r_F/L_o on the fiber mechanical response. *Left*: Along-the-chord modulus E_{eq} at $\varepsilon_F = 0$, normalized with respect to the initial fibril modulus $E_f(0)$ for different reference centerline geometries ($\omega = 0$: *continuous line*; $\omega = 8$: *dotted line*). *Right*: Along-the-chord modulus E_{eq} normalized with respect to the fibril modulus $E_f(\varepsilon_f)$ vs. the nominal fiber strain ε_F for $\omega = 4$. Parameters: $L_o = 100\,\mu\text{m}$, $\chi = 0.1$

transversally isotropic behavior with the symmetry axis coincident to the fiber-chord direction **t**. Due to the small values of the fiber flexural stiffness and of E_M [9], the Poisson ratio (ν_F), the tangent tensile (E_F) and the shear (G_F) elastic moduli of an equivalent fiber within the tissue are represented by:

$$E_{F,L}(\varepsilon_F) = E_{eq}(\varepsilon_F), \qquad E_{F,T} = E_M, \tag{21}$$

$$\nu_{F,LT} = \nu_{F,TT} = \nu_M, \qquad G_{F,LT} = E_M/[2(1+\nu_M)], \tag{22}$$

with $E_{eq}(\varepsilon_F)$ as in Eq. (12) and where the subscript L refers to the direction identified by **t**, whereas T to any direction orthogonal to **t**.

Once crimped collagenous fibers are reduced to equivalent straight fibers, through the previous homogenization step and accounting for geometrical and material non-linearities, standard arguments for fiber-reinforced composite materials can be employed in order to describe the macromechanics of soft collagenous tissues.

7.1 Uni-directional Tissues: Tendons and Ligaments

A further homogenization step at the macroscale is carried out by employing the mixture rule [60]. Accordingly, a uni-directional collagenous tissue is reduced to a homogeneous medium with a transversally isotropic behavior, the isotropy plane being orthogonal to **t**. Therefore, tangent equivalent elastic constants of the tissue at the along-the-chord fiber strain level ε_F result in:

$$E_L(\varepsilon_F) = V_f E_{F,L}(\varepsilon_F) + (1-V_f)E_M, \qquad E_T = \frac{E_{F,T}E_M}{E_{F,T}(1-V_f)+E_M V_f}, \tag{23}$$

$$G_{LT} = \left(\frac{V_f}{G_{F,LT}} + \frac{1-V_f}{G_M}\right)^{-1}, \qquad G_{TT} = \left(\frac{V_f}{G_{F,TT}} + \frac{1-V_f}{G_M}\right)^{-1}, \tag{24}$$

$$\nu_{LT} = V_f \nu_{F,LT} + (1-V_f)\nu_M, \qquad \nu_{TT} = \frac{E_T}{2G_{TT}} - 1, \tag{25}$$

where V_f is the fiber volume fraction. Referring to the standard Voigt notation, the tangent stiffness matrix $\tilde{\mathbb{C}}$ in the material coordinate system (**t**,**n**,**k**), with $\mathbf{k} = \mathbf{t} \times \mathbf{n}$, is:

$$\tilde{\mathbb{C}}(\varepsilon_F) = \begin{bmatrix} \mathbb{L}(\varepsilon_F) & 0 \\ 0 & \mathbb{M} \end{bmatrix}, \tag{26}$$

where

$$\mathbb{L}(\varepsilon_F) = \frac{1}{D}\begin{bmatrix} E_L(1-\nu_{TT}^2) & E_T\nu_{LT}(1+\nu_{TT}) & E_T\nu_{LT}(1+\nu_{TT}) \\ E_T\nu_{LT}(1+\nu_{TT}) & E_T(1-\kappa\nu_{LT}^2) & E_T(\nu_{TT}+\kappa\nu_{LT}^2) \\ E_T\nu_{LT}(1+\nu_{TT}) & E_T(\nu_{TT}+\kappa\nu_{LT}^2) & E_T(1-\kappa\nu_{LT}^2) \end{bmatrix}, \tag{27}$$

$$\mathbb{M} = \mathrm{diag}(G_{TT}, G_{LT}, G_{LT}), \qquad D(\varepsilon_F) = 1 - \nu_{TT}^2 - 2(1+\nu_{TT})\kappa\nu_{LT}^2, \tag{28}$$

with $E_L = E_L(\varepsilon_F)$ and $\kappa = \kappa(\varepsilon_F) = E_T/E_L(\varepsilon_F)$.

Accordingly, at the macroscale and in a global coordinate system, the tangent homogenized constitutive law for the tissue results in:

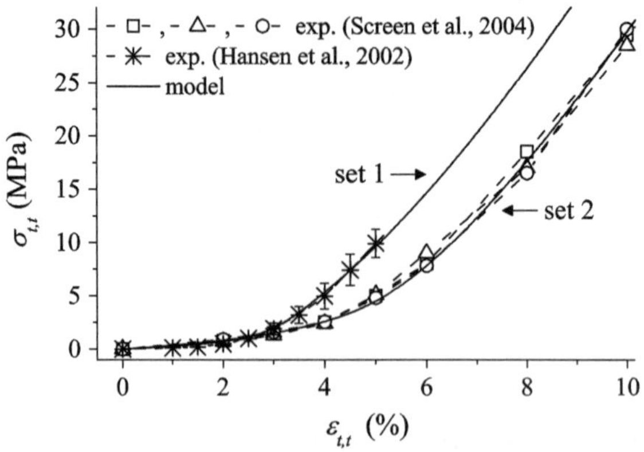

Fig. 11 Mechanics of uni-directional tissues: experimental along-**t** constitutive response of rat tail tendons [27, 61] compared with the numerical results obtained via present model by using two consistent sets of micro/macro model parameters (Table 1). $\sigma_{t,t}$ and $\varepsilon_{t,t}$ denote direct stress and strain components, respectively, along the fiber-chord direction **t**. The shape of the collagen fiber centerline in the reference configuration is defined as in Eq. (20) with $\chi = 0$. Values of nanoscale model parameters are set equal to: $\ell_p = 14.5$ nm, $\ell_c = 287$ nm, $\hat{E}_o = 1$ GPa, $\hat{E} = 80$ GPa, $\eta = 22.5$, $\varepsilon_o^h = 0.1$, $\ell_{kinks} = 14$ nm, $A_m = 1.41$ nm^2, $T = 310.15$ K, $\mu = 1$, $\lambda k_{cl} = 10$ pN/nm

Table 1 Values of micro- and macroscale parameters for tendons employed in numerical applications with relevant references

	H_o/L_o	r_F/L_o	L_o (μm)	V_f	E_M (MPa)	ν_M
Set 1	0.05	0.02	200	50 %	1	0.49
Set 2	0.0635	0.0325				
Ref.	[28]	[29]	[27, 28]	[30]	[62]	[62]

$$\dot{\boldsymbol{\sigma}}_t = \mathbb{C}(\varepsilon_F)\dot{\boldsymbol{\varepsilon}}_t = \hat{\mathbb{T}}_\sigma(\mathbf{t})\tilde{\mathbb{C}}(\varepsilon_F)[\hat{\mathbb{T}}_\varepsilon(\mathbf{t})]^{-1}\dot{\boldsymbol{\varepsilon}}_t, \tag{29}$$

where $\hat{\mathbb{T}}_\sigma(\mathbf{t})$ and $\hat{\mathbb{T}}_\varepsilon(\mathbf{t})$ are the stress and strain transformation matrix from the local to the global coordinate system, and $\dot{\boldsymbol{\varepsilon}}_t$, $\dot{\boldsymbol{\sigma}}_t$ are the increments of macro strain and stress vectors (in Voigt notation) for the equivalent homogeneous tissue, respectively.

In order to validate the present approach, uni-axial traction along the fiber-chord direction of tendinous tissues is addressed, comparing available experimental data for rat tail tendons with numerical results obtained via the proposed model. Figure 11 shows the excellent agreement between experimental results measured from two different tendinous specimens [27, 61] and the model's outcomes. These are obtained by using an incremental approach and considering two different sets of parameters (Table 1). The major issue is that, setting the same values for nanoscale parameters (since the collagen-related nanoscale features are similar among different healthy tissues), different tissue mechanical responses are clearly reproduced by considering differences in microscale fiber geometry.

7.2 Multi-directional Tissues: The Aortic Media Case

The proposed approach can be also employed to describe multi-layered tissues, comprising layers each having a uni-directional collagen fiber arrangement. This is the case of the arterial tunica media. In the following, reference will be made to the aorta.

Following experimental evidences [31, 32, 36] and well-established modeling approaches [47], aortic media is modelled as a multi-layered thick-walled cylinder (with internal radius r_i and thickness h_a) made up of N identical layers (media lamellar units MLUs, see Fig. 12). Such a multi-layered cylinder is assumed to have a length much greater than r_i, and to be loaded by a uniform internal pressure distribution p, undergoing prevailing membranal response with negligible flexural effects. Let the cylindrical coordinate system (z, φ, ρ) be introduced, where ρ is the radial coordinate, and φ and z the angular and axial coordinates, respectively.

Each MLU comprises an elastin rich elastic lamina (EL) h_e thick and an interlamellar substance (IL) with thickness h_{IL}.

The EL sub-layer is modelled as a two-phase substance comprising void (or very soft constituents with negligible stiffness) with volume fraction V_o, and elastin. The latter is assumed with a linearly elastic isotropic behavior and characterized by the Young's modulus E_e and the Poisson's ratio ν_e. Accordingly, EL can be reduced by the mixture rule to a homogeneous layer with equivalent isotropic elastic constants equal to $(1 - V_o)E_e$ and $(1 - V_o)\nu_e$.

The interlamellar substance, in turn, can be regarded as a multi-layered sub-structure, made up of concentrically fiber-reinforced layers, comprising elastin, muscle cells, and crimped collagenous fibers whose main direction is helically arranged around the vessel axis. In agreement with well-established histological in vivo measures [36] and as previously recalled (see Sect. 2.2.2), the wrapping angle θ_F of collagen fibers can be described as a function of the radial coordinate ρ (Fig. 12).

The kth MLU ($k = 1 \ldots N$) is reduced to a homogeneous layer, comprising an anisotropic (generally with a monoclinic symmetry) non-linearly elastic material characterized by a tangent equivalent stiffness matrix $\bar{\mathbb{C}}^k(\varepsilon_F)$ (and compliance matrix $\bar{\mathbb{S}}^k = (\bar{\mathbb{C}}^k)^{-1}$). The latter is obtained by accounting for the kth MLU-based microstructure through a homogenization step carried out via the standard laminate theory [60, 63]. To this aim, at the radial position ρ, the local tangent stiffness matrix $\mathbb{C}$ is computed referring to a uni-directional collagenous tissue with the fiber chord direction $\mathbf{t}(\rho)$ inclined by $\theta_F(\rho)$ with respect to the vessel axis z. Accordingly, $\mathbb{C} = \mathbb{C}(\theta_F, \varepsilon_F)$, where $\theta_F = \theta_F(\rho)$ and $\varepsilon_F = \varepsilon_F(\rho)$, and thereby $\mathbb{C} = \mathbb{C}(\rho)$ at each incremental step. Since $(h_e + h_{\text{IL}})/r_i \ll 1$, any curvature effect is disregarded. It is worth observing that if ρ identifies a position within a EL sub-layer, $\mathbb{C}$ is described in agreement with an isotropic response, that is by involving the elastic constants E_e and ν_e only.

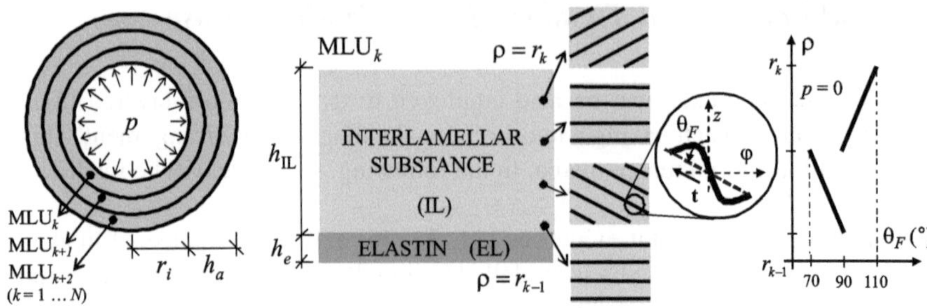

Fig. 12 Model of the multi-layered aortic tunica media. Notation. The function $\theta_F(\rho)$ has been assumed in agreement with evidences proposed in [36]

Indicating with $\bar{S}_{ij}^k$ (respectively, C_{ij} and S_{ij}) the components of the 6×6 compliance matrix $\bar{\mathbb{S}}^k$ (respectively of $\mathbb{C}(\rho)$ and $\mathbb{S}(\rho)$) expressed in the standard Voigt notation, the homogenized tangent compliance results from [60, 63]:

$$\begin{bmatrix} \bar{S}_{11}^k & \bar{S}_{12}^k & \bar{S}_{16}^k \\ \bar{S}_{21}^k & \bar{S}_{22}^k & \bar{S}_{26}^k \\ \bar{S}_{61}^k & \bar{S}_{62}^k & \bar{S}_{66}^k \end{bmatrix} = h_{\mathrm{IL}}\,(A^k)^{-1}, \tag{30}$$

$$\begin{bmatrix} \bar{S}_{31}^k & \bar{S}_{32}^k & \bar{S}_{36}^k \end{bmatrix} = (A^k)^{-1} \left\{ \int_{r_{k-1}}^{r_k} [S_{13}(\rho) \quad S_{23}(\rho) \quad S_{36}(\rho)]\, Q^k(\rho) d\rho \right\}, \tag{31}$$

$$\bar{S}_{33}^k = \frac{1}{h_{\mathrm{IL}}} \int_{r_{k-1}}^{r_k} \frac{1}{C_{33}(\rho)} \left\{ 1 - \bar{S}_{31}^k C_{13}(\rho) - \bar{S}_{32}^k C_{23}(\rho) - \bar{S}_{36}^k C_{63}(\rho) \right\} d\rho, \tag{32}$$

$$\begin{bmatrix} \bar{S}_{44}^k & \bar{S}_{45}^k \\ \bar{S}_{54}^k & \bar{S}_{55}^k \end{bmatrix} = \frac{1}{h_{\mathrm{IL}}} \int_{r_{k-1}}^{r_k} \begin{bmatrix} S_{44}(\rho) & S_{45}(\rho) \\ S_{54}(\rho) & S_{55}(\rho) \end{bmatrix} d\rho, \tag{33}$$

where

$$Q^k(\rho) = \begin{bmatrix} C_{11}(\rho) & C_{12}(\rho) & C_{16}(\rho) \\ C_{21}(\rho) & C_{22}(\rho) & C_{26}(\rho) \\ C_{61}(\rho) & C_{62}(\rho) & C_{66}(\rho) \end{bmatrix}, \qquad A^k = \int_{r_{k-1}}^{r_k} Q^k(\rho) d\rho, \tag{34}$$

and where r_{k-1} and r_k are the minimum and the maximum radial coordinate of the kth MLU, respectively.

Since non-fibrous components within interlamellar space are essentially elastin and muscle cells, and the stiffness of muscle cells can be retained as comparable to the elastin's [64], as a first approximation no distinction between elastin and muscle phases is assumed. Accordingly, the computation of $\mathbb{C}(\rho)$ via Eqs. (23)–(29) can be carried out by applying the mixture rule and by letting:

$$\frac{E_M}{E_e} = \frac{\nu_M}{\nu_e} = \left(1 - \frac{V_o}{1 - V_f} \right). \tag{35}$$

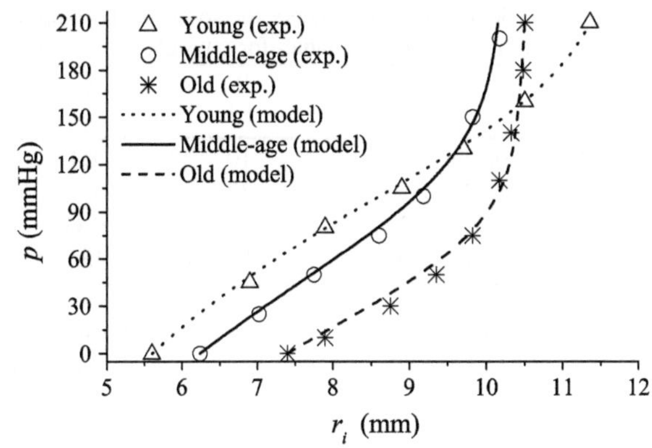

Fig. 13 Experimental pressure p vs. the internal-radius r_i for human aortas at different ages [34], compared with the numerical results obtained by means of the present multiscale model. The shape of the collagen fiber centerline in the reference configuration is defined as in Eq. (20) with $\chi = 0$. Age-dependent model parameters are summarized in Table 2, whereas constant parameters are set equal to: $\ell_p = 14.5$ nm, $\ell_c = 287$ nm, $\hat{E}_o = 1$ GPa, $\hat{E} = 80$ GPa, $\eta = 22.5$, $\varepsilon_o^h = 0.1$, $\ell_{kinks} = 14$ nm, $A_m = 1.41$ nm^2, $T = 310.15$ K, $\mu = 1$, and $L_o = 5\,\mu$m

Due to the previous symmetry assumptions and disregarding any possible end effect, problem reduces to an axisymmetric generalized plane strain problem. In the framework of an incremental approach, the incremental elastic equilibrium solution for such a composite multi-layered thick-walled structure is obtained at each pressure variation $\dot{p}$ as briefly summarized in the Appendix, updating at each step the fiber chord direction $\mathbf{t}(\rho)$ and the fiber strain $\varepsilon_F(\rho)$ as functions of the actual macroscale strain field [8].

In Fig. 13, p-r_i curves numerically obtained via present model are compared with the available experimental results proposed in [34]. In agreement with histological observations [31], the MTU number N has been set equal to 60. Moreover, since the experimental data have been obtained on excised dead aortas, both longitudinal pre-stretch (that is, the direct strain ε_o, see Appendix) and smooth muscle tone have been assumed to vanish, whereas geometrical radial pre-stretch effects have been taken into account referring the fiber morphological parameters to the experimentally-analyzed undamaged aortic cross-sections at zero-load state [36].

Proposed numerical results show the effects related to the variation of some model parameters, with the aim to reproduce the evolution of the tissue mechanics induced by age. The age-dependent values of model parameters have been assumed in agreement with the evidences proposed through the recent specialized literature [3, 34, 35, 57, 65], wherein well-established data and results give indication on the evolution of some histological, geometric and mechanical characteristics in healthy human aortas. In the lack of detailed experimental data, parameters have been set within well-established physiological ranges, and their age-dependent variation has been chosen in agreement with available qualitative evidences, disregarding possible differences between the thoracic and the abdominal aortic branches.

Table 2 Values of age-dependent model parameters for aortic media employed in numerical applications and relevant references (when available) for physiological ranges

Age (years)	λk_{cl} (pN/nm)	H_o/L_o (-)	r_F/L_o (-)	$V_f/(1-V_o)$ (%)	V_o (%)	E_e (kPa)	h_e (µm)	h_a (mm)	$r_i\|_{p=0}$ (mm)
20–24	5	0.425	0.065	15	10	30	1.4	0.588	5.6
36–42	10	0.295	0.032	20	30	80	1.5	0.630	6.2
71–78	15	0.225	0.022	25	50	120	1.65	0.693	7.4
Ref.		[36]	[36]	[66]	[66]	[35]	[35]	[35]	[34]

Experience-based age-dependent aortic features are reported in the following, together with the corresponding evolution trends for the physically-based parameters introduced in the present model:

- Unloaded aortic radius [34, 65] and media thickness [65] increase with age. Physiological ranges are about 5.5-8.65 mm and 0.44-0.76 mm, respectively. Accordingly, $r_i|_{p=0}$ and h_a increase with age.
- Elastin stiffness increases in men, due to increasing glycation with age [35], in the range 30–280 kPa. Thereby, E_e increases with age.
- The stiffness of collagenous constituents at high pressure (when the collagen fibers are straightened) increases with age [35], indicating a higher stiffness of the collagen fibrils, that is E_f. Since the genetic pattern of healthy people is practically constant in time, it is unlikely to believe that this stiffening behavior is related to variations in the nanoscale structure of collagen molecules. Thereby, variations in E_f are associated with the alteration of cross-link stiffness and/or occurrence, that is with the variation of λk_{cl}, which increases with age. This is fully in agreement with the experimental results from [57].
- With increasing age, collagen-induced stiffening effects occur at lower pressure levels during the cardiac cycle [35]. In agreement with present microscale results (see Fig. 10), this evidence can be associated with a reduced aspect ratio H_o/L_o of the collagen fibers, which then results to be inversely proportional to age.
- The stiffness of collagenous constituents at very low physiological pressure decreases with age [65]. As shown in Fig. 10, a reduction in stiffness of collagenous constituents at small strains can be associated with either more crimped or thinner fibers. In agreement with the previous evidences, the occurrence of fibers more crimped has been excluded, associating the present finding to a reduced value of the collagen fiber radius. Thereby, ratio r_F/L_o decreases with age.
- Dry weight of the aortic wall decreases with age (about four or five times, [66]) despite of the increase of the aortic thickness [65]. Therefore, aortic water content (non-solid matter) increases with age, leading to the decrease of the absolute amount of elastin and collagen. On the other hand, the increase (about three times) of the relative amount of collagen with respect to the non-collagenous solid matter has been reported [66]. Accordingly, V_o increases with age, as well as also $V_f/(1-V_o)$.

In agreement with previous considerations, parameters employed in proposed numerical applications are chosen as summarized in Table 2.

Figure 13 clearly shows that proposed model is able to reproduce the mean experimental p-r_i curves associated with young (20–24 years), middle-age (36–42 years), and old (71–78 years) subjects. Although values of model parameters are chosen only in agreement with well-established trends and there is no quantitative relationships with the available experiments, the model reveals to be attractive and potentially effective in relating tissue structure at different scales and its evolution to the tissue macroscopic mechanical behavior.

8 Conclusions

In the present paper, a structural multiscale elastic formulation for modeling soft collagenous tissues has been presented, based on the approach proposed and applied by authors in [8, 51]. To this aim, single-scale models of collagenous bio-structures at very different length scales have been introduced, addressing molecules (nanoscale), fibrils (mesoscale) and crimped fibers (microscale). Following a multiscale structural rationale, that allows to account for the hierarchical multiscale arrangement of the constituents within tissues, these models have been integrated by means of consistent inter-scale relationships and homogenization arguments, leading to an accurate macroscale description of soft collagenous tissues.

Soundness and effectiveness of the present approach have been proved by recovering a number of well-established evidences at different length scales, without the use of phenomenological descriptions. It is shown that proposed single-scale models are able to reproduce the transition from entropic-to-energetic mechanisms at the nanoscale, the effects of inter-molecular cross-links on fibril mechanics, and the coupling between material and geometric non-linearities affecting the deformation process of crimped collagen fibers. Moreover, available experimental data on the biomechanical response of tendons and aortas have been compared with the numerical results obtained by the macroscale model, highlighting an excellent agreement. Such a capability arises from the structural multiscale approach herein adopted, that allows to account for material and geometrical non-linearities at different scales, as well as for nanoscale mechanisms. In fact, as proved in [8], a structural approach involving only microscale mechanisms is generally not able to recover accurately, for different strain levels, the non-linear mechanical response of soft collagenous tissues.

Following the proposed approach, few and experimentally measurable model parameters are introduced. Thereby, the effects of altered histological features on tissue mechanics can be straightforwardly investigated. For instance, different microscale fiber geometric features allow to explain the variability in constitutive response experienced when data measured on different specimens of rat tail tendons are compared. Moreover, the age-dependent evolution of aortic mechanical behavior has been accurately reproduced, simply by incorporating the well-documented features of age-dependent tissue remodeling.

Therefore, multiscale structural approaches allow to implement numerical simulations, that are patient-specific not only for the geometric description of tissues and organs, but also in terms of tissue constitutive properties. This opens to the possibility of numerically investigating the effects of histological and biochemical rearrangement on the mechanics of an organ, as well as of estimating, in an inverse-like scheme, the values of histo-mechano-chemical features by means of non-invasive techniques. Accordingly, among the available modeling approaches for the analysis of soft tissues, the structural multiscale rationale can be retained as the most promising for conceiving and developing groundbreaking virtual tools, allowing to improve diagnosis and to assess customized clinical treatments.

Acknowledgments Authors would like to thank Professor Franco Maceri for valuable suggestions and fruitful discussions on this paper.

This work was developed within the framework of Lagrange Laboratory, a European research group comprising CNRS, CNR, the Universities of Rome "Tor Vergata", Calabria, Cassino, Pavia, and Salerno, Ecole Polytechnique, University of Montpellier II, ENPC, LCPC, and ENTPE.

Appendix

The incremental elastic equilibrium solution for the multi-layered aortic cylinder, comprising N identical media lamellar units (MLUs) and loaded with a uniform internal pressure increment $\dot{p}$, is herein briefly reported. Reference is made to the linearly elastic solution proposed in [67], considering the problem as an axisymmetric generalized plane strain problem, characterized by a given non-vanishing constant direct strain increment $\dot{\varepsilon}_o$ along the cylinder axis z. Accordingly, for the kth MLU, the incremental components of the displacement field in a cylindrical system of coordinates result in

$$\dot{u}_\rho^k = A_1^k \rho^{\alpha_k} + A_2^k \rho^{-\alpha_k} + \dot{\varepsilon}_o a_1^k \rho + B^k a_2^k \rho^2, \tag{36}$$

$$\dot{u}_\varphi^k = B^k \rho z, \tag{37}$$

$$\dot{u}_z^k = \dot{\varepsilon}_o z. \tag{38}$$

The non-trivial increments of strain components are

$$\dot{\varepsilon}_\rho^k = \frac{\partial \dot{u}_\rho^k}{\partial \rho}, \qquad \dot{\varepsilon}_\varphi^k = \frac{\dot{u}_\rho^k}{\rho}, \qquad \dot{\gamma}_{z\varphi}^k = B^k \rho, \tag{39}$$

and the components of stress increments are

$$\dot{\sigma}_\rho^k = A_1^k \beta_{\rho 1}^k \rho^{\alpha_k - 1} + A_2^k \beta_{\rho 2}^k \rho^{-\alpha_k - 1} + \dot{\varepsilon}_o \beta_{\rho 3}^k + B^k \beta_{\rho 4}^k \rho, \tag{40}$$

$$\dot{\sigma}_{\varphi}^{k} = A_1^k \beta_{\varphi 1}^k \rho^{\alpha_k - 1} + A_2^k \beta_{\varphi 2}^k \rho^{-\alpha_k - 1} + \dot{\varepsilon}_o \beta_{\varphi 3}^k + B^k \beta_{\varphi 4}^k \rho, \tag{41}$$

$$\dot{\sigma}_{z}^{k} = A_1^k \beta_{z 1}^k \rho^{\alpha_k - 1} + A_2^k \beta_{z 2}^k \rho^{-\alpha_k - 1} + \dot{\varepsilon}_o \beta_{z 3}^k + B^k \beta_{z 4}^k \rho, \tag{42}$$

$$\dot{\tau}_{z\varphi}^{k} = A_1^k \beta_{\gamma 1}^k \rho^{\alpha_k - 1} + A_2^k \beta_{\gamma 2}^k \rho^{-\alpha_k - 1} + \dot{\varepsilon}_o \beta_{\gamma 3}^k + B^k \beta_{\gamma 4}^k \rho, \tag{43}$$

where

$$\beta_{q1}^k = \bar{C}_{g2}^k + \alpha_k \bar{C}_{g3}^k, \qquad \beta_{q2}^k = \bar{C}_{g2}^k - \alpha_k \bar{C}_{g3}^k, \tag{44}$$

$$\beta_{q3}^k = \bar{C}_{g1}^k + (\bar{C}_{g2}^k + \bar{C}_{g3}^k) a_1^k, \qquad \beta_{q4}^k = \bar{C}_{g6}^k + (\bar{C}_{g2}^k + 2\bar{C}_{g3}^k) a_2^k, \tag{45}$$

with the index $q = z, \varphi, \rho, \gamma$ corresponding to the g-index values $g = 1, 2, 3, 6$, respectively, and with

$$\alpha_k^2 = \bar{C}_{22}^k / \bar{C}_{33}^k, \tag{46}$$

$$a_1^k = (\bar{C}_{12}^k - \bar{C}_{13}^k)/[\bar{C}_{33}^k(1 - \alpha_k^2)], \qquad a_2^k = (\bar{C}_{26}^k - 2\bar{C}_{36}^k)/[\bar{C}_{33}^k(4 - \alpha_k^2)], \tag{47}$$

$\bar{C}_{ij}^k$ denoting the (i, j) component of the stiffness matrix $\mathbb{C}^k$ for the kth MLU.

It is worth pointing out that during the deformation path both strain and stress components, and thereby also the strain-dependent MTU material properties, are z- and φ-independent.

Assuming the MLUs as perfectly bonded layers, the $3N$ unknown constants A_1^k, A_2^k, B^k (with $k = 1 \ldots N$) are solved by imposing the following $3(N-1)$ continuity conditions at the MLU's interfaces:

$$\dot{u}_{\rho}^k(r_k) = \dot{u}_{\rho}^{k+1}(r_k), \qquad \dot{u}_{\varphi}^k(r_k) - \dot{u}_{\varphi}^{k+1}(r_k), \qquad \dot{\sigma}_r^k(r_k) = \dot{\sigma}_r^{k+1}(r_k), \tag{48}$$

and the three equilibrium incremental relationships:

$$\dot{\sigma}_r(r_i) = -\dot{p}, \qquad \dot{\sigma}_r(r_i + h_a) = 0, \qquad 2\pi \sum_{k=1}^{N} \int_{r_{k-1}}^{r_k} \rho^2 \dot{\tau}_{z\varphi}^k d\rho = 0, \tag{49}$$

the latter prescribing the average torque related to the tangential stress increment $\dot{\tau}_{z\varphi}$ to be zero.

References

1. van Holde, K.E., Matthews, C.: Biochemistry. Benjamin/Cummings Publishing Company Inc., Menlo Park (1995)
2. Fratzl, P.: Collagen: Structure and Mechanics. Springer, New York (2008)
3. Bruel, A., Oxlund, H.: Changes in biomechanical properties, composition of collagen and elastin, and advanced glycation endproducts of the rat aorta in relation to age. Atherosclerosis **127**, 155–165 (1996)

4. Bruel, A., Ørtoft, G., Oxlund, H.: Inhibition of cross-links in collagen is associated with reduced stiffness of the aorta in young rats. Atherosclerosis **140**, 135–145 (1998)
5. Mao, J.R., Bristow, J.: The Ehlers–Danlos syndrome: on beyond collagens. J. Clin. Invest. **107**, 1063–1069 (2001)
6. Carmo, M., Colombo, L., Bruno, A., Corsi, F.R.M., Roncoroni, L., Cuttin, M.S., Radice, F., Mussini, E., Settembrini, P.G.: Alteration of elastin, collagen and their cross-links in abdominal aortic aneurysms. Eur. J. Vasc. Endovasc. Surg. **23**, 543–549 (2002)
7. Järvinen, T.A.H., Järvinen, T.L.N., Kannus, P., Józsa, L., Järvinen, M.: Collagen fibres of the spontaneously ruptured human tendons display decreased thickness and crimp angle. J. Orthop. Res. **22**, 1303–1309 (2004)
8. Maceri, F., Marino, M., Vairo, G.: A unified multiscale mechanical model for soft collagenous tissues with regular fiber arrangement. J. Biomech. **43**, 355–363 (2010)
9. Buehler, M.J., Wong, S.Y.: Entropic elasticity controls nanomechanics of single tropocollagen molecules. Biophys. J. **93**, 37–43 (2007)
10. Marko, J.F., Siggia, E.D.: Stretching DNA. Macromolecules **28**, 8759–8770 (1995)
11. MacKintosh, F.C., Käs, J., Janmey, P.A.: Elasticity of semiflexible biopolymer networks. Phys. Rev. Lett. **75**, 4425–4428 (1995)
12. Sun, Y.L., Luo, Z.P., Fertala, A., An, K.N.: Direct quantification of the flexibility of type I collagen monomer. Biochem. Biophys. Res. Commun. **295**, 382–386 (2002)
13. Wang, M.D., Yin, H., Landick, R., Gelles, J., Block, S.M.: Stretching DNA with optical tweezers. Biophys. J. **72**, 1335–1346 (1997)
14. Holzapfel, G.A., Ogden, R.W.: On the bending and stretching elasticity of biopolymer filaments. J. Elast. **104**, 319–342 (2010)
15. Maceri, F., Marino, M., Vairo, G.: Elasto-damage modeling of biopolymer molecules response. Comput. Model. Eng. Sci. (2012, to appear) ISSN:1526-1492
16. Gautieri, A., Buehler, M.J., Redaelli, A.: Deformation rate controls elasticity and unfolding pathway of single tropocollagen molecules. J. Mech. Behav. Biomed. Mat. **2**, 130–137 (2009)
17. Eyre, D.R., Weis, M.A., Wu, J.J.: Advances in collagen cross-link analysis. Methods **45**(1), 65–74 (2008)
18. Orgel, J.P.R.O., Irving, T.C., Miller, A., Wess, T.J.: Microfibrillar structure of type I collagen in situ. Proc. Nat. Acad. Sci. USA **103**, 9001–9005 (2006)
19. Petruska, J.A., Hodge, A.J.: A subunit model for the tropocollagen macromolecule. Proc. Natl Acad. Sci. USA **51**, 871–876 (1964)
20. Pins, G.D., Christiansen, D.L., Patel, R., Silver, F.H.: Self-assembly of collagen fibers. Influence of fibrillar alignment and decorin on mechanical properties. Biophys. J. **73**, 2164–2172 (1997)
21. Redaelli, A., Vesentini, S., Soncini, M., Vena, P., Mantero, S., Montevecchi, F.M.: Possible role of decorin glycosaminoglycans in fibril to fibril force transfer in relative mature tendons—a computational study from molecular to microstructural level. J. Biomech. **36**, 1555–1569 (2003)
22. Craig, A.S., Birtles, M.J., Conway, J.F., Parry, D.A.: An estimate of the mean length of collagen fibrils in rat tail-tendon as a function of age. Connect. Tissue Res. **19**, 51–62 (1989)
23. Provenzano, P.P., Vanderby, R.J.: Collagen fibril morphology and organization: implications for force transmission in ligament and tendon. Matrix Biol. **25**, 71–84 (2006)
24. Fessel, G., Snedeker, J.G.: Equivalent stiffness after glycosaminoglycan depletion in tendon—an ultra-structural finite element model and corresponding experiments. J. Theor. Biol. **268**, 77–83 (2011)
25. Martini, F.H., Timmons, M.J., Tallitsch, R.B.: Human Anatomy. Prentice Hall, Englewood Cliffs (1994)
26. Sasaki, N., Odajima, S.: Elongation mechanism of collagen fibrils and force–strain relations of tendon at each level of structural hierarchy. J. Biomech. **29**, 1131–1136 (1996)
27. Hansen, K.A., Weiss, J.A., Barton, J.K.: Recruitment of tendon crimp with applied tensile strain. J. Biomech. Eng. **124**, 72–77 (2002)

28. Yamamoto, E., Kogawa, D., Tokura, S., Hayashi, K.: Biomechanical response of collagen fascicles to restressing after stress deprivation during culture. J. Biomech. **40**, 2063–2070 (2007)
29. Kannus, P.: Structure of the tendon connective tissue. Scand. J. Med. Sci. Sports **10**, 312–320 (2000)
30. Silver, F.H., Freeman, J.W., Horvath, I., Landis, W.J.: Molecular basis for elastic energy storage in mineralized tendon. Biomacromolecules **2**, 750–756 (2001)
31. Wolinsky, H., Seymour, G.: A lamellar unit of aortic medial structure and function in mammals. Circ. Res. **20**, 99–111 (1967)
32. Clark, J.M., Glagov, S.: Transmural organization of the arterial media—the lamellar unit revisited. Arteriosclerosis **5**, 19–34 (1985)
33. Wolinsky, H., Seymour, G.: Comparison of abdominal and thoracic aortic medial structure in mammals. Circ. Res. **25**, 677–686 (1969)
34. Hallock, P., Benson, I.C.: Studies on the elastic properties of human isolated aorta. J. Clin. Invest. **16**, 595–602 (1937)
35. Åstrand, H., Stålhand, J., Karlsson, M., Sonesson, B., Länne, T.: In vivo estimation of the contribution of elastin and collagen to the mechanical properties in the human abdominal aorta: effects of age and sex. J. Appl. Phys. 110:176–187 (2011)
36. O'Connell, M.K., Murthy, S., Phan, S., Xu, C., Buchanan, J., Spilker, R., Dalman, R.L., Zarins , C.K., Denk, W., Taylor, C.A.: The three-dimensional micro- and nanostructure of the aortic medial lamellar unit measured using 3D confocal and electron microscopy imaging. Matrix Biol. **27**, 171–181 (2008)
37. Behmoaras, J., Osborne-Pellegrin, M., Gauguier, D., Jacob, M.P.: Characteristics of the aortic elastic network and related phenotypes in seven inbred rat strains. Am. J. Physiol. Heart. Circ. Physiol. **288**, 769–777 (2005)
38. Merrilees, M., Tiang, K.M., Scott, L.: Changes in collagen fibril diameters across artery walls including a correlation with glycosaminoglycan content. Connect. Tissue Res. **16**, 237–257 (1987)
39. Rachev, A., Stergiopulos, N., Meister, J.J.: Theoretical study of dynamics of arterial wall remodeling in response to changes in blood pressure. J. Biomech. **29**, 635–642 (1996)
40. Zhang, W., Herrera, C., Atluri, S.N., Kassab, G.S.: The effect of longitudinal pre-stretch and radial constraint on the stress distribution in the vessel wall: a new hypothesis. Mech. Chem. Biosyst. **2**, 41–52 (2005)
41. Kassab, G.S.: Biomechanics of the cardiovascular system: the aorta as an illustratory example. J. R. Soc. Interface **3**, 719–740 (2006)
42. Fung, Y.C.: Biorheology of soft tissues. Biorheology **10**, 199–212 (1973)
43. Yin, L., Elliott, D.M.: A biphasic and transversely isotropic mechanical model for tendon: application to mouse tail fascicles in uniaxial tendons. J. Biomech. **37**, 907–916 (2004)
44. Comninou, M., Yannas, I.V.: Dependance of stress–strain nonlinearity of connective tissues on the geometry of collagen fibers. J. Biomech. **9**, 427–433 (1976)
45. Lanir, Y.: A structural theory for the homogeneous biaxial stress–strain relationships in flat collagenous tissues. J. Biomech. **12**, 423–436 (1979)
46. Freed, A.D., Doehring, T.C.: Elastic model for crimped collagen fibrils. J. Biomech. Eng. **127**, 587–593 (2005)
47. Holzapfel, G.A., Gasser, T.C., Ogden, R.W.: A new constitutive framework for arterial wall mechanics and a comparative study of material models. J. Elast. **61**, 1–48 (2000)
48. Holzapfel, G.A., Gasser, T.C., Stadler, M.: A structural model for the viscoelastic behavior of arterial walls: continuum formulation and finite element analysis. Eur. J. Mech. A Solids **21**, 441–463 (2002)
49. Ciarletta, P., Micera, S., Accoto, D., Dario, P.: A novel microstructural approach in tendon viscoelastic modelling at the fibrillar level. J. Biomech. **39**, 2034–2042 (2006)
50. Tang, H., Buehler, M.J., Moran, B.: A constitutive model of soft tissue: from nanoscale collagen to tissue continuum. Ann. Biomed. Eng. **37**, 1117–1130 (2009)

51. Maceri, F., Marino, M., Vairo, G.: From cross-linked collagen molecules to arterial tissue: a nano-micro-macroscale elastic model. Acta Mech. Solida Sin. **23**(S1), 98–108 (2010)
52. Marino, M., Vairo, G.: Stress and strain localization in stretched collagenous tissues via a multiscale modeling approach. Comput. Methods Biomech. Biomed. Engin. (2012). doi:10.1080/10255842.2012.658043
53. Maceri, F., Marino, M., Vairo, G.: An insight on multiscale tendon modeling in muscle-tendon integrated behavior. Biomech. Model. Mechanobiol. **11**, 505–517 (2011)
54. Bozec, L., Horton, M.: Topography and mechanical properties of single molecules of type I collagen using atomic force microscopy. Biophys. J. **88**, 4223–4231 (2005)
55. Holmes, D.F., Graham, H.K., Trotter, J.A., Kadler, K.E.: STEM/TEM studies of collagen fibril assembly. Micron **32**, 273–285 (2001)
56. Buehler, M.J.: Nanomechanics of collagen fibrils under varying cross-link densities: atomistic and continuum studies. J Mech. Behav. Biomed. Mat. **1**, 59–67 (2008)
57. Bailey, A.J.: Molecular mechanisms of ageing in connective tissues. Mech. Ageing Dev. **122**, 735–755 (2001)
58. Couppé, C., Hansen, P., Kongsgaard, M., Kovanen, V., Suetta, C., Aagaard, P., Kjær, M., Magnusson, S.P.: Mechanical properties and collagen cross-linking of the patellar tendon in old and young men. J. Appl. Physiol. 107:880–886 (2009)
59. Marino, M., Vairo, G.: Equivalent stiffness and compliance of curvilinear elastic fibers. In: Maceri, F., Frémond, M. (eds.) Mechanics, Models and Methods in Civil Engineering. Lecture Notes in Applied & Computational Mechanics, vol. 61, pp. 309–332. Springer, Berlin (2011)
60. Kollar, L.P., Springer, G.S.: Mechanics of Composite Structures. Cambridge University Press, Cambridge (2003)
61. Screen, H.R.C., Lee, D.A., Bader, D.L., Shelton, J.C.: An investigation into the effects of the hierarchical structure of tendon fascicles on micromechanical properties. Proc. Inst. Mech. Eng. **218**, 109–119 (2004)
62. Lavagnino, M., Arnoczky, S.P., Caballero, O., Kepich, E., Haut R., C.: A finite element model predicts the mechanotrasduction of tendon cells to cyclic tensile loading. Biomech. Model. Mechanobiol. **7**, 405–416 (2008)
63. Auricchio, F., Sacco, E., Vairo, G.: A mixed FSDT finite element for monoclinic laminated plates. Comput. Struct. **84**, 624–639 (2006)
64. Zulliger, M.A., Rachev, A., Stergiopulos, N.: A constitutive formulation of arterial mechanics including vascular smooth muscle tone. Am. J. Physiol. Heart Circ. Physiol. **287**, 1335–1343 (2004)
65. Åstrand, H., Rydén-Ahlgren, Å ., Sandgren, T., Länne, T.: Age-related increase in wall stress of the human abdominal aorta: an in vivo study. J. Vasc. Surg. 42:926–931 (2005)
66. Cattell, M.A., Anderson, J.C., Hasleton, P.S.: Age-related changes in amounts and concentrations of collagen and elastin in normotensive human thoracic aorta. Clin. Chim. Acta **245**, 73–84 (1996)
67. Marklund, E., Varna, J.: Modeling the effect of helical fiber structure on wood fiber composite elastic properties. Appl. Compos. Mater. **16**, 245–262 (2009)

Multiscale Modeling of Ligaments and Tendons

Shawn P. Reese, Benjamin J. Ellis and Jeffrey A. Weiss

Abstract Ligaments and tendons are composed primarily of water and fibrillar type I collagen, which is hierarchically organized into complex structures that span multiple physical scales. Forces at the macroscopic joint level are transmitted via interactions at the mesoscale, microscale and nanoscale. Tissue repair and growth is mediated by fibroblasts and tenocytes, which are subjected to a unique microscale mechanical environment. The burgeoning field of multiscale modeling holds promise in filling the gaps in our understanding of structure–function relationships and mechanotransduction in these tissues, and these questions are difficult or impossible to address using experimental techniques alone. This article reviews the state of the art in multiscale modeling of ligaments and tendons, while providing sufficient background on the structure and function of these tissues to allow a reader who is new to the area to proceed without substantial outside reading. The multiscale structure of ligaments and tendons is described in detail. The available data on material characterization at different physical scales is reviewed as well. The final section of the chapter summarizes the efforts at developing and validating multiscale models that are relevant to ligament and tendon mechanics, and identifies future directions for research. Multiscale modeling of tendon and ligament holds considerable promise in advancing our understanding regarding the complex mechanisms of multiscale force transfer within these tissues.

J. A. Weiss (✉)
Departments of Bioengineering and Orthopedics, and Scientific Computing and Imaging Institute, University of Utah, Salt Lake City, UT, USA
e-mail: jeff.weiss@utah.edu

S. P. Reese · B. J. Ellis
Department of Bioengineering, and Scientific Computing and Imaging Institute, University of Utah, Salt Lake City, UT, USA

Stud Mechanobiol Tissue Eng Biomater (2013) 14: 103–147

DOI: 10.1007/8415_2012_157

Published Online: 22 November 2012

1 Introduction

Although the structure and mechanics of collagenous connective tissues have been studied for decades, a clear understanding of the relationships between hierarchical organization and material behavior is severely lacking. This can be attributed, at least in part, to an inability to integrate and couple mechanics between different physical scales. In the case of ligaments and tendons, this requires integration of information on structure, organization and material behavior of constituents between the nanoscale, microscale and mesoscale.

In theory, this integration can be accomplished using the theory of homogenization [37, 78, 129, 203, 256]. In homogenization, a microscale boundary value problem is used to determine the governing behavior at the macroscale. In linear theory, a homogenization yields the components of the so called "effective" elasticity tensor [172]. In nonlinear theory, the macroscale boundary value problem is solved simultaneously using the methods of computational mechanics, and a nested solution of two boundary value problems is obtained. Homogenization techniques have been applied to bone [165, 172, 176, 197, 223], cartilage [203], myocardium [146, 159], arteries [207], cells [97], connective tissues [4, 12, 37, 42, 43, 137, 149, 187] and biomaterials [43].

The function of the organizational motifs of collagen in connective tissues, and their mechanical interactions across scale levels, is of fundamental importance in understanding normal tissue function and the etiology and treatment of injury and disease, both acquired and inherited. In the case of injury and disease, changes to ECM organization are one of the most often observed effects on the tissue. It is well known that ligaments and tendons primarily heal by scar formation [10, 68, 69, 106, 140, 141, 238], and that the healed tissue is inferior to the normal tissue both in terms of structural organization and material properties [24, 54, 69, 106, 112, 238]. The reasons for the lack of a more "regenerative" healing response continue to evade us, but an improved understanding of the basic tissue structure will help to interpret the alterations in structure that are present in healing tissues. Many heritable diseases directly affect type I collagen structure and fibrillogenesis, resulting in varied phenotypes that alter the multiscale structure of collagen (e.g., osteogenesis imperfecta, Ehlers-Danlos syndrome, [193, 219]). These diseases cause relatively well-characterized alterations in structure/organization of type I collagen at the nanoscale (fibril) level, as well as other levels. Additionally, since collagen fibrillogenesis is regulated by other ECM components, alterations to these components can directly influence collagen structure. Examples include disorders that affect decorin, biglycan and elastin/fibrillin such as congenital stromal corneal dystrophy [36], periodontal disease [92] and Marfan Syndrome [161].

The overall objective of this chapter is to review the state of the art in multiscale modeling of ligaments and tendons, while providing sufficient background on the structure and function of these tissues to allow a reader who is new to the area to proceed without substantial outside reading. Section 2 reviews the multiscale structure and function of ligaments and tendons, including the constituents and their organization. Section 3 provides a detailed review of the available data on material characterization of ligaments and tendons at different scales. Section 4 reviews the mathematical fundamentals behind nonlinear continuum mechanics and homogenization theory. Finally, Sect. 5 critically reviews the state of the art in multiscale modeling of ligaments and tendons, and identifies directions for future research.

2 Ligament and Tendon Structure

2.1 Ligament

Ligaments are soft, fibrous tissues that connect bone to bone at the joints. They help to guide and limit the motion of the bones so that the joint articulates with no separation or only a limited separation of the bones. Ligaments are passive stabilizers and work in conjunction with other passive stabilizers, including the articulating surfaces of the bones and, in most diarthrodial joints (major joints—knee, hip and shoulder), other soft tissues such as the meniscus in the knee and the labrum in the shoulder and hip. In diarthrodial joints, ligaments are primarily banded or cordlike. For instance, the medial collateral ligament (MCL) of the knee is a banded ligament, while the anterior cruciate ligament (ACL) is a cordlike ligament. These knee ligaments resist motion along a single line of action and transmit tensile load, but also experience shear, transverse and compressive loads [5, 6, 20, 29, 33, 41, 49, 65, 75, 76, 233, 240, 242, 245]. The MCL, for example, primarily resists valgus knee motion, which loads the MCL in tension [3, 11, 87, 113, 115, 142, 157, 160]. However, articulation of the joint and contact with the bones will also generate shear, transverse and compressive loads [49, 65, 75, 76]. The ligaments in the shoulder and hip form thin, dense bands of tissue around the joint capsule and are known as capsular ligaments. The inferior glenohumeral ligament (IGHL), for example, is a capsular ligament in the shoulder. While it can be argued that capsular ligaments resist motion primarily in one direction, they are thought to constrain more complex motions than knee ligaments through their connection with the rest of the capsule [51, 52, 56, 57, 63, 64, 164].

Although ligaments are considered passive stabilizers, there are stresses in the tissue when the joint is in a neutral position [1]. These in situ stresses are responsible for the stability of the joint when muscle forces are not acting across the joint. Due to the difficulty in measuring in situ stresses, in situ strains are usually measured [75, 244]. Ligament in situ strains are inhomogenous, subject-specific, and vary depending on joint position [75, 244]. Previous research has shown that ligament in situ strains must be measured in order to accurately measure or predict ligament strains and stresses due to external loading [75].

Ligaments attach to bone at insertion sites. There are two types of insertion sites: direct and indirect insertions. Direct insertion sites occur over a distance of less than 1 mm [245], and consist of a distinct right-angle boundary where deep collagen fibrils extend out of the ground substance matrix and become fibrocartilage tissue, mineralized fibrocartilage tissue and then bone [47]. Indirect insertion sites occur over a larger area where superficial collagen fibers gradually blend into the periosteum at more acute angles. Deep collagen fibers also make attachments at indirect insertion sites, but the connections are fewer than at direct insertion sites, occur at more acute angles and without the fibrocartilagenous transitional zone observed in direct insertions [26]. Some ligaments have the same type of insertion site at both ends, while other ligaments have different types of insertion sites on opposing ends. The ACL in the knee, for example, has direct insertion sites on both ends, while the MCL in the knee attaches to the femur with a direct insertion, but attaches to the tibia with an indirect insertion.

2.2 Tendon

Tendons connect muscles to bones and transmit the forces generated by the muscles to the bones. Tendons generally stretch more than ligaments during use, with some tendons being very efficient at storing and recovering energy [23, 40, 48, 71, 100, 116, 139, 151, 155, 204]. These elastic properties allow tendons to passively modulate forces during locomotion, providing additional stability with no active work [71, 116, 119, 139]. The length of a tendon significantly contributes to these characteristics. Shorter tendons allow for more muscle mass [173, 178], but longer tendons provide more elastic recovery of stored energy [30, 71, 116, 152, 153, 155, 177, 204]. The primary function of tendons like the rotator cuff, finger tendons, and animal extensor tendons is to transfer load generated by their associated muscles [23, 27, 28, 123, 177, 178, 258]. In contrast, tendons like the Achilles tendon and animal flexor tendons store substantial amounts of energy and are thought to act like biological springs [30, 71, 116, 152, 153, 155, 177, 204]. Structure and composition vary between different tendons and between different locations within individual tendons [28, 121, 190]. Similar to ligaments, tendons connect to bones with either direct or indirect insertion sites. These attachments are complex, are often the site of injury and as such are an area of extensive continued research [8, 22, 60, 79–81, 150, 151, 214, 215, 246, 255].

2.3 Hierarchical Structure of Ligament and Tendon

Tendons and ligaments are multiphasic biological composites. The extracellular matrix (ECM) is composed of a fluid phase and a solid phase, with the bulk of the tissue consisting of the fluid phase (i.e. water). The solid phase consists primarily of type I collagen, which is organized into a complex hierarchy where tropocollagen monomers form fibrils at the nanoscale, fibrils form fibers at the microscale,

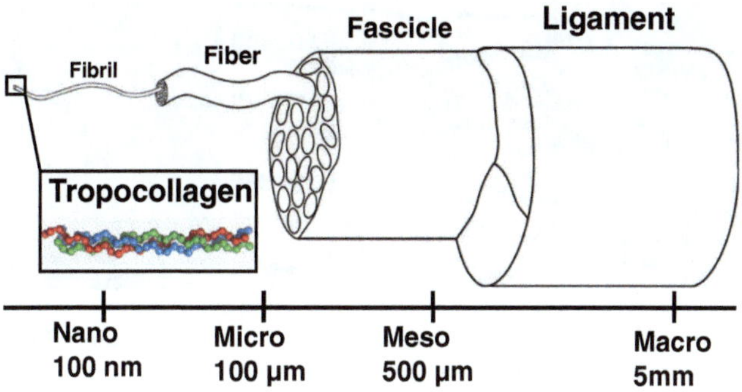

Fig. 1 Hierarchical organization of ligament from the molecular level to the joint level

fibers form fascicles at the mesoscale and fascicles form the whole tendon or ligament at the macroscale (Fig. 1).

The ECM of ligaments and tendons is formed by self-assembly of cell-secreted proteins and consists of approximately 70 % water [28]. The solid phase of these tissues is primarily composed of type I collagen (60–80 %), with the remainder consisting of elastin, proteoglycans and glycosaminoglycans (GAGs), other types of collagen (types III, IV, V, VI), fibrillin and other proteins [121, 125, 230]. Type I collagen exhibits different organizational motifs at each scale (Figs. 1 and 2) [121]. At the nanoscale, tropocollagen monomers are assembled to form **fibrils** (50–200 nm dia.), which display a characteristic d-banding period (67 nm) [121, 170, 171, 224]. Tropocollagen monomers are held together by a combination of hydrogen, ionic and covalent bonds [125, 230]. Fibrils are spaced regularly within healthy tissue and predominantly aligned in parallel [31, 209, 227]. Cross sectional TEM images reveal that fibrils are well organized and separated by a regular spacing within healthy tissue. At the microscale, fibrils are assembled into **fibers** (20–50 μm dia.) [50, 117]. Fibroblasts and tenocytes (10 μm width × 60 μm length) are located in the interfiber space [121, 125]. Fibroblasts and tenocytes are responsible for regulating the ECM in response to loading and injury, and mechanotransduction plays a major role in their function [229]. The characteristic crimp pattern is visible at the fiber level, with a period of 50–200 μm [112, 117]. Fibers are arranged in a largely parallel fashion [121]. At the mesoscale, fibers are assembled into **fascicles** (100–500 μm dia) [50, 121, 230]. To at least some extent, crimp is registered between fibers [124, 168]. Fascicles are organized in parallel [96]. Fascicles and fibers are surrounded by a thin fascia (referred to as endotenon) [85, 121, 206]. At the macroscale, groups of fascicles are organized into functional bands (100 μm–1 mm dia) [50].

Noncollagenous ECM constituents include proteoglycans (PGs) such as decorin (~1%/wt), biglycan (~0.5%/wt) and others (fibromodulin, lumican, aggrecan, versican), fibrillin [21, 114, 125, 143–145, 167] and elastin (1–2%/wt) [121, 125, 192, 206]. The large PGs (e.g., aggrecan) contribute to the apparent viscoelastic material behavior of these tissues by controlling water content and flux [114].

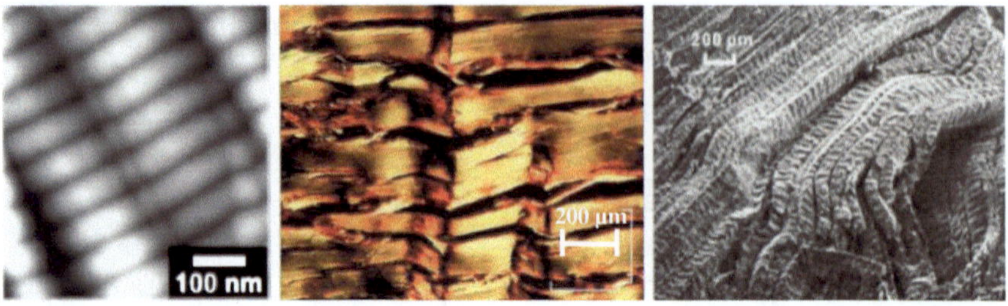

Fig. 2 Unique structural motifs exist at each scale level. At the nanoscale, 67 nm d-banding is observed (*left*), at the microscale fiber crimp is present (*middle*) and at the macroscale fascicles align in a parallel fashion (*right*). Note that crimp is generally in register within fascicles (reproduced with permission from [50, 67, 224])

The mechanical role of small PGs (decorin, biglycan) is debated [175, 259], but our own research demonstrated that their GAG side chains have a negligible contribution to tissue level mechanics in ligament [102, 104, 144, 145]. Elastin and fibrillin are thought to contribute to the toe region of the stress–strain curve in ligament and tendon [32, 136, 166, 220].

Structurally, ligaments and tendons share the same hierarchical organization and structural motifs. However, there are significant differences in fibril diameter distributions, fiber diameter and crimp morphologies, metabolic activity and the relative percentage of certain components such as water, proteoglycans and types I and III collagen [7, 121, 125, 194].

3 Multiscale Material Characterization

The normal mechanical function of ligament and tendon is an emergent property of complex interactions between physical scale levels (Fig. 3). In order to model these complex interactions, a firm experimental basis must first be obtained. This section reviews and summarizes the experimentally observed material behavior of ligament and tendon at the macroscale, mesoscale, microscale and nanoscale.

3.1 Elastic and Viscoelastic Behavior

When studying the material behavior of ligament and tendon, it is convenient to isolate the equilibrium elastic response from the dynamic, or time dependent response. The elastic response is experimentally measured using slow strain rates or stress relaxation and creep testing, whereby step displacements or loading are applied and equilibrium values are obtained [144]. The elastic response arises primarily from stretching and interactions of the solid phase components.

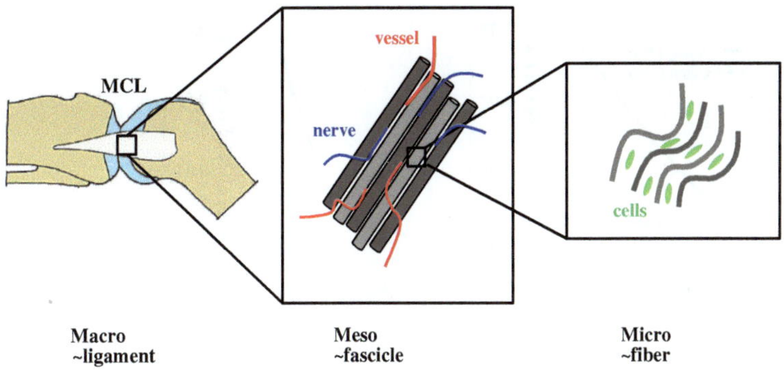

Fig. 3 Multiscale force transmission. Force transmission within a macroscale tissue structure (e.g., an MCL, shown on the *left*) is mediated at the mesoscale by fascicles (*center*) and at the microscale by fiber (*right*)

However, hydration also plays an important role in modulating equilibrium elastic response [45]. The viscoelastic response is experimentally measured by performing mechanical testing at varied strain rates, stress relaxation testing, creep testing and harmonic testing [34, 135, 145, 241]. The viscoelastic response is attributed to both fluid flow-dependent and fluid flow-independent effects. Flow-independent effects refer to an intrinsic viscoelasticity of the solid phase (e.g. a viscous sliding of fibrils), while flow-dependent effects refer to the pressure driven transport of free water through a permeable tissue (e.g. described using biphasic theory, discussed in Sect. 4.3) [46, 233, 235, 236, 253].

3.2 Macroscopic Material Characterization

For this chapter, the macroscale will be defined as tissue structures that are within the range of millimeters to centimeters, most commonly consisting of whole ligament and tendon preparations or subsamples that are dissected or punched out [27, 131, 144, 147, 249]. The 3D elastic response of ligament and tendon tissue is complex and difficult to fully characterize. The material response is highly dependent on the predominant alignment of the collagen fiber families. Although some capsular ligaments appear to have an isotropic fiber distribution (e.g. glenohumeral capsule [63, 64]), the material symmetry of most ligaments and tendons is reasonably described by transverse isotropy, with the collagen fibers predominantly aligned in the direction of in vivo loading [233, 235]. To fully characterize the elastic material response of these tissues, a combination of tensile, compression and shear testing must be performed in parallel and transverse directions to the predominate fiber family [34, 147, 232, 234].

Tensile testing in the fiber direction reveals a nonlinear stress–strain response consisting of a so called "toe region" and a linear region [34, 233] (Fig. 4, left).

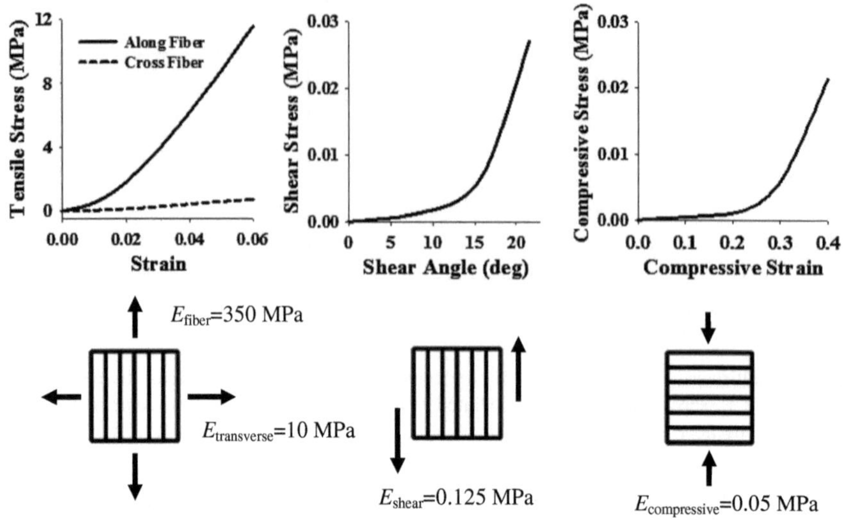

Fig. 4 The elastic behavior of ligament is anisotropic and nonlinear. The tensile stiffness along the fiber direction is an order of magnitude stiffer than in the transverse direction (*left*) [34, 183]. In shear, ligament is two orders of magnitude more compliant than in the transverse direction (*center*) [234]. In compression, ligament is over three times more compliant than in tension, indicating compression-tension nonlinearity (*right*) [231]. Approximate linear modulus (E) is shown for each test type

It has been hypothesized that the nonlinear toe region results from the uncrimping and/or successive recruitment of the aligned collagen fibers, while the linear region is understood as the tensile response of the collagen fibers. The reported linear modulus varies widely between tissue types, location and studies [19, 34, 147, 243, 245, 250]. The tensile response in the direction transverse to the fiber axis is nearly linear and an order of magnitude more compliant than the longitudinal response (Fig. 4, left). The shear response is nonlinear, with a tangent stiffness that is three orders of magnitude less than the fiber stiffness (Fig. 4, middle). Ligament stiffness is lowest when tested in unconfined compression transverse to the fiber axis, yielding a nonlinear response with a tangent compressive modulus nearly four orders of magnitude less than the tensile modulus of the fiber family (Fig. 4, right). The elastic volumetric response describes the change in volume in response to tensile or compressive loading. This has most commonly been reported as the Poisson's ratio (which is a linear measure of volume change), and less commonly as the Poisson's function (which is a nonlinear measure of volume change) [25, 187]. The Poisson's ratio (or function) is a kinematic measure that relates the axially applied strain to the laterally induced strain [25]. Experimentally measured Poisson's ratios for uniaxial tensile testing in the fiber direction have revealed values ranging from 1.0 in capsular ligament to 3.0 for flexor tendon [105, 147]. These values exceed the thermodynamic limit of 0.5 for isotropic tissues and are indicative of volume loss under tensile loading. This volume loss is generally understood to be the result of fluid exudation during tensile loading, which has been reported in the literature [94, 101]. The elastic response is not spatially homogenous. Strain measurements techniques such as speckle tracking and digital image correlation have shown that strains are highly inhomogeneous during tensile loading [53, 55, 133, 158]. Although the origins of this behavior are unclear, it may be due in part to variation in the tissue mechanical properties [163], clamping artifacts or an uneven fascicle stiffness and preload [126, 233]. It appears that spatial inhomogeneity within strain distribution may be an intrinsic property of ligament and tendon tissue.

The viscoelastic response of ligament and tendon is believed to play an important role in the normal function of these tissues [34]. It is experimentally manifested as stress relaxation under a step displacement, creep under a step loading, hysteresis, and a phase shift during harmonic loading [134]. Stress relaxation testing of ligament and tendon reveals a dependence of both the relaxation rate and magnitude of relaxation on the strain level [2, 131, 145, 179]. Similarly, the creep rate and creep magnitude are also strain dependent [218]. The tensile modulus is strain rate dependent, while the damping is relatively independent of strain rate [228, 233, 235, 240]. During high rate loading, the volumetric behavior of ligament and tendon appears to be incompressible [233]. Although viscoelastic testing is most commonly reported for uniaxial tensile testing in the fiber direction, both viscoelastic tensile testing in the transverse direction and in shear has been reported [34]. The magnitude of stress relaxation is relatively large for testing in the axial, transverse and shear directions, with times to equilibrium on the order of ten minutes or more [2, 131, 145, 147]. Because of the considerable importance of the fluid phase to tissue viscoelasticity, it comes as little surprise that the viscoelastic response is significantly altered by varied levels of tissue hydration [144]. There is also an observed effect of temperature on the elastic and viscoelastic response [45].

3.3 Mesoscale Material Characterization

Fascicles are the primary load bearing mesoscale structure found within ligament and tendon, and range in diameter from 100 to 500 μm [50, 121, 230]. Experimental studies at the mesoscale have interrogated the fascicle response by testing both isolated individual fascicles as well as fascicles in situ. Rat tail tendons have often been used in such studies because they are readily available, the tendons have large aspect ratios, and it is relatively easy to isolate the fascicles from the tendon. A number of studies have reported both elastic and viscoelastic properties of rat tail tendon fascicles [95, 184, 191]. The qualitative elastic and viscoelastic response is similar to that observed for macroscopic tissue, with a nonlinear toe region and a large stress relaxation.

Several studies have performed tensile testing on progressively divided tendons (e.g. into half and quarter sections) as well as individually isolated fascicles.

Uniaxial tensile testing was performed on whole tendons, split tendons and isolated individual fascicles from rabbit patellar tendons [250]. A comparison of quasi-static stress–strain curves indicated that the intact tendons were stiffer than split tendons, which were in turn stiffer than individually isolated tendon fascicles. Stress relaxation testing revealed that whole tendons had a larger stress relaxation magnitude and lower stress relaxation rate compared to individual fascicles. A similar study reported that whole porcine cruciate ligaments were stiffer than split ligaments and isolated fascicles [107]. The results of these studies seem counterintuitive, in that macroscopic structures were stiffer than the constituents. A parallel spring mechanism was proposed to explain this, but this awaits experimental verification [107].

The opposite trend was observed in similar studies. In one such study, human patellar tendons were sectioned into half, quarter and individual fascicles [16]. A comparison of the elastic modulus, relaxation magnitude and rate revealed a strong dependence on cross sectional area. As the sample cross sectional area decreased, the linear modulus increased and the rate and magnitude of stress relaxation decreased. This result is supported by another study in which macroscopic human Achilles tendon samples were clamped and subjected to multiple quasistatic tensile testing experiments [126]. For each test, a fascicular bundle was severed and another stress–strain test was performed. The construct stiffness increased as the cross sectional area decreased. Another finding from this study was that fascicles within the tissue did not appear to bear load evenly, with some fascicles carrying considerably more load than others. This may explain the macroscale observation of inhomogeneity in strain distribution.

The preceding paragraphs highlight a discrepancy in the literature regarding the variation of stiffness across scale levels. Some studies report increasing stiffness with increasing scale level (e.g. [107, 162, 250]), while other report the opposite (e.g. [16, 126]) Although the cause of this discrepancy is unclear, differences may arise from the use of animal versus human tissue, clamping methods and methods used for sectioning and separating the tendons. In either case, it is clear that the uniaxial tensile behavior in tendon depends on the physical scale.

It has also been shown that the shear behavior displays a scale dependence [96]. In this study two adjacent fascicles in human patellar tendon were isolated from the whole tendon. The preparations were subjected to repeated tensile loading. On the first cycle, both fascicles were intact. On the second cycle, a single fascicle was cut on one end. On the last cycle, the second fascicle was cut on the opposing end such that force could only be transmitted through an inter-fascicle shearing mechanism. The results indicated that very little load was transferred through shearing of adjacent fascicles, suggesting that fascicles are free sliding and largely independent in the transmission of tensile forces across the tendon.

3.4 Microscale Mechanical Characterization

The collagen fiber is the primary load bearing tissue constituent in ligaments and tendons at the microscale. Located between collagen fibrils are fibroblasts in ligaments and tenocytes in tendons, which are responsible for secreting collagen and other ECM materials in order to maintain the mechanical integrity of the tissue [230]. It is well established that fibroblasts respond to local strain fields via mechanotransduction, making the study of microscale force transfer particularly important [27]. Both direct (e.g. isolating individual fibers) and indirect (e.g. confocal imaging of loaded tissue) studies have been performed.

Only one study has directly examined the stress–strain response of individual fibers [162]. In this study, individual fibers (~ 1 µm in diameter) were isolated from rabbit patellar tendon and subjected to uniaxial tensile loading. The reported stiffness was compared to the fascicle and whole tendon data from a previous study [250]. The individual fibers were less stiff than both individual fascicles and whole tendons. This result implies that the macrostructures are stiffer than their constituents. As with the mesoscale fascicle test data discussed in Sect. 3.3, this result awaits a satisfactory explanation.

Confocal imaging of rat tail tendon fascicles has yielded considerable insight into the microscale strain environment of collagen fibers and tenocytes. In these studies single rat tail tendon fascicles were stained for collagen and cell nuclei, subjected to tensile loading and imaged using confocal microscopy [199, 201, 202]. These studies have revealed that the local strain field within fascicles is highly inhomogeneous. In response to uniaxial tensile loading, the predominant mode of microscale deformation is shearing, whereby individual fibers slide rel ative to adjacent fibers. As a result, local fiber strains are much smaller in magnitude than applied tensile strains. In one study, the local fiber strain was ~ 1 % in response to an applied strain of 6 % [201]. Resulting tenocyte strains were also inhomogeneous, with both tensile strains and large shearing strains being induced by tensile loading.

These experiments have also yielded insights into the microscale viscoelastic response. Two microscale mechanisms of viscoelasticity have been observed: a time dependent shearing of adjacent fascicles, and a time dependent stretching of individual fibers. The inter-fiber sliding response displayed a much larger magnitude of stress relaxation than the individual fibers, suggesting that microscale shearing may play an important role in the solid phase viscoelastic component of tendon. Although the source of the microscale strain inhomogeneity and large inter-fiber shear is still under investigation, it has been suggested that this may result from the uncrimping of the ubiquitous collagen crimping pattern [201, 202].

3.5 Nanoscale Mechanical Characterization

The collagen fibril is the primary load bearing structure at the nanoscale. A number of methods have been employed to study the behavior of ligament and tendon fibrils, including direct testing of isolated fibrils, atomic force microscopy of strained fibrils and the use of X-ray diffraction techniques. Numerous studies have isolated individual collagen fibrils and subjected them to uniaxial tensile testing (e.g. [222, 239, 251, 252]). The stiffness of fibrils varied between studies and was dependent on hydration, mounting method, crosslinking and strain rate. Single fibrils tested in this manner display viscoelastic behavior such as strain rate dependence, hysteresis on unloading and stress relaxation [212]. In another study, AFM was performed on strained fibrils within murine Achilles tendon tissue, revealing that the local fibril strain was considerably less than the applied macroscale strain ($\sim$2 % fibril strain for an applied 10 % macroscale strain). A large lateral contraction (corresponding to a Poisson's ratio of $\sim$0.8) was also observed [190]. Mechanical testing of single tropocollagen molecules has also been reported [35, 210]. In these studies, force-extension relationships were measured and analyzed by fitting the data to a worm-like chain elasticity model. The contour lengths reported ranged from approximately 200–300 nm.

It is believed that other nanoscale components may also contribute to the macroscopic mechanical behavior of ligament and tendon, including proteoglycans such as decorin, biglycan and others [114, 125, 144]. Although not a direct test of multiscale interactions, a number of knockout studies in mice have been performed that suggest macroscale effects from the altered expression of various nanoscale constituents [191, 259]. For instance, decorin deficient mice have been found to have mechanically inferior tendon fascicles [191]. In vitro studies have also been used to investigate the role of nanoscale constituents such as decorin. In a number of studies, samples of human MCL were subjected to tensile testing before and after decorin digestion (via incubation in chondroitinase ABC, ChABC) and no significant changes in mechanical behavior were found [131, 144, 145]. However, in similar studies that utilized single rat tail tendon fascicles, a change in mechanical behavior was found in response to digestion incubation in ChABC. This suggests that perhaps the mechanical function of certain proteoglycans may vary between tissue types and scale levels [198, 200]. Still, the changes in mechanical behavior were minimal. Although there were some trends towards incubation decreasing the stiffness and ultimate strength of the fascicles, the increased strain at the onset of visible fiber sliding was the only significant difference found in the tensile test data from both studies [198, 200].

In summary, it is clear that force is transmitted across scales in a complex manner. Although there are conflicting reports, it appears that tissue constituents are stiffer at lower scale levels than at higher scale levels (Fig. 5). In response to tensile loading in the fiber direction, microscale strains are less than the applied macroscale strain, often times by a large amount. Finally, strains are inhomogenously distributed at the macroscale, mesoscale and microscale. These findings are summarized in Table 1.

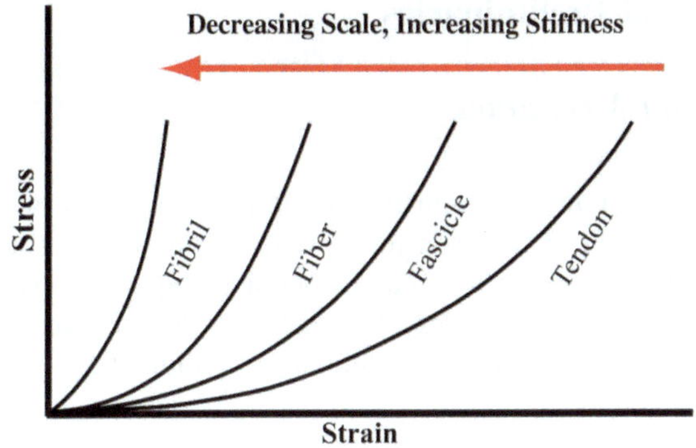

Fig. 5 Although there are conflicting reports, it appears that the stiffness of ligament and tendon tissue constituents increases with decreasing scale

Table 1 Mechanical testing of ligaments and tendons has been performed at multiple scales

Scale	Studies	Methods	Key Findings
Macro	[55, 133, 147, 234, 241, 250]	Tensile testing	Anisotropic, nonlinear mechanical behavior
		Shear testing	Large viscoelastic response
		Viscoelastic testing	Spatially inhomogeneous strain field
Meso	[16, 96, 107, 126, 250]	Subdivision of tendon	Scale dependent stiffness
		Isolation of single fascicles	Scale dependent viscoelasticity
		Severing fascicles within intact tissue	Weak shear coupling between fascicles
			Fascicles are loaded unevenly
Micro	[89, 162, 198–202]	Isolation of single fibers	Spatially inhomogeneous deformation
		Confocal imaging of strained tendon fascicles	Large inter-fiber sliding
			Micro strain less than macro strain
			Microscale mechanisms of viscoelasticity
Nano	[190, 222, 228, 239, 251]	Testing isolated fibrils	Fibrils are stiffer than macroscale tissue
		AFM of strained fibrils	Fibrils are viscoelastic
		X-ray diffraction	Fibril strain less than macroscopic strain

Summarized in this table are representative studies for each scale level, the test methods used and some key findings

4 Mathematical Preliminaries

4.1 Continuum Mechanics

The presence of a continuum assumes that the length scale of microstructures is infinitesimally small in comparison to the macroscale, and that the deformation gradient, and thus strain and stress, can be defined uniquely at every point within the domain. This latter point implies a homogeneous deformation map in which an infinitesimal line element $\boldsymbol{dX}$ in the reference configuration is mapped to the current configuration $\boldsymbol{dx}$:

$$\boldsymbol{dx} = \boldsymbol{F} \cdot \boldsymbol{dX} + \boldsymbol{X}_0, \tag{1}$$

in which $\boldsymbol{dX}$ is an infinitesimal material line element in the reference configuration, $\boldsymbol{dx}$ is the deformed version of the material line element, and $\boldsymbol{X_0}$ represents a rigid body translation vector. $\boldsymbol{F}$ is the nonsymmetric deformation gradient:

$$\boldsymbol{F} = \frac{\partial \boldsymbol{x}}{\partial \boldsymbol{X}}. \tag{2}$$

A number of second order strain measurement tensors are computed from the deformation gradient, including the right Cauchy deformation tensor ($\boldsymbol{C}$), the Green–Lagrange strain ($\boldsymbol{E}$) and the engineering or infinitesimal strain ($\boldsymbol{e}$):

$$\boldsymbol{C} = \boldsymbol{F}^T \boldsymbol{F}, \tag{3}$$

$$\boldsymbol{E} = \frac{1}{2}(\boldsymbol{C} - 1), \tag{4}$$

$$\boldsymbol{e} = \frac{1}{2}\left[(\boldsymbol{F} - 1) + (\boldsymbol{F} - 1)^T\right]. \tag{5}$$

The engineering strain is used extensively for linear elasticity, but is generally of limited use for the finite deformations seen in biological tissues. A useful concept in the study of aligned collagenous tissue is the notion of a unit vector to describe the fiber direction, which is denoted $\boldsymbol{a_0}$ in the reference configuration. The fiber vector is rotated and stretched by the deformation gradient, $\lambda \boldsymbol{a} = \boldsymbol{F} \cdot \boldsymbol{a}_0$, where λ is the fiber stretch. The concept of strain invariants is of particular importance in biosolid mechanics, since they provide an objective measure of strain that is invariant to rotation and rigid body motion [108, 208].

4.2 Continuum Based Constitutive Models

In order to compute a stress from the aforementioned strain measures, a constitutive model is required. In the case of linear elasticity, this constitutive model defines the Cauchy stress, $\boldsymbol{\sigma}$, to be the inner product of the engineering strain, $\boldsymbol{\varepsilon}$, and the fourth order elasticity tensor, $\mathbb{C}$, such that: $\boldsymbol{\sigma} = \mathbb{C} : \boldsymbol{\varepsilon}$. Due to the inherent nonlinearity of ligament and tendon tissue, strain energy approaches (referred to as hyperelasticity) based on the invariants of the deformation tensor (I_1, I_2, I_3, I_4, I_5), are commonly utilized. Such an approach is particularly attractive because it automatically satisfies a number of constraints, such that the formulation will be objective (i.e., invariant to rigid body rotation and displacement) and the tangent elasticity tensor (i.e., the linearization) will be positive definite for a polyconvex strain energy function [108]. In this approach a scalar strain energy function (W) is defined which is typically (but not necessarily) a function of the strain invariants. The Cauchy stress tensor is computed by taking the derivative of the strain energy function with respect to the right Cauchy deformation tensor $\boldsymbol{C}$:

$$\boldsymbol{\sigma} = \frac{2}{\boldsymbol{J}}\boldsymbol{F}\left(\frac{\partial \boldsymbol{W}}{\partial \boldsymbol{C}}\right)\boldsymbol{F}^T. \tag{6}$$

The fourth order elasticity tensor (necessary for the linearization and subsequent nonlinear analysis in numerical methods) is found by taking the second derivative:

$$\mathbb{C} = 4\frac{\partial^2 W}{\partial \boldsymbol{C} \partial \boldsymbol{C}}, \tag{7}$$

where $\mathbb{C}$ is the elasticity tensor in the material frame, which is pushed forward to the spatial frame in most practical implementations.

Hyperelastic, invariant-based, anisotropic continuum models have proved successful in modeling the macroscale behavior of ligament and tendon [233, 235]. One such formulation has been used to model the macroscopic stress–strain behavior of ligament [65, 75, 76, 232, 233, 235, 237]:

$$W = W_m(I_1, I_2) + W_f(\lambda) + U(J)\,. \tag{8}$$

Here, W is the total strain energy, W_m is the strain energy for the inter-fiber matrix, W_f is the strain energy for the collagen fibers, and U represents a volumetric strain energy. I_1 and I_2 are the invariants of the right Cauchy deformation tensor $\boldsymbol{C}$, λ is the stretch along the fiber direction, and $J = \det(\boldsymbol{F})$ is the volume ratio. The fiber strain energy term (W_f) is defined to capture the toe region and linear region of the stress–strain curve for ligaments, and to represent the relatively small compressive stiffness:

$$\lambda\frac{\partial W_f}{\partial \lambda} = \left\{\begin{array}{cc} 0 & \lambda \le 1 \\ c_2\left(e^{c_3(\lambda-1)} - 1\right) & 1 < \lambda < \lambda^* \\ c_4\lambda + c_5 & \lambda > \lambda^* \end{array}\right\}. \tag{9}$$

This formulation represents a structurally motivated constitutive model, as it specifies strain energy terms for the collagen fiber family and the inter-fiber matrix. Numerical implementation of hyperelastic constitutive models in finite element (FE) codes often make use of an additive decomposition of the strain energy into volumetric and deviatoric parts, based on the multiplicative decomposition of the deformation gradient $\boldsymbol{F}$ [108, 233, 235]. This requires a small modification of the equations above. The uncoupled strain energy equations are advantageous for representing these tissues in FE software because they can make use of element formulations that allow for nearly and fully incompressible material behavior without element locking [233, 235].

4.3 Constitutive Modeling of Viscoelasticity

Ligament and tendon viscoelasticity has most commonly been represented by quasilinear viscoelastic (QLV) [2, 62, 74, 181, 233, 235] and nonlinear viscoelastic constitutive models [135, 174, 179, 180]. The QLV theory postulates that the time response and the elastic response are independent [134]. The time response is described using a relaxation function, while the elastic stress response is typically described using a hyperelastic constitutive model [181]. The time dependent stress is then obtained by convolving the relaxation function with the elastic stress. According to QLV theory, the relaxation function is implicitly related to the creep function via a convolution [135]. Thus, an experimentally measured relaxation function should predict an experimentally measured creep function. Although a number of studies have applied the QLV theory successfully to describe the time- and rate-dependent material behavior of ligaments and tendons [62, 74, 118, 215, 216], several studies have suggested that these materials do not strictly behave as quasilinear viscoelastic materials [59, 61, 179, 217]. This has motivated the development of nonlinear viscoelastic models [174, 179, 180].

The apparent viscoelasticity of ligament and tendon can also be described using biphasic theory [137, 185, 236, 253]. Biphasic theory postulates an interaction between a porous, elastic solid phase and an incompressible fluid phase. Loading of the biphasic material induces volumetric changes in the elastic phase. This creates pressure gradients, which drive a time dependent fluid flux through the porous matrix. Diffusive drag and thus energy dissipation is induced by the local difference in velocity between the solid and fluid phases. Biphasic materials exhibit stress relaxation, creep and hysteresis. A necessary component of the field equations for the biphasic theory is the introduction of additional degrees of freedom related to the time and spatially varying fluid pressure field (or fluid velocity), thus making analytical solutions more difficult to compute than for standard viscoelastic constitutive models. Because of this, quasi-analytic solutions to biphasic problems have only been obtained for simplified geometries and loading scenarios [13, 46]. These include the confined and unconfined loading of a cylinder subjected to ramp loading, step loading and harmonic loading [13, 46] for linear material behavior, and for certain nonlinear materials [15, 110]. Both flow-dependent (e.g. biphasic material) and flow-independent mechanisms may be needed to accurately describe and predict the apparent viscoelasticity of some biological soft tissues [109, 110]. Poroviscoelastic formulations have been proposed that utilize a viscoelastic continuum model within the solid phase. These approaches have found utility in the field of cartilage mechanics [109, 110, 154].

4.4 Computational Modeling

Analytical solutions to the equations of motion for the mechanics of ligaments and tendons can only be obtained for simplified geometries and loading scenarios (e.g., uniaxial tension-compression). For complex geometries and loading patterns such as simulation of the mechanics of a ligament within an intact joint, the geometry and governing equations must be discretized and solved numerically [233, 235]. The FE method is by far the most commonly used numerical method in the field of biosolid mechanics. Commercial and freely available software packages support preprocessing, solution and postprocessing the nonlinear FE problems. Many studies in the literature have used FE methods for the simulation of ligament and tendon mechanics (e.g., [64, 65, 75, 235]). In addition to elastic problems, the FE method can also be used to solve viscoelastic problems and biphasic problems. In the past, addressing these types of problems was more difficult due to the lack of a FE framework specifically designed for biological applications. To address this issue, our lab developed FEBio, a nonlinear implicit finite element framework designed specifically for analysis in computational solid biomechanics (www.febio.org) [148].

4.5 Homogenization

Although continuum based constitutive models are useful for describing macroscopic behavior, they do not address the mechanical behavior that occurs at lower length scales and are not always useful for the study of structure–function relationships between the microscale and the macroscale. Because of the multiscale structure of ligaments and tendons, it is sometimes desirable to use models that can simultaneously describe both macroscale and microscale behavior. This is the goal of multiscale modeling in mechanics, and homogenization is part of the foundation of multiscale modeling. Homogenization is the process of obtaining a macroscopic stress–strain response from a material with a known heterogeneous microstructure [78, 122, 211]. It is based on the concept of a representative volume element (RVE), which can be considered representative of the continuum [82, 91, 120] (Fig. 6). An RVE must be large enough to be statistically representative of the material microstructure, but it must still satisfy the continuum assumption that its dimensions are much smaller than the macroscale dimension [82]. For the case of a perfectly periodic microstructure (e.g., a lattice of spherical particles), the RVE reduces to a unit cell [172]. In a homogenization, the RVE is subjected to the

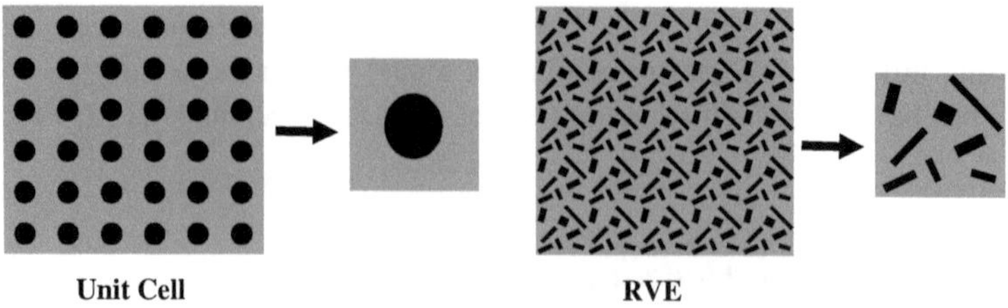

Fig. 6 Comparison of a unit cell and an RVE. For materials with a periodic microstructure, such as a lattice of spheres embedded in a matrix material (*left*), a unit cell (*middle left*) can be defined that describes the microscale geometry. For the case of media with random microstructures (*middle right*), a volume element representative of the microstructure, called an RVE (*right*), can be defined

appropriate boundary conditions and then simulated macroscopic loading is used to compute the effective material response. For a periodic unit cell, the exact homogenized effective material properties are obtained. If the RVE is statistically representative of the material microstructure, the "apparent material properties" are obtained [99, 172].

The concept of homogenization is based upon the Hill principle [98], which states that the volume averaged strain energy at the macroscale is equal to the volume averaged strain energy at the microscale (i.e., energy is conserved):

$$\langle \boldsymbol{\sigma} : \boldsymbol{\varepsilon} \rangle = \langle \boldsymbol{\sigma} \rangle : \langle \boldsymbol{\varepsilon} \rangle. \tag{10}$$

Special boundary conditions must be applied to satisfy the Hill condition. For a periodic unit cell, they are periodic boundary conditions [172, 247]. The periodic boundary conditions enforce the constraints that opposing faces of the unit cell must deform identically, and that the traction forces on opposing faces must be antiperiodic [172, 211, 247]:

$$\left.\begin{aligned} \boldsymbol{u}^{k+}(\boldsymbol{x}^{+}) - \boldsymbol{u}^{k-}(\boldsymbol{x}^{-}) &= \varepsilon_0(\boldsymbol{x}^{+} - \boldsymbol{x}^{-}) \\ \boldsymbol{t}^{k+}(\boldsymbol{x}^{+}) &= -\boldsymbol{t}^{k-}(\boldsymbol{x}^{-}) \end{aligned}\right\} \quad \text{on } \Gamma, \tag{11}$$

where $\boldsymbol{u}^{k+}$ and $\boldsymbol{u}^{k-}$ are the displacements on opposing faces and $\boldsymbol{t}^{k+}$ and $\boldsymbol{t}^{k-}$ are traction forces on opposing faces (both on the boundary Γ), ε_0 is the applied strain and $\boldsymbol{x}^{+}$ and $\boldsymbol{x}^{-}$ are the position vectors on opposing faces.

Historically, homogenizations have been primarily used to analyze linear material behavior, with the effective coefficients of the linear elasticity tensor being computed. Analytical methods, which obtain homogenized coefficients via closed form solutions, have been applied to problems that feature simple RVE geometries (e.g. a homogenization of annulus fibrosis [254], also refer to [122] for a summary of such methods in engineering applications). However, homogenization techniques based on analytical methods lack the ability to address the complex 3D microstructural features in ligament and tendon. Thus, methods based on FE discretization are particularly appealing.

For a properly discretized RVE, the FE method can be used to perform the homogenization. In the linear case, FE simulations are used to obtain the unknown coefficients of the elasticity tensor. To obtain a full set of homogenized material coefficients, a sufficient number of loading conditions must be applied. Depending on the type of homogenization and the underlying material symmetry, this may include simulated tensile testing in orthogonal directions and shear testing in orthogonal shearing directions. For a unit cell with an orthotropic symmetry, a total of 6 unique loading simulations must be performed to obtain the 9 independent coefficients in the elasticity tensor [77]. For a FE simulation, the periodic displacement boundary conditions (Eq. 11) must be enforced explicitly [247]. This can be achieved by converting the periodic boundary equations into a set of linear constraint equations (e.g., via a master node approach) within the FE solver. The application of periodic boundary conditions typically requires that the FE mesh has identical nodal distributions on opposing faces and edges (i.e., the faces and edges are conformal). For homogenizations that utilize a RVE that does not have conformal faces, other permissible boundary conditions must be used. These include kinematic boundary conditions, traction boundary conditions and mixed boundary conditions [172]. For these cases, the resulting homogenization is not exact.

Although homogenization methods have historically been applied to linear material behavior and kinematics, they can also be applied to nonlinear materials and nonlinear kinematics [98]. In the linear case, a finite number of loading scenarios can be used to solve for the unknown coefficients. In the nonlinear case, this methodology cannot be used because the functional form of the stress–strain response is unknown. For example, there is no combination of loading scenarios that can directly resolve whether a stress–strain response is quadratic, exponential or some other function. Strain energy based approaches have been suggested that curve fit an assumed functional response or populate a lookup table for interpolation [257]. However, they have yet to find widespread use.

An attractive alternative for nonlinear homogenizations is the use of a micromechanical model in combination with the appropriate boundary conditions. A micromechanical model can be subjected to loading scenarios that are of interest (e.g. uniaxial tensile loading of a tendon) in combination with periodic boundary conditions, and the homogenized response can be examined. This has proven useful in several studies that have sought to examine microscale forces and structure–function relationships in ligaments and tendons [186, 205]. See Sect. 5.2 for a discussion of these models.

However, micromechanical models are limited to very specific loading scenarios and do not provide a general homogenized response. In order to provide a more general homogenization, the FE^2 method has been proposed. FE^2 based homogenization utilizes a nested FE problem that consists of a macroscale boundary value problem and a microscale boundary value problem [66, 78, 122, 127, 128, 213, 256]. A macroscale mesh is defined in the normal fashion. When the stress or elasticity tensor for the macroscale model are needed during the nonlinear FE solution procedure, the macroscale deformation gradient is passed to the microscale problem and a homogenization is performed on a discretized RVE

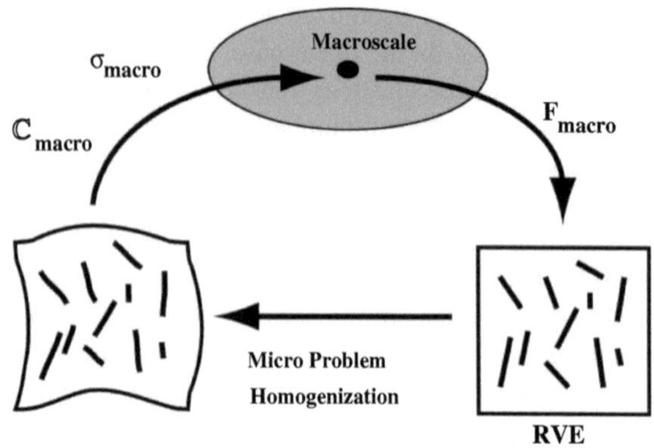

Fig. 7 FE^2 homogenization procedure. The FE^2 method solves a nested FE problem in which the deformation gradient $\mathbf{F}_{\mathbf{macro}}$ computed at a macroscale point is passed to the microscale RVE. A homogenization is performed using the deformation gradient coupled with suitable boundary conditions, and the resulting stress tensor ($\boldsymbol{\sigma}_{\mathbf{macro}}$) and elasticity tensor $\mathbb{C}_{\mathbf{macro}}$ are passed back to the macroscale. Figure adapted from [128]

(Fig. 7). Thus, the constitutive model is itself an FE problem. This framework provides a generalized strategy for performing a nonlinear homogenization on an arbitrarily defined RVE. The primary downside to this methodology is the substantial increase in computational demand. However, increased computing power coupled with parallelization methods will make future FE^2 considerably faster [169].

A fundamental assumption for the aforementioned homogenization approaches is that the RVE is infinitesimally small in comparison to the macroscale. Under this assumption, a homogenous deformation is described via the deformation gradient (Eq. 2). Since the deformation gradient is computed by taking a first order derivative, continuum based homogenization methods are referred to as 1st order methods. Some homogenization problems, however, feature RVEs that are not infinitesimally small in comparison to the macroscale. In these cases, microstructural size effects must be taken into account [39, 129]. This is particularly relevant in the study of damage initiation, as size effects play a critical role in this field [122, 225]. The most generalized approach to accounting for these size effects is to utilize 2nd order FE^2 strategies. In these methods, a microscale RVE problem is still used. However, the homogenization utilizes a quadratic version of the deformation map. In this method a Taylor series expansion is used to express the infinitesimal material line element $\boldsymbol{dx}$ as:

$$\boldsymbol{dx} = \boldsymbol{F} \cdot \boldsymbol{dX} + \frac{\boldsymbol{1}}{\boldsymbol{2}} \left(\boldsymbol{dX} \cdot {}^{3}\boldsymbol{G} \cdot \boldsymbol{dX} \right) + \boldsymbol{X}_0, \tag{12}$$

where the 3rd-order tensor ${}^{3}\boldsymbol{G} = \nabla \boldsymbol{F}$ is used. This method explicitly accounts for the length scale through the size of the RVE, and thus allows for computational homogenization of materials for which the assumptions of the 1st order method are not appropriate. Second order methods are particularly attractive for biological materials, as physical scales are often not separated sufficiently in size. As an example, the microstructures in ligament and tendon have similar physical dimensions to the macroscale. Fascicles, for instance, have a diameter of

Table 2 Summary of homogenization methods

Homogenization	Number of simulations	Purpose
Linear	Finite number, based on number of unknowns in linear elasticity tensor	Homogenized linear material coefficients
Micromechanical	Finite number, based on model application	Investigate nonlinear behavior and structure–function relationships
1st order FE^2	Simulation every time stress and tangent stiffness are evaluated within macro model	Nonlinear homogenization for infinitesimally small microstructures
2nd order FE^2	Simulation every time stress and tangent stiffness are evaluated within macro model	Nonlinear homogenization that incorporates size effects

The method of homogenization used is dependent on the intended application

~250 μm, which is only 1–2 orders of magnitude less than ligament and tendon widths, which are ~5–30 mm (refer to Sect. 2.3). For a comparison of homogenization techniques, refer to Table 2.

5 Multiscale Modeling of Ligament and Tendon Mechanics

5.1 Introduction

The field of multiscale mechanical modeling, and the multiscale modeling of ligament and tendon in particular, is in its relative infancy. For example, there seems to be no clear consensus on the very definition of multiscale modeling, as it has been applied to a large number of models that vary as to what scales are included and how the scales are linked. For the purposes of this section, a multiscale model is defined as a model that addresses two or more physical scales. This implies that both macroscale and microscale stress and strain are computed from a single simulation. Furthermore, this requires that there is some form of linking between scale levels. This linking is based on the appropriate application of boundary conditions, which may include periodic boundary conditions, prescribed boundary conditions, homogenous boundary conditions and a mixture thereof. This definition includes micromechanical models in which both a macroscale response and microscale response is described, as well as full nonlinear homogenizations (e.g., FE^2 methods). It is important to note that this definition does not include structurally motivated constitutive models, which may utilize the notion of microstructural features (e.g., uncrimping of collagen fibers [70, 88], fiber recruitment [111] or fiber families embedded within a ground substance [18]). This definition also precludes models that utilize generalized continuums, which address microscale size effects but do not specifically define microscale stress and strain [225]. This section reviews the state of the art in multiscale modeling and multiscale model validation as they relate to ligament and tendon.

5.2 Micromechanical Modeling of Ligament and Tendon

We define a micromechanical model as a 2D or 3D model that specifies a microscale geometry (e.g. a unit cell), applies macroscale boundary conditions (e.g. simulated tensile loading combined with periodic boundary conditions) and solves the governing equations over the simulation domain. Such a model yields both a microscale response (e.g. within the unit cell) and a macroscale response (e.g. reaction force of a unit cell subjected to tensile deformation). Since even simple 2D geometries do not generally have tractable solutions, these models almost exclusively rely on computational methods, most commonly the FE method. Models within the literature are primarily 2D and focused on equilibrium elasticity, although a biphasic model and 3D models have been proposed [137, 187]. Boundary conditions include fixed boundary conditions (e.g. zero displacement on a model edge), prescribed boundary conditions (e.g. a prescribed load or displacement on a model edge), periodic boundary conditions and a combination of these. Both linear and nonlinear micromechanical models have been proposed for biological tissues. Linear models (ubiquitous in the study of trabecular bone, e.g. [165, 172, 176, 197, 223]) generally seek to obtain homogenized coefficients of the linear elastic stiffness tensor. Due to the nonlinear nature of ligament and tendon, most micromechanical models for this application are nonlinear and seek to explore nonlinear behavior as the primary goal, although some studies have reported homogenized linear coefficients as well (e.g. [187].). The utility of microscale models is found in a number of ways. Some models are used to study certain structure function relationships (e.g. how certain microscale structures affect macroscale behavior) [84, 187]. Other models are used to study microscale damage mechanisms [205], and still others investigate microscale mechanotransduction [137].

Micromechanical models have been used to examine the structure–function relationship between fibril shape, fibril aspect ratio, the stiffness of fibrils and the inter-fibril matrix [83, 84]. In these studies, a 2D plane strain model was used to examine force transfer between adjacent collagen fibrils via an inter-fiber matrix [84]. A unit cell was created that consisted of a discretized fibril embedded within a matrix material (Fig. 8). The fibrils were given cylindrical or parabloidal (i.e. tapered) endpoints [83]. The unit cell was subjected to homogenous boundary conditions in which a displacement was applied to the sides of the model. The aspect ratio of the fiber and applied load were varied parametrically and their influence on the fibril stress, inter-fiber force transfer and strain was examined. Simulations revealed that fiber strain displayed a dependency on the end shape of the fibril, on the fibril aspect ratio and the ratio of the fibril stiffness to the matrix stiffness. The effect of tapered fibril ends was to decrease stress within fibers. The effect of increasing the stiffness of the inter fibril matrix was to increase load sharing between fibril and the matrix, which yielded decreased fibril strains. By utilizing a unit cell approach, this study was able to study the influence of structure–function relationships that would be difficult if not impossible to investigate

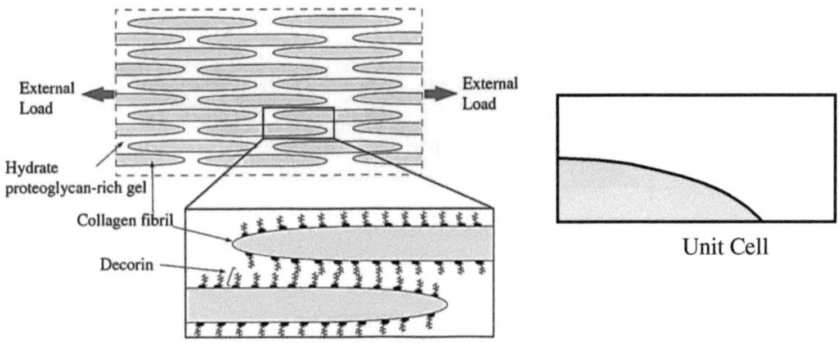

Fig. 8 Collagen fibril micromechanical model [84]. A model put fourth for collagen fibers consisted of fibrils of a finite length and with tapered tips connected via proteoglycan matrix material (*left*). A quarter symmetric micromechanical model was defined (*right*) and subjected to simulated loading. Figure adapted from [84]

using experimental or analytical approaches. Within these studies, the concept of an inter-fiber matrix material was utilized. The matrix material is thought to consist of proteoglycans, elastin and other ECM proteins that may mechanically couple collagen. Such a concept has been used in numerous studies (e.g. [17, 18, 132, 187, 232]) and is used to describe the substance that mechanically couples collagen fibrils and fibers within tendon and ligament.

One area that shows great promise in the field of multiscale modeling is in the study of stress and strain localization as it pertains to damage initiation. Although no studies have yet utilized micromechanical models to study damage initiation in tendon, they have been utilized in studying microscale strain patterns in the myotendinous junction (MTJ), which displays similarities to tendon and ligament tissue. In one such study, a 2D micromechanical model was used to explore microscale strain distributions within the MTJ, a common location for musculoskeletal injuries [205]. At the MTJ, muscle fibers taper as they insert into the tendon via the endomysium, creating a potential location for strain concentration and damage initiation. By utilizing a microscale unit cell model, this study sought to investigate strain concentrations within this region. The unit cell consisted of a single tapered muscle fiber inserting into tendon at a pennation angle of 37° (Fig. 9). The endomysium was given a transversely isotropic constitutive model similar to that presented in Sect. 4.2 and the muscle fiber was given an active contraction material model developed for muscle tissue [205]. The unit cell was subjected to prescribed displacement along the fiber direction corresponding to a 24 % strain. The edges of the unit cell were given periodic boundary conditions, which simulated a fiber embedded in macroscopic tissue. Simulations were run with both passive and active fiber recruitment. Model validation was performed by comparing the predicted fiber strains to those experimentally measured for relaxed and strained muscle fibers. More specifically, the deflection of the A-bands within the muscle fibers were experimentally measured and compared to those obtained from the FE models (refer to [205] for more detail regarding validation methods).

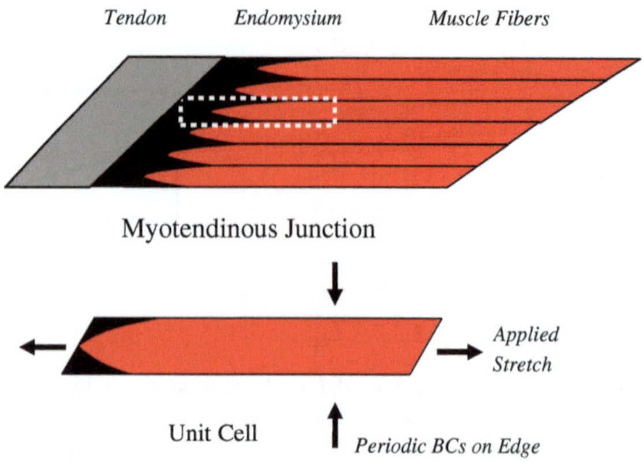

Fig. 9 Myotendinous junction micromechanical model [205]. (*Top*) The myotendinous junction (MTJ) consists of muscle fibers (*red*) that insert into tendon tissue (*gray*) via the endomysium (*black*). A micromechanical model was made of the MTJ by creating a unit cell (outlined in white *dashed box* and shown on *bottom*) and subjecting it to periodic boundary conditions on the edge and prescribed boundary conditions on the ends. Figure adapted from [205]

The most significant result of this study was that the FE micromechanical model predicted stress concentration and microscale strains that were significantly larger than the macroscale strains. This suggests that origins of damage mechanisms may initiate within the myotendinous junction, demonstrating the utility of micromechanical models in the study of damage initiation. The FE models predicted the experimentally measured deflection of the A-bands of muscle fibers. Although not a direct validation of the 2D strains within the unit cell, this validation provides evidence of the accuracy of the FE simulations. By creating micromechanical FE simulations driven by macroscopic loading, this study was able to utilize modeling as a means for investigating microscale strain concentrations, something that would have been difficult using experimental methods alone.

The aforementioned studies utilized 2D simulations. In our own research, we have used 3D micromechanical FE models to study structure function relationships in tendon and ligament tissue [187]. The aim of this research was to examine how fibril organization contributes to the elastic volumetric response. The volumetric response is quantified using the Poisson's ratio in linear theory and the Poisson's function in nonlinear theory. Experimentally observed Poisson's ratios range from 1.0 to 3.0 for tendon and ligament [105, 147], yet the structural underpinnings for these large values are not known. It was hypothesized that a planar, crimped arrangement of fibrils would not account for these large Poisson's ratios, while a helical organization of fibrils would.

To test this hypothesis, 3D unit cells were created that explicitly modeled collagen fibrils embedded within a matrix material (Fig. 10, top). The fibrils were given crimped, helical and combined crimped with a superhelical organization (Fig. 10, top). The models were given periodic boundary conditions and subjected to simulated tensile loading in the fiber direction, which yielded a homogenized

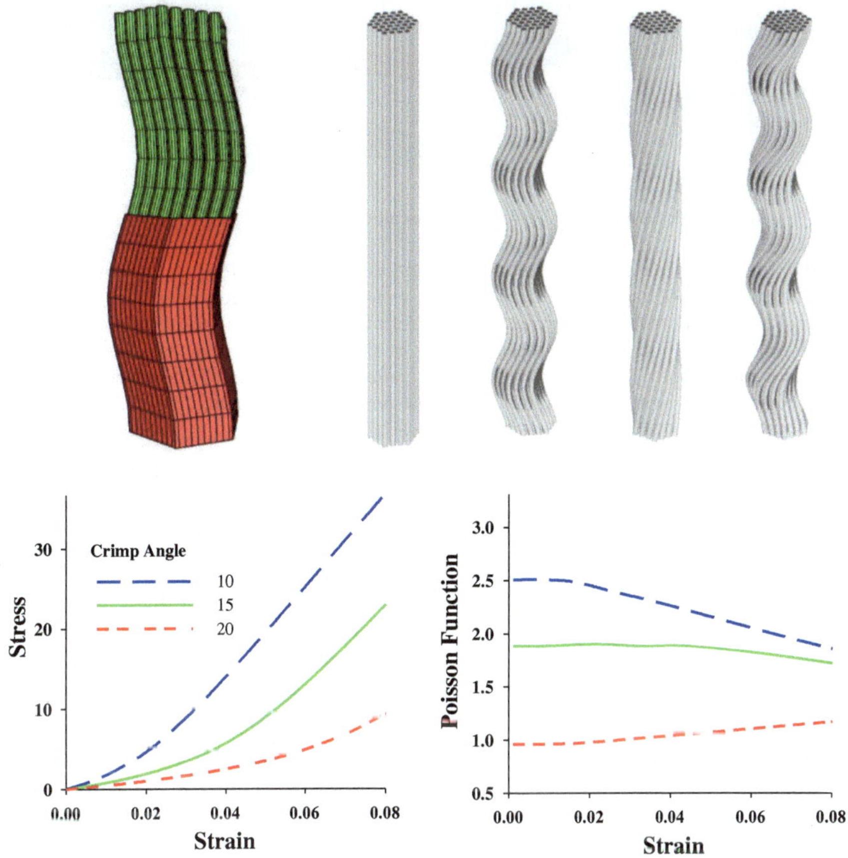

Fig. 10 3D micromechanical models with crimped and helical fiber organization [187]. *Top left* micromechanical FE model shows fibrils (*green*) embedded within the matrix (*red*). *Top right* straight, crimped, helical and helically crimped fibril organizations were modeled. *Bottom left* crimped models were able to reproduce the classic nonlinear behavior for tendon. *Bottom right* crimped models with a constant helical pitch predicted both the nonlinear stress–strain behavior and the large, nonlinear Poisson's function

macroscale stress–strain response and a homogenized Poisson's function. For a subset of models, tensile strains of 8 % were applied and the nonlinear stress–strain response and the Poisson's function were obtained (Fig. 10, bottom). For all other models, small strains (0.5 %) were applied and homogenized Poisson's ratios were obtained.

Models with planar crimp (both with and without a helical twist) could generate the classic nonlinear response, but only models with a helical twist could generate large Poisson's ratios (Fig. 10, bottom). This suggests that helical twisting of fibrils (which has been observed histologically [226, 248]) may contribute to the large experimentally measured Poisson's ratios. A parametric study which varied

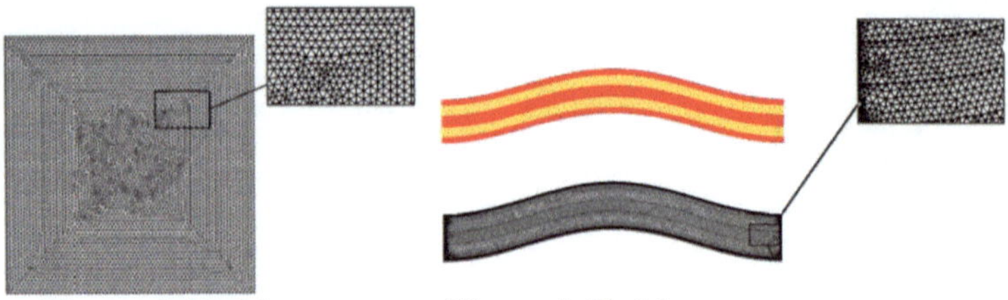

Fig. 11 Micromechanical stress and strain localization of tendon [156]. A macroscale model consisting of a cube of tendon tissue (mesh shown on *left*) was subjected to simulated loading and the macroscale deformation was computed. This deformation was then applied to the microscale model (*right*). The microscale model consisted of fibers (shown on the upper right in *red*) connected via an inter-fiber matrix (shown in *yellow*). A closeup of the macroscopic and microscopic model mesh is shown in the *breakout boxes*. Figure adapted from [156]

crimp angle, helical twist, the number of fibrils and the stiffness of the fibrils and matrix suggested that the large Poisson's ratios were predicted across a range of physiologically relevant values for these parameters. This study highlights the utility of homogenized micromechanical models in testing structure function based hypothesis that are otherwise difficult to address. Furthermore, it demonstrates the use of 3D until cells with a non-rectangular cross section for the use of nonlinear homogenization.

In the previously discussed studies, boundary conditions on the micromechanical model were applied a priori to the microscale models. In a recent study, micromechanical models were combined with a macroscale simulation to solve the localization problem [156]. In a localization problem, a macroscale deformation (generally computed from a continuum based FE simulation) is applied to an RVE, which is then solved in order to obtain the microscale stress and strain [122]. In this study, an analytically based homogenization was used as the constitutive model for a macroscopic FE simulation. Briefly, the analytical homogenization modeled collagen crosslinks at the nanoscale and collagen fiber uncrimping at the microscale to specify a macroscopic continuum response. The homogenization was not based on an explicit microstructural organization, therefore no microscale strains were computed. A macroscale cube was subjected to a tensile loading (Fig. 11, left), which yielded a macroscale deformation within each element. This macroscale deformation from an element in the interior of the macroscale mesh was applied to a microscale RVE (Fig. 11, right), which was solved using an FE simulation. The microscale FE results revealed a heterogenous distribution of stress and strain at the microscale. Although this study was primarily the presentation of a new method, it demonstrates the utility of micromechanical models in solving the localization problem by applying the results of a macroscale simulation to a microscale RVE.

The aforementioned models were limited to quasistatic elastic simulations only. However, models that can incorporate the time dependence and biphasic nature of these tissues are desirable. In one study, a macroscopic biphasic FE model was

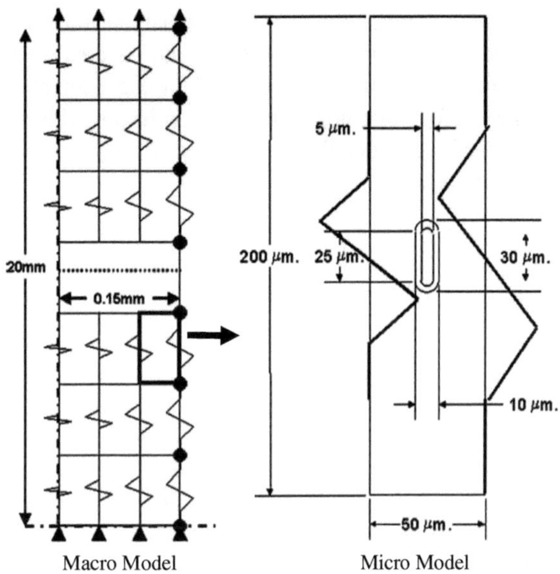

Fig. 12 Two level biphasic simulation of a tendon fascicle [137]. A biphasic axisymmetric macroscale model was created that contained nonlinear springs (Mesh, springs and model dimensions shown on *left*). The resulting strains and fluid flux was applied to a microscale model (*right*) that contained nonlinear springs and an ovoid shaped fibroblast. Figure adapted from [137]

combined with a microscopic biphasic FE model in order to examine the link between macroscale loading and microscale mechanotransduction [137]. To facilitate this, a macroscopic nonlinear biphasic model of a rat tail tendon fascicle was first solved. The macroscopic model utilized nonlinear springs to represent the transversely isotropic symmetry and nonlinear stress–strain behavior of tendon fascicles (Fig. 12, left). The springs were embedded in a porous matrix that had a transversely isotropic porosity. The model was subjected to uniaxial loading under a constant strain rate, and the computationally obtained macroscale stress–strain curve was validated against experimental stress–strain data of rat tail tendon fascicles. The deformation and fluid flux obtained from this model was then used to generate boundary conditions for the microscale model, which featured an ovoid shaped fibroblast aligned with the collagen matrix (Fig. 12, right). Finally, the predicted microscale fluid shear and microscale cell membrane deformation was correlated to collagenase mRNA levels that were experimentally measured in rat tail tendon fascicles. Briefly, fresh rat tail tendon fascicles were subjected to loading scenarios identical to those applied to the simulations. Four loading scenarios were utilized, including low strain, high strain, low strain rate and high strain rate. After each experiment, collagenase mRNA levels (MMP-13) were measured using real time quantitative PCR.

The macroscale models predicted a stress–strain response that was in good agreement with experimentally measured values. The microscale models predicted significant cell membrane strains and fluid shear stress on the embedded cell for the high strain rate and large strain models. The strains and shear stress correlated to a decreased expression of mRNA for collagenase. Experimentally, it can be concluded that decreased loading results in increased collagenase activity. By utilizing a computational micromechanical model, estimates of fluid shear stress on fibroblasts as well as cell membrane strains were able to be correlated to these experimental results. No tractable experimental methods currently have been proposed for obtaining microscale values of this sort, highlighting the important role that such modeling studies can play in extending our understanding of mechanotransduction within tendon tissue.

5.3 1st Order FE^2 Methods

The aforementioned micromechanical models report both macroscale and microscale measures, but are limited to a specific set of loading conditions. A more general approach is offered by the 1st order FE^2 homogenization method described in Sect. 4.5. To date, no FE^2 models have been proposed for the simulation of ligament and tendon. However, an FE^2 approach has been developed and applied to model collagen hydrogels, which have characteristics in common with connective tissues [43, 195, 196]. In this work, FE^2 models were made of type I collagen gels that were molded into a cruciform shape. The gels were seeded with fibroblasts and were allowed to undergo cell mediated compaction of the fibril network. The gels were subjected to biaxial testing and the macroscopic stress and strain were measured. Additionally, polarametric fiber alignment imaging (PFAI) was used to measure fibril orientations during testing. A quarter symmetric macroscopic FE model was constructed (Fig. 13, left) that mimicked the geometry and loading conditions of the experimentally tested cruciform gels. The fibril orientation (e.g. the angular distribution of fibrils) measured in the reference position was used in order to generate 3D RVE models that consisted of beam elements, which represented collagen fibrils within the gel (Fig. 13, middle). A constitutive model specific to collagen fibrils was developed for use in the RVEs. The macroscopic FE model utilized an FE^2 methodology described in Sect. 4.5, whereby the RVE homogenization was utilized as the constitutive model for the macroscopic simulation. The results of the FE simulations generated both macroscopic stress and strain as well as microscopic stress and strain within the fibril network. The FE predicted principle angles for fiber orientation were extracted from the microscale RVE simulations and compared to the experimentally obtained principle angles (via PFAI). There was good agreement between the predicted and measured orientation of the fibrillar network, which provided validation of the methodology. The microscale RVE simulations revealed realignment of the fibrillar network with applied strain. A considerable percentage of the fibrils were subjected to compressive buckling, revealing microscale inhomogeneity in response to a homogenous macroscale deformation. Collagen hydrogels are substantially different from tendon and ligament; however, this same microscale heterogeneity has been observed in tendon fascicles [201] (refer to Sects. 3.3 and 3.4). Although not directly applicable to tendon and ligament, this study provides a

Fig. 13 FE^2 simulation of collagen gel mechanics [195]. A quarter symmetric macroscale FE model was created of the cruciform gel sample (*left*). The RVE for the simulations consisted of beam elements with a specified angular distribution (*middle*). Models were validated using polarimetric fiber alignment imaging, which yielded fiber alignment vectors and contour plots of the principle angle of the fiber alignment (*right*). Figure adapted from [195]

template for how first order FE^2 methodologies could be applied to connective tissues such as tendon and ligament.

5.4 Multiscale Model Validation

Validation of computational models is fundamentally important if the models are to be interpreted in a meaningful way [9, 103]. Validation of macroscale computational models of ligaments and tendons consist of comparing macroscale FE simulations to experimentally measured metrics such as joint reactions forces and in situ tissue strains, with a number of examples being present in the literature (e.g. [75, 235]). Validation of multiscale models proceeds in a similar model, but ideally, experimental validation would occur at each scale level simulated (e.g., the macroscale and microscale). In a number of the aforementioned studies, validation of the mechanical simulations were performed at a single scale level (e.g., at the micro level for the unit cells of the myotendinous junction [205], and at the macro level for the biphasic tendon model [137]). To our knowledge, only one model relevant to tendons and ligaments has been validated at two scale levels [195]. This study, described in the preceding section, validated macroscale results via stress–strain data and validated microscale results via polarametric fiber alignment imaging.

Because of the important role validation plays in interpreting the results of computational studies, it is desirable to develop multiscale simulation strategies and experimental validation methods concurrently. Although macroscale validation methods have been described, microscale validation methods are still in need of improvement and development. In order to validate microscale models, data regarding stress and strain at lower scale levels (e.g. within fascicles, fibers and fibrils) must be obtained. One possible route involves isolating and mechanically testing individual tissue constituents, such as those described in Sects. 3.3 and 3.4. However, such methods have yielded widely variable results, likely due to the difficulty in consistently isolating substructures without causing tissue damage. Microscopic imaging studies, such as the confocal studies described in Sect. 3.4, provide considerable promise for use in validation of microscale models. However, the highly inhomogeneous strain fields and the complex microscale fiber structure make this a challenging starting point [199, 201, 202].

In order to address these challenges, our lab has developed a surrogate material for use as a physical model to aid in the development of multiscale modeling and validation methods. A physical model reduces the number of uncontrolled variables related to the structural organization of ligaments and tendons. To create the physical surrogates, dense (~25 % collage/wt), extruded collagen fibers were embedded within a collagen gel matrix (~0.5 % collagen/wt). Surrogates served as physical models to emulate features of ligament and tendon tissue in a controlled and reproducible manner. Two different colors of fluorescent beads were embedded in the fibers and gel matrix (Fig. 14, left) for use as microscopic fiducial markers. 3D micromechanical FE models of the surrogates were then constructed (Fig. 14, bottom). A constitutive model based on a continuous elliptical fiber distribution was used to describe the mechanical behavior of the collagen gel and embedded fibers [14]. This constitutive model emulated the reorganization of fibrils with applied strain. The model was curve fit to tensile testing data for isolated gel and extruded fiber samples and was found to accurately model both the uniaxial stress–strain behavior and the 2D strain behavior (i.e., the nonlinear Poisson's function). Micromechanical FE models were subjected to uniaxial strain, and the macroscale and microscale stress and 2D strain were determined. FEBio was used for all analysis (http://www.febio.org) [148]. To validate the FE models, the physical surrogates were subjected to tensile loading in a custom testing apparatus on an inverted confocal microscope. Confocal images were acquired at 6 strain increments at both 2.5X and 10X, while force was measured simultaneously. Texture correlation was used to measure strain at the macroscale and to measure strain within the fibers and strain in the inter-fiber matrix at the microscale [221].

The microscopic 2D strains were inhomogeneous, and the macroscopic 2D strain was not representative of the microscopic 2D strain (Fig. 14, right). The magnitude of the transverse strain in the fibers greatly exceeded the macroscopic transverse strain, while the magnitude of the transverse matrix strain was significantly less than the macroscopic strain. The macroscopically measured Poisson's ratio was 1.72 ± 0.26, which is comparable to experimentally measured values for tendon and ligament [105, 147]. The micromechanical FE model was able to simultaneously predict the macroscopic stress–strain behavior and the 2D macroscale and microscale strains (Fig. 14, right). The predicted macroscopic stress and macroscopic transverse strain closely matched the experimentally measured values with normalized root mean square (NRMSE) values of 0.015 and 0.085, respectively. The predicted microscopic transverse fiber strain was closely

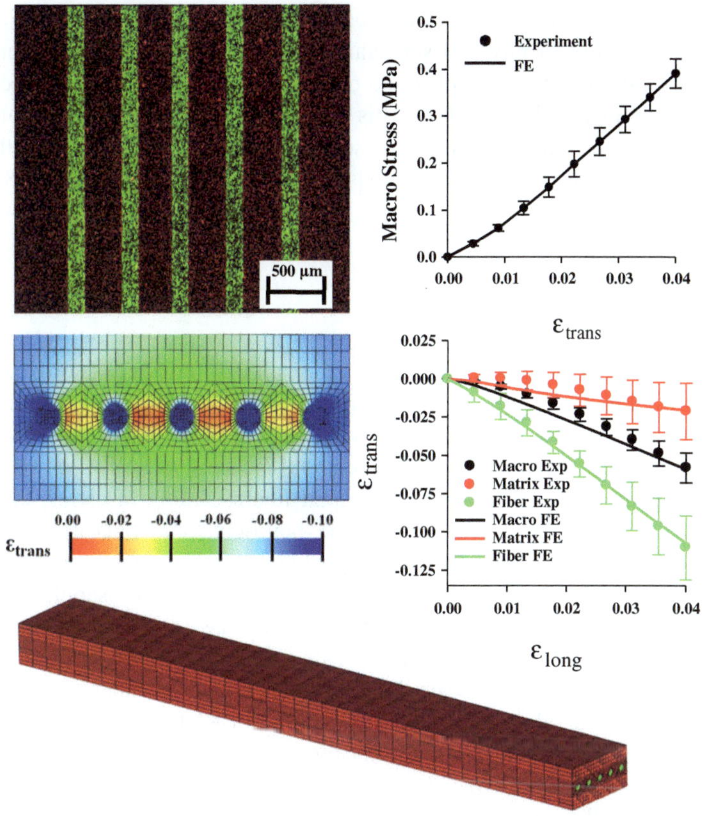

Fig. 14 Multiscale validation of a micromechanical model. (*Upper left*) A dual channel 4X confocal image of a surrogate construct shows the *red* fluorescent beads in the gel matrix and the *green* fluorescent beads in the fiber that were used for microscopic and macroscopic strain measurements. (*Upper right*) Micromechanical FE model stress–strain predictions were in excellent agreement with the experimental data. (*Middle left*) The constrained surrogate model displayed considerable heterogeneity in transverse strain. (*Middle right*) Microscopic and macroscopic strain measurement results show that the macroscopic transverse strain (*black line*) was not representative of the microscopic fiber strain (*green line*) or matrix strain (*red line*). The error bars represent the standard deviation computed for all samples. (*Bottom*) Micromechanical FE model of the surrogate. *Green* elements represent the fibers and *red* elements represent the gel matrix

matched by the experimentally measured values (NRMSE = 0.018), while predictions for the microscopic transverse matrix strain were reasonable but not as accurate (NRMSE = 0.190). When simulations were performed using coefficients that varied by a single standard deviation, all of the predictions were closely bounded by this uncertainty. A sensitivity study was then performed in which the inter-fiber spacing and the inter-fiber matrix material properties were varied.

The results of this work indicate that the micromechanical model was able to accurately predict the strains at both the macroscopic and microscopic level, demonstrating the utility of this approach for the study of fibrous biological composites. The use of physical surrogate materials provides a means for developing and validating more complex and physiologically relevant micromechanical and multiscale mechanical models. This study illustrates the feasibility of simultaneous validation at macro and micro scales that could be extended to the validation of micromechanical or FE^2 models of native ligament or tendon [188, 189].

5.5 Future Directions

There are a number of important areas of research that remain as future directions and opportunities for multiscale modeling of tendon and ligament. As of yet, no fully nested 1st or 2nd order FE^2 models have been proposed for tendon and ligament. This will require the development and meshing of RVEs that sufficiently characterize the microstructure (e.g. crimped fibers, sliding fascicles, etc.), constitutive models capable of describing the nanoscale and microscale response within the RVE (e.g., Eq. 9), software capable of performing the recursive FE^2 simulations using 1st and 2nd order procedures, as well as microscopic techniques for validating model simulations. Given the multiple scales of organization in these tissues (i.e. nanoscale fibrils, microscale fibers and mesoscale fascicles), it is unclear whether two levels of scale linking is adequate, or whether additional scale levels must be included. Furthermore, the criteria for RVE size in tendinous tissue and the importance of using of 2nd order methods has yet to be investigated. The use of physical surrogates may provide a controlled means for answering such questions, as well as provide a simplified approach for developing multiscale modeling and validation methods.

The development of validated multiscale models of tendons and ligaments will provide insight into structure function relationships and will be invaluable in the study of damage mechanisms, which appear to have microscale origins [58, 72, 73]. In the future, such methods may provide a means for creating scaffolds with a microstructure that can simultaneously address macroscale mechanical requirements of whole organs (e.g. an ACL graft) and the microscale mechanical requirements necessary for cellular proliferation. Although this chapter has focused primarily on the quasistatic elastic response, models that capture the viscoelastic and biphasic response of these tissues will be of substantial importance. Homogenization methods have been developed to account for viscoelasticity [38, 44, 90, 130], but these methods have yet to be applied to biological tissues. The next paradigm in multiscale modeling will likely address growth and remodeling of tissue (e.g. remodeling in bone [86, 93, 138]). Nested FE^2 multiscale methods for such simulations remain to be developed for aligned collagenous tissues.

References

1. Abdel-Rahman, E.: Three-dimensional dynamic anatomically based model of the human tibio-femoral joint. Ph.D., University of Toledo, Toledo (1995)
2. Abramowitch, S.D., Clineff, T.D., Withrow, J.D., Papageorgiou, C.D., Woo, S.L.: The quasilinear viscoelastic properties of the healing goat medial collateral ligament: an experimental & analytical approach. In: 23rd Annual Meeting of the American Society of Biomechanics (1999)
3. Abramowitch, S.D., Papageorgiou, C.D., Debski, R.E., Clineff, T.D., Woo, S.L.: A biomechanical and histological evaluation of the structure and function of the healing medial collateral ligament in a goat model. Knee Surg. Sports Traumatol. Arthrosc. **11**(3), 155–162 (2003)
4. Agoram, B., Barocas, V.H.: Coupled macroscopic and microscopic scale modeling of fibrillar tissues and tissue equivalents. J. Biomech. Eng. **123**(4), 362–369 (2001)
5. Ahmed, A.M., Burke, D.L., Duncan, N.A., Chan, K.H.: Ligament tension pattern in the flexed knee in combined passive anterior translation and axial rotation. J. Orthop. Res. **10**(6), 854–867 (1992)
6. Ahmed, A.M., Hyder, A., Burke, D.L., Chan, K.H.: In vitro ligament tension pattern in the flexed knee in passive loading. J. Orthop. Res. **5**(2), 217–230 (1987)
7. Amiel, D., Frank, C., Harwood, F., Fronek, J., Akeson, W.: Tendons and ligaments: a morphological and biochemical comparison. J. Orthop. Res. **1**(3), 257–265 (1984)
8. Andarawis-Puri, N., Kuntz, A.F., Ramsey, M.L., Soslowsky, L.J.: Effect of supraspinatus tendon repair technique on the infraspinatus tendon. J. Biomech. Eng. **133**(3), 031008 (2011)
9. Anderson, A.E., Ellis, B.J., Weiss, J.A.: Verification, validation and sensitivity studies in computational biomechanics. Comput. Methods Biomech. Biomed. Eng. **10**(3), 171–184 (2007)
10. Anderson, D.R., Weiss, J.A., Takai, S., Ohland, K.J., Woo, S.L.-Y.: Healing of the medial collateral ligament following a triad injury: a biomechanical and histological study of the knee in rabbits. J. Orthop. Res. **10**, 485–495 (1992)
11. Anderson, D.R., Weiss, J.A., Takai, S., Ohland, K.J., Woo, S.L.: Healing of the medial collateral ligament following a triad injury: a biomechanical and histological study of the knee in rabbits. J. Orthop. Res. **10**(4), 485–495 (1992)
12. Annovazzi, L., Genna, F.: An engineering, multiscale constitutive model for fiber-forming collagen in tension. J. Biomed. Mater. Res. A **92**(1), 254–266 (2010)
13. Armstrong, C.G., Lai, W.M., Mow, V.C.: An analysis of the unconfined compression of articular cartilage. J. Biomech. Eng. **106**(2), 165–173 (1984)
14. Ateshian, G.A., Rajan, V., Chahine, N.O., Canal, C.E., Hung, C.T.: Modeling the matrix of articular cartilage using a continuous fiber angular distribution predicts many observed phenomena. J. Biomech. Eng. **131**(6), 061003 (2009)
15. Ateshian, G.A., Warden, W.H., Kim, J.J., Grelsamer, R.P., Mow, V.C.: Finite deformation biphasic material properties of bovine articular cartilage from confined compression experiments. J. Biomech. **30**(11–12), 1157–1164 (1997)
16. Atkinson, T.S., Ewers, B.J., Haut, R.C.: The tensile and stress relaxation responses of human patellar tendon varies with specimen cross-sectional area. J. Biomech. **32**(9), 907–914 (1999)
17. Ault, H.K., Hoffman, A.H.: A composite micromechanical model for connective tissues: Part I—Theory. J. Biomech. Eng. **114**(1), 137–141 (1992)
18. Ault, H.K., Hoffman, A.H.: A composite micromechanical model for connective tissues: Part II—Application to rat tail tendon and joint capsule. J. Biomech. Eng. **114**(1), 142–146 (1992)
19. Babuska, I.: The finite element method with Lagrangian multipliers. Numer. Math. **20**, 179–192 (1973)

20. Bach, J.M., Hull, M.L., Patterson, H.A.: Direct measurement of strain in the posterolateral bundle of the anterior cruciate ligament. J. Biomech. **30**(3), 281–283 (1997)
21. Banos, C.C., Thomas, A.H., Kuo, C.K.: Collagen fibrillogenesis in tendon development: current models and regulation of fibril assembly. Birth Defects Res. C Embryo Today **84**(3), 228–244 (2008)
22. Barton, E.R., Gimbel, J.A., Williams, G.R., Soslowsky, L.J.: Rat supraspinatus muscle atrophy after tendon detachment. J. Orthop. Res. **23**(2), 259–265 (2005)
23. Batson, E.L., Paramour, R.J., Smith, T.J., Birch, H.L., Patterson-Kane, J.C., Goodship, A.E.: Are the material properties and matrix composition of equine flexor and extensor tendons determined by their functions? Equine Vet. J. **35**(3), 314–318 (2003)
24. Battlehner, C.N., Carneiro Filho, M., Ferreira Junior, J.M., Saldiva, P.H., Montes, G.S.: Histochemical and ultrastructural study of the extracellular matrix fibers in patellar tendon donor site scars and normal controls. J. Submicrosc. Cytol. Pathol. **28**(2), 175–186 (1996)
25. Beatty, M.F., Stalnaker, D.O.: The Poisson function of finite elasticity. J. Appl. Mech. **53**(4), 807–813 (1986)
26. Benjamin, M., Evans, E.J., Copp, L.: The histology of tendon attachments to bone in man. J. Anat. **149**, 89–100 (1986)
27. Benjamin, M., Kaiser, E., Milz, S.: Structure-function relationships in tendons: a review. J. Anat. **212**(3), 211–228 (2008)
28. Benjamin, M., Ralphs, J.R.: Tendons and ligaments—an overview. Histol. Histopathol. **12**(4), 1135–1144 (1997)
29. Berns, G.S., Hull, M.L., Patterson, H.A.: Strain in the anteromedial bundle of the anterior cruciate ligament under combination loading. J. Orthop. Res. **10**(2), 167–176 (1992)
30. Birch, H.L.: Tendon matrix composition and turnover in relation to functional requirements. Int. J. Exp. Pathol. **88**(4), 241–248 (2007)
31. Birk, D.E., Zycband, E.I., Woodruff, S., Winkelmann, D.A., Trelstad, R.L.: Collagen fibrillogenesis in situ: fibril segments become long fibrils as the developing tendon matures. Dev. Dyn. **208**(3), 291–298 (1997)
32. Black, L.D., Brewer, K.K., Morris, S.M., Schreiber, B.M., Toselli, P., Nugent, M.A., Suki, B., Stone, P.J.: Effects of elastase on the mechanical and failure properties of engineered elastin-rich matrices. J. Appl. Physiol. **98**(4), 1434–1441 (2005)
33. Blankevoort, L., Kuiper, J.H., Huiskes, R., Grootenboer, H.J.: Articular contact in a three-dimensional model of the knee. J. Biomech. **24**(11), 1019–1031 (1991)
34. Bonifasi-Lista, C., Lake, S.P., Small, M.S., Weiss, J.A.: Viscoelastic properties of the human medial collateral ligament under longitudinal, transverse and shear loading. J. Orthop. Res. **23**(1), 67–76 (2005)
35. Bozec, L., Horton, M.: Topography and mechanical properties of single molecules of type I collagen using atomic force microscopy. Biophys. J. **88**(6), 4223–4231 (2005)
36. Bredrup, C., Stang, E., Bruland, O., Palka, B.P., Young, R.D., Haavik, J., Knappskog, P.M., Rodahl, E.: Decorin accumulation contributes to the stromal opacities found in congenital stromal corneal dystrophy. Invest. Ophthalmol. Vis. Sci. **51**(11), 5578–5582 (2010)
37. Breuls, R.G., Sengers, B.G., Oomens, C.W., Bouten, C.V., Baaijens, F.P.: Predicting local cell deformations in engineered tissue constructs: a multilevel finite element approach. J. Biomech. Eng. **124**(2), 198–207 (2002)
38. Brinson, L.C., Lin, W.: Comparison of micromechanics methods for effective properties of multiphase viscoelastic composites. Compos. Struct. **41**(3–4), 353–367 (1998)
39. Buechner, P.M., Lakes, R.S.: Size effects in the elasticity and viscoelasticity of bone. Biomech. Model. Mechanobiol. **1**(4), 295–301 (2003)
40. Butler, D.L., Kay, M.D., Stouffer, D.C.: Comparison of material properties in fascicle-bone units from human patellar tendon and knee ligaments. J. Biomech. **19**(6), 425–432 (1986)
41. Butler, D.L., Noyes, F.R., Grood, E.S.: Ligamentous restraints to anterior-posterior drawer in the human knee. J. Bone Joint Surg. (Am) **62**, 259–270 (1980)
42. Cetinkaya, M., Xiao, S., Markert, B., Stacklies, W., Grater, F.: Silk fiber mechanics from multiscale force distribution analysis. Biophys. J. **100**(5), 1298–1305 (2011)

43. Chandran, P.L., Barocas, V.H.: Deterministic material-based averaging theory model of collagen gel micromechanics. J. Biomech. Eng. **129**(2), 137–147 (2007)
44. Cherraf-Schweyer, C., Maurice, G., Taghite, M., Taous, K.: An experimental and theoretical approach of elasticity and viscoelasticity of compact and spongy bone with periodic homogenization. Comput. Methods Biomech. Biomed. Eng. **10**(3), 195–207 (2007)
45. Ciccone, W.J., Bratton, D.R., Weinstein, D.M., Elias, J.J.: Viscoelasticity and temperature variations decrease tension and stiffness of hamstring tendon grafts following anterior cruciate ligament reconstruction. J. Bone Joint Surg. (American volume) **88**(5), 1071–1078 (2006)
46. Cohen, B., Lai, W.M., Mow, V.C.: A transversely isotropic biphasic model for unconfined compression of growth plate and chondroepiphysis. J. Biomech. Eng. **120**(4), 491–496 (1998)
47. Cooper, R.R., Misol, S.: Tendon and ligament insertion. A light and electron microscopic study. J. Bone Joint Surg. (Am) **52**, 1–20 (1970)
48. Cui, L., Maas, H., Perreault, E.J., Sandercock, T.G.: In situ estimation of tendon material properties: differences between muscles of the feline hindlimb. J. Biomech. **42**(6), 679–685 (2009)
49. Daniel, D.M., Akeson, W.H., O'Connor, J.J.: Knee Ligaments: Structure, Function, Injury and Repair. Raven Press, New York (1990)
50. Danylchuk, K.D., Finlay, J.B., Krcek, J.P.: Microstructural organization of human and bovine cruciate ligaments. Clin. Orthop. Relat. Res. **131**, 294–298 (1978)
51. Debski, R.E., Weiss, J.A., Newman, W.J., Moore, S.M., McMahon, P.J.: Stress and strain in the anterior band of the inferior glenohumeral ligament during a simulated clinical examination. J. Shoulder Elbow Surg./Am. Shoulder Elbow Surg. [et al.], **14**(1 Suppl S), 24S–31S (2005)
52. Debski, R.E., Wong, E.K., Woo, S.L., Sakane, M., Fu, F.H., Warner, J.J.: In situ force distribution in the glenohumeral joint capsule during anterior-posterior loading. J. Orthop. Res. **17**(5), 769–776 (1999)
53. Defrate, L.E., van der Ven, A., Boyer, P.J., Gill, T.J., Li, G.: The measurement of the variation in the surface strains of Achilles tendon grafts using imaging techniques. J. Biomech. **39**(3), 399–405 (2006)
54. deVente, J.E., Lester, G.E., Trotter, J.A., Dahners, L.E.: Isolation of intact collagen fibrils from healing ligament. J. Electron Microsc. **46**(4), 353–356 (1997)
55. Doehring, T.C., Kahelin, M., Vesely, I.: Direct measurement of nonuniform large deformations in soft tissues during uniaxial extension. J. Biomech. Eng. **131**(6), 061001 (2009)
56. Drury, N.J., Ellis, B.J., Weiss, J.A., McMahon, P.J., Debski, R.E.: Finding consistent strain distributions in the glenohumeral capsule between two subjects: implications for development of physical examinations. J. Biomech. **44**(4), 607–613 (2011)
57. Drury, N.J., Ellis, B.J., Weiss, J.A., McMahon, P.J., Debski, R.E.: The impact of glenoid labrum thickness and modulus on labrum and glenohumeral capsule function. J. Biomech. Eng. **132**(12), 121003 (2010)
58. Duenwald-Kuehl, S., Kondratko, J., Lakes, R.S., Vanderby Jr, R.: Damage mechanics of porcine flexor tendon: mechanical evaluation and modeling. Ann. Biomed. Eng. **40**(8), 1692–1707 (2012)
59. Duenwald, S.E., Vanderby Jr, R.: Lakes RS: viscoelastic relaxation and recovery of tendon. Ann. Biomed. Eng. **37**(6), 1131–1140 (2009)
60. Edelstein, L., Thomas, S.J., Soslowsky, L.J.: Rotator cuff tears: what have we learned from animal models? J. Musculoskelet. Neuronal Interact. **11**(2), 150–162 (2011)
61. Einat, R., Yoram, L.: Recruitment viscoelasticity of the tendon. J. Biomech. Eng. **131**(11), 111008 (2009)
62. Elliott, D.M., Robinson, P.S., Gimbel, J.A., Sarver, J.J., Abboud, J.A., Iozzo, R.V., Soslowsky, L.J.: Effect of altered matrix proteins on quasilinear viscoelastic properties in transgenic mouse tail tendons. Ann. Biomed. Eng. **31**(5), 599–605 (2003)

63. Ellis, B.J., Debski, R.E., Moore, S.M., McMahon, P.J., Weiss, J.A.: Methodology and sensitivity studies for finite element modeling of the inferior glenohumeral ligament complex. J. Biomech. **40**(3), 603–612 (2007)
64. Ellis, B.J., Drury, N.J., Moore, S.M., McMahon, P.J., Weiss, J.A., Debski, R.E.: Finite element modelling of the glenohumeral capsule can help assess the tested region during a clinical exam. Comput. Methods Biomech. Biomed. Eng. **13**(3), 413–418 (2010)
65. Ellis, B.J., Lujan, T.J., Dalton, M.S., Weiss, J.A.: Medial collateral ligament insertion site and contact forces in the ACL-deficient knee. J. Orthop. Res. **24**(4), 800–810 (2006)
66. Feyel, F.: A multilevel finite element method (FE2) to describe the response of highly non-linear structures using generalized continua. Comput. Methods Appl. Mech. Eng. **192**(28–30), 3233–3244 (2003)
67. Franchi, M., Fini, M., Quaranta, M., De Pasquale, V., Raspanti, M., Giavaresi, G., Ottani, V., Ruggeri, A.: Crimp morphology in relaxed and stretched rat Achilles tendon. J. Anat. **210**(1), 1–7 (2007)
68. Frank, C., McDonald, D., Bray, D., Bray, R., Rangayyan, R., Chimich, D., Shrive, N.: Collagen fibril diameters in the healing adult rabbit medial collateral ligament. Connect. Tissue Res. **27**(4), 251–263 (1992)
69. Frank, C., Woo, S.L., Amiel, D., Harwood, F., Gomez, M., Akeson, W.: Medial collateral ligament healing. A multidisciplinary assessment in rabbits. Am. J. Sports Med. **11**(6), 379–389 (1983)
70. Freed, A.D., Doehring, T.C.: Elastic model for crimped collagen fibrils. J. Biomech. Eng. **127**(4), 587–593 (2005)
71. Fukunaga, T., Kubo, K., Kawakami, Y., Fukashiro, S., Kanehisa, H., Maganaris, C.N.: In vivo behaviour of human muscle tendon during walking. Proc. R. Soc. Biol. Sci. **268**(1464), 229–233 (2001)
72. Fung, D.T., Sereysky, J.B., Basta-Pljakic, J., Laudier, D.M., Huq, R., Jepsen, K.J., Schaffler, M.B., Flatow, E.L.: Second harmonic generation imaging and Fourier transform spectral analysis reveal damage in fatigue-loaded tendons. Ann. Biomed. Eng. **38**(5), 1741–1751 (2010)
73. Fung, D.T., Wang, V.M., Laudier, D.M., Shine, J.H., Basta-Pljakic, J., Jepsen, K.J., Schaffler, M.B., Flatow, E.L.: Subrupture tendon fatigue damage. J. Orthop. Res. **27**(2), 264–273 (2009)
74. Funk, J.R., Hall, G.W., Crandall, J.R., Pilkey, W.D.: Linear and quasi-linear viscoelastic characterization of ankle ligaments. J. Biomech. Eng. **122**(1), 15–22 (2000)
75. Gardiner, J.C., Weiss, J.A.: Subject-specific finite element analysis of the human medial collateral ligament during valgus knee loading. J. Orthop. Res. **21**, 1098–1106 (2003)
76. Gardiner, J.C., Weiss, J.A., Rosenberg, T.D.: Strain in the human medial collateral ligament during valgus loading of the knee. Clin. Orthop. **391**, 266–274 (2001)
77. Garnich, M.R., Karami, G.: Finite element micromechanics for stiffness and strength of wavy fiber composites. J. Compos. Mater. **38**(4), 273 (2004)
78. Geers, M.G.D., Kouznetsova, V.G., Brekelmans, W.A.M.: Multi-scale computational homogenization: trends and challenges. J. Comp. Appl. Math. **234**, 2175–2182 (2010)
79. Gimbel, J.A., Mehta, S., Van Kleunen, J.P., Williams, G.R., Soslowsky, L.J.: The tension required at repair to reappose the supraspinatus tendon to bone rapidly increases after injury. Clin. Orthop. Relat. Res. **426**, 258–265 (2004)
80. Gimbel, J.A., Van Kleunen, J.P., Lake, S.P., Williams, G.R., Soslowsky, L.J.: The role of repair tension on tendon to bone healing in an animal model of chronic rotator cuff tears. J. Biomech. **40**(3), 561–568 (2007)
81. Gimbel, J.A., Van Kleunen, J.P., Williams, G.R., Thomopoulos, S., Soslowsky, L.J.: Long durations of immobilization in the rat result in enhanced mechanical properties of the healing supraspinatus tendon insertion site. J. Biomech. Eng. **129**(3), 400–404 (2007)
82. Gitman, I.M., Askes, H., Sluys, L.J.: Representative volume: existence and size determination. Eng. Fract. Mech. **74**(16), 2518–2534 (2007)

83. Goh, K.L., Meakin, J.R., Aspden, R.M., Hukins, D.W.: Influence of fibril taper on the function of collagen to reinforce extracellular matrix. Proc. R. Soc. Biol. Sci. **272**(1575), 1979–1983 (2005)
84. Goh, K.L., Meakin, J.R., Aspden, R.M., Hukins, D.W.: Stress transfer in collagen fibrils reinforcing connective tissues: effects of collagen fibril slenderness and relative stiffness. J. Theor. Biol. **245**(2), 305–311 (2007)
85. Gotoh, T., Murashige, N., Yamashita, K.: Ultrastructural observations on the tendon sheath of the rat tail. J. Electron. Microsc. (Tokyo) **46**(3), 247–252 (1997)
86. Graham, J.M., Ayati, B.P., Ramakrishnan, P.S., Martin, J.A.: Towards a new spatial representation of bone remodeling. Math. Biosci. Eng. **9**(2), 281–295 (2012)
87. Grood, E.S., Noyes, F.R., Butler, D.L., Suntay, W.J.: Ligamentous and capsular restraints preventing straight medial and lateral laxity in intact human cadaver knees. J. Bone Joint Surg. (Am) **63**, 1257–1269 (1981)
88. Grytz, R., Meschke, G.: Constitutive modeling of crimped collagen fibrils in soft tissues. J. Mech. Behav. Biomed. Mater. **2**(5), 522–533 (2009)
89. Gupta, H.S., Seto, J., Krauss, S., Boesecke, P., Screen, H.R.: In situ multi-level analysis of viscoelastic deformation mechanisms in tendon collagen. J. Struct. Biol. **169**(2), 183–191 (2010)
90. Gururaja, S., Kim, H.J., Swan, C.C., Brand, R.A., Lakes, R.S.: Modeling deformation-induced fluid flow in cortical bone's canalicular-lacunar system. Ann. Biomed. Eng. **33**(1), 7–25 (2005)
91. Gusev, A.A.: Representative volume element size for elastic composites: a numerical study. J. Mech. Phys. Solids **45**(9), 1449–1459 (1997)
92. Hakkinen, L., Strassburger, S., Kahari, V.M., Scott, P.G., Eichstetter, I., Lozzo, R.V., Larjava, H.: A role for decorin in the structural organization of periodontal ligament. Lab. Invest. **80**(12), 1869–1880 (2000)
93. Hambli, R., Katerchi, H., Benhamou, C.L.: Multiscale methodology for bone remodelling simulation using coupled finite element and neural network computation. Biomech. Model. Mechanobiol. **10**(1), 133–145 (2011)
94. Hannafin, J.A., Arnoczky, S.P.: Effect of cyclic and static tensile loading on water content and solute diffusion in canine flexor tendons: an in vitro study. J. Orthop. Res. **12**, 350–356 (1994)
95. Hansen, K.A., Weiss, J.A., Barton, J.K.: Recruitment of tendon crimp with applied tensile strain. ASME **124**, 72 (2002)
96. Haraldsson, B.T., Aagaard, P., Qvortrup, K., Bojsen-Moller, J., Krogsgaard, M., Koskinen, S., Kjaer, M., Magnusson, S.P.: Lateral force transmission between human tendon fascicles. Matrix Biol. **27**(2), 86–95 (2008)
97. Hartmann, D.: A multiscale model for red blood cell mechanics. Biomech. Model. Mechanobiol. **9**(1), 1–17 (2010)
98. Hazanov, S.: Hill condition and overall properties of composites. Arch. Appl. Mech. **68**, 385–394 (1998)
99. Hazanov, S.: On apparent properties of nonlinear heterogeneous bodies smaller than the representative volume. Acta Mech. **134**(3–4), 1619–6937 (1998)
100. Heinemeier, K.M., Kjaer, M.: In vivo investigation of tendon responses to mechanical loading. J. Musculoskelet. Neuronal Interact. **11**(2), 115–123 (2011)
101. Helmer, K.G., Nair, G., Cannella, M., Grigg, P.: Water movement in tendon in response to a repeated static tensile load using one-dimensional magnetic resonance imaging. J. Biomech. Eng. **128**(5), 733–741 (2006)
102. Henninger, H.B., Maas, S.A., Shepherd, J.H., Joshi, S., Weiss, J.A.: Transversely isotropic distribution of sulfated glycosaminoglycans in human medial collateral ligament: a quantitative analysis. J. Struct. Biol. **165**(3), 176–183 (2009)
103. Henninger, H.B., Reese, S.P., Anderson, A.E., Weiss, J.A.: Validation of computational models in biomechanics. Proc. Inst. Mech. Eng. H **224**(7), 801–812 (2010)

104. Henninger, H.B., Underwood, C.J., Ateshian, G.A., Weiss, J.A.: Effect of sulfated glycosaminoglycan digestion on the transverse permeability of medial collateral ligament. J. Biomech. **43**(13), 2567–2573 (2010)
105. Hewitt, J., Guilak, F., Glisson, R., Vail, T.P.: Regional material properties of the human hip joint capsule ligaments. J. Orthop. Res. **19**(3), 359–364 (2001)
106. Hildebrand, K.A., Frank, C.B.: Scar formation and ligament healing. Can. J. Surg. Journal Canadien de Chirurgie **41**(6), 425–429 (1998)
107. Hirokawa, S.: An experimental study of the microstructures and mechanical properties of swine cruciate ligaments. JSME Int. J. **46**(4), 1417–1425 (2003)
108. Holzapfel, G.A.: Nonlinear Solid Mechanics. Wiley, Chichester (2000)
109. Huang, C.Y., Mow, V.C., Ateshian, G.A.: The role of flow-independent viscoelasticity in the biphasic tensile and compressive responses of articular cartilage. J. Biomech. Eng. **123**(5), 410–417 (2001)
110. Huang, C.Y., Soltz, M.A., Kopacz, M., Mow, V.C., Ateshian, G.A.: Experimental verification of the roles of intrinsic matrix viscoelasticity and tension-compression nonlinearity in the biphasic response of cartilage. J. Biomech. Eng. **125**(1), 84–93 (2003)
111. Hurschler, C., Loitz-Ramage, B., Vanderby Jr, R.: A structurally based stress-stretch relationship for tendon and ligament. J. Biomech. Eng. **119**(4), 392–399 (1997)
112. Hurschler, C., Provenzano, P.P., Vanderby Jr, R.: Scanning electron microscopic characterization of healing and normal rat ligament microstructure under slack and loaded conditions. Connect. Tissue Res. **44**(2), 59–68 (2003)
113. Ichiba, A., Nakajima, M., Fujita, A., Abe, M.: The effect of medial collateral ligament insufficiency on the reconstructed anterior cruciate ligament: a study in the rabbit. Acta Orthop. Scand. **74**(2), 196–200 (2003)
114. Ilic, M.Z., Carter, P., Tyndall, A., Dudhia, J., Handley, C.J.: Proteoglycans and catabolic products of proteoglycans present in ligament. Biochem. J. **385**(Pt 2), 381–388 (2005)
115. Inoue, M., McGurk-Burleson, E., Hollis, J.M., Woo, S.L.: Treatment of the medial collateral ligament injury. I: The importance of anterior cruciate ligament on the varus-valgus knee laxity. Am. J. Sports Med. **15**(1), 15–21 (1987)
116. Ishikawa, M., Komi, P.V., Grey, M.J., Lepola, V., Bruggemann, G.P.: Muscle-tendon interaction and elastic energy usage in human walking. J. Appl. Physiol. **99**(2), 603–608 (2005)
117. Jarvinen, T., Jarvinen, T.L., Kannus, P., Jozsa, L., Jarvinen, M.: Collagen fibres of the spontaneously ruptured human tendons display decreased thickness and crimp angle. J. Orthop. Res. **22**(6), 1303–1309 (2004)
118. Johnson, G.A., Tramaglini, D.M., Levine, R.E., Ohno, K., Choi, N.Y., Woo, S.L.: Tensile and viscoelastic properties of human patellar tendon. J. Orthop. Res. **12**(6), 796–803 (1994)
119. Jung, H.J., Fisher, M.B., Woo, S.L.: Role of biomechanics in the understanding of normal, injured, and healing ligaments and tendons. Sports Med. Arthroscopy Rehabil. Therapy Technol. (SMARTT) **1**(1), 9 (2009)
120. Kanit, T., Forest, S., Galliet, I., Mounoury, V., Jeulin, D.: Determination of the size of the representative volume element for random composites: statistical and numerical approach. Int. J. Solids Struct. **40**(13–14), 3647–3679 (2003)
121. Kannus, P.: Structure of the tendon connective tissue. Scand. J. Med. Sci. Sports **10**(6), 312–320 (2000)
122. Kanouté, P., Boso, D., Chaboche, J., Schrefler, B.: Multiscale methods for composites: a review. Arch. Comput. Methods Eng. **16**(1), 31–75 (2009)
123. Karas, V., Cole, B.J., Wang, V.M.: Role of biomechanics in rotator cuff pathology: North American perspective. Med. Sport Sci. **57**, 18–26 (2012)
124. Kastelic, J., Galeski, A., Baer, E.: The multicomposite structure of tendon. Connect. Tissue Res. **6**(1), 11–23 (1978)
125. Kjaer, M.: Role of extracellular matrix in adaptation of tendon and skeletal muscle to mechanical loading. Physiol. Rev. **84**(2), 649–698 (2004)

126. Komolafe, O.A., Doehring, T.C.: Fascicle-scale loading and failure behavior of the Achilles tendon. J. Biomech. Eng. **132**(2), 021004 (2010)
127. Kouznetsova, V., Brekelmans, W.A.M., Baaijens, F.P.T.: An approach to micro-macro modeling of heterogeneous materials. Comput. Mech. **27**(1), 37–48 (2001)
128. Kouznetsova, V., Geers, M.G.D., Brekelmans, W.A.M.: Multi-scale constitutive modelling of heterogeneous materials with a gradient-enhanced computational homogenization scheme. Int. J. Numer. Meth. Eng. **54**(8), 1235–1260 (2002)
129. Kouznetsova, V.G., Geers, M.G.D., Brekelmans, W.A.M.: Multi-scale second-order computational homogenization of multi-phase materials: a nested finite element solution strategy. Comput. Methods Appl. Mech. Eng. **193**(48–51), 5525–5550 (2004)
130. Labed, N., Turbe, N.: Computation of homogenized coefficients for a viscoelastic composite reinforced with spherical inclusions. J. Compos. Mater. **32**(14), 1297–1310 (1998)
131. Lafortune, M.A., Cavanagh, P.R., Sommer III, H.J., Kalenak, A.: Three-dimensional kinematics of the human knee during walking. J. Biomech. **25**, 347–357 (1992)
132. Lake, S.P., Barocas, V.H.: Mechanical and structural contribution of non-fibrillar matrix in uniaxial tension: a collagen-agarose Co-Gel model. Ann. Biomed. Eng. **39**, 1891–1903 (2011)
133. Lake, S.P., Miller, K.S., Elliott, D.M., Soslowsky, L.J.: Effect of fiber distribution and realignment on the nonlinear and inhomogeneous mechanical properties of human supraspinatus tendon under longitudinal tensile loading. J. Orthop. Res. **27**(12), 1596–1602 (2009)
134. Lakes, R.S.: Viscoelastic Materials. Cambridge University Press, New York (2009)
135. Lakes, R.S., Vanderby, R.: Interrelation of creep and relaxation: a modeling approach for ligament. J. Biomech. Eng. **121**(6), 612–615 (1999)
136. Lanir, Y.: Structure-strength relationships in mammalian tendon. Biophys. J. **24**, 541–554 (1978)
137. Lavagnino, M., Arnoczky, S.P., Kepich, E., Caballero, O., Haut, R.C.: A finite element model predicts the mechanotransduction response of tendon cells to cyclic tensile loading. Biomech. Model. Mechanobiol. **7**(5), 405–416 (2008)
138. Li, J., Li, H., Shi, L., Fok, A.S., Ucer, C., Devlin, H., Horner, K., Silikas, N.: A mathematical model for simulating the bone remodeling process under mechanical stimulus. Dent. Mater. **23**(9), 1073–1078 (2007)
139. Lichtwark, G.A., Wilson, A.M.: In vivo mechanical properties of the human Achilles tendon during one-legged hopping. J. Exp. Biol. **208**(Pt 24), 4715–4725 (2005)
140. Linder, L.H., Sukin, D.L., Burks, R.T., Haut, R.C.: Biomechanical and histological properties of the canine patellar tendon after removal of its medial third. Am. J. Sports Med. **22**(1), 136–142 (1994)
141. Liu, S.H., Panossian, V., al-Shaikh, R., Tomin, E., Shepherd, E., Finerman, G.A., Lane, J.M.: Morphology and matrix composition during early tendon to bone healing. Clin. Orthop. Relat. Res. **339**, 253–260 (1997)
142. Loitz-Ramage, B.J., Frank, C.B., Shrive, N.G.: Injury size affects long-term strength of the rabbit medial collateral ligament. Clin. Orthop. Relat. Res. **337**, 272–280 (1997)
143. Lujan, T.J., Dalton, M.S., Thompson, B.M., Ellis, B.J., Weiss, J.A.: Effect of ACL deficiency on MCL strains and joint kinematics. J. Biomech. Eng. **129**(3), 386–392 (2007)
144. Lujan, T.J., Underwood, C.J., Henninger, H.B., Thompson, B.M., Weiss, J.A.: Effect of dermatan sulfate glycosaminoglycans on the quasi-static material properties of the human medial collateral ligament. J. Orthop. Res. **25**(7), 894–903 (2007)
145. Lujan, T.J., Underwood, C.J., Jacobs, N.T., Weiss, J.A.: Contribution of glycosaminoglycans to viscoelastic tensile behavior of human ligament. J. Appl. Physiol. **106**(2), 423–431 (2009)
146. Lumens, J., Delhaas, T., Kirn, B., Arts, T.: Modeling ventricular interaction: a multiscale approach from sarcomere mechanics to cardiovascular system hemodynamics. Pac. Symp. Biocomput. **13**, 378–389 (2008)

147. Lynch, H.A., Johannessen, W., Wu, J.P., Jawa, A., Elliott, D.M.: Effect of fiber orientation and strain rate on the nonlinear uniaxial tensile material properties of tendon. J. Biomech. Eng. **125**(5), 726–731 (2003)
148. Maas, S.A., Ellis, B.J., Ateshian, G.A., Weiss, J.A.: FEBio: finite elements for biomechanics. J. Biomech. Eng. **134**(1), 011005 (2012)
149. Maceri, F., Marino, M., Vairo, G.: A unified multiscale mechanical model for soft collagenous tissues with regular fiber arrangement. J. Biomech. **43**(2), 355–363 (2010)
150. Maganaris, C.N., Narici, M.V., Almekinders, L.C., Maffulli, N.: Biomechanics and pathophysiology of overuse tendon injuries: ideas on insertional tendinopathy. Sports Med. **34**(14), 1005–1017 (2004)
151. Maganaris, C.N., Narici, M.V., Maffulli, N.: Biomechanics of the Achilles tendon. Disabil. Rehabil. **30**(20–22), 1542–1547 (2008)
152. Maganaris, C.N., Paul, J.P.: Tensile properties of the in vivo human gastrocnemius tendon. J. Biomech. **35**(12), 1639–1646 (2002)
153. Magnusson, S.P., Narici, M.V., Maganaris, C.N., Kjaer, M.: Human tendon behaviour and adaptation, in vivo. J. Physiol. **586**(1), 71–81 (2008)
154. Mak, A.F.: Unconfined compression of hydrated viscoelastic tissues: a biphasic poroviscoelastic analysis. Biorheology **23**(4), 371–383 (1986)
155. Malvankar, S., Khan, W.S.: Evolution of the Achilles tendon: the athlete's Achilles heel? Foot **21**(4), 193–197 (2011)
156. Marino M, Vairo G: Stress and strain localization in stretched collagenous tissues via a multiscale modelling approach. Comput. Methods Biomech. Biomed. Eng. 2012. doi:10.1080/10255842.2012.658043
157. Markolf, K.L., Mensch, J.S., Amstutz, H.C.: Stiffness and laxity of the knee—the contributions of the supporting structures. A quantitative in vitro study. J. Bone Joint Surg. (Am) **58**, 583–594 (1976)
158. Matsumoto, H., Suda, Y., Otani, T., Niki, Y., Seedhom, B.B., Fujikawa, K.: Roles of the anterior cruciate ligament and the medial collateral ligament in preventing valgus instability. J. Orthop. Sci. **6**(1), 28–32 (2001)
159. May-Newman, K., McCulloch, A.D.: Homogenization modeling for the mechanics of perfused myocardium. Prog. Biophys. Mol. Biol. **69**(2–3), 463–481 (1998)
160. Mazzocca, A.D., Nissen, C.W., Geary, M., Adams, D.J.: Valgus medial collateral ligament rupture causes concomitant loading and damage of the anterior cruciate ligament. J. Knee Surg. **16**(3), 148–151 (2003)
161. McKusick, V.A.: The defect in Marfan syndrome. Nature **352**(6333), 279–281 (1991)
162. Miyazaki, H., Kozaburo, H.: Tensile tests of collagen fibers obtained from the rabbit patellar tendon. Biomed. Microdevices **2**(2), 151–157 (1999)
163. Mommersteeg, T.J., Blankevoort, L., Kooloos, J.G., Hendriks, J.C., Kauer, J.M., Huiskes, R.: Nonuniform distribution of collagen density in human knee ligaments. J. Orthop. Res. **12**(2), 238–245 (1994)
164. Moore, S.M., Ellis, B., Weiss, J.A., McMahon, P.J., Debski, R.E.: The glenohumeral capsule should be evaluated as a sheet of fibrous tissue: a validated finite element model. Ann. Biomed. Eng. **38**(1), 66–76 (2010)
165. Müller, R., Rüegsegger, P.: Three-dimensional finite element modelling of non-invasively assessed trabecular bone structures. Med. Eng. Phys. **17**(2), 126–133 (1995)
166. Nakagawa, H., Mikawa, Y., Watanabe, R.: Elastin in the human posterior longitudinal ligament and spinal dura. A histologic and biochemical study. Spine **19**(19), 2164–2169 (1994)
167. Neame, P.J., Kay, C.J., McQuillan, D.J., Beales, M.P., Hassell, J.R.: Independent modulation of collagen fibrillogenesis by decorin and lumican. Cell. Mol. Life Sci. **57**(5), 859–863 (2000)
168. Niven, H., Baer, E., Hiltner, A.: Organization of collagen fibers in rat tail tendon at the optical microscope level. Coll. Relat. Res. **2**(2), 131–142 (1982)

169. Okada, J-i, Washio, T., Hisada, T.: Study of efficient homogenization algorithms for nonlinear problems. Comput. Mech. **46**(2), 247–258 (2010)
170. Ottani, V., Martini, D., Franchi, M., Ruggeri, A., Raspanti, M.: Hierarchical structures in fibrillar collagens. Micron **33**(7–8), 587–596 (2002)
171. Ottani, V., Raspanti, M., Ruggeri, A.: Collagen structure and functional implications. Micron **32**(3), 251–260 (2001)
172. Pahr, D.H., Zysset, P.K.: Influence of boundary conditions on computed apparent elastic properties of cancellous bone. Biomech. Model. Mechanobiol. **7**(6), 463–476 (2008)
173. Payne, R.C., Crompton, R.H., Isler, K., Savage, R., Vereecke, E.E., Gunther, M.M., Thorpe, S.K., D'Aout, K.: Morphological analysis of the hindlimb in apes and humans. I. Muscle architecture. J. Anat. **208**(6), 709–724 (2006)
174. Pena, E., Pena, J.A., Doblare, M.: On modelling nonlinear viscoelastic effects in ligaments. J. Biomech. **41**(12), 2659–2666 (2008)
175. Pins, G.D., Christiansen, D.L., Patel, R., Silver, F.H.: Self-assembly of collagen fibers. Influence of fibrillar alignment and decorin on mechanical properties. Biophys. J. **73**(4), 2164–2172 (1997)
176. Podshivalov, L., Fischer, A., Bar-Yoseph, P.Z.: 3D hierarchical geometric modeling and multiscale FE analysis as a base for individualized medical diagnosis of bone structure. Bone **48**(4), 693–703 (2011)
177. Pollock, C.M., Shadwick, R.E.: Allometry of muscle, tendon, and elastic energy storage capacity in mammals. Am. J. Physiol. **266**(3 Pt 2), R1022–R1031 (1994)
178. Pollock, C.M., Shadwick, R.E.: Relationship between body mass and biomechanical properties of limb tendons in adult mammals. Am. J. Physiol. **266**(3 Pt 2), R1016–R1021 (1994)
179. Provenzano, P., Lakes, R., Keenan, T., Vanderby Jr, R.: Nonlinear ligament viscoelasticity. Ann. Biomed. Eng. **29**(10), 908–914 (2001)
180. Provenzano, P.P., Lakes, R.S., Corr, D.T., Vanderby Jr, R.: Application of nonlinear viscoelastic models to describe ligament behavior. Biomech. Model. Mechanobiol. **1**(1), 45–57 (2002)
181. Puso, M.A., Weiss, J.A.: Finite element implementation of anisotropic quasi-linear viscoelasticity using a discrete spectrum approximation. J. Biomech. Eng. **120**(1), 62–70 (1998)
182. Puso, M.A., Weiss, J.A.: Finite element implementation of anisotropic quasi-linear viscoelasticity. J. Biomech. Eng. **120**(1), 62–70 (1998)
183. Quapp, K.M., Weiss, J.A.: Material characterization of human medial collateral ligament. J. Biomech. Eng. **120**, 757–763 (1998)
184. Race, A., Amis, A.A.. The mechanical properties of the two bundles of the human posterior cruciate ligament. J. Biomech. **27**(1), 13–24 (1994)
185. Rausch, M.K., Reese, S., Maas, S., Weiss, J.A.: Can poroelasticity predict the cyclic tensile viscoelastic behavior of ligament? In: Proceedings of the 55th Annual Meeting of the Orthopaedic Research Society (2009)
186. Reese, S., Weiss, J.: Measurement of Poisson's ratio and transverse strain in rat tail tendon during stress relaxation. In: Proceedings of the 56th Annual Meeting of the Orthopaedic Research Society, New Orleans, LA (2010)
187. Reese, S.P., Maas, S.A., Weiss, J.A.: Micromechanical models of helical superstructures in ligament and tendon fibers predict large Poisson's ratios. J. Biomech. **43**(7), 1394–1400 (2010)
188. Reese, S.P., Maas, S.A., Weiss, J.A.: A bottom-up approach to construction and validation of multiscale models for aligned collagenous tissues. In: Workshop on Microscale Modeling in Biomechanics and Mechanobiology (2011)
189. Reese, S.P., Weiss, J.A.: Multiscale micromechanical model of a collagen-based composite: development and validation. In: Multiscale Methods and Validation in Medicine and Biology (2012)

190. Rigozzi, S., Stemmer, A., Muller, R., Snedeker, J.G.: Mechanical response of individual collagen fibrils in loaded tendon as measured by atomic force microscopy. J. Struct. Biol. **176**(1), 9–15 (2011)
191. Robinson, P.S., Huang, T.F., Kazam, E., Iozzo, R.V., Birk, D.E., Soslowsky, L.J.: Influence of decorin and biglycan on mechanical properties of multiple tendons in knockout mice. J. Biomech. Eng. **127**(1), 181–185 (2005)
192. Rosenbloom, J.: Elastin: biosynthesis, structure, degradation, and role in disease processes. Connect. Tissue Res. **10**, 73–91 (1982)
193. Royce, P.M., Steinmann, B.: Connective Tissue and its Heritable Disorders: Molecular, Genetic, and Medical Aspects. Wiley, New York (2002)
194. Rumian, A.P., Wallace, A.L., Birch, H.L.: Tendons and ligaments are anatomically distinct but overlap in molecular and morphological features—a comparative study in an ovine model. J. Orthop. Res. **25**(4), 458–464 (2007)
195. Sander, E.A., Stylianopoulos, T., Tranquillo, R.T., Barocas, V.H.: Image-based multiscale modeling predicts tissue-level and network-level fiber reorganization in stretched cell-compacted collagen gels. Proc. Natl. Acad. Sci. U.S.A. **106**(42), 17675–17680 (2009)
196. Sander, E.A., Tranquillo, R.T., Barocas, V.H.: Image-based multiscale structural models of fibrous engineered tissues. In: Conference Proceedings: Annual International Conference of the IEEE Engineering in Medicine and Biology Society. IEEE Engineering in Medicine and Biology Society. Conference, 2009, pp. 4270–4272 (2009)
197. Sanz-Herrera, J.A., Garcia-Aznar, J.M., Doblare, M.: On scaffold designing for bone regeneration: a computational multiscale approach. Acta Biomater. **5**(1), 219–229 (2009)
198. Screen, H.R., Chhaya, V.H., Greenwald, S.E., Bader, D.L., Lee, D.A., Shelton, J.C.: The influence of swelling and matrix degradation on the microstructural integrity of tendon. Acta Biomater. **2**(5), 505–513 (2006)
199. Screen, H.R., Lee, D.A., Bader, D.L., Shelton, J.C.: An investigation into the effects of the hierarchical structure of tendon fascicles on micromechanical properties. Proc. Inst. Mech. Eng. [H] **218**(2), 109–119 (2004)
200. Screen, H.R., Shelton, J.C., Chhaya, V.H., Kayser, M.V., Bader, D.L., Lee, D.A.: The influence of noncollagenous matrix components on the micromechanical environment of tendon fascicles. Ann. Biomed. Eng. **33**(8), 1090–1099 (2005)
201. Screen, H.R.C., Bader, D.L., Lee, D.A., Shelton, J.C.: Local strain measurement within tendon. Strain **40**, 157–163 (2004)
202. Screen, H.R.C., Cheng, V.W.T.: The micro-structural strain response of tendon. J. Mater. Sci. **19**, 1–2 (2007)
203. Sengers, B.G., Van Donkelaar, C.C., Oomens, C.W., Baaijens, F.P.: The local matrix distribution and the functional development of tissue engineered cartilage, a finite element study. Ann. Biomed. Eng. **32**(12), 1718–1727 (2004)
204. Shadwick, R.E.: Elastic energy storage in tendons: mechanical differences related to function and age. J. Appl. Physiol. **68**(3), 1033–1040 (1990)
205. Sharafi, B., Ames, E.G., Holmes, J.W., Blemker, S.S.: Strains at the myotendinous junction predicted by a micromechanical model. J. Biomech. **44**(16), 2795–2801 (2011)
206. Smith, K.D., Vaughan-Thomas, A., Spiller, D.G., Innes, J.F., Clegg, P.D., Comerford, E.J.: The organisation of elastin and fibrillins 1 and 2 in the cruciate ligament complex. J. Anat. **218**(6), 600–607 (2011)
207. Speirs, D.C., de Souza Neto, E.A., Peric, D.: An approach to the mechanical constitutive modelling of arterial tissue based on homogenization and optimization. J. Biomech. **41**(12), 2673–2680 (2008)
208. Spencer, A.J.M.: Continuum Mechanics. Dover Publications, New York (1980)
209. Starborg, T., Lu, Y., Huffman, A., Holmes, D.F., Kadler, K.E.: Electron microscope 3D reconstruction of branched collagen fibrils in vivo. Scand. J. Med. Sci. Sports **19**(4), 547–552 (2009)
210. Sun, Y.L., Luo, Z.P., Fertala, A., An, K.N.: Direct quantification of the flexibility of type I collagen monomer. Biochem. Biophys. Res. Commun. **295**(2), 382–386 (2002)

211. Suquet, P.M.: Elements of homogenization theory for inelastic solid mechanics. In: Sanchez-Palencia, E., Zaoui, A. (eds.) Homogenization Techniques for Composite Media. Springer, Berlin (1985)
212. Svensson, R.B., Hassenkam, T., Hansen, P., Peter Magnusson, S.: Viscoelastic behavior of discrete human collagen fibrils. J. Mech. Behav. Biomed. Mater. **3**(1), 112–115 (2010)
213. Tawhai, M., Bischoff, J., Einstein, D., Erdemir, A., Guess, T., Reinbolt, J.: Multiscale modeling in computational biomechanics. IEEE Eng. Med. Biol. Mag. **28**(3), 41–49 (2009)
214. Thomopoulos, S., Hattersley, G., Rosen, V., Mertens, M., Galatz, L., Williams, G.R., Soslowsky, L.J.: The localized expression of extracellular matrix components in healing tendon insertion sites: an in situ hybridization study. J. Orthop. Res. **20**(3), 454–463 (2002)
215. Thomopoulos, S., Williams, G.R., Gimbel, J.A., Favata, M., Soslowsky, L.J.: Variation of biomechanical, structural, and compositional properties along the tendon to bone insertion site. J. Orthop. Res. **21**(3), 413–419 (2003)
216. Thomopoulos, S., Williams, G.R., Soslowsky, L.J.: Tendon to bone healing: differences in biomechanical, structural, and compositional properties due to a range of activity levels. J. Biomech. Eng. **125**(1), 106–113 (2003)
217. Thornton, G.M., Oliynyk, A., Frank, C.B., Shrive, N.G.: Ligament creep cannot be predicted from stress relaxation at low stress: a biomechanical study of the rabbit medial collateral ligament. J. Orthop. Res. **15**(5), 652–656 (1997)
218. Thornton, G.M., Oliynyk, A., Frank, C.B., Shrive, N.G.: Ligament creep cannot be predicted from stress relaxation at low stresses: a biomechanical study of the rabbit medial collateral ligament. J. Orthop. Res. **15**, 652–656 (1997)
219. Tsipouras, P., Ramirez, F.: Genetic disorders of collagen. J. Med. Genet. **24**(1), 2–8 (1987)
220. Uitto, J.: Biochemistry of the elastic fibers in normal connective tissues and its alterations in diseases. J Invest Dermatol **72**(1), 1–10 (1979)
221. Upton, M.L., Gilchrist, C.L., Guilak, F., Setton, L.A.: Transfer of macroscale tissue strain to microscale cell regions in the deformed meniscus. Biophys. J. **95**(4), 2116–2124 (2008)
222. van der Rijt, J.A., van der Werf, K.O., Bennink, M.L., Dijkstra, P.J., Feijen, J.: Micromechanical testing of individual collagen fibrils. Macromol. Biosci. **6**(9), 697–702 (2006)
223. Varga, P., Dall'ara, E., Pahr, D.H., Pretterklieber, M., Zysset, P.K.: Validation of an HR-pQCT-based homogenized finite element approach using mechanical testing of ultra-distal radius sections. Biomech. Model. Mechanobiol. **10**, 431–444 (2010)
224. Venturoni, M., Gutsmann, T., Fantner, G.E., Kindt, J.H., Hansma, P.K.: Investigations into the polymorphism of rat tail tendon fibrils using atomic force microscopy. Biochem. Biophys. Res. Commun. **303**(2), 508–513 (2003)
225. Vernerey, F.J., Kabiri, M.: An adaptive concurrent multiscale method for microstructured elastic solids. Comput. Methods Appl. Mech. Eng. **241–244**, 52–64 (2012)
226. Vidal, B.C.: Crimp as part of a helical structure. C R Acad. Sci. III **318**(2), 173–178 (1995)
227. Vidal, D.C.: Image analysis of tendon helical superstructure using interference and polarized light microscopy. Micron **34**(8), 423–432 (2003)
228. Waggett, A.D., Ralphs, J.R., Kwan, A.P., Woodnutt, D., Benjamin, M.: Characterization of collagens and proteoglycans at the insertion of the human Achilles tendon. Matrix Biol. **16**(8), 457–470 (1998)
229. Wang, C.J., Walker, P.S., Wolf, B.: The effects of flexion and rotation on the length patterns of the ligaments of the knee. J. Biomech. **6**, 587–596 (1973)
230. Wang, J.H.: Mechanobiology of tendon. J. Biomech. **39**(9), 1563–1582 (2006)
231. Weiss, J.A.: Behavior of human medial collateral ligament in unconfined compression. In: Othopaedic Research Society 46th Annual Meeting, ORS, Orlando, FL (2000)
232. Weiss, J.A.: A constitutive model and finite element representation for transversely isotropic soft tissues. Ph.D., University of Utah, Salt Lake City (1994)
233. Weiss, J.A., Gardiner, J.C.: Computational modeling of ligament mechanics. Crit. Rev. Biomed. Eng. **29**(3), 303–371 (2001)

234. Weiss, J.A., Gardiner, J.C., Bonifasi-Lista, C.: Ligament material behavior is nonlinear, viscoelastic and rate-independent under shear loading. J. Biomech. **35**, 943–950 (2002)
235. Weiss, J.A., Gardiner, J.C., Ellis, B.J., Lujan, T.J., Phatak, N.S.: Three-dimensional finite element modeling of ligaments: technical aspects. Med. Eng. Phys. **27**(10), 845–861 (2005)
236. Weiss, J.A., Maakestad, B.J.: Permeability of human medial collateral ligament in compression transverse to the collagen fiber direction. J. Biomech. **39**(2), 276–283 (2006)
237. Weiss, J.A., Maker, B.N., Govindjee, S.: Finite element implementation of incompressible, transversely isotropic hyperelasticity. Comp. Meth. Appl. Mech. Eng. **135**, 107–128 (1996)
238. Weiss, J.A., Woo, S.L.-Y., Ohland, K.J., Horibe, S., Newton, P.O.: Evaluation of a new injury model to study medial collateral ligament healing: Primary repair versus nonoperative treatment. J. Orthop. Res. **9**, 516–528 (1991)
239. Wenger, M.P., Bozec, L., Horton, M.A., Mesquida, P.: Mechanical properties of collagen fibrils. Biophys. J. **93**(4), 1255–1263 (2007)
240. Woo, S.L.-Y.: Biomechanics of tendons and ligaments. In: Schmid-Schonbein, G.W., Woo, S.L.-Y., Zweifach, B.W. (eds.) Frontiers in Biomechanics, pp. 180–195. Springer, New York (1986)
241. Woo, S.L.-Y.: Mechanical properties of tendons and ligaments I. Quasi-static and nonlinear viscoelastic properties. Biorheology **19**, 385–396 (1982)
242. Woo, S.L-Y., An, K.-N., Arnoczky, S.P., Wayne, J.S., Fithian, D.C., Myers, B.S.: Anatomy, biology, and biomechanics of tendon, ligament and meniscus. In: Simon, S. (ed.) Orthopaedic Basic Science. American Academy of Orthopaedic Surgeons, pp. 47–74 (1994)
243. Woo, S.L.-Y., Hollis, J.M., Adams, D.J., Lyon, R.M., Takai, S.: Tensile properties of the human femur-anterior cruciate ligament-tibia complex. The effects of specimen age and orientation. Am. J. Sports Med. **19**, 217–225 (1991)
244. Woo, S.L.-Y., Weiss, J.A., Gomez, M.A., Hawkins, D.A.: Measurement of changes in ligament tension with knee motion and skeletal maturation. J. Biomech. Eng. **112**(1), 46–51 (1990)
245. Woo SL-Y, Weiss JA, MacKenna DA, 1990, "Biomechanics and morphology of the medial collateral and anterior cruciate ligaments," Biomechanics of Diarthrodial Joints, Volume 1, V. a. R. Mow, A and Woo, SL-Y, ed., pp. 63-103
246. Wurgler-Hauri CC, Dourte LM, Baradet TC, Williams GR, Soslowsky LJ: Temporal expression of 8 growth factors in tendon-to-bone healing in a rat supraspinatus model. J. Shoulder Elbow Surg./Am. Shoulder Elbow Surg. [et al.], **16**(5 Suppl), S198–203 (2007)
247. Xia, Z., Zhou, C., Yong, Q., Wang, X.: On selection of repeated unit cell model and application of unified periodic boundary conditions in micro-mechanical analysis of composites. Int. J. Solids Struct. **43**(2), 266–278 (2006)
248. Yahia, L.H., Drouin, G.: Microscopical investigation of canine anterior cruciate ligament and patellar tendon: collagen fascicle morphology and architecture. J. Orthop. Res. **7**(2), 243–251 (1989)
249. Yamamoto, E., Hayashi, K., Yamamoto, N.: Effects of stress shielding on the transverse mechanical properties of rabbit patellar tendons. J. Biomech. Eng. **122**(6), 608–614 (2000)
250. Yamamoto, E., Hayashi, K., Yamamoto, N.: Mechanical properties of collagen fascicles from the rabbit patellar tendon. J. Biomech. Eng. **121**(1), 124–131 (1999)
251. Yang, L., van der Werf, K.O., Fitie, C.F., Bennink, M.L., Dijkstra, P.J., Feijen, J.: Mechanical properties of native and cross-linked type I collagen fibrils. Biophys. J. **94**(6), 2204–2211 (2008)
252. Yang, L., van der Werf, K.O., Koopman, B.F., Subramaniam, V., Bennink, M.L., Dijkstra, P.J., Feijen, J.: Micromechanical bending of single collagen fibrils using atomic force microscopy. J. Biomed. Mater. Res. A **82**(1), 160–168 (2007)
253. Yin, L., Elliott, D.M.: A biphasic and transversely isotropic mechanical model for tendon: application to mouse tail fascicles in uniaxial tension. J. Biomech. **37**(6), 907–916 (2004)
254. Yin, L., Elliott, D.M.: A homogenization model of the annulus fibrosus. J. Biomech. **38**(8), 1674–1684 (2005)

255. Yokota A, Gimbel JA, Williams GR, Soslowsky LJ: Supraspinatus tendon composition remains altered long after tendon detachment. J. Shoulder Elbow Surg./American Shoulder Elbow Surg. [et al.] **14**(1 Suppl S), 72S–78S (2005)
256. Yuan, Z., Fish, J.: Toward realization of computational homogenization in practice. Int. J. Numer. Meth. Eng. **73**(3), 361–380 (2008)
257. Yvonnet, J., Gonzalez, D., He, Q.C.: Numerically explicit potentials for the homogenization of nonlinear elastic heterogeneous materials. Comput. Methods Appl. Mech. Eng. **198**(33–36), 2723–2737 (2009)
258. Zajac, F.E.: How musculotendon architecture and joint geometry affect the capacity of muscles to move and exert force on objects: a review with application to arm and forearm tendon transfer design. J. Hand Surg. **17**(5), 799–804 (1992)
259. Zhang, G., Ezura, Y., Chervoneva, I., Robinson, P.S., Beason, D.P., Carine, E.T., Soslowsky, L.J., Iozzo, R.V., Birk, D.E.: Decorin regulates assembly of collagen fibrils and acquisition of biomechanical properties during tendon development. J. Cell. Biochem. **98**(6), 1436–1449 (2006)

Multiscale Modelling of Lymphatic Drainage

Tiina Roose and Gavin Tabor

Abstract In this chapter we will describe the latest developments in the area of lymphatic modelling. The lymphatic system is one of the key elements of the human circulation, serving the dual functions of draining interstitial fluid and returning this to the general blood circulation, together with processing this lymph fluid which is a key component of the body's immune response system. Compared to the main cardiovascular system however, remarkably little modelling has been attempted. At the same time, the distribution of pumping activity (contractile lymphangions coupled with simple valves) throughout the system, passive primary lymphatics and complex lymph nodes combining to form an active network, makes the system a prime candidate for multiscale modelling.

1 Introduction

Even though the existence of lymphatic vessels has been known since the seventeenth century, until very recently not much was known about their functioning and development. This was due to our failure to understand their importance in the proper functioning of tissues. However, in the last ten years, lymphatics have come to the forefront of biomedical research largely due to findings highlighting their importance to cancer growth and metastasis [48]. Thus there are now a large

T. Roose (✉)
Bioengineering Sciences Research Group, Faculty of Engineering and Environment, University of Southampton, Southampton, UK
e-mail: t.roose@soton.ac.uk

G. Tabor
College of Engineering, Maths and Physical Sciences, University of Exeter, Exeter, UK
e-mail: g.r.tabor@ex.ac.uk

Stud Mechanobiol Tissue Eng Biomater (2013) 14: 149–176
DOI: 10.1007/8415_2012_148

Published Online: 31 October 2012

number of experimental studies of molecular and micromechanical factors controlling lymphatics. Although these studies have greatly increased our knowledge of different aspects of the lymphatic system there is a need to integrate these studies to produce quantitative models of the functioning of the lymphatic system. In particular such models should aim to explain the effect of the lymphatics on the macroscale, i.e., organ and body scale, in terms of the system's microscopic and molecular properties. As well as providing a framework for interpreting experimental results and highlighting interesting avenues for new experimental studies, in the long term such modelling can lead to new ways of treating important medical conditions.

The fundamental role of the lymphatic system is to collect excess interstitial fluids, tissue waste products and plasma proteins from tissue and return them to the blood. Tissues require fluid and nutrients and these are supplied to them by blood vessels. After the fluid has performed its function it is reabsorbed either by the lymphatics or by postcapillary venules, with the dominant contribution in most cases being due to the lymphatics. Thus when the lymphatic system is unable to function the interstitial pressure becomes heightened and the tissues become swollen. In addition, as in most cases reabsorption drives the flow of fluids through tissues, failure of the lymphatic system results in much slower, mainly diffusive, movement of nutrients into tissues and the accumulation of waste products in the tissues. In order to perform its important task, the lymphatic system has evolved into an elaborate, highly branched, highly valved, unidirectional drainage network. We can distinguish three main parts of the lymphatic system; the primary lymph system, comprising passive microcapiliary ducts which drain the interstitial tissue and feed into the secondary lymphatic ducts or lymphangions, larger vessels with active, contractile walls and interspersed with simple valves, forming a distributed pumping system. Secondary lymph vessels combine to feed larger vessels draining whole areas of the body. Finally, lymph nodes are distributed through the system at intervals, whose function is to process the lymph fluid as it passes through. In addition to its primary task the lymphatic system also plays an important part in the immune system. In particular, lymphocytes that reside and multiply in the lymphatic system clean lymph fluid of bacteria and other contaminants. Thus diseases of the lymphatic system often result in compromised immune competence. The lymph nodes play a critical part in the body's response to infection, and in the spread of cancer and HIV.

Many medical conditions have now been linked to a malfunctioning of the lymphatic system, for example lymphedema, Melkersson-Rosenthal-Meischer syndrome, Kaposi sarcoma, lymphatic filariasis. Lymphatic filariasis is a parasitic disease that is thought to be globally the second leading cause of permanent and long-term disability [20]. In recent years, the lack of lymphatic function in solid tumours has been identified as one cause for hindered delivery of chemotherapeutic drugs to solid tumours [19]. Tumour metastasis is also thought to involve lymphatics as one, if not the primary, pathway [7, 47, 48].

2 Lymphatic Physiology

2.1 Primary Lymphatics

Primary lymphatics (initial lymphatics, prelymphatics or lymph capillaries) form typically hexagonal or pentahexagonal capillary plexuses which are covered by a layer of lymphatic endothelial cells. These cells are attached to the interstitium at one end whilst the other end is free to move, but overlaps the adjacent lymphatic endothelial cells (the end which is stuck to the intersititum). The suggestion is that this construction creates a series of miniature fluid valves which ensure that the fluid can move from the interstitium into the lymphatic lumen, but it cannot move from the lymphatic lumen back to the interstitium [31]. The assumption about there being a long thin channel between the overlapping endothelial cells is somewhat debatable since in most of the histological, microscopical, and anatomical representations of single primary lymphatic valves [23] there does not appear to be any indication of this. However, clearly there can be exceptions and possibly more thorough anatomical and experimental investigations into this issue are needed. Irrespective of the details, the primary lymphatic vessels act to drain the interstitial fluid and supply it to the larger, secondary lymphatics. Unlike the secondary lymphatics, the primary lymphatics have no pumping action and so this drainage must be a purely passive process (Fig. 1).

2.2 Secondary Lymphatics

The primary lymphatics feed lymph fluid into the next level, that of the secondary lymphatics (lymph vessels or collecting lymphatics). The secondary lymphatics are vessels with walls comprising smooth muscle, collagen and elastin fibres, lined by endothelial cells and bound to surrounding tissue with fibrous adventitia [1] (see Fig. 2). At intervals, simple valves occur, and the periodic contraction of the smooth muscle in the walls interacting with the valves produces a pumping action which transports the lymph fluid, despite any adverse pressure gradients, through the system. This pumping action is thought to be enhanced by the proximity of the secondary lymphatics to muscles, arteries and veins, which lead to a natural mechanical forcing on the walls [42]. The ultimate destination of the lymph fluid is the venous system, and as under normal conditions the fluid pressure in the veins is higher than the fluid pressure in most tissues it implies that the lymphatic system, i.e., the secondary lymphatics, can generate a reasonably large pumping force. However there are conditions in which the reverse is true, such as edema, or cases where limbs have been raised or subjected to compression for medical purposes.

The basic unit of a lymphatic vessel between two valves is referred to as the lymphangion. As the lymph vessel accumulates more and more lymph from the capillaries and connected lymphangions it swells. At intervals along the secondary

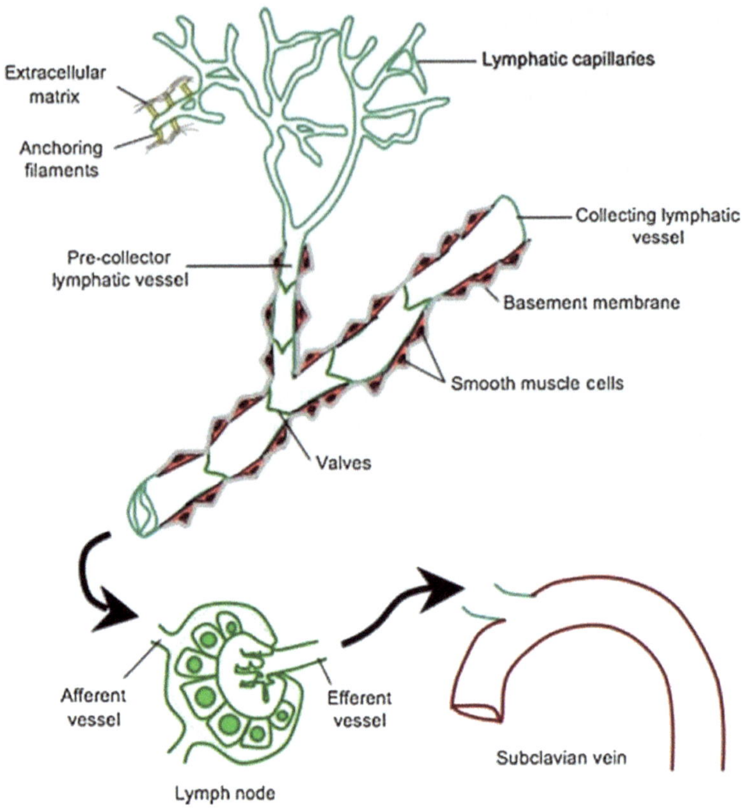

Fig. 1 Schematics showing lymphatic flows. Reproduced from [18]

lymphatic network reside complex structures known as lymph nodes that are fed by an upstream afferent lymph vessel. The lymph node itself can be basically divided into an outer cortex and an inner medulla. The surrounding fibrous capsule extends into the interior structure of the node forming trabeculae. Thin reticular fibres form an interior supporting network called the reticular network. Lymph fluid passing through the node experiences a degree of chemical and biological processing, in particular through interaction with B-cells and T-cells in the cortex, which represent a significant element of the body's immune response. However the variety of entities present in a lymph node makes it difficult to isolate the role played by each specific component in the overall behaviour of the node. Lymph fluid passing through the node drains into a single efferent lymph vessel, which itself may be afferent to another node further downstream. All lymph vessels ultimately drain into one of two lymph ducts; the right lymph duct which drains the upper right side of the trunk, neck and head together with the right arm; and the left or thoracic duct which drains the rest of the body. These ducts return lymph to the blood stream, emptying into the subclavian veins.

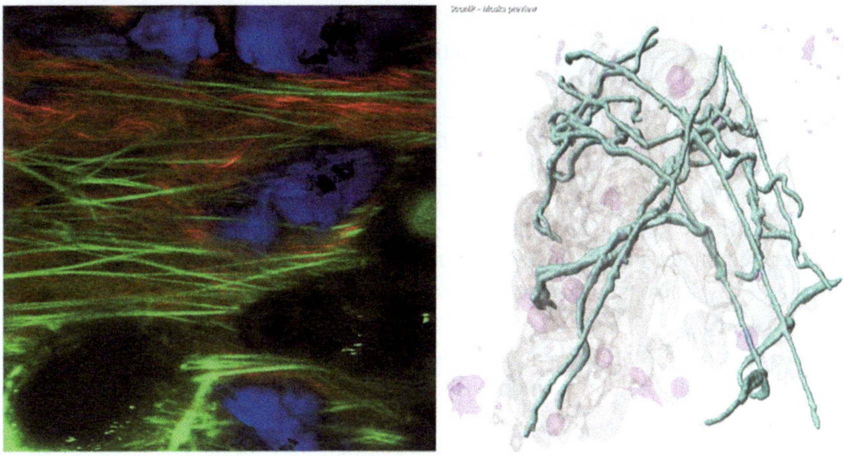

(a) (b)

Fig. 2 **a** Collecting lymph wall imaged using various two photon microscopy techniques superimposed to create a composite image [1]. Lymph wall cells are shown in *blue*, with elastin fibres (*green*) and collagen (*red*). **b** Data processed using image based meshing package ScanIP to generate 3D representation of structure, focussing on the elastin fibres. Image courtesy of Simpleware Ltd (www.simpleware.com)

3 Modelling Primary Lymphatics

3.1 Derivation of Macroscale Primary Lymphatic Fluid Flow

To describe the fluid flow in the primary lymphatic structure we first have to write down the fundamental equations describing these flows and then use multiple scale homogenisation to derive the tissue scale equations. This homogenisation procedure relies on the fact that the primary lymphatics have a very simple, repeating periodic hexagonal structure and because of the periodicity is very amenable to such analysis.

3.1.1 Fundamental Equations

We begin by describing the fluid flow inside the lymphatic capillary lumen. Our starting point is to use the full Navier–Stokes equations and supplement these with suitable boundary conditions that relate to the fluid drainage. Thus in the lymphatic capillary domain Ω^C we have

$$\rho[\partial_t \mathbf{u}^C + (\mathbf{u}^C \cdot \nabla)\mathbf{u}^C] = -\nabla p^C + \mu \nabla^2 \mathbf{u}^C, \quad \text{and} \quad \nabla \cdot \mathbf{u}^C = 0, \tag{1}$$

where ρ is the lymph fluid density and μ is the viscosity. From a consideration of the valve dynamics we get a relationship for the flux of fluid from the interstitium into the primary lymphatics and this relationship is usually a function of the pressure difference. Thus, as a lymphatic capillary surface $\partial\Omega$ boundary condition we take

$$\mathbf{n}\cdot\mathbf{u}^C = -\alpha(p^I - p^C) \quad \text{on the boundary,} \tag{2}$$

where p^I is the interstitial fluid pressure, $\mathbf{n}$ is the unit inward pointing normal to the capillary surface and α is the lymphatic capillary permeability, i.e., we include the primary lymphatic valve function via this effective capillary lumen permeability parameter. We also supplement this with a no-slip boundary condition on the lumen surface.

To describe the fluid flow in the interstitial space we use Darcy's law, i.e. we will assume that the fluid flux in the interstitium is related to the intersititial pressure gradient, i.e.,

$$\mathbf{u}^I = -\frac{k}{\mu}\nabla p^I, \tag{3}$$

where k is the hydraulic permeability (cm^2) and μ is the intersititial fluid viscosity. We will consider the viscosity of the interstitial fluid and lymphatic fluid to be the same. Combining (3) with the fluid conservation equation $\nabla\cdot\mathbf{u}^I = 0$ we get

$$\nabla^2 p^I = 0, \tag{4}$$

when k/μ is constant.

Clearly, on the lymphatic capillary lumen boundary the fluid flux out of the interstitium should be the same as fluid flux into the interstitium, i.e. on $\partial\Omega^C$ we have (looking from the interstitial side)

$$\mathbf{n}\cdot\mathbf{u}^C = \mathbf{n}\cdot\mathbf{u}^I = \mathbf{n}\cdot\left(-\frac{k}{\mu}\nabla p^I\right) = -\alpha(p^I - p^C). \tag{5}$$

3.1.2 Dimensionless Equations

Our next task in the homogenisation procedure is to non-dimensionalise the Navier–Stokes and Darcy equations (1) and (4) with suitable scales. We begin by defining the standard Navier–Stokes scalings, i.e.,

$$\mathbf{x}\sim d, \quad p^C\sim P, \quad p^I\sim P, \quad t\sim d/U \quad \text{and} \quad \mathbf{u}^C\sim U, \tag{6}$$

where d is the length scale of the typical single lymphatic structure (i.e. the so called unit cell scale), U is the typical fluid flow velocity in the lymphatic capillaries, and the classical pressure scale P is given by $P = \mu U/d$. Thus, the dimensionless Navier–Stokes equations are

$$Re[\partial_t \mathbf{u}^C + (\mathbf{u}^C \cdot \nabla)\mathbf{u}^C] = -\nabla p^C + \nabla^2 \mathbf{u}^C, \quad \nabla \cdot \mathbf{u}^C = 0, \tag{7}$$

where $Re = \rho dU/\mu$ is the Reynolds number based on the microscopic length scale. Clearly, the interstitial fluid flow equation (4) remains unchanged by the pressure scaling. The dimensional boundary condition (2) becomes the following dimensionless boundary condition

$$\mathbf{n} \cdot \mathbf{u}^C = -\widetilde{R}(p^I - p^C) \quad \text{on the boundary,} \tag{8}$$

where $\widetilde{R} = \alpha\mu/d$.

The microscale Reynolds number $Re = \rho dU/\mu$ is based on the microscopic (unit) length scale d. However, we expect the dominant fluid pressure variation within lymphatic capillaries to occur over the macroscale, and therefore it makes sense to rescale pressures with $1/\varepsilon$, where $\varepsilon = d/L$ is the ratio of the microscale (i.e. the unit cell scale) d to macroscale (i.e. tissue scale) L. After such rescaling the dimensionless Navier–Stokes equations become

$$\varepsilon Re[\partial_t \mathbf{u}^C + (\mathbf{u}^C \cdot \nabla)\mathbf{u}^C] = -\nabla p^C + \varepsilon\nabla^2 \mathbf{u}^C, \quad \nabla \cdot \mathbf{u}^C = 0, \tag{9}$$

with the boundary condition on the capillary surface $\partial\Omega^C$ given by

$$\mathbf{n} \cdot \mathbf{u}^C = -R(p^I - p^C) \quad \text{on the boundary,} \tag{10}$$

where $R = \widetilde{R}/\varepsilon = \alpha\mu L/d^2$ is the dimensionless modified lymphatic capillary permeability.

The dimensionless interstitial equations (4) and associated boundary condition (5) are

$$\nabla^2 p^I = 0, \tag{11}$$

$$\mathbf{n} \cdot \nabla p^I = \Psi(p^I - p^C) \quad \text{on the boundary,} \tag{12}$$

where $\Psi = \alpha\mu d/k$.

3.1.3 Parameter Values

There is a high level of uncertainty about some of the parameters, such as lymphatic wall permeability α, but the other parameters such as interstitial permeability and geometry are reasonably well known. We present estimates for all the parameters required for the model in Table 1.

Based on the values in Table 1 the values of the dimensionless parameters are as follows: $\varepsilon \sim 10^{-2}$, $Re \sim 10^{-4}$ to 10^{-3}, $R \sim 2 \times 10^{-6}$ and $\Psi \sim 0.5 \times (1 \text{ to } 10^{-3})$. Thus, since ε and Re are always small we can always neglect the inertial terms in the Navier–Stokes equations and use Stokes equations. Evidently also, the other parameters are comparable to ε, i.e., we can rewrite them as $R = \hat{R}\varepsilon^3$ and $\Psi = \hat{\Psi}\varepsilon^\beta$

Table 1 Parameter values for all primary lymphatic fluid flow parameters

Parameter	Value	Unit	Description	Reference
d	500	μm	Length of a single lymphatic structure	[6]
L	5	cm	Tissue length scale	–
k/μ	10^{-8}(normal) to 10^{-4}(tumour)	$\text{cm}^2\,\text{mmHg}^{-1}\,\text{s}^{-1}$	Interstitial hydraulic conductivity	[52]
μ	1.5	cP	Interstitial and lymph fluid viscosity	[52]
α	2×10^{-7}	$\text{cm mmHg}^{-1}\,\text{s}^{-1}$	Lymphatic capillary surface permeability	[53]
ρ	1	kg L^{-1}	Lymph fluid density	–
U	1–50	$\mu\text{m s}^{-1}$	Average lymph fluid flow velocity	[4, 43]

where $0 \leq \beta \leq 2$ and $\hat{R}$ and $\hat{\Psi}$ are both $O(1)$ quantities. $\beta = 2$ is the case which is the most relevant.

3.1.4 Multiscale Homogenisation

Multiple scale homogenisation is applicable to systems with structure on more than one characteristic length scale, i.e. in the case of lymphatics the length of an uninterrupted lymphatic vessel and the tissue length scale. The method is based upon the observation that the functions which represent physical variables in such systems tend to contain components which vary over the same length scales, i.e. in the case of lymphatics that some parts of the solution will vary significantly on the length scale of a section of lymphatic vessel whereas other parts of the solution will only vary significantly on the tissue scale. The method seeks solutions of this type and makes them easier to identify by inventing a new spatial variable $\mathbf{y}$ in addition to the true position variable $\mathbf{x}$ where rapidly varying functions are functions of $\mathbf{x}$ and slowly varying functions are functions of $\mathbf{y}$. Clearly there is only one spatial variable and thus $\mathbf{y}$ is a function of $\mathbf{x}$ and in general $\mathbf{y}$ is chosen to be given by $\mathbf{y} = \varepsilon\mathbf{x}$, where $\varepsilon = d/L$ is a number of order the ratio of the small d and large L length scales and is significantly less than 1. The use of the second spatial variable means that the spatial derivatives are given by $\nabla \rightarrow \nabla_x + \varepsilon\nabla_y$ and $\nabla^{\rightarrow}\nabla_x^2 + \varepsilon(\nabla_x \cdot \nabla_y + \nabla_y \cdot \nabla_x) + \varepsilon^2\nabla_y^2$. After introduction of this two space scale concept, a standard perturbation expansion solution in ε is sought for the problem. Before we proceed with this we will present the equations that reflect the two scales. The dimensionless equations used for the perturbation expansion are

$$\begin{aligned} -(\nabla_x + \varepsilon\nabla_y)p^C + \varepsilon(\nabla_x^2 + \varepsilon(\nabla_x \cdot \nabla_y + \nabla_y \cdot \nabla_x) + \varepsilon^2\nabla_y^2)\mathbf{u}^C &= 0, \\ (\nabla_x + \varepsilon\nabla_y) \cdot \mathbf{u}^C &= 0, \end{aligned} \tag{13}$$

$$(\nabla_x^2 + \varepsilon(\nabla_x \cdot \nabla_y + \nabla_y \cdot \nabla_x) + \varepsilon^2\nabla_y^2)p^I = 0, \tag{14}$$

with boundary conditions

$$\mathbf{n}\cdot\mathbf{u}^C = -\varepsilon^3\hat{R}(p^I + p^C) \quad \text{and} \quad \mathbf{n}\cdot(\nabla_x + \varepsilon\nabla_y)p^I = \varepsilon^2\hat{\Psi}(p^I - p^C) \tag{15}$$
on the boundary.

We use the perturbation expansions given by

$$\mathbf{u}^C = \mathbf{u}_0^C(\mathbf{x},\mathbf{y}) + \varepsilon\mathbf{u}_1^C(\mathbf{x},\mathbf{y}) + \varepsilon^2\mathbf{u}_2^C(\mathbf{x},\mathbf{y}) + \cdots, \tag{16}$$

$$p^C = p_0^C(\mathbf{x},\mathbf{y}) + \varepsilon p_1^C(\mathbf{x},\mathbf{y}) + \varepsilon^2 p_2^C(\mathbf{x},\mathbf{y}) + \cdots, \tag{17}$$

$$p^I = p_0^I(\mathbf{x},\mathbf{y}) + \varepsilon p_1^I(\mathbf{x},\mathbf{y}) + \varepsilon^2 p_2^I(\mathbf{x},\mathbf{y}) + \cdots. \tag{18}$$

Substituting Eqs. (16)–(18) into Eqs. (13)–(15) we find the first order, i.e., $O(\varepsilon^0)$, capillary flow equations to be

$$-\nabla_x p_0^C = 0, \quad \nabla_x\cdot\mathbf{u}_0^C = 0, \tag{19}$$

with $\mathbf{n}\cdot\mathbf{u}_0^C = 0$ on the boundary. This equation essentially says that $p_0^C = p_0^C(\mathbf{y})$, i.e., at the leading order the pressure depends only on the macroscopic space scale $\mathbf{y}$ and has no rapid local variations.

The first order interstitial flow equation is

$$\nabla_x^2 p_0^I = 0, \tag{20}$$

with $\mathbf{n}\cdot\nabla_x p_0^I = 0$ on the boundary and thus similarly to the capillary flow problem $p_0^I = p_0^I(\mathbf{y})$, i.e., the interstitial pressure is also only varying on the macroscale.

The $O(\varepsilon)$ capillary flow equation together with $O(1)$ continuity equation gives us two equations for $\boldsymbol{u}_0^C$ and p_1^C, i.e.,

$$-\nabla_x p_1^C - \nabla_y p_0^C + \nabla_x^2\mathbf{u}_0^C = 0, \quad \nabla_x\cdot\mathbf{u}_0^C = 0, \tag{21}$$

with $\mathbf{n}\cdot\mathbf{u}_0^C = 0$ on the boundary. The equations suggest that we can look for separable solutions, i.e.,

$$\mathbf{u}_0^C(\mathbf{x},\mathbf{y}) = -\xi^j(\mathbf{x})\frac{\partial p_0^C}{\partial y_j}, \quad p_1^C = -\pi^j(\mathbf{x})\frac{\partial p_0^C}{\partial y_j}, \tag{22}$$

where $\xi^j(\mathbf{x})$ and $\pi^j(\mathbf{x})$ are called local corrector functions that only depend on the microscale variable $\mathbf{x}$. The local corrector functions are defined/determined by the specific local solution that depends on the specific microstructure, i.e.,

$$\nabla_x\cdot\xi^j = 0, \quad \nabla_x\pi^j = \nabla_x^2\xi^j + \mathbf{e}_j, \tag{23}$$

with $\xi^j = \mathbf{0}$ on the internal microstructure surface, and all variables periodic on the external (on the unit square/rectangle) surfaces; $\mathbf{e}_j$ is the unit vector in the j coordinate direction. For any specific geometric situation one would in general need to solve these numerically, although for some simplified geometries (i.e. infinitely thin lymphatic capillaries etc.) analytic solutions may exist.

We are really interested in average velocity of fluid in the capillaries and thus defining the average fluid velocity over the volume of the unit cell as

$$\mathbf{U}^C = \frac{1}{V}\int_\Omega \mathbf{u}_0^C dV, \tag{24}$$

we find by integrating (22) that the average velocity in the capillaries is given by

$$\mathbf{U}^C = -\mathbf{K}\cdot\nabla_y p_0^C, \tag{25}$$

where

$$\mathbf{K} = \frac{1}{V}\int_\Omega \xi_i^j dV. \tag{26}$$

K is essentially the Darcy permeability that contains all the necessary information about the microstructure in it via the direct computation of ξ using equations (23) on specific microstructure. With modern computational packages, like Comsol Multiphysics, it is relatively straightforward and easy to determine ξ and subsequently **K**.

In the same spirit, substituting (22) into the $O(\varepsilon)$ continuity equation $\nabla_x\cdot\mathbf{u}_1^C + \nabla_y\cdot\mathbf{u}_0^C = 0$ and then integrating over the unit cell whilst taking into account the internal microstructure boundary condition $\mathbf{n}\cdot\mathbf{u}_1^C = 0$ and the periodic boundary conditions on the outside surfaces, we get the following macroscale conservation equation

$$\nabla_y\cdot\mathbf{U}^C = 0, \quad \text{i.e.} \quad \nabla_y\cdot(\mathbf{K}\cdot\nabla_y p_0^C) = 0, \tag{27}$$

which is the macroscale conservation equation for average capillary flow, i.e., macroscale capillary law is essentially described by the Darcy's law with permeability **K**. All the information about the microstructure is in **K**. In principle, for any specific geometric configuration **K** needs to be calculated independently, but it is crucial to point out that **K** does not depend on the macroscopic pressure variations, only on the local specific microstructure.

The $O(\varepsilon)$ interstitial flow equations are

$$\nabla_x^2 p_1^I + \nabla_x\cdot\nabla_y p_0^I + \nabla_y\cdot\nabla_x p_0^I = 0, \tag{28}$$

with $\mathbf{n}\cdot(\nabla_x p_1^I + \nabla_y p_0^I) = 0$ on the boundary. This suggests that the solution is again scale separable (i.e., function of **x** and **y** separately) and given by

$$p_1^I = -\omega_j(\mathbf{x})\frac{\partial p_0^I}{\partial y_j}, \tag{29}$$

where the local interstitial corrector function ω_j satisfies

$$\nabla_x^2 \omega_j = 0, \tag{30}$$

with $\mathbf{n} \cdot \nabla_x \omega_j = \mathbf{n} \cdot \mathbf{e}_j$ on the internal microstructure surfaces and periodic on the outer surfaces.

The order $O(\varepsilon^2)$ interstitial equations are

$$\nabla_x^2 p_2^I + \nabla_x \cdot \nabla_y p_1^I + \nabla_y \cdot \nabla_x p_1^I + \nabla_y^2 p_0^I = 0, \tag{31}$$

with boundary condition

$$\mathbf{n} \cdot (\nabla_x p_2^I + \nabla_y p_1^I) = \hat{\Psi}(p_0^I - p_0^C) \quad \text{on} \quad \partial\Omega. \tag{32}$$

As for the capillary flow equation, substituting (29) into the equation above and integrating over the whole interstitial domain we obtain

$$\nabla_y \cdot (\mathbf{E} \cdot \nabla_y p_0^I) = \frac{S\hat{\Psi}}{V}(p_0^I - p_0^C), \tag{33}$$

where S is the surface area of lymphatic capillaries within the unit cell, V is the volume of the unit cell, and hence S/V is the lymphatic surface area density in the tissue. The effective tissue interstitial permeability is given by

$$E_{ij} = \frac{V_I}{V}\delta_{ij} + \frac{1}{V}\int_{\partial\Omega} \omega_j n_i dS, \tag{34}$$

where V_I is the volume of interstitial space in the unit cell, V is the volume of the unit cell, and $\delta_{ij} = 1$ if $i = j$ and zero otherwise.

3.1.5 Summary of the Macroscale Equations for Primary Lymphatic Drainage

The capillary flow problem is described as

$$\nabla_y \cdot (\mathbf{K} \cdot \nabla_y p_0^C) = 0, \tag{35}$$

whilst the interstitital flow problem is described as

$$\nabla_y \cdot (\mathbf{E} \cdot \nabla_y p_0^I) = \frac{S\hat{\Psi}}{V}(p_0^I - p_0^C). \tag{36}$$

These equations can now be solved in a fast and efficient manner with suitable boundary conditions either analytically (1D equations) or numerically in higher dimensions.

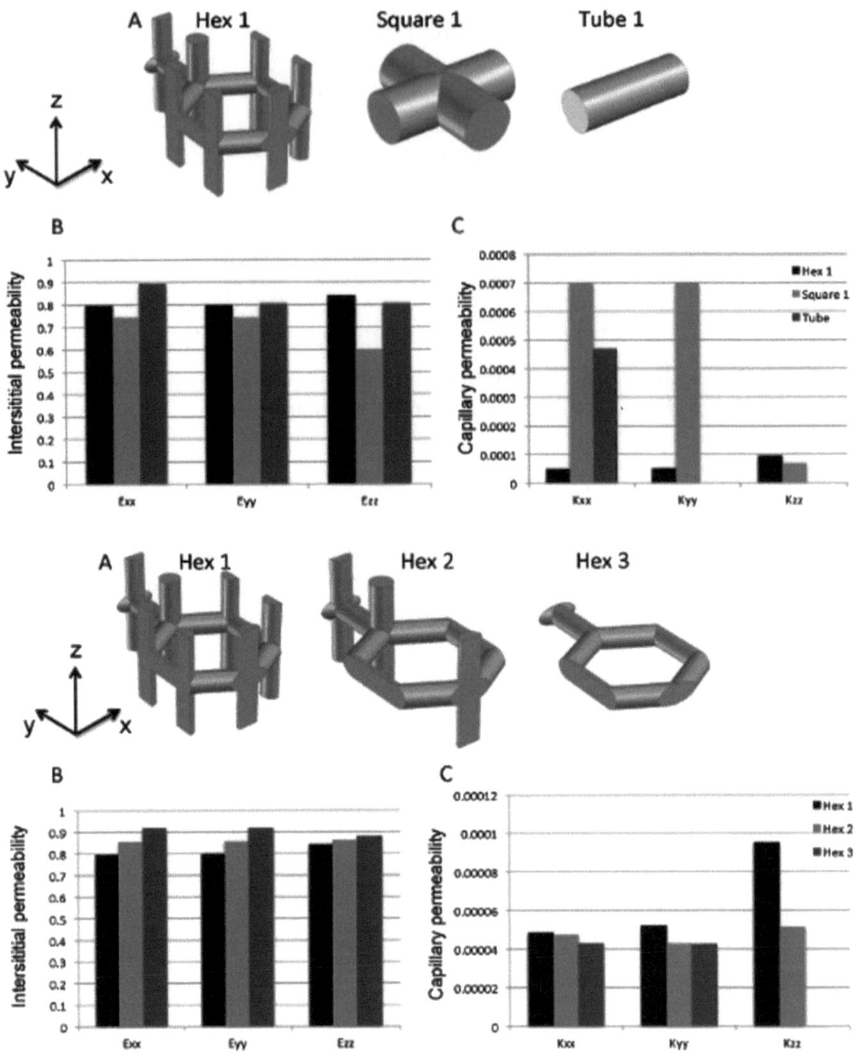

Fig. 3 Comparison of capillary and intestitial permeabilities calculated using Comsol Multiphysics. Reproduced from [40]

The cell problems that enable comparison between different microstructures need to be solved only once and in general using numerical packages such as Comsol Multiphysics, Abacus, Fluent etc. This was specifically done in [40] where different surface area matched (i.e. S/V constant) lyphatic capillary configurations were compared. The results are reproduced on Fig. 3.

4 Modelling Collecting Lymphatics

Moving downstream within the lymphatic system, primary lymphatics become larger and acquire valves along the lumen, morphing into the secondary or collecting lymphatic system. These vessels are somewhat larger than the primary lymphatics, with Reynolds numbers around $Re \sim 1$. Modelling of the individual lymphangions has typically taken one of two forms; development of zero dimensional or lumped-parameter models often based on a circuit analog, or one-dimensional models based on the Navier Stokes equations in 1D in a tube with contractile, elastic walls. Both approaches have required experimental input to determine specific parameters of the lymph vessels such as elastance. In a typical experiment [30] an appropriate lymphangion is excised from the body (e.g. bovine mesentry from an abbatoir), kept in a warmed bath of Krebs solution, and cannulated with fine glass tubes which can be connected to fluid reservoirs. This in vitro experiment can then be manipulated to measure physical parameters of the lymphangion, for example valve opening and closure pressures, elasticity and elastance of the walls, and even flow rates for different pumping effects. A typical lymphangeon and experimental setup is shown in Fig. 4. Models of individual lymphangions, together with simplistic valve models, can be (and have been) linked together to investigate the system properties of short chains of lymphangions, for instance to answer questions related to the degree of coordination of contractile behaviour between lymphangions. Higher dimensional modelling has not been widely investigated. Detailed modelling of other elements in the system, particularly detailed valve modelling or modelling of the fluid dynamics of the lymph nodes, has not been undertaken, although there are agent-based models of the immune system within lymph nodes which do take account of the drainage of fluid through the individual nodes. Accordingly in this section we will outline the modelling of individual and chains of lymphangions, whilst Sect. 5 will discuss other aspects of lymph system modelling.

4.1 Zero Dimensional Models

Flows in complicated hydraulic networks have often been represented by means of lumped-parameter components, where the details of flow through any particular section of duct are replaced by a simple relation linking pressure and volumetric flow rate. Such relations are functionally equivalent to electric networks, with pressure and volumetric flow rate being identified with electric voltage and current, whilst wall compliance, viscosity and inertia are equivalent to capacitance, resistance and inductance respectively. The network is divided into a number of sections (the number determined by intrinsic features of the network such as nodes and bifircations, or by resolution requirements). Lumped-parameter models have been extensively used in a variety of contexts including cardiovascular networks

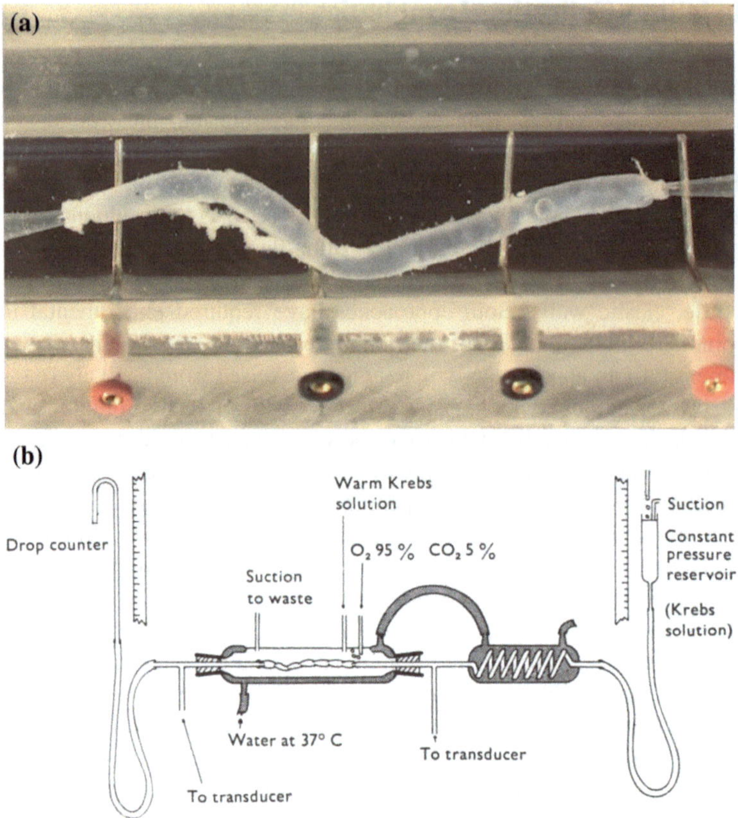

Fig. 4 **a** Cannulated lymph vessel in vitro. **b** Typical experimental setup for measuring flow and physical properties, reproduced from [30]

Table 2 Components of the zero dimensional lymphatic models

Fluid device		Electrical component	
Compliant vessel	$Q = C\frac{dp}{dt}$	$q = CV \Rightarrow \frac{dq}{dt} = I = C\frac{dV}{dt}$	Capacitor
Viscosity	$\Delta p = QR$	$V = IR$	Resistor
Inertia	$\Delta p = L\frac{dQ}{dt}$	$V = -L\frac{dI}{dt}$	Inductor

[45]; in the lymph network they have been applied by Lambert and Benoit[21] and extensively by Quick, Venugopal and colleagues [35, 56].

A simple deformable pipe can be modelled through a combination of the three individual elements detailed in Table 2; with the circuit capacitance representing the wall compliance, resistance representing the fluid viscosity and self-inductance representing fluid inertia in the system. There are numerous possible arrangements of components which can be used to model the deformable pipe in arbitrary detail; the simplest model is the Windkessel circuit shown in Fig. 5. This can be used to

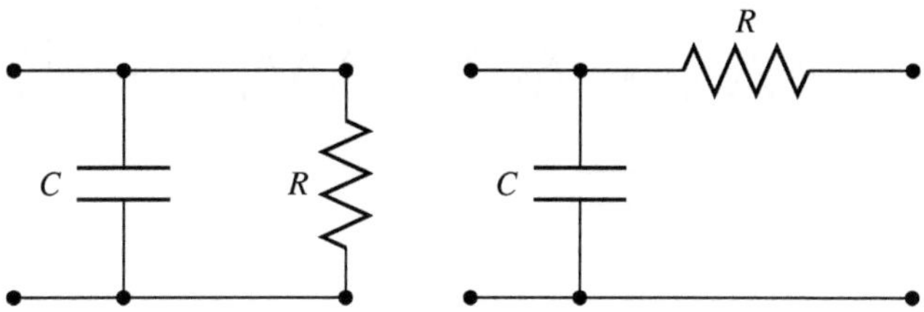

Fig. 5 Windkessel circuit. The left circuit represents a complete system modelled as a windkessel circuit; the right a single element in a chain

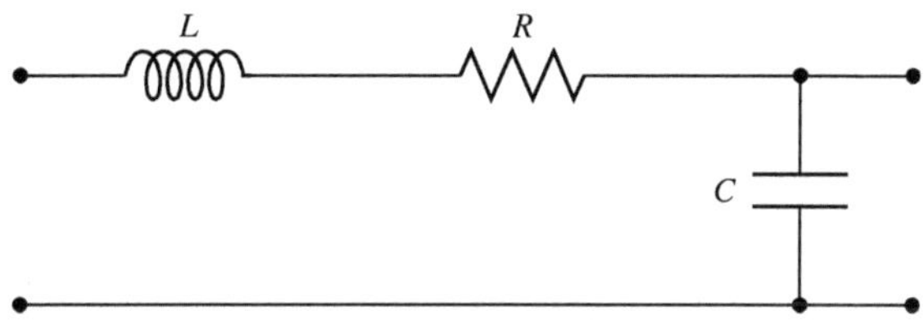

Fig. 6 Transmission line description of a blood vessel

represent an entire system (left figure) or an element in a chain (right figure). Fluid flowing into the system can flow out the other side, with associated pressure drop (the circuit resistance) or the vessel can distend, storing fluid (equivalent to the capacitor). Applying Kirchoff's laws to the circuit on the left, for instance, we can derive the equation

$$Q(t) = \frac{p(t)}{R} + C\frac{dp(t)}{dt}, \tag{37}$$

relating volumetric flow rate and pressure for the system. With only one possible time constant this circuit cannot represent all possible frequencies in a problem, leading to the development of more sophisticated circuits in order to improve the detailed modelling [45]. For instance, the transmission line model represented in Fig. 6 which includes the effect of fluid inertia via self-inductance L. All that remains with all of these circuits is to identify values or models for the terms R, C and so forth.

The earliest attempt to represent the lymphatic system in terms of an equivalent circuit was that of Drake et al. [9], as a simple resistor/diode arrangement. The value of R was determined from in vitro experimental measurements on a dog lung lymph vessel. This was cannulated and the volumetric flow rate measured for various back pressures with the slope of Q vs. p_{out} giving the resistance R. However this takes no account of the pumping action. In terms of this modelling, lymph vessels differ from arteries in one very important respect; whilst arterial walls are compliant they are also passive, so the response of the system can be modelled by finding suitable fixed values for L, C and R. Lymphangion walls

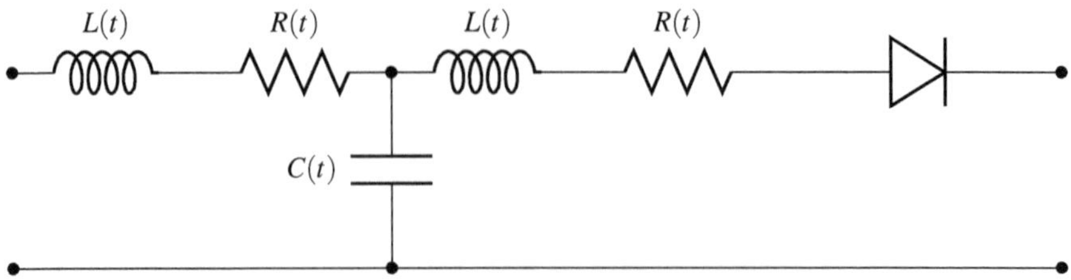

Fig. 7 Transmission line description of lymphangeon used by Quick, Venugopal and collaborators [35, 56]

are actively contractile, contributing to the pumping; and in fact the determination and modelling of this is a significant part of modelling the system as a whole.

Venugopal, Quick and collaborators [34, 35, 55–57] developed a more sophisticated lumped parameter representation of the lymph system using the circuit given in Fig. 7 with the parameters being time-varying. The circuit also includes a diode to simulate the non-return property of the valve. Representation of the pumping effect uses time-varying elastance theory, originally developed as a simple model of the heart [49–51]. The elastance of the pumping vessel (heart or lymphangion) is defined as

$$E(t) = \frac{p(t)}{V(t) - V_0}. \tag{38}$$

The chamber volume $V(t)$ varies as a result of the pressure $p(t)$; V_0 is the theoretical volume for zero pressure, or dead volume. If the chamber were passive this would be a constant as it inflated and deflated; however the physiological changes in the wall through the pumping cycle result in changes in the physical stiffness, i.e. changes in E. Experimental measurements on cannulated lymphangions were used to determine $p - V$ data through multiple cycles. As with other pumps (e.g. the heart), the $p - V$ data forms a loop, albiet in the case of the lymphatic pump, a somewhat eratic one [24]. Quick and collaborators developed an empirical model of the time-varying elastance from in vitro pressure and volume measurements on a pumping, cannulated valveless section of bovine mesenteric lymphangion [35]; the same paper also presented valve behaviour data for different pressure gradients. With the assumption of a cylindrical shape and Poiseuille's law, the time-varying resistance and self-inductance can also be calculated:

$$R(t) = \frac{\Delta p}{Q} = \frac{8\mu l}{\pi r^4}, \tag{39}$$

$$L(t) = \frac{\Delta p}{dQ/dt} = \frac{\rho l}{\pi r^2}. \tag{40}$$

Valve resistance is specified as a small fixed value for open valves ($\Delta p_{valve} > 0$) switching to infinite for a closed valve ($\Delta p_{valve} < 0$), whilst the inlet and outlet boundaries are simulated with additional fixed resistances [56]. The model has been validated against experimental results for a single lymphangion [35], and expanded to explore networks of four lymphangions [56] and branching networks [55].

This lumped-parameter approach has proved very successful, allowing investigation of individual pumping behaviour [35], for example demonstrating that whilst lymphatic pumping action is beneficial under normal (positive) pressure gradients, it is counterproductive for reverse (negative) gradients. Such reverse gradients occur for cases of edema, external compression or limb elevation, where the interstitial fluid pressure is artificially raised, leading to the permanent opening of the lymphangion valves and the lymphangion acting as a simple conduit [34]. It has also allowed the investigation of the effect of coordination of pumping between successive lymphangions in series [56]. Introducing a phase change into the time-periodic functions ($E(t)$, $R(t)$ etc.) allows the creation of contractile waves propagating along the series, both orthograde (in the direction of flow) and retrograde (opposing). The authors found this coordination of the contraction to have very little impact upon the pumping, and the orthograde and retrograde waves had similar effects, suggesting that individual lymphangions are able to function independently and thus adapt to local conditions as necessary. It has also proved possible to use the model to examine optimal structures of the lymph system [55]. The branching structure of the arterial system is well predicted by Murray's law [33]: in an optimal system, where an artery bifurcates, the cubes of the radii of the daughter vessels must sum to a constant value. This result is based on fundamental physical principles (minimum energy loss in the system) and has been well validated experimentally. Venugopal and collaborators demonstrate a similar effect for lymphatic networks, showing that a ratio of 1.26 for the upstream and downstream *lengths* optimises lymph flow for symmetric networks. The same authors have also investigated the effect of linear and non-linear contractile behaviour on the pumping effect [57], a significant issue in understanding the response of lymphangions to increasing transmural pressure, for example in response to edema.

4.2 One Dimensional Models

Lumped parameter models assume a uniform distribution of the dependent variables throughout the individual elements of the model at a particular instant in time. Although they are computationally efficient and so can be applied to large networks, and, as demonstrated, can produce very good results, the individual elements of the model are empirical and require fine tuning. A more fundamental approach is to directly solve the basic equations of fluid dynamics, the Navier–Stokes equations, for flow through the lymphangions. This has been carried out in 1D using purpose-built codes, discussed in this section, and could potentially be investigated further in 2D and 3D (see Sect. 3). These models allow for the spatial variation of the dependent variables; the main challenge is including the mechanics of the wall, its compliance and contractility. Experimental input to determine physical parameters such as elastic modulii as well as for model validation is still important, but the flow model itself is more directly derived.

The earliest attempt to model the collecting lymphatics was that of Reddy and co-workers [37, 38]. The Navier–Stokes equations can be reduced to the following form in cylindrical polar coordinates:

$$\frac{\partial u_x}{\partial t} + u_x \frac{\partial u_x}{\partial x} + u_r \frac{\partial u_x}{\partial r} = -\frac{1}{\rho}\frac{\partial}{\partial x}[p + \rho g h] + \mu\left[\frac{\partial^2 u_x}{\partial x^2} + \frac{1}{r}\frac{\partial}{\partial r}\left(r\frac{\partial u_x}{\partial r},\right)\right] \tag{41}$$

This represents the flow of fluid in a cylindrical tube oriented along the x-axis; with the possibility of including head differences ($\rho g h$ term) which will be irrelevant in laboratory in vitro studies but may be quite important for a chain of lymphangions across a significant section of the human body. Assuming no radial variation in the flow, i.e.

$$\frac{\partial p}{\partial r} = 0 \qquad u_r = 0, \tag{42}$$

and also assuming the flow to be low-Reynolds creeping flow;

$$u_x \frac{\partial u_x}{\partial x} = 0, \tag{43}$$

we can simplify (Eq. 41) as follows:

$$\frac{\partial u_x}{\partial t} = -\frac{1}{\rho}\frac{\partial}{\partial x}[p + \rho g h] + \frac{\nu}{r}\frac{\partial}{\partial r}\left(r\frac{\partial u_x}{\partial r}.\right) \tag{44}$$

Integrating this over the area of the tube ($\int \ldots 2\pi r\, dr$) gives

$$\frac{\partial Q}{\partial t} = -\frac{\pi a^2}{\rho}\frac{\partial}{\partial x}[p + \rho g h] + \frac{2\pi a}{\rho}\tau\bigg|_{r=a}, \tag{45}$$

where τ is the wall shear stress, a the tube radius and Q the volumetric flow. For Poiseuille flow it is straightforward to show that the shear stress at the wall is

$$\tau\bigg|_{r=a} = \frac{4\mu}{\pi a^3} Q, \tag{46}$$

(see eg. [11]), giving

$$\frac{\partial Q}{\partial t} = -\frac{\pi a^2}{\rho}\frac{\partial}{\partial x}[p + \rho g h] + \frac{8\mu}{\rho a^2} Q. \tag{47}$$

Finally, an additional term can be introduced to represent the effect of the valve, switching between finite and infinite resistance to represent the closed valve.

For a flexible tube (i.e. one whose radius can change) it is straightforward to show that

$$\frac{\partial A}{\partial t} + \frac{\partial}{\partial x}(u_x A) = 0, \tag{48}$$

represents the equation of continuity for an incompressible fluid [16, 44], where A is the cross sectional area. Rewriting this in terms of the radius ($A = \pi a^2$) and Q,

$$\frac{\partial a}{\partial t} = -\frac{1}{2\pi a}\frac{\partial Q}{\partial x}. \tag{49}$$

To close this system we need an expression for the pressure p. For an incompressible fluid within an elastic tube, the pressure, or rather the pressure difference between outside and inside, is dictated by the dilation of the wall. Reddy used a standard expression for a thin-walled tube for this:

$$p - p_{ext} = \frac{h}{a}\left(\Sigma_{hoop} + \Sigma_{act}\right), \tag{50}$$

where p_{ext} is the external imposed pressure, h the vessel wall thickness, a the radius, Σ_{hoop} the hoop stress (due to the elastic properties of the wall) and Σ_{ext} the additional stress due to the active contraction of the wall. This was represented as a combination of a simple sinusoidal function with a fixed period and a limiting contractile stress representing the minimum condition necessary to activate contraction.

With this, the equation set can now be discretised using standard finite difference techniques. Reddy and collaborators used upwind differencing and created a simple explicit algorithm with one grid point per lymphangion, essentially creating a lumped-parameter model based directly on the Navier–Stokes equations [37]. This was subsequently extended to model the complete left-side lymphatic network for the human body, i.e. the larger subsystem draining into the thoracic duct, including the effects of elevation in the system and including an external pressure term representing the effect of breathing [38]. The authors took care to validate their modelling as far as was possible by comparing average flow rates with known values, and were able to demonstrate behaviour in the theoretical model consistent with observed pumping behaviour in real lymphatics, but did observe somewhat intermittent and random flow and pressure behaviour.

MacDonald and collaborators [25, 26, 28] used the same basic framework to construct a more spatially refined model of an individual lymphangion and of short series of lymphangions [27], using 4–6 grid points to resolve the axial distribution of flow and wall properties within each lymphangion. In addition this work attempted to improve some of the detailed modelling, in particular by introducing a more complex model for the wall:

$$p - p_{ext} = \Phi\left(\frac{A}{A_0}\right) - \frac{T}{D_0}\frac{\partial^2 A}{\partial x^2} + \gamma\frac{\partial A}{\partial t}, \tag{51}$$

where A is the cross-sectional area, (A_0 the relaxed area), T the longitudinal wall tension and γ a damping factor. Φ represents the circumferential wall stress, which was expressed using the thick wall model [3] often used in arterial mechanics:

$$\Phi(\Delta a_{ext}) = E\Delta a_{ext}^2 \frac{(a_{ext}^2 - a_{int}^2)}{2(1-\sigma^2)a_{int}^2 a_{ext}}, \tag{52}$$

where a_{ext} and a_{int} are the external and internal radii, Δa_{ext} the change in external radius caused by the pressure difference Φ, and σ is Poisson's ratio. E is the Young's modulus for the wall material. To simulate the effect of pumping, either the natural (resting) radius or E can be changed; both approaches were investigated. The computational work was supported by experimentation to determine appropriate values for coefficients such as T and γ. Experimentation was also used to determine the pumping model for E,

$$E(t) = E_{relaxed} + (E_{contracted} - E_{relaxed})\xi(t), \tag{53}$$

and a similar model for the zero-pressure radius a_0; for the pumping function $\xi(t) \in [0, 1]$, two formulations were used, a simple sine wave with period t_p, and a shorter sine wave with pause at relaxation (total time period t_p). Phase differences between the pumping function at adjacent nodes enabled the simulation of orthograde and retrograde contractile waves. Results for these simple pumping functions were entirely consistent with experimentally observed results, generating a sawtooth output which matched observed radius variations. Pumping worked best with contraction being nearly simultaneous in all nodes within a lymphangion, and differences between orthograde and retrograde contractile waves were minor. For sufficiently high reverse gradients (i.e. edema), pumping merely served to reduce the flow rate, in agreement with the computational results of [56] reported earlier. Numerically the explicit algorithm placed a stringent constraint on the computational timestep which could be used, governed by the propagation of pressure/contractile waves in the system; the additional terms T and γ also acted to smooth the solution somewhat, producing less erratic results.

4.3 Higher Dimensions

Full solution of the Navier Stokes equations in 2D or 3D, except in very limited cases, is only really possible numerically, a fact which has led to the development of the subject of Computational Fluid Dynamics or CFD. CFD has had a major impact on biomechanics, in both cardovascular and respiratory [8] areas. Commonly-used is the Finite Volume method (FVM), in which the domain of interest is divided into a multitude of small *cells* or *control volumes*; the assembly of these is referred to as a *mesh*. The Navier–Stokes equations are integrated over each cell, and this, combined with application of Gauss' theorem to convert the transport term in the Navier–Stokes equation into fluxes across the cell faces, converts these partial differential equations into difference equations which can be solved, generally through an implicit, matrix-inversion-based algorithm. The big advantage of the finite volume method is that the cells can be any shape necessary, thus adapting to the needs of the geometry. Additional modelling may be introduced, e.g. turbulence, non-Newtonian flow, and particle tracking. Prewritten codes are available, both commercial (e.g. ANSYS Fluent, Star-CCM) and open source (e.g. OpenFOAM [58]). Alternative to the finite volume method is the functionally equivalent Finite Element method (FEM), in which the domain is represented by a collection of vertices connected by edges, and solution proceeds through numerical minimisation of a cost function. Again, preexisting codes are available (e.g. Comsol Multiphysics).

The major issue with CFD modelling of lymphatic vessels is that of treating the wall compliance. Flow of fluid in a duct which can deform is known as fluid-structure interaction (FSI). The FVM or FEM approaches can both handle this by permitting the motion of boundary patches and allowing the interior mesh structure to adapt accordingly. A full FSI calculation for a lymphangion would require the solution of the flow in the lymphangion and the resultant wall forces (pressures and shear stresses), together with a stress calculation in the lymphangion wall, with coupling acting in both directions. This is theoretically achievable but computationally costly [29, 46] and has not currently been attempted. However if we can determine the motion of the wall (and this may be more appropriate given its active state) the boundary motion may be imposed as a condition, simplifying the calculation considerably, Rahbar and Moore [36] have done this, creating an idealised model of a lymphangion using the commercial code StarCCM+, of length 1000 μm (contracting section 500 μm) and a variety of contractions (ranging from radii contracting from 60 to 40 μm to 120 contracting to 60 μm. Sinusoidal and skewed-sinusoidal wall motions were imposed with a period of 3.24 s, and steady and unsteady inlet velocity profiles imposed. The aim of the research was to ascertain whether the commonly-used assumption of Poiseuille flow, appropriate for steady flow in a straight, rigid tube, is appropriate for the case of flow in lymph vessels. In general, the authors found that it was, with discrepancies from strict Poiseuille flow of less than 4 % for wall shear stress, and parabolic velocity profiles strongly suggesting Poiseuille flow throughout the whole pumping cycle.

5 Other Modelling Related to Lymphatics

In addition to the two fluid flow modelling areas discussed above there are additional emerging fields that deal with modelling the detailed nature of how the lymphatic primary and secondary valves function and how the lymphatic system develops. We will discuss both very briefly below. In addition there are models that deal with the immune response within the lymph nodes [2, 17, 32], but since these do not usually have any fluid dynamics included in them we will not discuss them in this review.

5.1 Valve and Node Modelling

Valves are obviously an important element of the overall lymphatic system, but are even less well researched than the systems behaviour of fluid flows within lymphatics. Both lumped-parameter and Navier–Stokes based models described above use empirically based valve models based on experimental measurements of valve opening and closing pressure drops, and resistance values for the opened valve. Usually these models have exhibited simple diodic on/off behaviour, although Macdonald and collaborators [28] developed a more complicated model with ramping of the valve resistance between the two states. The situation is made more complex by there being at least two different types of valve in operation in the system. In the initial lymphatics, overlapping endothelial cells in the vessel wall form a primitive non-return valve for fluid transiting from the intrastitial space into the vessel [41, 54], often referred to as the primary valves; the vessels themselves however are non-pumping, being simple drainage ducts, and so do not possess internal valves. Mendoza and Schmid-Schönbein [31] attempted to develop a model for the primary valves by treating the overlapping cell as a flat flexible plate subject to a deflection $w(x)$ with governing equation

$$\frac{d^4 w(x)}{dx^4} = \frac{1}{EI}\Delta p(x), \tag{54}$$

with E as the elastic modulus and I the moment of inertia; the pressure drop across the plate $\Delta p(x)$ being related to Stokes flow through the gap, giving

$$\frac{dp}{dx} = -\frac{12Q\mu}{w(x)^3 T}, \tag{55}$$

for tissue thickness T. The equations were solved numerically using Mathematica. Results showed that, mechanistically, such simple structures could indeed operate as a valve, opening and closing in appropriate timescales and under appropriate pressures, and gave a likely estimate of the size of the opening.

Mendoza and Schmid-Schonbein's model [31] represents the situation where the valve leaflets are overlapping by a large distance, and hence the primary impedance for the lymphatic drainage is the movement of fluid in the narrow channel between two overlapping lymphatic endothelial cells. Galie and Spilker's model [12] is essentially similar to that of Mendoza and Schmid-Schonbein with the difference that they make the fluid dynamics around the valve more sophisticated. Both of these models also neglect the overall curvature of the lymphatic lumen.

Downstream in the network, as the vessels become larger and form the collecting lymphatic system, we find valve structures between individual lymphangions. However their exact structure is difficult to identify, being largely known from 2D micrographs such as that shown in Fig. 8. Two distinct structures have been proposed; a bicuspid valve analogous to the tricuspid shape of the heart valve

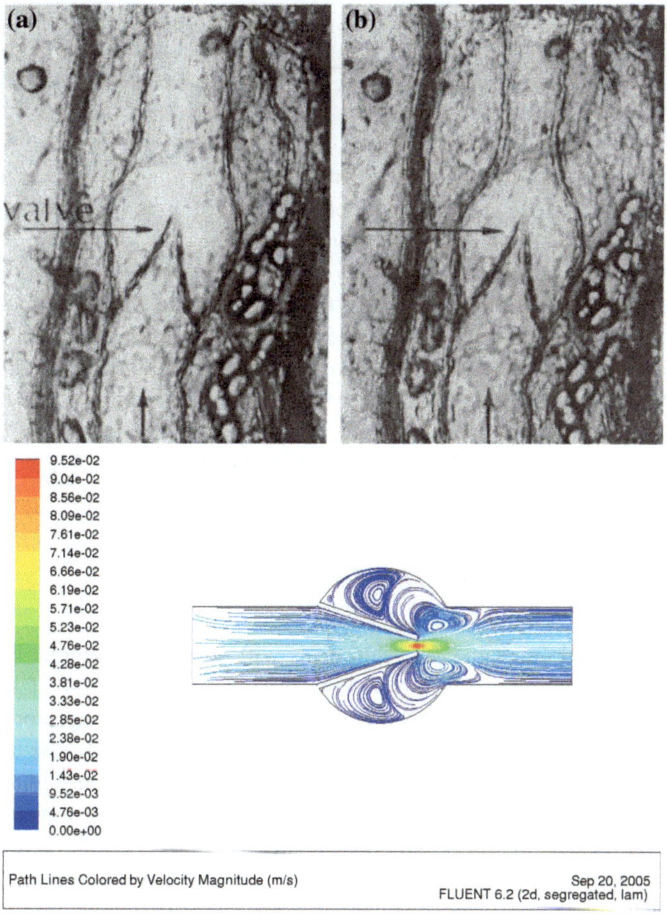

Fig. 8 *Top*: cat mesenteric valve (reproduced from [61]). *Bottom*: 2D simulation of opening lymphatic valve using Fluent

[13–15] or a funnel-like structure [22]. Meanwhile, recent work based on 3D reconstruction of images from confocal multiphoton microscopy suggests a bileaflet structure anchored at the walls [60]. Macdonald [27] attempted to model a valve as a 2D bileaflet with the commercial CFD code Fluent; the valve leaflets were represented by rigid structures with predefined motion, enabling the investigation of the response of the flow to the opening and closing of the leaflets. This, combined with experimental data on the valve behaviour, formed the basis for the empirical valve model used in [28]. Typical results are shown in Fig. 8, from which the flow resistance for various degrees of opening could be calculated.

Recently, Bertram et al. [5] have attempted to refine the valve modelling within a lumped-parameter chain model. This modelled the valve behaviour more smoothly than a simple on–off diode, with a resistance which ramped smoothly:

$$R_V = R_{V_{min}} + R_{V_{max}} \left(\frac{1}{1 + \exp(s_{open}(\Delta p - p_{open}))} + \frac{1}{1 + \exp(-s_{fail}(\Delta p - p_{fail}))} - 1 \right). \tag{56}$$

The two exponential terms here represent the opening resistance and valve failure, respectively. The authors developed a model of up to five lymphangions in a chain using Matlab; representative parameters for a small lymph vessel were used, and the effect of various conditions and coordination between lymphangions explored. Most significantly, this model enabled a more detailed exploration of the failure mode for adverse pressures; at the highest adverse pressures pumping was found to fail through one of two failure modes; through leakback through the valve or through the valve simply failing to open. Flow rates were also found to be sensitive to the conditions, with benefits to coordinated pumping and maximum flow rates and pressures sensitive to the contractile state.

5.2 Models of Lymphatic Development

There is a new and emerging area that deals with modelling lymphatic development. This is an undoubtedly an exciting area, but suffers from a scarcity of systematic experimental knowledge to build truly encompassing models. Thus, only two, very specialised models have been published [10, 39]. Roose and Fowler [39] developed a model to describe the fluid flow channelisation within collagen gels. This channelisation is thought to precede and guide the development of mature lymph vessels in mouse tail. The model essentially considered the collagen gel to be a multiphase rubber like material, and thus Flory-Huggins polymer theory was used for modelling. This resulted in a macroscopic model in the form of Cahn–Hilliard equation that was shown to exhibit experimentally consistent results of collagen patterning. However, clearly, not enough is known about all the processes to confirm all the model findings.

Lolas and Friedman [10] developed a model for lymphangiogenesis during solid tumour growth. The model consists of eight reaction-diffusion equations that include the effects of different growth factors (VEGF-C, plasmin etc.) on the development of lymphatic endothelial cell and cancer density, in addition to the effect of extracellular matrix.

All these models are clearly only first step in the longer research programs that are likely to grow in importance as the applied developmental questions related to lymphatics are addressed by, for example, tissue engineers who are looking to create artificial tissue scaffolds with not only functioning vasculature, but also with working lymphatics.

6 Conclusions and Challenges

The lymphatic system acts to drain and return interstitial fluids to the main circulatory system, pumping it from the low-pressure interstitial environment and returning it to the (usually higher pressure) venous system via the lymph ducts. This is accomplished by means of a complex structure of primary and secondary lymphatics and lymph nodes spanning three orders of magnitude in scale. In addition to its role in fluid transport, it provides a significant part of the body's defense against disease, particularly through the intermission of the lymph nodes. Medically, issues arise when compromised through surgery or infection, and it also plays a role in the spread of cancer through the body.

The range of scales involved indicates the necessity to invoke multiscale modelling; simulating individual components to investigate their behaviour, but also using this investigation to create simplified models which can act as building blocks for the next level up. In the case of the primary lymphatics, this has taken the form of multiple scale perturbation expansion and has had success in explaining the nature of fluid flow coupling between the interstitium and lymphatic capillaries.

In the case of the secondary lymphatics, two basic approaches have been followed; lumped-parameter modelling using electrical circuit analogies, and direct solution of the Navier–Stokes equations in 1D. Both have been highly successful, generating results which have been broadly in agreement both with each other and with experiment, and have been used to discover aspects of the system behaviour which would have been difficult to identify otherwise, such as coordination issues and the non-pumping behaviour for the case of edema. A more extensive multiscale approach is starting to evolve within this area with detailed modelling of valves beginning to inform the development of the network model. In all cases, a synergistic approach involving experiment and modelling has been successfully followed, partly as a pragmatic response to the lack of background information available about physical parameters of the system.

Further modelling of the primary lymphatic system within the context of detailed molecular factors influencing its development is clearly still needed. There are two preliminary models [10, 39], but many more detailed modelling studies performed as part of experimental investigations are still needed. In the secondary lymphatic system, advanced microscopy techniques [1, 60] should be more widely used to probe the structure of the secondary lymphatic vessel walls. Further CFD modelling can investigate the detailed flow through valves and lymphangions, possibly employing the techniques of Image Based Meshing to link with the microscopy [59]. Coordination and integration of the modelling with the primary lymphatics is also important, with the primary lymphatics providing input boundary conditions for the secondary lymphatic models. The final aim must be a complete, detailed model of the whole lymph system which can be used to investigate not just drainage issues but propagation of components such as malignant cancer cells through the system.

Acknowledgments Tiina Roose acknowledges the award of a Royal Society University Research Fellowship to fund her research.

References

1. Arkill, K.P., Moger, J., Winlove, C.P.: The structure and mechanical properties of collecting lymphatic vessels: an investigation using multimodal nonlinear microscopy. J. Anat. **216**, 547–555 (2010)
2. Baldazzi, V., Paci, P., Bernaschi, M., Castiglione, F.: Modelling lymphocyte homing and encounters in lymph nodes. BMC Bioinform. **10**, 387–398 (2009)
3. Bergel, D.H.: The static elastic properties of the arterial wall. J. Physiol. **156**, 445–457 (1961)
4. Berk, D.A., Swartz, M.A., Leu, A.J., Jain, R.K.: Transport in lymphatic capillaries. II. Microscopic velocity measurement with fluorescence photobleaching. Am. J. Physiol. Heart Circ. Physiol. **270**, H330–H337 (1996)
5. Bertram, C.D., Macaskill, C., Moore, J.E.: Simulation of a chain of collapsible contracting lymphangions with progressive valve closure. J. Biomech. Eng. **133**, 011008-1–011008-10 (2011)
6. Boardman, K.C., Swartz, M.A.: Interstitial flow as a guide for lymphangiogenesis. Circ. Res. **92**, 801–808 (2003)
7. Cassella, M., Skobe, M.: Lymphatic vessel activation in cancer. Ann. N. Y. Acad. Sci. **979**, 120–130 (2002)
8. Collins, T.P., Tabor, G.R., Young, P.G.: A computational fluid dynamics study of inspiratory flow in orotracheal geometries. Med. Biol. Eng. Comput. **45**(9), 829–836 (2007)
9. Drake, R.E., Allen, S.J., Katz, J., Gabel, J.C., Laine, G.A.: Equivalent circuit technique for lymph flow studies. Am. J. Physiol. Heart Circ. Physiol. **251**, H1090–H1094 (1986)
10. Friedman, A., Lolas, G.: Analysis of a mathematical model of tumor lymphangiogeneis. Math. Models Meth. Appl. Sci. **15**, 95–107 (2005)
11. Fung, Y.C.: Biomechanics: Circulation. 2nd edn. Springer, New York (1997)
12. Galie, P., Spilker, R.L.: A two-dimensional computational model of lymph transport across primary lymphatic valves. J. Biomech. Eng. **131**, 1297–1307 (2009)
13. Gnepp, D.R.: Lymphatics. In: Staub, N.C., Taylor, A.E. (eds) Edema, pp. 263–298. Raven Press, New York (1984)
14. Gnepp, D.R., Green, F.H.: Scanning electron microscopy of collecting lymphatic vessels and their comparison to arteries and veins. Scan. Electron Microsc. **3**, 756–762 (1979)
15. Gnepp, D.R., Green, F.H.Y.: Scanning electron-microscopic study of canine lymphatic vessels. Lymphology **13**(2), 91–99 (1980)
16. Grotberg, J.B., Jensen, O.E.: Biofluid mechanics in flexible tubes. Ann. Rev. Fluid Mech. **36**, 121–147 (2004)
17. Guo, Z., Sloot, P.M.A., Tay, J.C.: A hybrid agent-based approach for modeling microbiological systems. J. Theor. Biol. **255**, 163–175 (2008)
18. Hajjami, H.M.-E., Petrova, T.V.: Developmental and pathological lymphangiogenesis: from models to human disease. Histochem. Cell Biol. **130**, 1063–1078 (2008)
19. Jain, R.K.: Delivery of molecular and cellular medicine to solid tumors. Adv. Drug Deliv. Rev. **46**, 149–168 (2001)
20. Jussila, L., Alitalo, K.: Vascular growth factors and lymphangiogenesis. Physiol. Rev. **82**, 673–700 (2002)
21. Lambert, M.W., Benoit, J.N.: Mathematical model of intestinal lymph flow and lymphatic pumping. FASEB J. **6**(5), A2078 (1992)
22. Lauweryns, J.M.: Stereomicroscopic funnel-like architecture of pulmonary lymphatic valves. Lymphology **4**(4), 125–132 (1971)

23. Leak, L.V.: Electron microscopic observations on lymphatic capillaries and the structural components of the connective tissue-lymph interface. Microvasc. Res. **2**, 361–391 (1970)
24. Li, B., Silver, I., Szalai, J. P., Johnson, M. G.: Pressure–volume relationships in sheep mesenteric lymphatic vessels in situ: response to hypovolemia. Microvasc. Res. **56**, 127–138 (1998)
25. Macdonald, A., Tabor, G., Winlove, C.P., Arkill, K., McHale, N.: Computational and experimental analysis of lymphatic valves. J. Biomech. **39**(Suppliment 1), S295 (2006)
26. Macdonald, A., Tabor, G., Winlove, P., Arkill, K., McHale, N.: The fluid dynamics of lymphatic vessels. Poster presentation at Cardiovascular Haemodynamics and Modelling, 25th–27th September 2005, Edinburgh (2005)
27. Macdonald, A.J.: The fluid dynamics of lymphatic vessels. PhD thesis, University of Exeter (2007)
28. Macdonald, A.J., Arkill, K.P., Tabor, G.R., McHale, N.G., Winlove, C.P.: Modeling flow in collecting lymphatic vessels: one-dimensional flow through a series of contractile elements. Am. J. Physiol. Heart Circ. Physiol. **295**, H305–H313 (2008)
29. Marzo, A., Luo, X.Y., Bertram, C.D.: Three-dimensional collapse and steady flow in thick-walled flexible tubes. J. Fluids Struct. **20**, 817–835 (2005)
30. McHale, N.G., Roddie, I.C.: The effect of transmural pressure on pumping activity in isolated bovine lymphatic vessels. J. Physiol. **261**, 255–269 (1976)
31. Mendoza, E., Schmid-Schönbein, G.W.: A model for mechanics of primary lymphatic valves. J. Biomech. Eng. 125:407–414 (2003)
32. Mirski, H.P., Miller, M.J., Linderman, J.J., Kirschner, D.E.: Systems biology approaches for understanding cellular mechanisms of immunity in lymph nodes during infection. J. Theor. Biol. **287**, 160–170 (2011)
33. Murray, C.D.: The physiological principle of minimum work: 1. The vascular system and the cost of blood volume. Proc. Natl Acad. Sci. USA **12**(3), 207–214 (1926)
34. Quick, C.M., Ngo, B.L., Venugopal, A.M., Stewart, R.H.: Lymphatic pump-conduit duality: contraction of postnodal lymphatic vessels inhibits passive flow. Am. J. Physiol. Heart Circ. Physiol. **296**, H662–H668 (2009)
35. Quick, C.M., Venugopal, A.M., Gashev, A.A., Zawieja, D.C., Stewart, R.H.: Intrinsic pump-conduit behavior of lymphangions. Am. J. Physiol. Regul. Integr. Comp. Physiol. **292**, R1510–R1518 (2007)
36. Rahbar, E., Moore, J.E.: A model of a radially expanding and contracting lymphangeon. J. Biomech. **44**, 1001–1007 (2011)
37. Reddy, N.P., Krouskop, T.A., Newell, P.H.: Biomechanics of a lymphatic vessel. Blood Vessels **12**, 261–278 (1975)
38. Reddy, N.P., Krouskop, T.A., Newell, P.H.: A computer model of the lymphatic system. Comp. Biol. Med. **7**, 181–197 (1977)
39. Roose, T., Fowler, A.C.: Network development in biological gels: role in lymphatic vessel development. Bull. Math. Biol. 70(6), 1772–1789 (2008)
40. Roose, T., Swartz, M.A.: Multiscale modeling of lymphatic drainage from tissues using homogenization theory. J. Biomech. **45**, 107–115 (2012)
41. Schmid-Schönbein, G.: The second valve system in lymphatics. Lymphat. Res. Biol. **1**, 25–29 (2003)
42. Schmid-Schönbein, G.W.: Microlymphatics and lymph flow. Physiol. Rev. **70**(4), 987–1028 (1990)
43. Schmid-Schonbein, G.W.: Microlymphatics and lymph flow. Physiol. Rev. **70**, 987–1026 (1990)
44. Sherwin, S.J., Franke, V., Peiró, J., Parker, K.: One-dimensional modelling of a vascular network in space-time variables. J. Eng. Math. **47**, 217–250 (2003)
45. Shi, Y., Lawford, P., Hose, R.: Review of zero-D and 1-D models of blood flow in the cardiovascular system. Biomed. Eng. OnLine **10**, 33 (2011)
46. Shim, E.B., Kamm, R.D.: Numerical simulation of steady flow in a compliant tube or channel with tapered wall thickness. J. Fluids Struct. **16**(8), 1009–1027 (2002)

47. Skobe, M., Hawighorst, T., Jackson, D.G., Prevo, R., Janes, L., Velasco, P., Riccardi, L., Alitalo, K., Claffey, K., Detmar, M.: Induction of tumor lymphangiogenesis by VEGF-C promotes breast cancer metastasis. Nat. Med. **7**(2), 192–198 (2001)
48. Stacker, S.A., Achen, M.G., Jussila, L., Baldwin, M.E., Alitalo, K.: Lymphangiogenesis and cancer metastasis. Nat. Rev. Cancer **2**, 573–583 (2002)
49. Suga, H., Sagawa, K.: Mathematical interrelationship between instantaneous ventricular pressure–volume ratio and myocardial force–velocity relation. Ann. Biomed. Eng. **1**, 160–181 (1972)
50. Suga, H., Sagawa, K.: Instantaneous pressure–volume relations and their ratio in the excised, supported canine left ventricle. Circ. Res. **35**, 117–126 (1974)
51. Suga, H., Sagawa, K., Shoukas, A.A.: Load independence of the instantaneous pressure–volume ratio of the canine left ventricle and effects of epinephrine and heart rate on the ratio. Circ. Res. **32**, 314–322 (1973)
52. Swartz, M.A., Fleury, M.E.: The physiology of the lymph system. Ann. Rev. Biomed. Eng. **9**, 229–256 (2007)
53. Swartz, M.A., Kaipainen, A., Netti, P.A., Boucher, Y., Grodzinsky, A.J., Jain, R.K.: Mechanics of interstitial-lymphatic fluid transport: theoretical foundation and experimental validation. J. Biomech. **32**, 1297–1307 (1999)
54. Trzewik, J., Mallipattu, S.K., Artmann, G.M., Delano, F.A., Schmid-Schönbein, G.W.: Evidence for a second valve system in lymphatics: endothelial microvalves. FASEB J. **15**, 1711–1717 (2001)
55. Venugopal, A.M., Quick, C.M., Laine, G.A., Stewart, R.H.: Optimal postnodal lymphatic network structure that maximizes active propulsion of lymph. Am. J. Physiol. Heart Circ. Physiol. **296**, H303–H309 (2009)
56. Venugopal, A.M., Stewart, R.H., Laine, G.A., Dongaonkar, R.M., Quick, C.M.: Lymphangion coordination minimally affects mean flow in lymphatic vessels. Am. J. Physiol. Heart Circ. Physiol. **293**, H1183–H1189 (2007)
57. Venugopal, A.M., Stewart, R.H., Laine, G.A., Quick, C.M.: Nonlinear lymphangion pressure–volume relationship minimizes edema. Am. J. Physiol. Heart Circ. Physiol. **299**, H876–H882 (2010)
58. Weller, H.G., Tabor, G., Jasak, H., Fureby, C.: A tensorial approach to computational continuum mechanics using object orientated techniques. Comput. Phys. **12**(6), 620–631 (1998)
59. Young, P.G., Beresford-West, T.B.H., Coward, S.R.L., Notarberardino, B., Walker, B., Abdul-Aziz, A.: An efficient approach to converting 3D image data into highly accurate computational models. Phil. Trans. R. Soc. A **366**, 3155–3173 (2008)
60. Zawieja, D.C.: Contractile physiology of lymphatics. Lymphat. Res. Biol. **7**(2), 87–96 (2009)
61. Zweifach, B.W., Prather, J.W.: Micromanipulation of pressure in terminal lymphatics in the mesentery. Am. J. Physiol. **288**(5), 1326–1331 (1975)

Part II
Muscle, Nerve, Epithelium and Endothelium

A Model of Electromechanical Coupling in the Small Intestine

Peng Du, Jeelean Lim and Leo K. Cheng

Abstract The motility of the intestines is partly governed by a bioelectrical activity termed intestinal slow wave activity; however, the dynamics of the electromechanical relationship have remained poorly defined. With the recent advances in continuum-based multi-scale modeling techniques, we present a modeling framework to investigate the electromechanical coupling in a segment of small intestine. The overall modeling framework included three parts: (i) an anatomical model describing the geometry and makeup of the smooth muscle fibers, (ii) an electrical model describing the slow wave propagation; and (iii) a mechanical model describing the active and passive tension laws during contraction. The resultant intraluminal pressure was approximated using Lamé's thick-walled cylinder equation. This modeling framework demonstrates the potential to be used in investigating the effects of intestinal slow wave dysrhythmias on the motility of the small intestine, and may be extended in the future to incorporate additional regulatory factors and pathways.

1 Introduction

Digestion of food and absorption of nutrients are two important aspects to maintaining the normal physiological functions of the body. Under the normal circumstances, the food we eat is moved through a long and convoluted muscular tubular organ called the gastrointestinal tract. The key physiological function of the gastrointestinal tract is the uptake of nutrients and water that are necessary for

P. Du (✉) · J. Lim · L. K. Cheng
Auckland Bioengineering Institute, The University of Auckland,
Auckland, New Zealand
e-mail: peng.du@auckland.ac.nz

Stud Mechanobiol Tissue Eng Biomater (2013) 14: 179–207
DOI: 10.1007/8415_2012_153

Published Online: 29 September 2012

the maintenance of life. We now understand much more about the remarkable mechanisms of digestion and absorption, thanks to over a century of accumulative research of modern gastrointestinal physiology [51]. Furthermore, as the focus of modern healthcare is shifting toward preventative medicine, it is also becoming clear that maintaining a healthy digestive system is a key strategy in the management of increasingly prevalent illnesses in the developed and developing worlds alike, such as diabetes, cardiovascular diseases, and obesity.

Even though the general topology of the gastrointestinal tract is a continuous long tube, different parts of the tract are unique in both anatomy and function, in order to specialize in different aspect of digestion and/or absorption. For example, the stomach is where most of the digestion takes place, with aid of digestive enzymes and gastric contractions; whereas the small intestine is where most of the nutrient absorption occurs. Partially digested food, called chyme, is emptied into the small intestine at a controlled rate from the stomach, through a muscular valve called the pylorus, and the nutrients in the chyme are absorbed through the blood vessels embedded in the villi of the intestinal wall via passive diffusion, as the food travels down the intestine. The adult human small intestine is approximately 6 m in length and 30 mm in diameter [7]. In order to ensure that the chyme is continuously processed through the entire intestine, a series of motor patterns slowly transport the food content down the small intestine over a period of approximately 8 h [20]. The intestinal content then becomes more compacted in the colon, after which the content is evacuated from the body. The transit time of food through the intestine is an important clinical indicator of digestive health. Abnormally quick transit through the intestine could result in diarrhea, which in severe and/or chronic cases can lead to malnourishment and dehydration. Conversely, abnormally slow transit could result in constipation, which is often accompanied with bloating, pain, and a general reduction in quality of life [22].

Since the advent of modern medical imaging techniques such as the X-ray, ultrasound, and magnetic resonance imaging, the motor patterns in the small intestine have been classified into a variety of modes depending on the physiological state of the body. This in turn is governed by a cohort of biological systems, such as the myogenic, neurogenic, and hormonal systems [51]. Out of the many types of motor patterns in the intestine, the most studied motor pattern is a series of contraction and relaxation movements known as peristalsis. Generally, the term "peristaltic contraction" refers to any constriction which travels along the intestine, while the "peristaltic reflex" is a subset of peristaltic contractions, and describes the neurally-mediated contractile reflex of the intestine to a food bolus [31].

Peristalsis is partly mediated by a propagating bioelectrical activity in the small intestine called the intestinal slow wave activity. In 1914, intestinal motor patterns were first studied in rabbits [3], and eight years later in 1922, slow wave activity was recorded for the first time from the intestines in a number of species [4]. Subsequent studies have established that the slow wave activity is actively generated by a special class of pacemaker cells known as the interstitial cells of Cajal. The slow wave activity then passively conducts to the surrounding smooth muscle cells, driving their motility. The interstitial cells of Cajal were in fact first described by the Spanish

Nobel Prize recipient, Santiago Ramón y Cajal, in 1911 [11]. However, it took decades before the pacemaking role of the interstitial cells of Cajal in electromechanical coupling activity in the gastrointestinal tract was established [21]. The clinical significance of interstitial cells of Cajal has been demonstrated by multiple studies showing a strong correlation between loss of the cells and intestinal motility disorders such as chronic intestinal pseudo-obstruction [22]. In the muscle layers of the small intestine, the smooth muscle cells are arranged in two sheets, defined as the circular and longitudinal layers, which refer to the principal direction of alignment of the individual muscle cells in each layer [9]. Since two of the components responsible for contraction, the actin and myosin filaments, are aligned roughly parallel to the longitudinal axis of the cell body, the alignment of the cells within the muscle layer ensures coordinated contraction and efficient force generation in the principal direction of the layer [9].

Peristaltic motor patterns involve contraction of both the circular and longitudinal muscle layers in the intestine, accomplished by voltage-mediated calcium entry into the smooth muscle cells and calcium release from intracellular stores [31]. In addition, the electromechanical sensitivity of the smooth muscle cells is further influenced by the resting membrane potential which varies across localised regions as well as across the intestinal wall [52], and changes in the amount of depolarization required to activate peristalsis in different segments of intestine. This intrinsic control of excitability also plays an important role in the formation of many other specific motor patterns in different physiological conditions and at different locations along the intestine.

In summary, there are multiple co-regulatory pathways imposing on the motility of the small intestine. These pathways integrate to generate the complex contractile patterns in response to slow wave activity in the small intestine. In this chapter, we focus on the electromechanical coupling between the intestinal slow wave activity and peristalsis. However, even with the current level of understanding we have of the underlying mechanisms of peristalsis, it is still difficult to evaluate this complicated response in vivo. This is where a mathematical modeling framework can offer a highly attractive *in silico* environment to understand peristalsis. Here, we present the underlying theories with simple examples of a modeling framework that can be used to investigate the electromechanical coupling in the small intestine. The framework offers the flexibility to modify or add new components as more information is established. Furthermore, the output variables of the electromechanical model are in physically meaningful quantities that can be validated experimentally. This additionally enables the model can be used to evaluate current data, and to help generate new hypotheses for subsequent experimental testing.

2 Geometric Modeling

The sophistication of multi-scale modeling work discussed in this chapter partly depends on the anatomy of the small intestine. It is therefore important to first incorporate an appropriate representation of the anatomy of intestinal tissue.

Generally, the smooth geometry of biological tissue has been represented by defining geometric finite element mesh with appropriate basis function classes. The concept of basis functions is used to discretize the physical geometric domain into multiple geometric elements that can be iteratively fitted to the shape of one part of the organ using the finite element method. For example, the geometric field value in a one-dimensional geometric field can be interpolated by a linear Lagrange basis function over a local coordinate system denoted by ξ, as follows,

$$u(\xi) = (1 - \xi)u_1 + \xi u_2, \tag{1}$$

where the field value, u is varied linearly between the nodal values (u_1 at $\xi = 0$; u_2 at $\xi = 1$). Higher order geometric basis functions such as quadratic or cubic functions can also be used to represent the geometric field. Even though the Lagrange basis functions provide continuity of u across elemental boundaries, they do not maintain higher order continuity, i.e., the derivatives on the curvature in terms of geometry, are not continuous, and therefore the smoothness of the biological tissue is not represented accurately using the Lagrange basis function. One approach that has been adopted is to define these additional derivative parameters, such as $\frac{\partial u}{\partial \xi}$, and ensure that both the field value and the derivative are continuous across the elemental boundaries, e.g., in the form of cubic Hermite basis functions as follows,

$$u(\xi) = u_1 + u_1'\xi + (3u_2 - 3u_1 - 2u_1' - u_2')\xi^2 + (u_1' + u_2' + 2u_1 - 2u_2)\xi^3, \tag{2}$$

where u_1, u_1' and u_2, u_2' are the nodal values and derivatives, respectively. A previous study has adopted the cubic Hermite basis functions to track the center-line of the small intestine based on imaging evidence from the Visible Human Project [37]; the wall of the intestinal model was then projected outward with a constant radius to create an intestinal lumen space (Fig. 1a).

Even though the model constructed from the Visible Human Project successfully generated the gross anatomy of the human small intestine, a more detailed representation of the intestinal micro-structure, in particular the muscle layers, would be required for a more realistic electromechanical model. Morphological studies of segments of small intestine have found that the radius and wall thickness are fairly uniform within each segment. Therefore, a simple hollow cylindrical tube can be used to represent a relatively short segment of the small intestine, in order to reduce the computation required to apply the anatomical model in an electromechanical modeling framework.

As the thickness and radial dimensions of intestines vary considerably between species. We chose to focus on a 40 mm segment of a rat small intestine as the rat species has been used in a number of experimental studies. Based on imaging and morphometric evidence [1, 19], the rat intestine has an external radius of 2.3 mm and a thickness of 0.9 mm. Figure 1b shows an idealized geometric model of the small intestine, defined by 216 nodes and 128 geometric elements with a combination of different types of basis functions. In a three-dimensional space, the

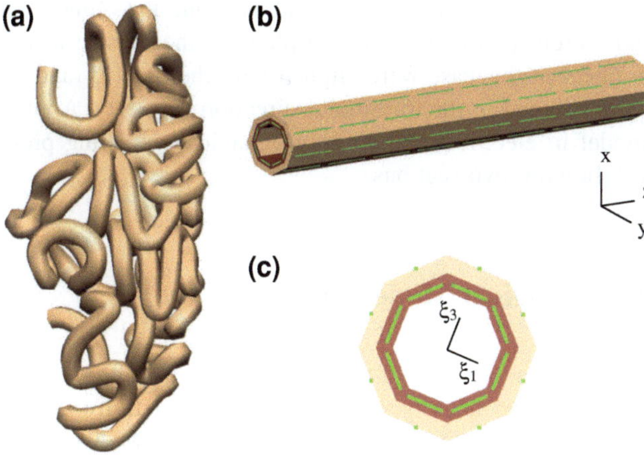

Fig. 1 Anatomical models of the small intestine. **a** An anatomical model constructed from the Visible Human Project using the cubic Hermite basis functions [37]. **b** A more structurally detailed linear model with idealized representation of the inner circular (*brown*) and outer longitudinal muscle layers (*light brown*). The green lines represent the orientations of the smooth muscle fibers in each geometric element. **c** A end-view of the linear intestinal anatomical model

idealized intestinal model was represented using the Cartesian coordinates (x, y, z) as the global coordinate system. Alternative coordinate systems such as the polar cylindrical coordinate system could also be used in this case. A local coordinate system (ξ_1, ξ_2, ξ_3) is usually used in the finite element method to represent the geometry within each local element. The local geometric coordinates were interpolated over the element using linear Lagrange basis functions in the circumferential (ξ_1) and transmural (ξ_3) directions, and a cubic Hermite basis function in the longitudinal (ξ_2) direction. This basis functions scheme represented a balance between computational efficiency and accuracy, since the physical dimension in the transmural and circumferential directions was smaller than the geometry in the longitudinal direction.

The wall of the intestine anatomical model was further divided into two smooth muscle layers of equal thickness, to represent the inner circular and outer longitudinal muscle layers (Fig. 1b). To represent the different alignment of the muscle fibers in the longitudinal and circular directions, a fiber coordinate system was prescribed at each node, and subsequently interpolated over the whole element, in this case using a trilinear basis function scheme. Note the other anatomical layer of the small intestine, e.g., the submucosal layer, was not explicitly defined in this initial model, because they do not contribute significantly to the active mechanical properties of the intestine.

Anisotropic material properties, such as electrical conductivities and the parameters for the constitutive equations, were specified along material axes which are referred to the fiber directions. These axes specify three directions: (i) 'fiber', which was aligned with the direction of the fibers; (ii) "sheet", which was perpendicular to the muscle fiber, but parallel to the plane of the muscle layer; and

(iii) "sheet-normal", which was orthogonal to both the "fiber" and "sheet" directions, and therefore orthogonal to the plane of the muscle layer. The orientations of the fibers in this case were aligned with the local ξ_1 and ξ_2 directions, i.e., no fiber was aligned in the transmural direction (Fig. 1c). Application of the anatomical model in electrical and electromechanical modeling process are discussed in the following two sections.

3 Electrophysiological Modeling

This section presents an overview of the mathematical models we used to simulate the intestinal slow wave activity, i.e., the electrical activity generated by the interstitial cells of Cajal and smooth muscle cells. The slow wave activity was then integrated in a continuum setting, i.e., across multiple cells in a spatially averaged electrical syncytium. The electrical control of the intestinal tissue is somewhat unique with respect to its involvement of two separate levels of control by these two different types of cell models. Earlier intestinal cell models utilized coupled oscillators to phenomenologically reproduce the periodicity of slow waves [2]. Even though these phenomenological models offer a computationally efficient way of simulating slow waves, they lack the sophistication required to link the electrical activity to mechanical activity. More recently biophysically-based models of these two types of cells have been developed. These models are generally based on the gated ion conductance approach pioneered by Hodgkin and Huxley [30]. These biophysically-based cell models have the advantage of relating physically meaningful parameters, e.g., ion concentrations, to physiologically realistic processes, e.g., voltage-dependent ion transport. In this section, we use a biophysically-based smooth muscle cell model as an example of how slow waves can be simulated using a system of ordinary differential equations.

A biophysically-based model of gastrointestinal smooth muscle cell was developed in 2007 [14]. This model adopts the Hodgkin and Huxley formulation to describe the eight types of commonly known ion channels in the smooth muscle cell membrane. These eight ion channels, also referred to as conductances, in combination with a pacemaking current term (i.e., the pacemaking current from the interstitial cells of Cajal), specify depolarization of the smooth muscle cell, through,

$$\frac{\partial V_m}{\partial t} = -\frac{1}{C_m}(I_{ion} - I_{ICC}), \tag{3}$$

where

$$I_{ion} = I_{CaL} + I_{LVA} + I_{Kr} + I_{Ka} + I_{BK} + I_{Kb} + I_{Na} + I_{NSCC}, \tag{4}$$

specifically, I_{CaL} represents Ca^{2+} current through voltage-dependent, dihydropyridine (DHP)-sensitive, L-type channels which are expressed in all gastrointestinal smooth muscle cells, and have therefore been referred to as the backbone of electromechanical coupling in the gut [49]. I_{LVA} describes a low voltage-activated, DHP-insensitive, T-type calcium current. I_{Kr} and I_{Ka} represent potassium currents through outward delayed rectifier channels and through outward fast-inactivating, Ca^{2+}-independent channels respectively. I_{BK} represents Ca^{2+}-activated potassium conductance, and I_{Kb}, an experimentally-determined, background potassium current. I_{Na} is a sodium conductance and I_{NSCC} is current through non-selective cation channels, which have been found in the gastrointestinal tract of several species. The stimulus current, I_{ICC}, represents the slow wave from the interstitial cells of Cajal which depolarizes the surrounding smooth muscle cells.

A generic expression of the conductance of ion x is in the form,

$$I_x = G_x(V_m - E_x)g \tag{5}$$

where I_x is the ionic current of an arbitrary ion species, G_x is the maximum channel conductance, and E_x is the Nernst potential of x. The voltage dependency of the ion conductance comes from the gating variable, g, which has a generic form,

$$\frac{dg}{dt} = \frac{g_\infty - g}{\tau_\infty} \tag{6}$$

Equation (6) specifies the rate of change of the gating variable, g, as a function of voltage-dependent variables, g_∞ and τ_∞, both of which can be fitted to experimental measurements.

Another important cellular process which contributes to the behavior of these ion channels, and which is of particular interest to the intestinal electromechanics, is the regulation of intracellular calcium during the depolarization of the smooth muscle cells. In this cell model, the increase in $[Ca^{2+}]_i$ is balanced by a calcium extrusion current (I_{CaEXT}), which includes re-uptake by the sarcoplasmic reticulum and mitochondria, as well as extrusion from the cell through Ca^{2+} pumps. These processes were combined and modeled using the following expression,

$$I_{CaExt} = 0.317\left(\left[\mathrm{Ca}^{2+}\right]_i^{1.34}\right), \tag{7}$$

based on observation of normal Ca^{2+} levels after each slow wave activity. The smooth muscle cell model was applied to successfully simulate slow wave membrane potential traces which have been validated against experimental recordings. Activity of the individual ion conductance was also simulated (Fig. 2). Most important to the electromechanical model, the corresponding Ca^{2+} transient had an amplitude of 300 pM (Fig. 2 top), as contractions of the smooth muscle cells are related to the amplitudes of Ca^{2+} transients [44].

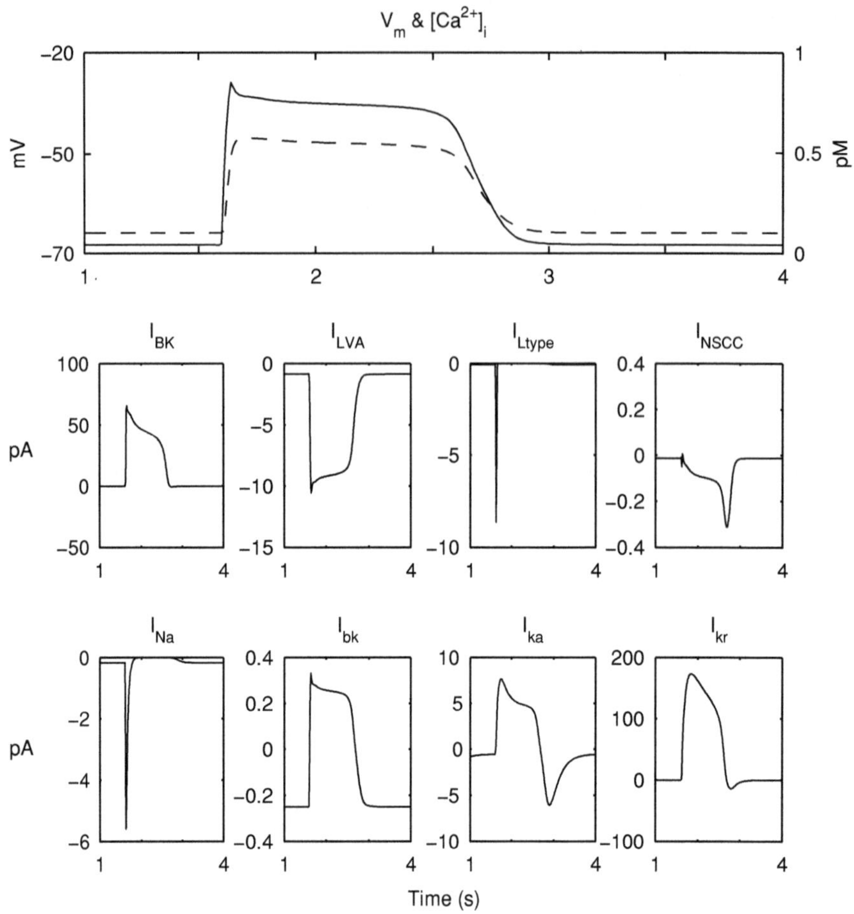

Fig. 2 A simulated trace of intestinal slow wave activity (V_m, measured in mV), Ca^{2+}, and ion conductances using the smooth muscle cell model (Eq. 3). I_{CaL} and I_{LVA} are calcium conductances; I_{BK}, I_{kr}, I_{ka}, I_{bk} are potassium conductances; I_{Na} is a sodium conductance; I_{NSCC} is a non-selective conductance, which is also Ca^{2+}-dependent

3.1 Continuum Model

In a continuum model of intestinal electrophysiology, the mathematical descriptions of slow wave propagation incorporate activities of scales larger than that of a single cell. In one popular approach, individual cells are modeled and integrated over an electrical syncytium. In the early stages of cardiac modeling, such a discrete approach was used to model networks of individual cardiac cells over minute scales (0.5 by 1.5 mm^2) but involved significant computational expense (4.5 h to simulate 520 ms of electrical activity) [40]. Even with the significant advances in computational power since, the discrete approach is still challenging to apply to intestinal slow wave propagation problems that need to be studied over seconds, if not minutes. In another approach, the basic unit of an electrical continuum is treated in a spatially averaged sense, i.e., the conductive medium is modeled as continuous rather than consisting of discrete cells [45]. The spatially-averaged treatment of the conductive medium reduces the size of the model and thereby reduces the computation time required to solve the model.

The bidomain model is one of the main continuum electrical models employed to simulate intestinal slow wave propagation. The principle of the bidomain equations is to model the electrical current flow between two inter-penetrating domains: (i) the intracellular domain (usually denoted by subscript *i*); and (ii) the extracellular domain (usually denoted by subscript *e*). The two domains are divided by a continuum cell membrane. The current fluxes across the cell membrane are described by an I_{ion} term denoting the ionic current term in Eq. (3). This critical step links the bidomain equations to the cell model. More specifically, the bidomain formulation involves two equations,

$$\nabla \cdot (\sigma_i \nabla V_m) = -\nabla \cdot ((\sigma_i + \sigma_e) \nabla \phi_e), \tag{8}$$

$$A_m \left(C_m \frac{\partial V_m}{\partial t} + I_{ion} \right) - \nabla (\sigma_i \nabla \phi_e) = \nabla \cdot (\sigma_i \nabla V_m), \tag{9}$$

where the σ terms denote tissue conductivity tensors, with subscript *i* denoting the intracellular domain and subscript *e* denoting the extracellular domain. The I_{ion} denotes the current flow through the cell membrane, as described in the cell models. Equation (8) describes the relationship between V_m and ϕ_e. Equation (9) is a reaction-diffusion equation in terms of the V_m, where the sum of ion conductances from cell models provides the non-linear reaction term [45]. The bidomain model presented here is a voltage dependent system, in line with the standard of previous simulation studies of both cardiac and gastrointestinal electrical activity [45].

In the numerical solution process, the bidomain equations can be approximated using a weighted residual approach, similar to that used for the finite deformation approximations (described in the following section). The bidomain equations approximated over a physical solution domain can therefore be expressed as,

$$\int_\Omega [\nabla \cdot (\sigma_i \nabla V_m)] w d\Omega = 0, \tag{10}$$

$$\int_\Omega \left[\nabla \cdot (\sigma_i \nabla V_m) + \nabla (\sigma_i \nabla \phi_e) - A_m \left(C_m \frac{\partial V_m}{\partial t} + I_{ion} \right) \right] w d\Omega = 0, \tag{11}$$

where *w* is the weighting factor, specified using the interpolating basis function. The basis functions are also used to interpolate parameter variables over the local coordinate of the geometric elements. The time derivative $\frac{\partial V_m}{\partial t}$ can be approximated using a difference method, for example,

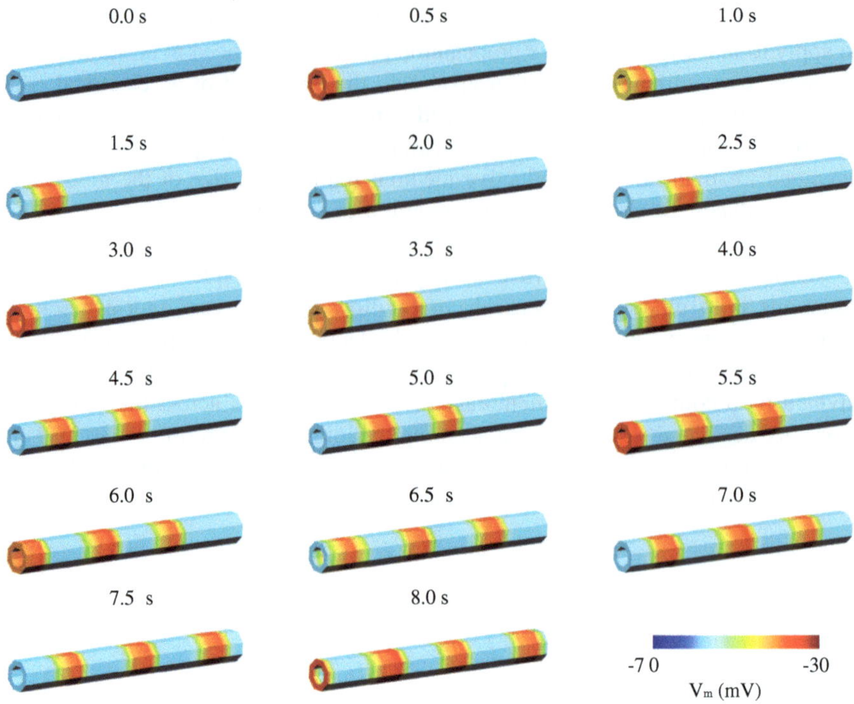

Fig. 3 Simulated intestinal electrical activity. Visualization of slow wave membrane potential at regular intervals in an idealized intestinal model, from 0 to 8 s. The *color bar* indicates membrane potential values in mV, ranging from −70 (*blue*) to −30 mV (*red*)

$$\frac{\partial V_m}{\partial t} = \frac{V_m^{t+1} - V_m^t}{\Delta t} \tag{12}$$

The discretized bidomain equations can be solved sequentially, with the V_m term from Eq. (11) is used to update Eq. (10) at each time step.

In our simulation, as an example of intestinal slow wave propagation, the smooth muscle cell model was solved at tissue-level using the bidomain formulation implemented within a grid-based (solution points) finite element framework. Four grid points were assigned in each ξ-direction of the segment model in Fig. 1b, to make a total of 64 grid points per element, and 8,192 in the whole geometry. The intestinal slow wave propagation was simulated for 2.8 s and visualized over the intestinal geometry (Fig. 3). Boundary conditions of zero current flux through the cell membrane boundary condition were assigned to the model. Anisotropic tissue conductivities assigned to the fiber, sheet and sheet-normal directions were 1, 0.5 and 0.019 mS mm^{-1} respectively in the intracellular domain and 1, 0.5 and 0.236 mS mm^{-1} in the extracellular domain.

The frequency of the simulated intestinal slow wave activity was 23 cpm, with a propagating velocity of 5 mm s^{-1}. In this initial simulation, the slow wave propagation in the circular direction was assumed to be instantaneous, so that all the points at the same axial location were activated simultaneously.

4 Electromechanical Coupling

The tension generated in intestinal smooth muscle tissue can be classified into passive and active components. Passive tension, or "tone", is mainly attributed to extracellular connective tissue opposing stretch in the muscle tissue [39]. Active tension is generated by the intracellular mechanisms the intestinal smooth muscle cells that are responsible for contractions. Active contractions can be further classified as either "tonic contractions", which maintain the steady-state force over long periods of time and generate "specific tone" in the tissue [26], or "phasic contractions", which are characterized by a rise in developed tension followed by relaxation [39]. Multiple phasic contractions may also fuse together, generating "tetanic tone" in the intestinal tissue [26].

During periods of contraction, the myosin units the smooth muscle cells are phosphorylated [49]. The activation allows interaction between actin and myosin elements in the cell. These interactions are regulated by intracellular Ca^{2+}, so initiation of contraction requires an increase in the $[Ca^{2+}]_i$. This increase in $[Ca^{2+}]_i$ can be induced by multiple factors, but under normal physiological conditions, electromechanical coupling is widely considered as the predominant mechanism [44, 49]. Slow wave activation of the smooth muscle cell depolarizes the membrane potential, and this depolarization leads to an increase in the influx of Ca^{2+} through voltage-gated Ca^{2+} membrane channels, e.g., I_{LVA} and I_{Ltype}. Normally, the increase in $[Ca^{2+}]_i$ triggers release of more Ca^{2+} from intracellular Ca^{2+} stores in the smooth muscle cell, e.g., the sarcoplasmic reticulum.

The next step in smooth muscle contraction is the binding of intracellular Ca^{2+} to calmodulin. There are four Ca^{2+} binding sites on calmodulin; at least three need to be filled before the Ca^{2+}-calmodulin complex is able to activate myosin-light-chain-kinase [33]. This partially explains the requirement for a high level of $[Ca^{2+}]_i$ for activating contraction. Myosin-light-chain-kinase catalyses the phosphorylation of the light chain subunit in myosin, allowing myosin cross-bridges to attach to actin. This process of activating cross-bridges through a phosphorylation switch is one of the major differences between striated and smooth muscle (in striated muscle, Ca^{2+} removes the inhibition of actin–myosin interaction instead [39]).

This section presents the principal theory and modeling technique used to simulate electromechanical coupling in the small intestine. Figure 4 summarizes the simulation process and highlights the main components of the electromechanical model, which are individually discussed in the following subsections. In brief, an anatomical model (Fig. 1b) was created and the initial timing of the electrical activation specified to match slow wave propagation velocity.

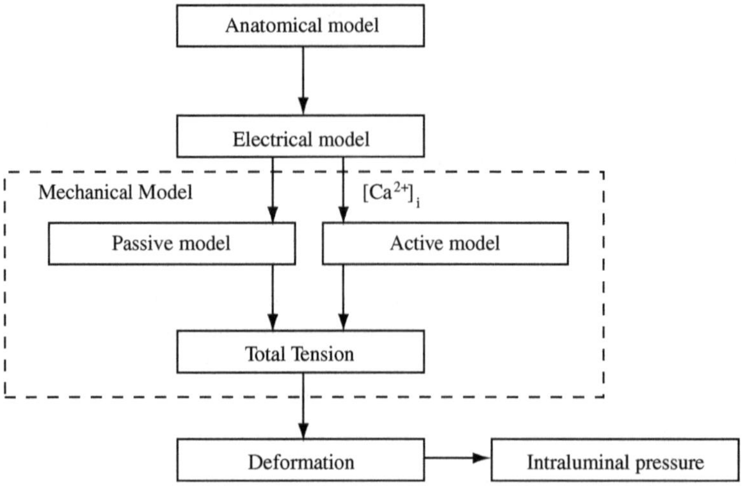

Fig. 4 Schematic of the electromechanical coupling process used in this simulation. At each solution step, the electrical component is solved first, and the relevant variables in the cell models, i.e., $[Ca^{2+}]_i$, is linked to the active component of the mechanical model. The active component is solved together with the passive component to update the deformation of the geometric model

The electrical model was solved for $[Ca^{2+}]_i$ values (Fig. 3), which are then fed to the active mechanics component. Resultant active tension was incorporated into the finite deformation equations that, together with the passive constitutive law, specified deformation of the geometry. The mesh was updated with this new geometry before the next solution step. Stress and radial measurements along the small intestine were used to approximate the intraluminal pressures at various axial locations along the anatomical model.

4.1 Finite Deformation Theory

In the small intestine, peristalsis can occlude more than 70 % of the diameter of the lumen [1]. Such large and nonlinear deformation presents a challenge to the conventional linear and small-strain deformation theory. In this case, a more fundamental approach of finite elasticity is needed when using the mechanical behavior of elastic materials which undergo large strains. Stress is another important measure in mechanical deformation, and it can also vary greatly between the undeformed and deformed states as the smooth muscle fiber elongates under large strains.

Formulation of finite elasticity is a relatively more complex procedure than linear elasticity formulations, as it involves "tracking" the deforming material. In finite deformation theory, particles in the material are "tracked" using a moving coordinate system called the *material* coordinate system, with coordinates denoted by (X_1, X_2, X_3). This coordinate system moves throughout the deformation to ensure that the coordinates of each particle remains constant. Another set of coordinates, the *spatial* coordinate system, is used as reference. This coordinate system marks coordinate points, denoted by (x_1, x_2, x_3) which remain constant throughout the deformation; the spatial coordinate system therefore remains static in time. The deformation of the material can be quantified by the changes in direction and length of the lines that connect two adjacent material particles. It is calculated from the differences in material and spatial coordinates and specified using the deformation gradient tensor (F),

$$F = \frac{\partial x}{\partial X}. \tag{13}$$

Removing the directional components of this tensor, i.e., removing the rotational dependency of the deformation gradient tenor, leaves the right Cauchy–Green deformation tensor (C),

$$C = F^T F, \tag{14}$$

which can be related to a strain tensor in the form of the Lagrangian finite strain tensor (E) through the following relationship,

$$E = \frac{1}{2}(C - I), \tag{15}$$

where I is the identity matrix. The right Cauchy–Green deformation tensor can be used to specify three invariants (I_1, I_3, I_3), i.e., quantities which remain constant under coordinate rotations, which are used for expressing several passive constitutive relationships (Sect. 4.2). The invariants can alternatively be specified in terms of the principal stretch ratios of C: λ_1, λ_2 and λ_3, such that,

$$I_1 = tr(C) = \lambda_1^2 + \lambda_2^2 + \lambda_3^2, \tag{16}$$

$$I_2 = \frac{1}{2}\left[tr(C)^2 - tr(C^2)\right] = \lambda_1^2\lambda_2^2 + \lambda_2^2\lambda_3^2 + \lambda_3^2\lambda_1^2, \tag{17}$$

$$I_3 = det(C) = \lambda_1^2\lambda_2^2\lambda_2^2. \tag{18}$$

Both the force and the unit surface area in a material are required in order to calculate stress. In a large deformation problem, both the force and area are different in the undeformed and deformed configurations. There are therefore various definitions for the stress tensor. An important definition, useful for representing passive material behavior, is the second Piola–Kirchhoff stress tensor (T), which represents the force measured per unit undeformed area, acting on a local geometric element in the undeformed configuration,

$$T = JF^{-1}\Sigma(F^T)^{-1}, \tag{19}$$

where J is the Jacobian matrix required for the transformation from undeformed to deformed coordinates. The Jacobian matrix is the determinant of the deformation gradient tensor,

$$J = det(F) = \lambda_1\lambda_2\lambda_3. \tag{20}$$

The second Piola–Kirchhoff stress tensor is particularly useful as it directly relates to the strain energy function used in formulating the constitutive models that specify passive material properties.

In order to specify anisotropic material properties, these tensors are additionally referred to the locally varying fiber coordinate system, v_α (i.e., the fiber (f), sheet (s) and sheet-normal (n) directions). By applying the *principle of virtual work*, stress equilibrium of the system, in the Einstein notation, yields,

$$\int_V T^{\alpha\beta}F^j_\beta \delta u_{j,\alpha} dV = \int_V \rho(b^j - f^j)\delta u_j dV + \int_S s^j \delta u_j dS, \tag{21}$$

$T^{\alpha\beta}$ and F^j_β are the second Piola–Kirchhoff stress tensor and deformation gradient tensor respectively, expressed with respect to the fiber coordinate system, so that the virtual displacements are expressed in the reference coordinate system and differentiated with respect to the v_α coordinate system [41]. The density of the material (ρ) is in the reference system; b^j and f^j are the body force and acceleration vectors over the body volume; V, and s^j are the surface traction vectors representing external surface forces acting on the deformed surface, S [41]. For the finite deformation problem, the finite element basis functions are used to interpolate the virtual displacement yields, δu_j in Eq. (21). The integrals in Eq. (21) are subsequently transformed to the local coordinate space. Due to the nonlinearity of the formulation, the resulting integrals form can be evaluated numerically using, for example, the Gaussian quadrature scheme, which approximates the integrals using a weighted sum of integrand evaluations,

$$\int_0^1 f(\xi)d\xi = \sum_{i=1}^{I} w_i f(\xi^i) + E_i, \tag{22}$$

where w_i are the weighting factors, often specified using the interpolating basis functions, and ξ^i are the locations of the solution points (gauss points) where the integrand is to be evaluated. E_i is the error in the approximation and I is the order of the Gaussian quadrature scheme. By incorporating the passive constitutive equations (described in Sect. 4.2), second Piola–Kirchhoff stress and Lagrangian strains may be calculated at the gauss points. These integral approximations are subsequently combined, together with boundary constraints, to form a global system of nonlinear equations in terms of the unknown displacements of each node in the mesh. The system can then be solved using an appropriate numerical method.

4.2 Modeling Passive Tension

Passive tension is an intrinsic material property and refers to the stiffness in the material that resists deformation. In smooth muscle tissue, this stiffness is largely due to inelastic collagen fibers, which prevent further deformation at high wall tensions [26]. It is important to include the effects of passive tension in a mechanics model of the intestine, because the ease with which a segment of the intestine is actively distended depends on passive stress, which in turn influences the flow and mixing within the intestinal lumen [26].

Specifying the passive behavior of smooth muscle tissue in response to stress requires two important specifications. These are the compressibility constraint and the constitutive behavior of the tissue. The intestinal tissue can be assumed to be incompressible, as smooth muscles consist mainly of incompressible fluids. Popular constitutive laws used to model isotropic material behavior include the Neo-Hookean and Mooney−Rivlin laws. However, the Mooney–Rivlin relationship may not be appropriate to model the intestinal tissue because the layered muscle structure, i.e., a hyperelastic and anisotropic constitutive law may be required and large strains [6]. Another important aspect of constitutive law modeling is viscoelasticity; although, viscoelasticity is challenging to model because of the time-dependency. However, an important assumption can be made in the case of smooth muscle tissue. Following a period of pre-stretching, after the muscle undergoes repeated loading–unloading cycles, the stress–strain relationship becomes relatively independent of strain rate; the response becomes constant and predictable. Fung termed this approximation "pseudoelasticity" [24]. The existence of a unique stress–strain relationship means that the tissue has an associated strain energy function, and can therefore be modeled as a hyperelastic material [24]. In general, the stress can be related to the strain energy function (W) and the strain (E) tensor of the hyperelastic material,

$$T = \frac{\partial W(E)}{\partial E}. \tag{23}$$

The general form of the strain energy function can be modeled using many different mathematical formulations, each with parameters that can be adjusted to fit data from material tests and yield the constitutive equations for the tissue. One such constitutive law is the Fung-type model, based on a previous biaxial study of intestinal tissue by Bellini et al. [6],

$$W = \frac{C}{2}\left(e^{Q} - 1\right) \tag{24}$$

where,

$$Q = a_1 E_{11}^2 + a_2 E_{22}^2 + a_3 E_{11} E_{22}, \tag{25}$$

and E_{11} and E_{22} are the strains in the circumferential and longitudinal directions respectively; C, a_1, a_2 and a_3 are parameters fitted from the experimental data. It was found that the direction of maximum stiffness at different location in the small intestine is not consistent across all the tissue samples from that location [6]. Not all, but a majority of the duodenal and ileum samples are stiffer in the circumferential direction, while a majority of the jejunal samples are stiffer in the longitudinal direction [6]. A lumped model was therefore developed in order to average out some of the anisotropic behaviors in the tissue.

4.2.1 Measurement of Passive Tension

Passive tension should technically be measurable between contractions when the muscle is at rest. However, this is challenging in practice, due to the spontaneous contractile nature of the muscle cells, as well as the presence of specific tone in the tissue, which is difficult to isolate from pure passive properties [26]. In their efforts to obtain the passive tension–length relationship in arterial smooth muscle, Herlihy and Murphy developed a protocol which involves reversibly stretching muscle strips [29]. Their results closely match measurements obtained from muscle strips which were irreversibly equilibrated in Ca^{2+}-free solution, a process which they believed inactivates the contractile apparatus. However, although various mechanisms have been found to arrest spontaneous contractions, the mechanical behavior of the "resting" smooth muscle varies with the type of mechanism used [24]. This means that there is an additional active component in the smooth muscle, i.e., tonic tension, which is still present, but to varying degrees depending on experimental conditions, even when the phasic behavior has been inactivated.

Assuming a relatively inert intestinal tissue sample can be obtained, passive stiffness in the material can be investigated using planar biaxial tests. The tissue sample is loaded simultaneously in the longitudinal and circumferential directions. By tracking the movement of markers on the surface of the sample, one can observe the dependence of the response in one direction on the material properties of the other direction. Assuming the material is incompressible, the response in the third direction, i.e., the transmural direction, can be approximated. The stress–strain responses are incorporated into a constitutive law, which can then be used to predict the behavior of the tissue under any general loading state. The review by Sacks provides a detailed explanation of the biaxial testing technique [48].

Another commonly used technique for measuring passive tension in tubular organs is balloon manometry. A probe is inserted into the intestine and a balloon attached to the probe is inflated to apply an outward pressure onto the intestinal wall [46]. The radius of the tube is measured using imaging techniques. The tension in the wall is then approximated from applied pressure, radius and wall thickness. This method has the advantage of being able to directly test on the intact intestinal tissue, rather than on muscle strips. However, this is essentially a uniaxial method, which is limited by the stiffness data that assumes isotropic material behavior in all three orthogonal material directions [48]. Recent adaption of the fiber optic technology also allowed intraluminal pressure to be measured at high-resolution using a pressure-sensing catheter placed in the colon [16, 18]. This method has the potential to resolve movements and pressure developments in multiple directions.

4.3 Modeling Active Tension

Active tension is generated by the contractile apparatus in the smooth muscle cells in response to the increase in $[Ca^{2+}]_i$ caused by slow wave depolarization of the cell. Smooth muscles are capable of generating longer sustained contractions than striated muscle with the similar level of myosin phosphorylation, because of the ability of smooth muscle to form latch-bridges. There are several models of the active behavior of the intestinal smooth muscle [25, 27, 47]. Many of these models describe detailed mechanisms of the cross-bridge cycling in the smooth muscle cells. Another simplified approach is to adapt the steady-state tension–length–calcium relationship (SS-T-L-Ca^{2+}) relationship developed by Hunter et al. [32], and use it to describe the active tension generated in response to the increase in $[Ca^{2+}]_i$ following electrical activation. The model assumes cardiac-specific kinetics of Ca^{2+}, including binding of Ca^{2+} to Troponin-C in cardiac cells. The Ca^{2+}-Troponin-C complex binds to tropomyosin and unblocks the binding sites, and the model provides a phenomenological description for the steady-state proportion of available binding sites (z_{ss}), as follows,

$$z_{ss} = \frac{\left[Ca^{2+}\right]_i^h}{\left[Ca^{2+}\right]_i^h + C_{50}^h}. \tag{26}$$

where C_{50} is the value of $[Ca^{2+}]_i$ required to achieve 50 % availability of actin binding sites and h determines the steepness of the sigmoidal curve. The tension developed by the muscle fiber at steady-state length is directly proportional to z_{ss}, and was fitted to cardiac isometric tension data using the following expression,

$$T = T_{ref}(1 + \beta_O[\lambda - 1]) \cdot z_{ss}, \tag{27}$$

where λ is the extension ratio of the muscle fiber, T_{ref} is the tension recorded at an extension ratio of 1, and β_O relates to myofilament cooperativity, a phenomenon which describes how activation of a single cardiac muscle cell facilitates activation of neighboring cells [53].

Although originally specified from cardiac muscle data, the parameters of the SS-T-L-Ca^{2+} model can be readily adjusted to simulate smooth muscle behavior. Figure 5 shows the tension-$[Ca^{2+}]_i$ relationship with parameters β_O = 1.45, C_{50} = 0.65, and h = 2.5 to match a set of normalized tension over a range of strains.

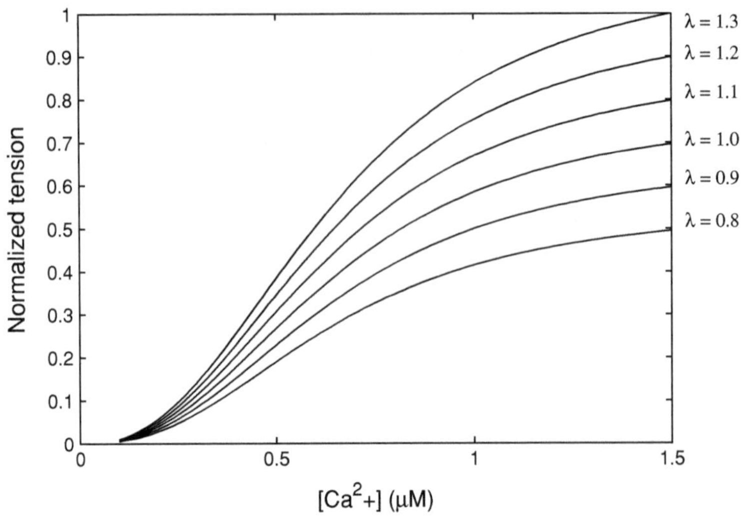

Fig. 5 Steady-state Ca^{2+}-tension-extension relationship. Simulated isometric tension-$[Ca^{2+}]_i$ relationship at different extension-ratios (λ)

Visualization of the electromechanical simulation using the SS-T-L-Ca^{2+} relationship is shown in Fig. 5.

4.4 Boundary Conditions and Numerical Solutions

Boundary conditions were chosen to restrict torsion and whole body movements of the intestine on one face of the model, but should allow contractions in the longitudinal and radial directions. In the anatomical model (Fig. 1c), the geometric elemental nodes on the left hand side face were fixed in the longitudinal direction. In addition, two nodes on the inner circular layer were fixed in the vertical circumferential direction, and the two nodes orthogonal to the previous two nodes were fixed in the horizontal direction. This is representative of physical constraints applied in several experimental studies on whole segments of small intestine [28, 50], in which only one end was constricted in the axial direction, so that longitudinal contractions could be measured. Boundary conditions also influence the degrees of freedom of a deformation problem. The degrees of freedom for the mechanical solution is an important indication of the computational time required for solving the model. The degrees of freedom may be measured by the number of dependent variables (deformed geometric coordinates) required at each geometric node, omitting the variables that are fixed due to boundary constraints. For example, the present mechanical model described in this chapter contained 1,240

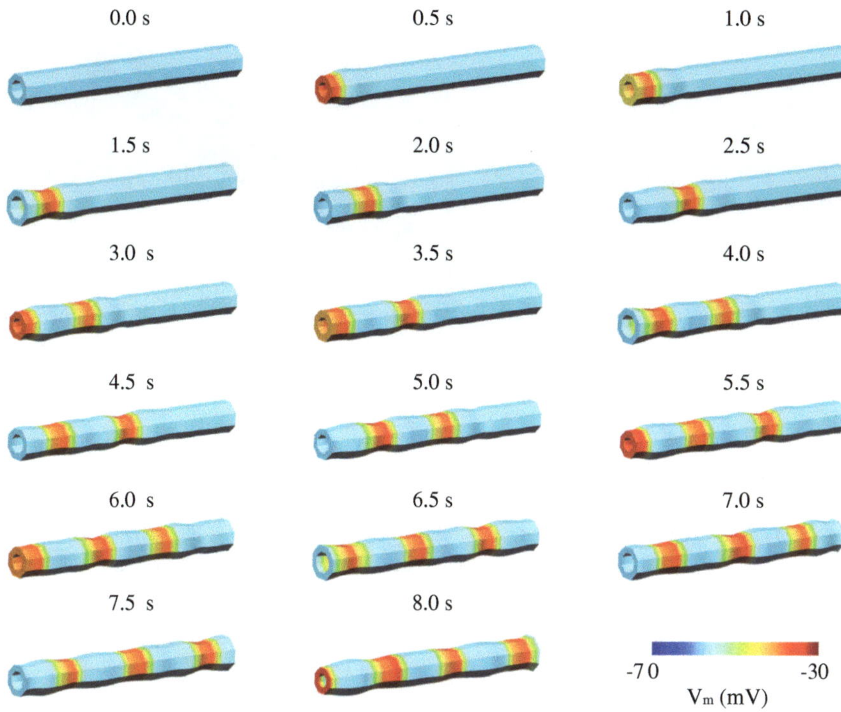

Fig. 6 Simulated intestinal electromechanical activity. Mechanical deformation is seen as occlusion of the intestinal lumen. Visualization of slow wave membrane potential at regular intervals in an idealized intestinal model, from 0 to 8 s. The *color bar* indicates membrane potential values in mV, ranging from −70 (*blue*) to −30 mV (*red*)

degrees of freedom in total. The solution procedure for the electromechanically coupling model involved solving for deformed geometric coordinates, which were then used to update the model geometry, including the coordinates of the electrical solution points, before the next solution step. At each mechanical solution point i.e., gauss point, the following variables were defined as outputs: (i) deformed nodal coordinates, (ii) Lagrangian strains, (iii) active Piola–Kirchhoff stresses, and (iv) total Piola–Kirchhoff stresses.

More specifically in this example, the electromechanically coupled model was solved for a simulated period of 8 s, and solutions output at time steps of 0.5 s (Fig. 6). The electrical component was solved first. The $[Ca^{2+}]_i$ values from the grid points were interpolated and used to update the Ca^{2+} variable at each mechanical gauss point. A trilinear interpolation scheme was used for this update step. These $[Ca^{2+}]_i$ values were subsequently input into the active mechanics model. For the mechanical component, the nonlinear finite deformation equations were linearized using the Newton–Raphson method, then numerically solved using UMFPACK. The solution converged within five iterations in most cases, with each iteration taking between 50 and 80 s to solve. Within each mechanical solution

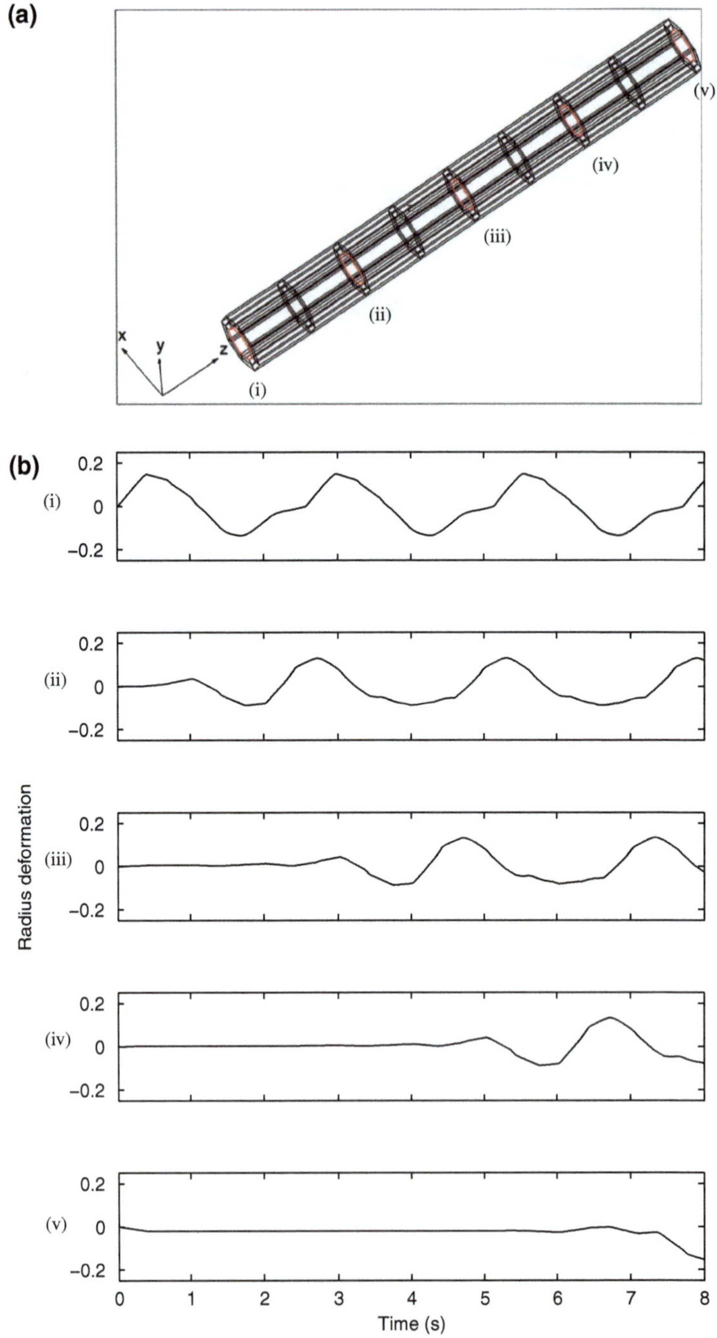

Fig. 7 Radial deformation over time. **a** The deformations were calculated at selected axial locations marked in *red* (i–v). **b** The deformed radii at each time step were calculated as ratios of the original radius

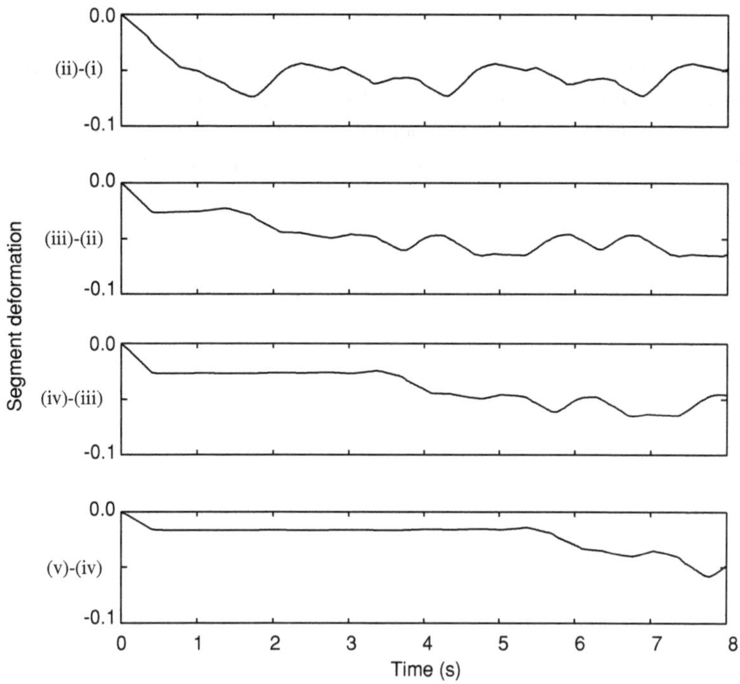

Fig. 8 Longitudinal deformation over time. The deformed segment strains at each time step were calculated as ratios of the original segmental length

step, the active tension model was solved using the improved Euler method. The entire simulation was implemented in a multi-scale modeling computation package called CMISS [12], on a single 2.8 GHz processor, and required 85.6 h of computational time. Based on the simulation results in Fig. 6, the maximum active stress generated along the fiber direction in each slow wave (electrical) cycle was approximately 937 Pa. Total stress in the fiber direction, i.e., the linear combination of passive and active stresses, was smaller and much more dependent on axial location, with maximum recorded values of 464 Pa. We were also able to quantify the circular contractions by calculating the deformed radius as a ratio of the original, relaxed radius (14 mm) at each axial node. The three transient decreases in radial measurement correspond to the three cycles of slow wave and contractile activity that passed this point during the simulation. The maximum radial deformation was 0.276 of the original radius. A plot of the radial deformation against time (calculated at the axial locations marked in red) is shown in Fig. 7. On the other hand, the longitudinal contraction (shortening) at each axial location was similarly quantified as the ratio of the original length of the axial segment (Fig. 8).

4.5 Intraluminal Pressure Calculation

Validation of the model is an important part of any multi-scale modeling study. Therefore, it is important for the model to have an output that can be measured experimentally. Clinically, antroduodenal manometry catheter is a commonly applied method of recording to assess the patient's intestinal motility. The assessment involves inserting a tube with pressure sensors through the stomach and into the small intestine, sometimes for hours at a time. In order to compare the model with experimental manometry recordings, we calculated the intraluminal pressure due to the mechanical contraction of the intestinal model. For this initial model, instead of attempting to deploy a computational fluid dynamics simulation within the complex movement of the geometry, a relatively simple approximation using Lamé's theory of stresses within thick-walled cylinders was applied. The thick-wall formulation of Lamé's theory is typically used if the cylinder has a ratio of wall thickness to internal diameter that is larger than 0.1 [38]; the ratio in this model geometry was 0.3. The theory approximates the circumferential, or hoop, stresses (σ_h) in the wall of the cylinder using the internal (p_i) and external (p_o) fluid pressures and the internal (r_i) and external (r_o) radial values,

$$\sigma_h = \frac{r_i^2 p_i - r_o^2 p_o}{r_o^2 - r_i^2} + \frac{r_i^2 r_o^2 (p_i - p_o)}{r^2 \left(r_o^2 - r_i^2\right)}, \tag{28}$$

where r is the coordinate corresponding to the radial location of this stress value. These parameters are visualized in Fig. 9. The derivation of Lamé's equation can be found in [42].

Assuming no external pressure, rearranging for internal pressure gives,

$$p_i = \sigma_h \frac{r^2 \left(r_o^2 - r_i^2\right)}{r_i^2 \left(r_o^2 + r^2\right)} \tag{29}$$

Stress values in the fiber, sheet and sheet-normal directions can be calculated at each gauss point and at each time step of the simulation. The stresses in the fiber direction at the inner surface of the intestinal model correspond to circular wall stresses at $r = r_i$ because the fibers on the inner surface of the intestine model are aligned in the circumferential direction. Intraluminal pressures calculated using stresses at $r = r_i$ can therefore be expressed as,

$$p_i = \sigma_{gauss} \frac{\left(r_o^2 - r_i^2\right)}{\left(r_o^2 + r_i^2\right)} \tag{30}$$

where σ_{gauss} are the fiber stresses at the innermost gauss points in the simulation. Radius values can be calculated by taking the distance between the innermost gauss point and the outermost gauss point at corresponding circumferential coordinates.

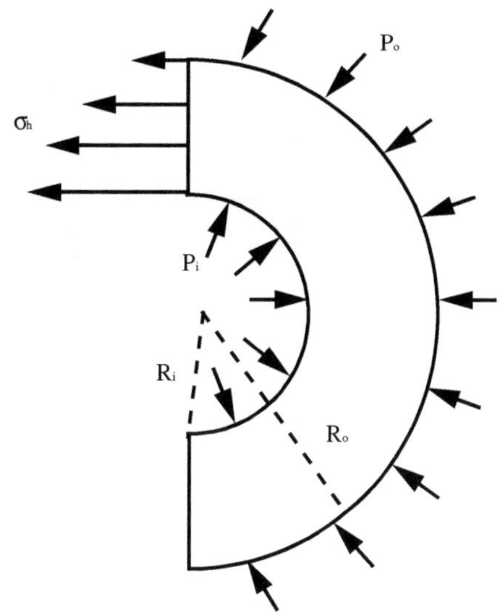

Fig. 9 Cross-sectional diagram of intestine illustrating variables used for intraluminal pressure calculation in Eq. (28): r_i and r_o are the internal and external radii respectively, r is the radial coordinate corresponding to the location of this stress value, and p_i and p_o are the internal and external pressures respectively

To make the output of the simulation consistent with measurements obtained from manometry, particularly high-resolution manometry [18], we calculated pressures at points spaced approximately 10 mm apart in the longitudinal direction on the inner surface of the intestinal model. Figure 10 demonstrates a sample of the pressure traces obtained from the axial locations marked in Fig. 7a. The top-most trace corresponds to the point at the most left-most location of the geometric model and the bottom trace to the most right-most location of the geometric model. We found that pressure varied depending on location within each element, as seen in the different shapes of the pressure transients in each plot in Fig. 10. The transient peaks in pressure correlated with periods of contractile activity. The maximum intraluminal pressure calculated was 148 Pa.

5 Perspectives

This chapter presents a preliminary framework for an electromechanically coupled simulation of the rat small intestine using multi-scale modeling. In this chapter, we have highlighted several features and limitations of the methods and models used. We have also proposed potential extensions that may be added to the modeling framework as we continue to learn more about the electromechanical dynamics in the small intestine. Due to the modular nature of this simulation framework, the model may be readily updated in the future when more specific information on intestinal motility has been understood. This is a first step toward creating a fully coupled model that can be applied in the clinical setting as the link between minimally-invasive electrical measurements and pressure recordings, in order to improve diagnosis and treatment of intestinal motility disorders.

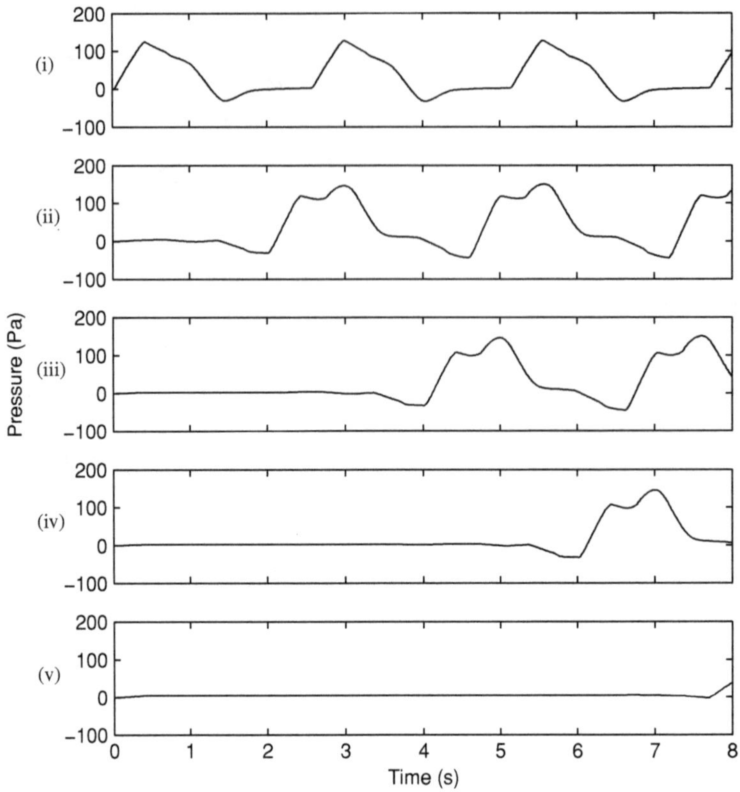

Fig. 10 Intraluminal pressure traces approximated using Lamé's theory of stresses within thick-walled cylinders. Each subplot illustrates the pressures calculated at the gauss point locations marked in Fig. 7a)

The modeling framework proposed in this chapter consists of three main parts: (i) geometric model; (ii) electrical model; and (iii) electromechanical model. Many simplifications were made in order to create a computationally feasible model. For example, a simplification applied to the model geometry is the two-layered wall structure representing the circular and longitudinal muscle layers. This idealized geometry ignores the effects of other anatomical layers in the small intestine. While the missing layers do not generate active tension, they contain structural features, such as collagen fibers, which contribute to the passive mechanical properties of the intestinal wall. Additionally for simplicity, we assumed equal thickness of the longitudinal and circular muscle layers, although morphometric studies have shown that the circular muscle layer of the rat small intestine is approximately 1.2 times thicker than the longitudinal layer [19]. Assuming muscle fiber density is consistent in each layer, this unequal thickness ratio may affect the active tension generated.

The geometric model was limited to contain mostly linear basis functions in order to minimize the degrees of freedom and therefore computation time. However, it would be of interest to evaluate how geometry and mechanical deformation are affected by the order of the interpolation functions in the simulation, e.g., using tri-cubic Hermite basis functions. Computing time may be reduced in the future by using distributed memory multiprocessing for solving the finite deformation problem. For example, the time consuming computation of the element stiffness matrices can potentially be sped up by utilizing parallel computation techniques.

The electrical activity of intestinal motility is not limited to slow wave only. For example, the changes in intestinal contractility also appear to be co-regulated by other electrophysiological mediating mechanisms, such as spike activity is voltage-dependent and rapid oscillations during the plateau phase of the intestinal slow waves, which apparently play an important role in mediating contractility during certain stages of intestinal transit, e.g., during fasting [15]. However, as there is a strong temporal relationship between slow waves, spikes, and contractions, mechanical activation by slow waves could justifiably be treated in the temporal domain as spike activation in this initial simulation. Nevertheless, it is still important to include an intermediate spiking mechanism to realistically simulate the autonomous control of a wider range of motility patterns in the small intestine. In a recent study [34], spike patches were shown to propagate faster in the opposite direction than the slow wave propagation, and spontaneously terminate spikes at variable distances. At present, however, the cellular mechanisms behind spike activity are poorly understood, and there are no existing models that simulate intestinal spike activity. It would be interesting to incorporate these different characteristics in future modeling studies to investigate how they affect intestinal motility.

Even though voltage-dependent entry of Ca^{2+} has been postulated as one of the main mechanisms of electromechanical coupling in the small intestine [44, 49], the enteric nervous system is also understood as another important regulator of gastrointestinal motility [31]. Incorporation of these neurological pathways will be an important future direction towards an integrated mechanical model. Recently, an extension to the bidomain model has been proposed as a possible framework to bridge the present electromechanical modeling framework and future neural models [10]. Much details of the neurologically mediated regulators of gastrointestinal motility still remain under investigations, and future modeling work will be required to study these mechanisms as further details become clear.

The passive and active mechanical models implemented in this simulation framework were developed by Bellini et al. [6] and adapted from Hunter et al. [32]. $[Ca^{2+}]_i$ acts as the intermediary between the electrical component and the active mechanical component. Active tension is combined with the passive tension to obtain the total tension at each gauss point. This is incorporated into the finite deformation formulation to determine the resultant deformed geometry. However, recent modeling studies have proposed more biophysically based descriptions of the relationship between $[Ca^{2+}]_i$ and cellular biomechanical response [25]. It will be necessary to incorporate a biophysical, gastrointestinal specific model that can accurately predict and explain the biophysical process that lead to the development of active tension.

The present simulation framework is a loosely coupled model, meaning it does not contain a mechanoelectrical feedback component. We assumed that deformation produces nominal changes in the electrophysiological description. However, feedback from the mechanical to the electrophysiology component should be considered in order to investigate the length-dependent Ca^{2+}-sensitivity of smooth muscle cells and tissues. Additionally, strain feedback would be required for describing mechanical transduction mechanisms such as mechanosensitive ion channels. For example, recently a voltage-dependent sodium channel ($Na_v1.5$ encoded by SCN5A) has been shown to exhibit mechanosensitivity [8]. Furthermore, the physical properties of the luminal contents may also exert a degree of influence on mixing, efficiency of digestion and, and absorption of the intestine [35]. Therefore, adding a mechanoelectrical component to the model framework is an important extension to consider in the future.

Intraluminal pressure approximated from Lamé's theory of stress in thick-walled cylinders provided a separate validation for the simulation results. However, this approximation does not take into account time-dependent behavior, e.g., viscosity of the fluid. The necessity to compare the model output with clinical manometry recordings highlights the importance of a complete numerical fluid analysis of the intestinal flow [23, 36]. Nevertheless, the procedure of inputting deformed nodal coordinates, solving the fluid–structure interaction and the flow equations at each time step, while maintaining a stable and coupled state throughout, is a challenging problem at present.

Minimally invasive methods that are capable of recording intestinal slow wave and mechanical contractions are being developed, e.g., endoscopic recordings of gastric mucosa [13] and laparoscopic measurements of the small intestinal serosal surfaces [43]. Additionally, recent developments in high resolution pressure recordings promise effective and minimally invasive diagnostic procedures [5, 17]. These pressure recordings can potentially be used with this simulation framework to investigate the sources of abnormal motility, without the need for imaging tools to visualize contractile activity. However, work is still needed to fully understand the relationships among electrical activity, intraluminal pressure and contractile activity in the small intestine.

In conclusion, we have presented the underlying theory of continuum-based electromechanical modeling, with a particular emphasis on electrical and finite elasticity modeling in the intestine. This framework can be used as a platform for integrating independent experimental measurements from multiple modalities, for example, electrical recording, image capture, and manometric analysis, and provides analysis on the mechanistic relationship between these recordings. A definitive future direction of gastrointestinal biomechanics is detailed constitutive laws that can sufficiently capture the material behaviours of the gastrointestinal tissues. Another important aspect of work in this field is the detailed mechanism of the electromechanical coupling and mechanical feedback to the cells. The progress in gastrointestinal electromechanical modeling is therefore only at the beginning, and more much work will be required before the work can be translated to a better understanding and treatment of gastrointestinal motility disorders in a clinical setting.

References

1. Ailiani, A.C., Neuberger, T., Brasseur, J.G., Banco, G., Wang, Y., Smith, N.B., Webb, A.G.: Quantitative analysis of peristaltic and segmental motion in vivo in the rat small intestine using dynamic MRI. Magnet. Reson. Med. **62**(1), 116–126 (2009)
2. Aliev, R.R., Richards, W., Wikswo, J.P.: A simple nonlinear model of electrical activity in the intestine. J. Theor. Biol. **204**(1), 21–28 (2000)
3. Alvarez, W.C.: Functional variations in contractions of different parts of the small intestine. Am. J. Physiol. **35**, 177–193 (1914)
4. Alvarez, W.C., Mahoney, L.J.: Action currents in stomach and intestine. Am. J. Physiol. **58**, 476–493 (1922)
5. Arkwright, J.W., Blenman, N.G., Underhill, I.D., Maunder, S.A., Szczesniak, M.M., Dinning, P.G., Cook, I.J.: In-vivo demonstration of a high resolution optical fiber manometry catheter for diagnosis of gastrointestinal motility disorders. Opt. Express. **17**, 4500–4508 (2009)
6. Bellini, C., Glass, P., Sitti, M., Di Martino, E.S.: Biaxial mechanical modeling of the small intestine. J. Mech. Behav. Biomed. Mater. **4**(8), 1727–1740 (2011)
7. Berne, R.M., Levy, M.N., Koeppen, B.M., Stanton, B.A.: Physiology, 4th edn. Mosby, St. Louis (1998)
8. Beyder, A., Rae, J.L., Bernard, C., Strege, P.R., Sachs, F., Farrugia, G.: Mechanosensitivity of Nav1.5, a voltage-sensitive sodium channel. J. Physiol. **588**(24), 4969–4985 (2010)
9. Bolton, T.B., Prestwich, S.A., Zholos, A.V., Gordienko, D.V.: Excitation contraction coupling in gastrointestinal and other smooth muscles. Annu. Rev. Physiol. **61**, 85–115 (1999)
10. Buist, M.L., Poh, Y.C.: An extended bidomain framework incorporating multiple cell types. Biophys. J. **99**(1), 13–18 (2010)
11. Cajal, S.R.: Histologie du Systeme Nerveux de l'homme et des Vertebretes. Maloine, Paris (1911)
12. CMISS. http://www.cmiss.org/
13. Coleski, R., Hasler, W.L.: Directed endoscopic mucosal mapping of normal and dysrhythmic gastric slow waves in healthy humans. Neurogastroenterol. Motil. **16**(5), 557–555 (2004)
14. Corrias, A., Buist, M.L.: A quantitative model of gastric smooth muscle cellular activation. Ann. Biomed. Eng. **35**(9), 1595–1607 (2007)
15. Costa, M., Sanders, K.M., Schemann, M., Smith, T.K., Cook, I.J., De Giorgio, R., Dent, J., Grundy, D., Shea-donohue, T., Tonini, M. et al.: A teaching module on cellular control of small intestinal motility. Neurogastroenterol. Motil. **17**, 4–19 (2005)
16. Davidson, J.B., O'Grady, G., Arkwright, J.W., Zarate, N., Scott, S.M., Pullan, A.J., Dinning, P.J.: Anatomical registration and three-dimensional visualization of low and high-resolution pan-colonic manometry recordings. Neurogastroenterol. Motil. **23**(4), e171 (2011)
17. Dinning, P.G., Arkwright, J.W., Costa, M., Wiklendt, L., Hennig, G., Brookes, S.J.H., Spencer, N.J.: Temporal relationships between wall motion, intraluminal pressure, and flow in the isolated rabbit small intestine. Am. J. Physiol. Gastr. L **300**(4), G577–G585 (2011)
18. Dinning, P.G., Zarate, N., Hunt, L.M., Fuentealba, S.E., Mohammed, S.D., Szczesniak, M.M., Lubowski, D.Z., Preston, S.L., Fairclough, P.D., Lunniss, P.J., Scott, S.M., Cook, I.J.: Pancolonic spatiotemporal mapping reveals regional deficiencies in, and disorganization of

colonic propagating pressure waves in severe constipation. Neurgastroenterol. Motil. **22**, e340–e349 (2010)
19. Dou, Y., Zhao, J., Gregersen, H.: Morphology and stress–strain properties along the small intestine in the rat. J. Biomed. Eng. **125**, 266 (2003)
20. Fallingborg, J., Pedersen, P., Jacobsen, B.A.: Small intestinal transit time and intraluminal pH in ileocecal resected patients with Crohn's disease. Dig. Dis. Sci. **43**(4), 702–705 (1998)
21. Farrugia, G.: Ionic conductances in gastrointestinal smooth muscles and interstitial cells of Cajal. Annu. Rev. Physiol. **61**, 45–84 (1999)
22. Farrugia, G.: Interstitial cells of Cajal in health and disease. Neurogastroenterol. Motil. **20**(Suppl 1), 54–63 (2008)
23. Ferrua, M.J., Singh, R.P.: Modeling the fluid dynamics in a human stomach to gain insight of food digestion. J. Food Sci. **75**(7), R151–R162 (2010)
24. Fung, Y.: Biomechanics: mechanical properties of living tissues. vol. 12, Springer, New York (1993)
25. Gajendiran, V., Buist, M.L.: A quantitative description of active force generation in gastrointestinal smooth muscle. Int. J. Numer. Method Biomed. Eng. **27**(3), 450–460 (2011)
26. Gregersen, H., Kassab, G.: Biomechanics of the gastrointestinal tract. Neurogastroenterol. Motil. **8**(4), 277–297 (1996)
27. Hai, C., Murphy, R.: Cross-bridge phosphorylation and regulation of latch state in smooth muscle. Am. J. Physiol. Cell Physiol. **254**(1), C99–C106 (1988)
28. Hennig, G.W., Costa, M., Chen, B.N., Brookes, S.J.: Quantitative analysis of peristalsis in the guinea-pig small intestine using spatio-temporal maps. J. Physiol. **517**(2), 575 (1999)
29. Herlihy, J., Murphy, R.: Length–tension relationship of smooth muscle of the hog carotid artery. Circ. Res. **33**(3), 275–283 (1952)
30. Hodgkin, A.L., Huxley, A.F.: A quantitative description of membrane current and its application to conduction and excitation in nerve. J. Physiol. **117**(4), 500 (1952)
31. Huizinga, J.D., Lammers, W.J.E.P.: Gut peristalsis is coordinated by a multitude of cooperating mechanisms. Am. J. Physiol. Gastrointest. Liver Physiol. **296**(1), 1–8 (2009)
32. Hunter, P.J., McCulloch, A.D., Ter Keurs, H.: Modelling the mechanical properties of cardiac muscle. Prog. Biophys. Mol. Biol. **69**(2–3), 289–331 (1998)
33. Kamm, K.E., Stull, J.T.: The function of myosin and myosin light chain kinase phosphorylation in smooth muscle. Annu. Rev. Pharmacol. **25**(1), 593–620 (1985)
34. Lammers, W., Stephen, B., Slack, J.R., Dhanasekaran, S.: Anisotropic propagation in the small intestine. Neurogastroenterol. Motil. **14**(4), 357–364 (2002)
35. Lentle, R.G., Janssen, P.W.: Physical characteristics of digesta and their influence on flow and mixing in the mammalian intestine: a review. J. Comp. Physiol. B **178**(6), 673–690 (2008)
36. Lentle, R.G., Janssen, P.W., Goh, K., Chambers, P., Hulls, C.: Quantification of the effects of the volume and viscosity of gastric contents on antral and fundic activity in the rat stomach maintained ex vivo. Dig. Dis. Sci. **55**, 3349–3360 (2010)
37. Lin, A.S., Buist, M.L., Smith, N.P., Pullan, A.J.: Modelling slow wave activity in the small intestine. J. Theor. Biol. **242**, 356–362 (2006)
38. Lindeburg, M.R.: Mechanical engineering reference manual for the PE exam. Professional Publications, Inc., Belmont (2006)
39. Meiss, R.A.: Mechanical properties of gastrointestinal smooth muscle. Compr. Physiol. 273–329 (1989)
40. Muller-Borer, B.J., Erdman, D.J., Buchanan, J.W.: Electrical coupling and impulse propagation in anatomically modeled ventricular tissue. IEEE Trans. Biomed. Eng. **41**(5), 445–454 (1994)
41. Nash, M.: Mechanics and material properties of the heart using an anatomically accurate mathematical mode. Thesis, The University of Auckland (1998)
42. Negi, L.: Strength of Materials. Tata McGraw-Hill Education, New Delhi (2007)
43. O'Grady, G., Du, P., Egbuji, J.U., Lammers, W.J.E.P., Wahab, A., Pullan, A.J., Cheng, L.K., Windsor, J.A.: A novel laparoscopic device for measuring gastrointestinal slow-wave activity. Surg. Endosc. **23**(12), 2842–2848 (2009)

44. Ozaki, H., Stevens, R.J., Blondfield, D.P., Publicover, N.G., Sanders, K.M.: Simultaneous measurement of membrane potential, cytosolic Ca^{2+}, and tension in intact smooth muscles. Am. J. Physiol. Gastr. L 260(5):C917–C925 (1991)
45. Pullan, A.J., Buist, M.L., Cheng, L.K.: Mathematically modelling the electrical activity of the heart: from cell to body surface and back again. World scientific publishing Co Pte Ltd, Singapore (2005)
46. Rao, S.S., Singh, S.: Clinical utility of colonic and anorectal manometry in chronic constipation. J. Clin. Gastroenterol. **44**(9), 579–609 (2010)
47. Rembold, C., Murphy, R.: Latch-bridge model in smooth muscle: $[Ca^{2+}]_i$ can quantitatively predict stress. Am. J. Physiol. Cell Physiol. 259(2):C251–C257 (1990)
48. Sacks, M.: Biaxial mechanical evaluation of planar biological materials. J. Elasticity **61**(1), 199–246 (2000)
49. Sanders, K.M.: Regulation of smooth muscle excitation and contraction. Neurogastroenterol. Motil. **20**, 39–53 (2008)
50. Seerden, T.C., Lammers, W., Winter, D., De Man, J.G., Pelckmans, P.A.: Spatiotemporal electrical and motility mapping of distension-induced propagating oscillations in the murine small intestine. Am. J. Physiol. Gastr. L **289**(6), G1043–G1051 (2005)
51. Szurszewski, J.H.: A 100-year perspective on gastrointestinal motility. Am. J. Physiol. 274(3 Pt 1): G447–G453 (1998)
52. Szurszewski, J.H., Farrugia, G.: Carbon monoxide is an endogenous hyperpolarizing factor in the gastrointestinal tract. Neurogastroenterol. Motil. 16(Suppl 1):81–85 (2004)
53. VanBuren, P., Palmer, B.M.: Cooperative activation of the cardiac myofilament. Circulation **121**(3), 351–353 (2010)

Multiscale Computational Modeling in Vascular Biology: From Molecular Mechanisms to Tissue-Level Structure and Function

Heather N. Hayenga, Bryan C. Thorne, Phillip Yen, Jason A. Papin, Shayn M. Peirce and Jay D. Humphrey

Abstract Blood vessels exhibit a remarkable ability to adapt in response to sustained alterations in hemodynamic loads and diverse disease processes. Although such adaptations typically manifest at the tissue level, underlying mechanisms exist at cellular and molecular levels. Dramatic technological advances in recent years, including sophisticated theoretical and computational modeling, have enabled significantly increased understanding at tissue, cellular, and molecular levels, yet there has been little attempt to integrate the associated models across these length and time scales. In this chapter, we suggest a new paradigm for identifying strengths and weaknesses of models at different scales and for establishing congruent models that more completely predict vascular adaptations. Specifically, we show the importance of linking intracellular with cellular models and cellular models with tissue level models. In this way, we propose a new approach for

H. N. Hayenga
Department of Bioengineering, University of Maryland, College Park, MD, USA
e-mail: hnhayenga@gmail.com

B. C. Thorne · P. Yen · J. A. Papin · S. M. Peirce
Department of Biomedical Engineering, University of Virginia, Charlottesville, VA, USA
e-mail: bct3d@virginia.edu

P. Yen
e-mail: py4wg@virginia.edu

J. A. Papin
e-mail: papin@virginia.edu

S. M. Peirce
e-mail: shayn@virginia.edu

J. D. Humphrey (✉)
Department of Biomedical Engineering, Yale University, 55 Prospect Street, MEC 212, New Haven, CT 06520, USA
e-mail: jay.humphrey@yale.edu

Stud Mechanobiol Tissue Eng Biomater (2013) 14: 209–240
DOI: 10.1007/8415_2012_147

Published Online: 16 September 2012

incorporating events across these three levels, thus providing a means to predict phenomena that can only emerge from a system of integrated interactions.

1 Introduction: Vascular Biology as a Complex System

Vascular development, adaptations to altered hemodynamics, the progression of disease, and responses to injury or clinical treatment—in each of these cases, one can identify tissue-level changes in geometry, structure, function, and properties that result from altered cellular phenotypes, which in turn depend on changes in intracellular signaling pathways. Indeed, our knowledge of the complex web of signals that underpin vascular function, homeostasis, growth, and remodeling at different levels of biological scale is growing exponentially as more sophisticated experimental models, techniques for analysis, and tools for integrating data become available. Recent technological developments in molecular biology and bioinformatics have thus made high-throughput analyses commonplace and "–omics" data widely accessible. The empirical tools we can use to manipulate and measure vascular structure, function, and adaptation in vivo—ranging from inducible genetic manipulations in mice to non-invasive intravital microscopy with single-cell resolution—are more flexible and precise than ever before. Parallel advances in systems biology, agent based modeling, continuum biomechanics, and computational methods have enabled significantly increased understanding of vascular biology at molecular, cellular, and tissue levels.

Despite all of these advances, critical questions in vascular mechanobiology—that is, many of the big questions that impact the care of thousands of patients each year—remain unanswered. For example, how do medial vascular smooth muscle cells (SMC) transduce mechanical stimuli in a way that impacts their production and secretion of proteases that degrade the extracellular matrix (ECM)? How, in turn, does degradation of ECM liberate growth factors that impact the proliferation of adventitial fibroblasts? What mechanical stimuli induce endothelial-to-mesenchymal transition and how does this perturb homeostasis? Similarly, many questions remain regarding interactions between wall mechanics and pharmacological treatments. What impact, for example, would a calcium channel blocker have on the stiffness of an arterial wall in the presence of a stiff atherosclerotic plaque that occludes 50 % of the lumen? While it is no small task to study such questions in isolation, the prospect of conceptualizing how these phenomena interact in space and time to create an emergent response is even more daunting. Indeed, the difficulty in answering these questions arises not from our ignorance of the individual components that are relevant in this complex system, but rather from the way in which they integrate to produce emergent outcomes. Even when we understand singular cause-and-effect relationships between two components, we face the challenge of integrating sets of relationships across different spatiotemporal scales in these complex systems. We submit that achieving a more holistic understanding of how blood vessels develop, maintain homeostasis, and respond to diseases requires an approach that couples experiments with theoretical and computational models and integrates processes across multiple length and time scales in a way that ultimately captures the emergent behaviors of the complex system that we know as a blood vessel.

This chapter will first provide an overview of vascular wall structure and many of the well characterized molecular signals that underpin functional and dysfunctional cellular behaviors in vascular tissues. We will then briefly describe some theoretical approaches, including continuum biomechanics, agent-based modeling, and models of intracellular signaling, that have been applied at different spatial scales to the study vascular function and adaptation. Examples from the literature will be highlighted to depict how modeling has been fruitful in producing new understanding at each level of spatial scale. We will then present our fundamental premise: that integration of processes across scales, as enabled by truly integrated, multiscale computational models, can result in a new understanding of emergent behaviors in vascular biology. This multiscale modeling approach, in turn, is expected to reveal new categories of questions that can be posed—questions that embrace multi-dimensional cause-and-effect relationships. We have developed the conceptual basis for such a multiscale model and begun integration efforts for a combined tissue-level continuum mixture and multi-cell agent-based model as well as for a combined agent-based and intracellular signaling model. We will conclude with a summary of our goal to integrate models from continuum to intracelluar scales and to highlight some of the opportunities and challenges posed by integrative multiscale modeling of complex systems.

2 Background

2.1 Vascular Wall Structure

The microstructure of arteries and veins varies with species, age, disease, and location along the vascular tree [30], yet the normal wall in maturity is characterized by three primary layers—the intima, media, and adventitia (Fig. 1). The intima, or inner layer, consists of a monolayer of endothelial cells (EC) and an underlying basal lamina composed of mesh-like type IV collagen and adhesion molecules such as laminin. In addition to being a smooth, nonthrombogenic interface between the blood and contents of the wall, the endothelium is biologically active. In response to chemical and mechanical stimuli, ECs produce a host of vasoactive molecules (which control vascular dilatation or constriction), growth factors (which promote cell replication or synthesis of proteins), proteases (which degrade proteins), and factors that regulate local immune responses and clotting processes. The endothelium also modulates transport of substances into the wall (e.g., white blood cells or lipids), and thereby plays important roles in diseases

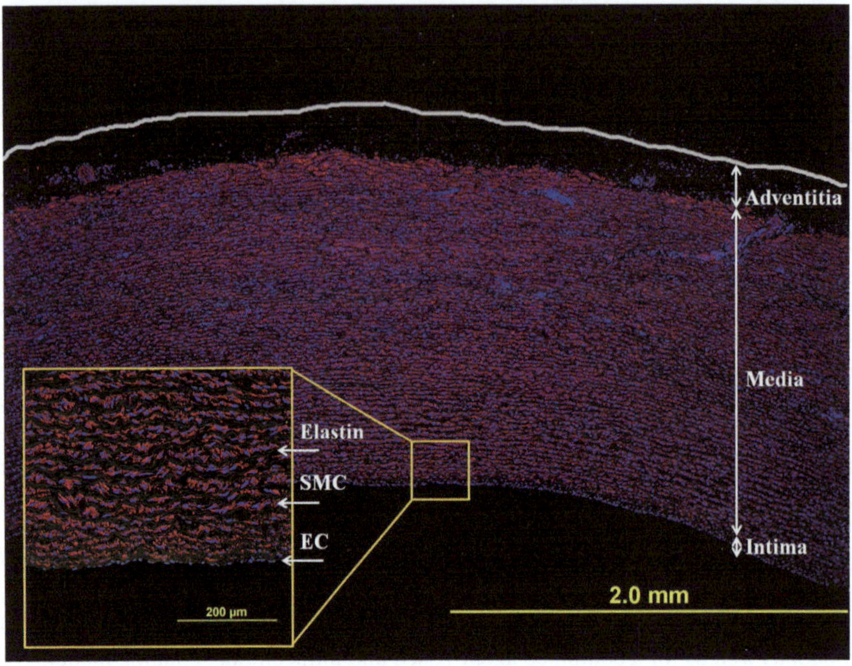

Fig. 1 Porcine aorta displaying the three arterial layers: thin intima, media, and thin adventitia. The monolayer of endothelial cells is revealed by the *blue lining* (cell nuclei) in the large image. The smooth muscle cells (SMC) are shown in *red* (alpha smooth muscle actin staining) and the layered elastin in *green*, which separates each SMC layer. One vessel of the vasa vasorum is seen in the *upper left* of the large image as a round collection of SMCs

such as atherosclerosis. Realization that many functions of the endothelium correlate with changes in hemodynamic loads provided important guidance for treating many vascular disorders and renewed interest in the biomechanics even though this layer contributes little to the overall structural integrity of the wall. Flow induced wall shear stresses tend to be on the order of 1.5 Pa in arteries and 0.15 Pa in veins; the mean value of this stress can be estimated in vivo from measurements of viscosity, volumetric flow-rate, and luminal radius.

The media, or middle layer, consists primarily of SMCs embedded in ECM consisting of elastic fibers, various types of collagen (I, III, V, etc.), and proteoglycans (Fig. 1). In general, the closer these vessels are to the heart the more elastin (the main constituent of elastic fibers), and the farther away the more smooth muscle. Regardless, the mean circumferential wall stress tends to be on the order of 100 kPa, which can be estimated in vivo via measurements of transmural pressure, inner radius, and wall thickness. Whereas SMCs primarily synthesize proteins of the ECM during development and disease, they endow the normal mature wall with an ability to constrict or dilate and thereby regulate blood flow locally. Smooth muscle contraction, hypertrophy (increase in size), hyperplasia (increase in number), apoptosis (cell suicide), and migration (often from the media to subintima) play essential roles in diseases such as aneurysms, atherosclerosis, and hypertension. Loss of matrix proteins, particularly elastin, similarly plays key roles in the formation of aneurysms or dissections.

The adventitia, or outer layer, often merges with the perivascular tissue. It consists primarily of fibroblasts and axially oriented type I collagen, but may include admixed elastic fibers, nerves, and its own small vasculature, the vasa vasorum, when the thickness of the wall is too great to allow sufficient transmural diffusion of oxygen directly from the blood. Fibroblasts are responsible primarily for regulating the matrix, particularly collagen, but they can migrate, proliferate, and differentiate. Indeed, there is growing evidence that migrating fibroblasts play significant roles in many diseases. Nevertheless, the normal adventitia is thought to serve, in large part, as a protective sheath that prevents over-distension of the media; like all muscle, smooth muscle contracts maximally at a certain length, above and below which the contractions are less forceful. Finally, the adventitia is typically demarcated from the media by an external elastic lamina (except in cerebral arteries); the media is similarly demarcated from the intima by an internal elastic lamina, which is a fenestrated, cylindrical sheet of elastin.

Cross-linked elastin is one of the most stable proteins in the body; it endows vessels with considerable elasticity over finite deformations (e.g., nearly linear stress response over stretches of 150–200 %) and it helps control the phenotype of the SMCs. Specifically, cross-linked elastin encourages a quiescent, contractile phenotype characteristic of maturity. This is in contrast to effects of the elastin precursor, tropoelastin, which is not cross-linked, contributes little to the structural integrity, and encourages smooth muscle migration, proliferation, and synthesis of extracellular matrix, particularly in development. The collagens are the primary family of structural proteins in the body, with fibrillar types I and III endowing tissues with significant tensile stiffness. Collagen fibers turn over continuously and thereby play key roles in homeostasis and remodeling. They can be on the order of microns in diameter and are often undulated slightly in the normal physiologic state; they manifest their high stiffness when straightened. Proteoglycans represent a large class of molecules having diverse functions. Structurally, they tend to be most important in sequestering water within the tissue, which, as in cells, typically accounts for $\sim$70 % of the total mass. To provide a better idea of relative distributions of these various constituents, the media of the thoracic aorta (cf. Fig. 1) consists of, by dry weight, $\sim$37 % collagen, 33 % SMCs, 24 % elastin, and 6 % other constituents whereas the adventitia consists of $\sim$78 % collagen, 9 % fibroblasts, 2 % elastin, and 11 % other constituents. Nevertheless, each vessel has different distributions of constituents within each layer and different relative thicknesses of the media and adventitia; overall structural integrity is thus differentially controlled by balances and imbalances in cell and matrix turnover (Fig. 2). For more on vascular ECM and its relation to mechanics, see [62].

Vascular structure, function, and material properties are dictated by the three primary cell types of the wall (endothelial, smooth muscle, and fibroblasts) and in some cases cells from the blood stream (e.g., platelets, monocytes, progenitor

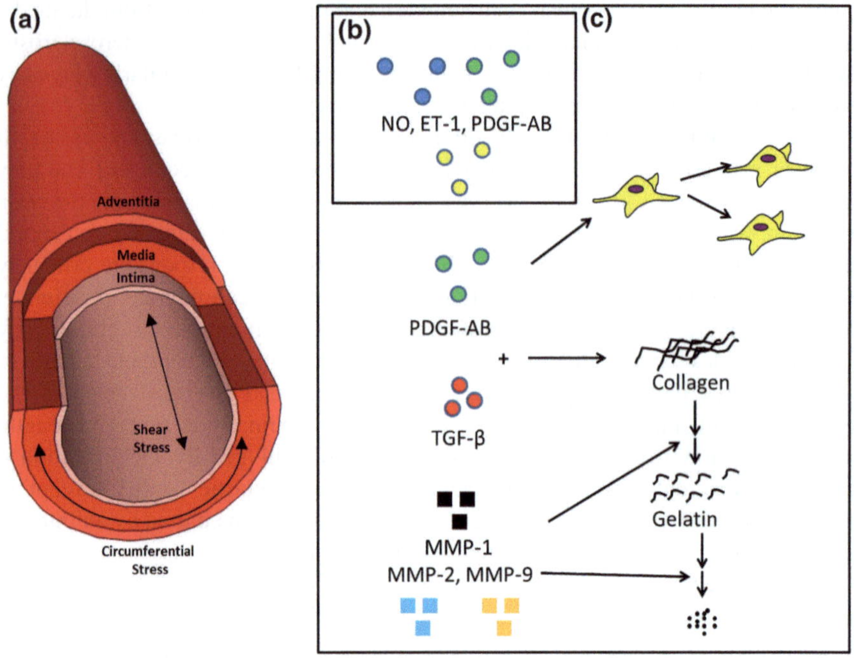

Fig. 2 Illustrative arterial mechanics and molecular interactions. **a** The three layers of an artery wall. The Intima is composed of endothelial cells and a basement membrane; the Media, smooth muscle cells and primarily elastin and collagen; the Adventitia, fibroblasts and extracellular matrix. Shear stress typically orients in the axial direction, while intramural stresses are circumferential and axial in orientation. **b** Factors governed by shear stress and produced by endothelial cells include NO, ET-1, and PDGF-AB. **c** Factors governed by circumferential stress and produced by smooth muscle cells include PDGF-AB, TGF-β, and MMPs. Illustrative effects of these growth factors and MMPs are shown to the *right*. PDGF-AB and TGF-β act through complex intracellular pathways to promote collagen production and cell proliferation or phenotypic switching, while MMPs degrade collagen, gelatin, and elastin

cells). That ECs respond directly to changing mechanical loads was first noted ~35 years ago and is now a well-documented example of mechanotransduction—that is, the sensing and converting of mechanical stimuli into a signal that controls gene expression and hence cellular activities. For example, ECs increase their production of vasodilators (e.g., nitric oxide (NO) and prostacyclin (PGI_2)) in response to increases in wall shear stress; this allows the vessel to dilate, thereby decreasing the resistance to flow, and promotes endothelial proliferation to cover the increased surface area of the dilated vessel. Conversely, ECs increase production of vasoconstrictors (e.g., endothelin-1 (ET-1) and angiotensin-II (ANG-II)) in response to decreases in flow or increases in pressure. Endothelial cells also produce a host of growth regulatory molecules (including vascular endothelial growth factor (VEGF), platelet derived growth factors (PDGF), fibroblast growth factors (FGF)), adhesion molecules (including vascular cell adhesion molecule (VCAM-1) and intercellular adhesion molecule (ICAM-1)), cytokines and chemokines (e.g., interleukin-1, IL-1, and monocyte chemoattractant protein, MCP-1), and clotting factors (e.g., tPA)—all in response to changing mechanical stimuli including local stresses or trauma. See Refs. [19, 28] for more on endothelial mechanobiology.

The structure of smooth muscle differs from that of skeletal and cardiac muscle, but its contractility also depends on a calcium dependent actin-myosin interaction. Vascular smooth muscle can generate contractile forces comparable to those of striated muscles while maintaining the contraction for longer periods and at a lower expenditure of energy. This feature allows blood vessels to maintain a "basal tone" from which they can dilate or constrict further. Like ECs, SMCs respond to changes in their mechanical environment; for example, SMCs alter their synthesis of collagen in response to changes in mechanical loading. Mechanical damage to elastin can also induce phenotypic changes in smooth muscle that promote migration, proliferation, and apoptosis in addition to synthesis of matrix. This causality appears to be fundamental to the response of the arterial wall to clinical interventions such as balloon angioplasty and stenting and likewise to the response of the venous wall to its clinical use as an arterial by-pass graft. See Refs. [42, 68] for more on the mechanobiology of smooth muscle.

Fibroblasts are primarily responsible for regulating the extracellular matrix in the adventitia, as, for example, via synthesis and degradation of collagen. Degradation is accomplished via ingestion by cells (phagocytosis) or the release of enzymes, including the matrix metalloproteinases (MMPs). Fibroblasts play an important role in regulating the ECM in many soft tissues (from the eye to the skin to heart tissue) and are easily studied in vitro. For these reasons, there is a considerable literature on the mechanobiology of fibroblasts and myofibroblasts (e.g., [59]). Macrophages are scavenger cells; in response to a local injury, they enter the vessel wall from the blood (actually blood borne monocytes adhere to the wall and transform into macrophages while inside the wall) and act primarily via phagocytosis or the release of MMPs. They, too, are responsive to changes in mechanical stimuli [64]. Platelets also circulate within the bloodstream; they play a key role in coagulation, but also release growth factors (e.g., PDGF) and vasoconstrictors (e.g., serotonin, 5-HT, and thromboxane, TXA_2) that affect both ECs and SMCs. Platelet derived vasoconstrictors play a particularly damaging role following the rupture of intracranial aneurysms, causing nearby vessels to constrict and cause distal strokes. More specifics of the molecular biology of blood borne cells can be found in general textbooks.

2.2 Key Signaling Pathways in Vascular Adaptation

Myriad signaling pathways play important roles in vascular homeostasis and adaptation in both large and small vessels. Many of these molecules are homologous across species and play important roles in mediating homeostasis and growth in other organ systems. For example, VEGF is a highly conserved family

Table 1 Some of the key molecules produced and/or expressed by vascular cells in response to altered hemodynamic loading, disease, and injury

Vasoactive molecules	Cytokines
Nitric oxide	IL-1
Endothelin-1	IL-6
Angiotensin-II	IL-8
Serotonin	IL-10
Thromboxane	SDF-1
Thrombin	
Growth factors/receptors	**Proteinases and modulators**
PDGF-BB/PDGFR-alpha, beta	MMP-2
VEGF/Flk-1, Flt-1, neuropilin	MMP-9
TGF-β1/TGFBRI/II	MMP-1
bFGF/FGF-R	TIMP-1
EphrinB2/EphB4	TIMP-2
Cell–cell adhesion	**ECM proteins**
P-selectin/PSGL-1	Type I collagen
E-selectin/PSGL-1	Type III collagen
CD-34/L-selectin	Type IV collagen
VCAM-1/VLA-4	Elastin/Microfibrils
ICAM-1/LFA-1 & MAC-1	Laminin/Fibronectin

of molecules present in zebrafish, mice, and humans, and is a pivotal regulator of growth and patterning in both the circulatory and nervous systems.

The key signaling pathways in the vascular tree can be broadly lumped into functional categories based on their ability to mediate vascular tone (vasoactive molecules), activate cells (cytokines), induce growth (growth factors), form and impact the extracellular milieu (ECM molecules and proteinases), and orchestrate cell adhesion (Table 1). While a comprehensive review of all of these key signaling pathways is beyond the scope of this chapter, we highlight some of the most widely studied molecules that play a diverse set of context-specific roles in the vasculature. We thus focus on a small sub-set of signaling pathways that are particularly important in mediating vascular responses to physiological and pathological alterations and those that are most relevant in our proposed multiscale model of hypertension, which is discussed in subsequent sections.

NO is one of the most widely studied signaling molecules in the vasculature. It exerts its effects at both systemic and cellular levels throughout the microcirculation and in larger vessels throughout the body. This highly diffusible and short-lived free radical gas is synthesized via nitric oxide synthases [35], a family of enzymes that convert L-arginine to NO (for review of this process in ECs see [54]). Impaired NO activity, due to decreased synthesis or increased degradation, is a hallmark of endothelial dysfunction and has been observed in a host of conditions and diseases ranging from aging to atherosclerosis, hypertension, and diabetes [47]. NO is a potent vasodilator [12] and thus regulates vascular SMC tone. Both exogenous NO [48, 72] and over-expression of endothelial nitric oxide synthase (eNOS) promote angiogenesis [56]. Interestingly, NO and eNOS also upregulate and are synergistic with pro-angiogenic growth factors, such as VEGF and Angiopoietin-1 [10]. This relationship supports an intriguing mechanistic linkage between hemodynamic alterations due to vasodilation and growth factor mediated angiogenesis. NO also plays a role in mediating microvascular permeability (for review see [22]) and inflammation by inhibiting platelet adhesion, aggregation, and leukocyte adhesion [24].

NO production via eNOS has long been associated with the mechanical shear stress experienced by the vascular endothelium of large vessels (e.g., during sustained changes in blood flow) and is critical for blood flow-dependent adaptive remodeling of the media [50], yet a direct molecular linkage between NO and SMC remodeling (proliferation and apoptosis) was only recently discovered by Yu et al. [69]. They demonstrated that abnormal flow-dependent remodeling in eNOS knockout mice is associated with activation of the PDGF signaling pathway and downstream inhibition of apoptosis. Moreover, they showed that NO negatively regulates PDGF-induced cell proliferation in vascular SMCs. Hence, this signaling module represents yet another example of the mechanistic linkages between mechanical forces experienced by blood vessels, diffusible signals and morphogens secreted by cells, and defined cellular behaviors that have important consequences on long-term vascular tissue structure and function.

As mentioned above, PDGF is a family of growth factors synthesized and secreted by vascular ECs and SMCs in homodimeric (e.g., PDGF-AA and PDGF-BB) and heterodimeric (e.g., PDGF-AB) forms [20]. PDGF homodimers and heterodimers bind to dimeric tyrosine kinase receptors, PDGFR-alpha and PDGFR-beta, with different affinities. PDGF is a potent mitogen for SMCs and fibroblasts, stimulating proliferation, migration, and preventing apoptosis [41] (Fig. 2). Dysfunction of the PDGF signaling pathway has been implicated in a number of diseases, including pulmonary hypertension [51], cancer [38], renal disease [41], and diabetic retinopathy [66]. There is extensive evidence that implicates PDGF in inhibiting SMC differentiation, and the extensive intracellular machinery (e.g., gene promoters and repressors) that enact its ability to shift SMC phenotypes from differentiated to synthetic/proliferative are well described [67]; for a review, see Ref. [36]. How the phenotypic states of a collection of SMCs within the medial wall, in turn, impact the mechanical stiffness of that tissue, which may further be influenced by regional NO levels, is less well understood and requires the type of multiscale modeling that we will focus on in subsequent sections of this chapter.

Thus far in this section, we have highlighted how small molecule signals (e.g., NO) and growth factors (e.g., PDGF) impact vascular adaptation, but we would be remiss to leave out the impact that ECM and its modifiers have on vascular growth and remodeling. While various extracellular proteins and glycoproteins (e.g., elastin, fibrillins, and fibulins: see Ref. [62]) provide a substrate for vascular cell assembly and stability, proteolytic enzymes such as the MMPs critically impact vessel homeostasis and adaptation by degrading the ECM (Fig. 2) and by mediating intercellular signaling (for review see Ref. [46]). The MMP family includes collagenases, gelatinases, stromelysins, matrilysins, and membrane-type MMPs. ProMMPs are cleaved into active forms, which degrade ECM proteins, and their effects are balanced by tissue inhibitors of metalloproteinases (TIMPs) that prevent excessive proteolytic ECM degradation.

As noted in subsequent sections, MMP-2 and MMP-9 (gelatinases A and B) are upregulated during sustained hypertension and contribute to ECM reorganization, SMC proliferation and migration, and vascular hypertrophy in large vessels. Increased MMP-2 levels, in particular, have been associated with impaired NO-mediated vasorelaxation, arterial wall hypertrophy, and excessive collagen and elastin deposition. Therapeutic MMP inhibition with doxycycline has been proposed as a pharmacological strategy to attenuate SMC proliferation and hypertrophy during hypertension (for review see Ref. [15]) as well as the treatment of aneurysms. MMPs can also liberate and activate matrix-bound growth factors, such as TGF-beta [13], which may have opposing influences on SMC differentiation. Thus it is important to quantitatively assess these interactions with spatial and temporal resolution in order to resolve issues of therapeutic dose and timing.

3 Modeling Foundations and Current Models

3.1 Continuum Biomechanics and Illustrative Vascular Models

Continuum biomechanics has proven to be an important contributor to our understanding of physiology and pathophysiology as well as to the design of medical devices, biomaterials, and tissue engineered constructs. It is fundamental, for example, to many analyses of vascular biology and pathophysiology that are based on clinically available information such as blood pressure, local blood flow, and complex geometry [30]. Continuum biomechanics is founded upon five basic postulates: balance of mass, linear momentum, and energy as well as balance of angular momentum and the entropy inequality. Whereas the first three types of relations provide partial differential equations of motion, the last two provide important restrictions on the forms of the constitutive relations (i.e., descriptors of individual material behaviors). An underlying assumption is that one can compute at each macroscopic point (or location) and each instant a meaningful "continuum average" of properties or physical quantities of interest; a general guideline is that the continuum assumption is reasonable if the characteristic length scale of the microstructure is much less than the characteristic length scale of the physical problem. For example, continuum biomechanics can be equally applicable to studying an arterial wall (wherein diameters of collagen and elastic fibers are on the order of μm and overall vessel diameter is on the order of mm or cm) or an isolated cell (wherein diameters of cytoskeletal filaments are on the order of nm and overall cell dimensions are on the order of μm).

To date, the two primary vascular applications of the continuum approach have been to compute pressure and velocity fields in blood flow (i.e., hemodynamics) and to compute stress and strain fields within the vascular wall (i.e., wall mechanics), each of which requires explicit solution of mass and linear momentum balance. Although cells cannot sense continuum metrics such as stress and strain, these quantities have proven useful in correlating mechanobiological responses by cells to diverse loads [31]. For example, simple parallel plate flow experiments demonstrate that ECs are very responsive to changes in wall shear stress, which is calculated using the continuum approach; simple organ culture experiments on straight segments of arteries and arterioles demonstrate that SMCs are very responsive to changes in pressure and extension, which induce intramural changes in stress and strain. Given the complexity of the microstructure of cells and tissues down to the level of molecular interactions, it is inconceivable that one would attempt to use a purely molecular dynamics simulation to study problems that manifest at a physiological or clinical scale. That is, continuum biomechanics is much more appropriate to study problems involving, for example, changes in the structural stiffness of the arterial wall in hypertension, the effects of evolving vascular diseases such as atherosclerosis or aneurysms, or the design of novel interventional devices such as intravascular stents, heart valves, or left ventricular assist devices.

Notwithstanding past successes, until recently continuum biomechanics had focused primarily on material behaviors at a particular time, not how they evolve. Moreover, most studies had assumed that the tissue (or cell) is materially uniform. Yet, all tissues are materially non-uniform, consisting of different types of cells and matrix that turnover, and so too cells consist of different organelles and cytoskeletal proteins that change over time. In an attempt to address these complexities, Humphrey and Rajagopal [33] proposed a Constrained Mixture Model (CMM) that allows one to track evolving changes in the properties, turnover rates, and natural (i.e., stress-free) configurations of individual structural constituents that comprise a tissue or cell. Computations have shown that this approach can capture salient features of diverse vascular adaptations and disease processes (cf. [6, 61]). Briefly, a CMM of arterial growth and remodeling consists of full mixture equations for mass balance plus a single equation for overall linear momentum balance that is solved for the net stress field. The associated two classes of constituents are: structurally insignificant but soluble constituents, such as vasoactive molecules, growth factors, cytokines, and proteases, and structurally significant but insoluble constituents, such as elastin, collagen, and muscle. Linear momentum balance is solved via a rule-of-mixtures constitutive relation for the structurally significant constituents. The need for but a single linear momentum equation stems from the assumption that negligible momentum exchanges exist between structurally significant constituents, which appear to deform together with the mixture. Because inertial loads are often negligible in the calculation of arterial wall stresses, even during transient loading, we assume further that structurally significant constituents experience quasi-static loading. The primary constitutive relations thus reduce to equations for the production and removal of structurally insignificant and significant constituents (the former via classical reaction–diffusion type relations) and stored-energy functions for the structurally significant constituents. Each of these relations must be formulated based on appropriate experimental data. For example, data reveal that NO not only causes vasodilatation (thus affecting the stress–strain behavior), it also affects the synthesis of collagen (production relation) and inhibits inflammation (which influences the removal relation).

3.2 *Agent Based Modeling and Illustrative Vascular-Specific Models*

Multi-cell biological phenomena, such as the assembly of cells into tissues in response to environmental cues, can be modeled using either continuum or discrete approaches. A primary focus of multi-cell modeling has been to quantify how patterns emerge in tissues and whether they arise from diffusible biochemical factors, mechanical forces, cell–cell interactions, cell–matrix interactions, or any combinations thereof. While most continuum-based models approximate individual cells as a series of similar units, discrete multi-cell models explicitly represent individual cells as distinct entities capable of exhibiting individual behaviors, which provides increased generality. Amongst the different approaches to discrete cell modeling, the most common are Agent-Based Models (ABMs), Cellular Potts Models (CPMs) [25], and statistical models. CPMs generalize an approach from statistical mechanics called the Ising model, and simulate biological cells by mapping them to domains on a lattice. Cell behaviors are described by effective energies and elastic constraints, and cellular dynamics, such as migration and cell shape changes, are guided by principles of energy minimization [16, 17, 34, 40, 70]. Statistical approaches, such as Monte Carlo simulations, model biological cells as discrete objects, and their behaviors are dictated by purely probabilistic rules [21].

3.2.1 Agent Based Modeling

ABMs represent a computational approach that has been used extensively in the social sciences and ecology [27], but only recently has it been employed in biomedical research to study multi-cell phenomena such as tumorigenesis [71], angiogenesis [44], inflammation [7], and arterial wall remodeling in hypertension [58]. This technique rests on the idea that local interactions between members of a population can result in complex higher-level emergent phenomena. The key components of an ABM include the agents themselves, their behaviors within their environment (i.e., simulation space), the rules that govern their behaviors, and the simulation inputs and outputs.

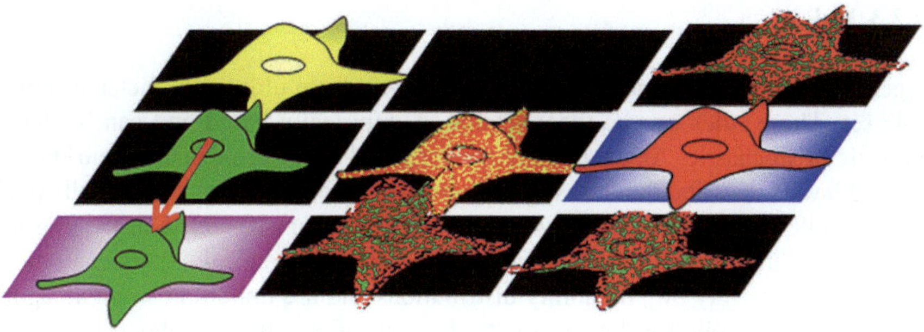

Fig. 3 Components of an agent-based model. The grid of squares represents the simulation space, which can hold concentrations of extracellular factors, denoted here by colored squares. Agents/cells reside at discrete points on the grid, and are colored to represent cell phenotype or activity. Agents perform behaviors according to their rule set. For example, the *yellow* cell is currently static, with no outside influences, while the *green* cell is following a rule instructing it to migrate up a gradient of a chemokine (*purple*). A third cell (*red*) is affecting the phenotype of its neighbors through paracrine secretion of a growth factor (*blue*)

3.2.2 Agents

Agents are discrete entities that perform a defined set of behaviors according to a set of rules. Agents representing cells, for example, are programmed to exhibit biologically-relevant behaviors, including migration, proliferation, differentiation, and apoptosis [3–5, 43, 44] as shown in Fig. 3. Rules governing an agent's decision to perform these behaviors may take into account, among other things, the agent's past state history and factors present in the local simulated tissue environment; these rules can thus be stochastic or deterministic. Agents can detect biochemical and biomechanical stimuli in their environment and respond by exhibiting a particular behavior. Agents can also affect their environment by secreting diffusible growth factors (e.g., PDGF-BB) or by producing ECM and its modifiers, such as collagen or MMP-9. Finally, agents can interact with one another. For example, neighboring agents with engaged cadherins can interact physically much in the same way as neighboring cells in a tissue would interact—by signaling directly to one another, transmitting a mechanical force, and so forth. All of these actions and interactions will impact an agent's state, and the ABM can record the state histories of each agent at each time step, facilitating individual cell lineage tracking. To simulate more complex phenomena, multiple cell types can be represented using multiple types of agents. For example, an ABM modeling the progression of an atherosclerotic plaque might contain agents that represent individual ECs, SMCs, macrophages, foam cells, or fibroblasts. The aggregate of agent behaviors and interactions over space and time produces emergent phenomena that could not be predicted by modeling a single cell or using models making a continuum assumption.

3.2.3 Rules

Agent-specific actions and interactions in an ABM are ultimately dictated by the rule set. These rules can be theoretically or empirically based and can be deterministic or stochastic. An example of a theoretically based rule is the use of Fick's second law to describe the diffusion of a growth factor that affects other cells [44]. An example of an empirically based rule is a dose–response curve describing the speed of cellular migration as a function of chemokine (e.g., IL-8) concentration. Stochastic rules rely on probability distributions and are important when an agent state does not explicitly dictate a behavior, but affects the likelihood of such a behavior. Rules are most often derived or estimated from the literature; the organization and presentation of these rules is facilitated by a table or flow chart accompanying the ABM (cf. [7]). Rule sets can vary from fewer than ten rules [26] to as many as 200 [7]. The composition, content, and accuracy of a rule set have a profound impact on the output of an ABM. Slight modifications to a single rule can dramatically alter the output of even the simplest ABM. Therefore, in designing, constructing, and implementing ABM rule sets, caution is urged to ensure that the rules are accurate, non-redundant, and necessary.

The validity of a rule set can be checked by contrasting model outputs against experimental data [39, 44, 55], and by performing a sensitivity analysis, where rules are systematically removed or adjusted incrementally to determine their contribution to the overall ABM output [26]. Because the outputs of an ABM are highly dependent on empirical rules, it is necessary to couple models with experiments at all stages of model development and to validate an ABM's rule set by performing iterative in silico and in vivo/in vitro experimentation. For a review of the integration of experimental data with ABMs, see Ref. [57], and for a method of assessing the quality of a model's rule set, see Ref. [58].

3.2.4 Inputs and Outputs

Most biological ABMs simplify tissue geometry by simulating cell behaviors in a quasi-two-dimensional simulation space that reduces model complexity and speeds up simulations. For example, one can use a one-cell thick axial slice of vessel to model arterial adaptations to hypertension [58]. This simplification enabled measurement of vascular wall thickness and cellularity and was sufficient to enable calculation of the concentration and diffusion of extracellular proteins as well as to facilitate direct comparisons with experimental data. An ABM simulation space can have closed [39], open [7], or periodic boundary conditions [53], and the positioning and state assignments for the agents at the start of a simulation are specified by initial conditions that are frequently derived from microscopic images obtained at a starting time [7, 39]. Setting initial conditions in this way enables direct comparisons with the experimental data at later time-points for model validation. The initial agent states are assigned based on empirical observations (e.g., histology) that describe baseline conditions for agent states. The time steps of biological ABMs can span a range of scales, from milliseconds to hours [14, 44], and simulation times can span years [1].

Outputs of ABMs include the spatial arrangements of agents within their simulation space, their internal state, and the state of the environment. Spatial patterns of agent organization may emerge during the simulation, and these can be analyzed quantitatively. For example, in an ABM simulating angiogenesis, the pattern of new microvessel growth can be assessed by measuring the new vascular length, counting branch points of vessel trees, and quantifying the number of new capillary sprouts that have developed over the course of the simulation time window. Analyzing agent patterns using metrics that are also used to quantify biological phenomena experimentally facilitates direct comparison of ABM predictions to experimental data, which enables rigorous model validation. Because ABMs generalize intracellular processes, but are more fine-grained (discrete) than is suitable for continuum analysis, they are uniquely suited to bridging disparate biological scales.

3.3 Signaling Pathways and Illustrative Vascular-Specific Models

Intracellular signaling pathways govern basic cell functions and allow cells to adapt to their microenvironments. These signaling pathways consist of a complex network of interacting molecules that give rise to a diverse range of cell functions such as proliferation and differentiation. Often, an extracellular ligand binds to a cell-surface receptor and triggers a cascade of intracellular interactions between signaling molecules and second messengers that ultimately results in a change in transcriptional activity, metabolism, or other regulatory function.

Because network elements of signaling pathways often overlap, the causal relationship between input and output is not always explained by a linear series of events. Furthermore, network motifs, such as positive and negative feedback loops, make it difficult to deduce the relationships between the network elements solely by intuition [2]. To better study and understand intracellular signaling pathways and networks, a combination of experimental and mathematical approaches have been used to disentangle the functions of the highly interconnected components. Mathematical and computational Intracellular Signaling Models (ISM) are used to contextualize experimental data and predict possible emergent behaviors that are difficult to realize by experimentation alone. Through cycles of model refinement and experimental validation, one can begin to understand and probe the complexities of the signaling network and make predictions about how specific cellular functions arise.

These computational models predict dynamic behaviors of biochemical reactions by using mathematical relations to describe the underlying molecular interactions. Traditionally, ordinary differential equations (ODEs) are used to model the concentration profiles of the different signaling molecules over time in response to a stimulus. Mass action and Michaelis–Menten kinetics are common ways to represent the reaction kinetics. These approaches have been used to demonstrate how even relatively small and focused modules can exhibit emergent behavior through feedback mechanisms [11]. If information about the spatial distributions of the molecules is desired, partial differential equations (PDEs) are used. ODEs and PDEs can then be implemented deterministically or stochastically. Deterministic systems have no element of randomness while stochastic systems incorporate probabilities in the evolution of the system output over time. The choice of how the model is formulated depends upon how the modeler views the system and the biological question being answered.

While modeling of intracellular signaling networks pertaining to blood vessels is still a relatively young field, an increasing number of computational models are being developed to study vascular function. To date, a few aspects of vascular intracellular signaling have been prominently modeled. Proliferation of ECs and the formation of vessels is one such area. For example, several intracellular signaling network models have been developed to study vessel formation in the context of vasculogenesis during embryonic development and angiogenesis in both physiologic and pathophysiological cases [37, 45, 52]. In terms of particular signaling pathways, those involving NO and calcium have received much attention in recent computational models [60, 65]. We will summarize some of the published models of vascular signaling networks in the following paragraphs and highlight the new understanding they produced.

As noted above, among other functions, NO is a key signaling molecule that regulates tissue-level vasodilation by affecting cell-level (SMC) contraction. One group modeled the NO/cGMP signaling pathway in vascular SMCs and reproduced NO/cGMP-induced smooth muscle relaxation effects, such as intracellular Ca^{2+} concentration reduction and Ca^{2+} desensitization of myosin phosphorylation and force generation [65]. The authors of this model proposed a cGMP feedback-controlled soluble guanylate cyclase (sGC) decay from its activated to its basal form and predicted that the intermediate form of sGC is the dominant steady-state form of sGC under physiological NO stimulation. This model thus suggested that the sGC desensitization by cGMP feedback may limit cGMP production over a wide range of NO concentration, which may contribute to the robustness of the response of vascular SMCs to small perturbations in NO. The different mathematical models used to investigate the role of NO in microcirculatory physiology have been reviewed in [60], and one of the overarching themes is that intracellular signaling models can provide valuable insights into roles of NO in physiology, especially because experimentally measuring tracer amounts and signaling events of NO in biological tissue with the appropriate spatial and temporal resolutions is difficult.

Calcium signaling is largely coupled to NO signaling. In the arterioles, SMCs and ECs are coupled via the exchange of Ca^{2+} along with other ions and the paracrine diffusion of NO. The vascular response to the nonlinear interactions of subcellular components and processes including Ca^{2+} signaling have been studied in rat mesenteric arterioles via mathematical modeling [35]. This ODE model predicted that in the rat mesenteric arterioles, ECs exert a stabilizing effect on intracellular vascular SMC Ca^{2+} oscillations, which synchronize with oscillations in vessel tone to cause vasomotion. This stabilization allows Ca^{2+} oscillations to be maintained over a wider range of agonist concentrations. Models related to intracellular signaling include those that simulate the dynamics of ionic flows across the membrane. One group modeled the dynamics of the Na^{+}/Ca^{2+} exchanger (NCX) in vascular SMCs; this exchanger regulates the reloading of the sarcoplasmic reticulum and the maintenance of Ca^{2+} oscillations in activated SMCs [23]. Although this model is not of an actual signaling network, it gives insights into the mechanism of the coupling between Na^{+} entry via TRPC6 (a non-specific cation channel) and the NCX. This implicates the concentration of Ca^{2+} in the vascular SMC, which affects the signaling pathways involving Ca^{2+}. The model incorporates a stochastic element to simulate the movement of single Na^{+} ions in the nanospace between the plasma membrane and the sarcoplasmic reticulum. This model predicted that in order to have a Na^{+} concentration transiently elevated in the plasma membrane/sarcoplasmic reticulum nanospace, there must be physical obstructions to Na^{+} motion, which form a relatively impermeable barrier around the TRPC6 channel. NCX must also be localized near TRPC6 within such barrier in order to sense the high Na^{+} concentration, reverse, and allow Ca^{2+} into the sarcoplasmic reticulum. As the details of individual intracellular signaling pathways become better understood, models can be evolved to incorporate additional pathways such that they combine to form an interconnected signaling network that increasingly describes the physiological system with more accuracy.

3.4 Limitations of Singular Modeling Approaches

3.4.1 CMM Limitations

Continuum models have proved very useful in vascular research, including helping to reveal the existence of residual stresses and their effects on the transmural distribution of stress that led to a fundamental mechanobiological hypothesis [32]. Continuum models also continue to be very successful in explaining and predicting a wide variety of nontrivial aspects of vascular physiology and pathophysiology, including the adaptation of arteries to sustained changes in blood pressure or flow as well as the rupture of aneurysms. Nevertheless, limitations remain. For example, the continuum approach assumes material is distributed continuously over particular length scales, which can mask specific mechanisms of mechanotransduction that result from cell–matrix interactions at discrete sites (e.g., focal adhesions) but otherwise help to drive overall tissue-level adaptations. Current models also do not account for the details of particular matrix–matrix interactions, including interactions between collagen fibers and proteoglycans. Perhaps most importantly, however, CMMs cannot exploit directly the growing knowledge-base of molecular vascular biology.

3.4.2 ABM Limitations

Agent-based models can explicitly include complex cell–cell interactions, but are limited when it comes to modeling multiscale phenomena. For example, current ABMs do not satisfy some of the important constraints on vascular behavior that stem from classical physics (e.g., conservation of linear momentum). Simulating whole tissues would also require an extremely large number of individual agents, which quickly becomes prohibitive computationally, particularly for complex rule-sets that may include stochastic rules. While stochastic rules can produce a population of results that more closely resembles actual experimental data, including stochasticity necessitates an even larger number of model runs to converge on an average output. Finally, single-scale, cell-level ABMs treat the agent as a black box; intracellular interactions are often included implicitly when rules are derived from cell-level experiments. When intracellular processes are simple, this is less of a problem, but as intracellular interactions grow more complex, experimental approaches to rule development can become intractable. Of course, as with any model, the ABM rule-set is only as good as the data from which it is derived. Often the kind of data needed to develop rules for cellular behaviors with complex multifactorial inputs are not available in the literature.

3.4.3 ISM Limitations

Models of intracellular signaling networks are useful tools in predicting the effects of signal transduction in a cell in response to a stimulus. These predictions are made at the cellular level, and may require coupling with higher level models to infer tissue or organ level function. For example, Ca^{2+} signaling in vascular smooth muscle cells is studied to make predictions about vasomotion in blood vessels. Vasomotion requires the synchronization of oscillations in the concentration of Ca^{2+} in a large group of vascular smooth muscle cells, and gap junctions are believed to play an important role in this process [18]. Hence, elements of cell–cell interactions often need to be coupled to an ISM to convincingly bridge the gap between single cell and tissue level predictions. Lastly, as with the other types of mathematical models, determination of the scope and complexity of an ISM during the developmental stages is not a trivial task. In some cases, the paucity and confidence of relevant experimental data to be used as model parameters can directly limit the scope and usefulness of an ISM.

4 Linking Intracellular, Cellular, and Tissue Level Models

4.1 An Integrative Approach

Single-scale models of the types described above all have their own unique advantages but also limitations. In vivo changes in vascular wall structure depend critically on interactions at all scale levels, from intracellular signaling to whole vessel biomechanics. A more complete understanding of the responses of blood vessels to changes in hemodynamics requires detail at the cellular level that CMMs alone cannot provide, while many of the functions of SMCs (such as proliferation and collagen production) result from the integration of so many signals that experiments to develop rules for a single-scale, cellular level ABM, are not feasible. Such an integration of signals could be carried out by an ISM, however. A multiscale, multi-model approach is thus in order, which promises to build on the strengths of each type of model while compensating for limitations by offloading them to another model better suited for those specific types of tasks.

Coupling models at different spatial and temporal scales brings forth a whole new set of challenges, however. In the next two sections, we focus on challenges unique to each coupling (intracellular to cellular-level, and cellular-level to tissue-level). We suggest ideas to consider when attempting this approach, and provide examples from the literature where these techniques have been successful.

4.2 Coupling Intracellular Signaling Models with Agent Based Models

Rather than building ABMs whose rules treat the cell as a black box, which provides specific outputs and behaviors given certain combinations of inputs, it is possible (experimental data permitting) to couple an ABM with an ISM. In this case, each agent monitors concentrations of extracellular signaling molecules and runs its own ISM to decide on a particular behavior, including amounts of proteins to produce. Such a coupling could thereby enable more fine-grained simulations that more closely replicate in vivo intracellular processes.

One of the areas where cellular-level ABMs can break down is in the combination of multiple influences on a single output. For example, both TGF-β and PDGF-AB influence collagen production by SMCs [58]. Each biomolecule's influences can be tested experimentally by varying concentrations applied in culture, but mapping out responses by SMCs to combinations of growth factors increases the complexity of the experiment exponentially. Indeed, addition of a third or fourth growth factor could make the experiment intractable. Hence, these kinds of experiments are rarely performed in vitro even though they are exactly the kind needed to inform a cell-level ABM that treats the cell as a black box. For example, development of a rule for the combined influences of growth factors on collagen production currently requires extrapolation from a few data points or parameterization of the ABM [58]. Conversely, if the appropriate intracellular signaling pathways were well understood, experimentally inferred rules could be improved by coupling the determination of collagen production to an ISM resulting in more physiological model output.

The scenario described above, however, would require each agent to run its own ISM or to query a single ISM running separately at every timestep. As the number of agents increases, this could become computationally expensive. Therefore, it is advisable to carefully select which rules or outputs should receive this treatment. A thorough sensitivity analysis of the ABM assists in this process by determining which cellular outputs are most important to model predictions. In contrast, another way to approach this coupling would be to pre-calculate outputs of the ISM within physiological ranges of the input parameters and then store results in a look-up table to be accessed by agents during the ABM simulation.

Other things to consider when determining how to couple cell-level with intracellular models is that temporal and spatial scales may differ by orders of magnitude. While the ideal time-step for an ABM may be on the order of minutes to days (e.g., depending on whether one is simulating cell migration or proliferation), intracellular protein interactions may occur on the order of nanoseconds to seconds. The majority of vascular models using some form of ABM-ISM integration to date have been in the field of angiogenesis. For example, Bauer et al. [8] developed a model of tumor induced angiogenesis using a cellular Potts model. In order to understand contributions of cadherins and ECM binding integrins to VEGF signaling, and the associated decision of a cell to migrate, proliferate, or apoptose, the authors went on to develop a Boolean network model incorporating the crosstalk between these three intracellular signaling pathways [9]. This model could potentially be used in conjunction with the CPM to dictate cellular behavior during angiogenesis. Likewise, Scianna [52] developed a hybrid approach coupling a CPM of vasculogenesis with an ISM using reaction–diffusion equations to couple VEGF signaling, arachidonic acid, and NO with calcium entry into the cell. These reactions occur with ten diffusion time steps per main time step.

The issue of very small time-scales for the ISM is much more easily addressed than the converse: when the signaling pathway being modeled contains transcription and translation, critical functions may occur over multiple ABM time-steps. In this case, the ISM might need to be capable of integrating changes in growth factor concentrations as a cell migrates, which may influence protein outputs 30 or more ABM time-steps later. While this has been addressed in previous ABMs by assuming that changes in protein levels happen instantaneously, or within one time-step [7], this may not always be possible when predicting sensitive outputs that require sufficient physiological detail. To our knowledge, this problem has not yet been satisfactorily solved and is something that needs to be seriously considered in the design phase of any integrated model.

4.3 Coupling Agent Based Models with Constrained Mixture Models

In this section we focus on coupling an ABM with a CMM, but the overall procedure can be applied to the coupling of many types of models. Any model is only as good as the data upon which it is built. Hence, we suggest that the first step in a multistep process of coupling different models should be to substantiate the goodness of the published data used to build both models. This can be done, in part, by evaluating the credibility of each data set based on four categories: (1) agreement with other published findings, (2) physiological conditions (e.g., in vitro versus in vivo), (3) appropriateness to the computational model (this metric assesses how well the data match the particular situation, that is, same cell type, organ system, species, environmental condition, and so forth) and (4) data precision (e.g., data measured directly and quantified numerically are better than extrapolated or theoretical); please see Ref. [58] for more details. After performing this data assessment one can then use the highest scored findings to update or create a data-driven computational model with more confidence.

The second step involves further model verification through stability and parameter sensitivity analyses. After the governing equations, parameters, and outputs have been defined for each model, one should confirm that the model is stable over known situations. For example, constituents within the vascular wall are synthesized and turned over at different rates; collagen may be secreted by the cell in less than an hour and has a half-life of $\sim$30–70 days, whereas elastin is primarily produced during the perinatal period and remains for the majority of one's lifetime. Despite these changes, the average geometry of a healthy, mature artery remains fairly constant over long periods. Therefore, it is expected that, under homeostatic conditions, the ABM and CMM should predict no net change in geometry, microstructure, mechanical properties, or biological response over such periods. Moreover, one should confirm that the model can capture acute reactions to transient perturbations. For example, a 10 % increase in pressure over a few hours can lead to transient spikes in growth factor production and yet no net changes in collagen or SMC content. In order to assess the sensitivity of each model to parameters that influence production and removal rates, it is important to conduct a one-dimensional, and if possible, a two-dimensional sensitivity analysis. A one-dimensional analysis will reveal to what degree a single parameter can be increased or decreased before outputs diverge from what is physiologically expected. For example, in a recently reported ABM, the production of new SMCs depended in part on a gain-type parameter multiplied by the concentration of PDGF; if this parameter increased, then SMC content increased and so too wall thickness, which could decrease the circumferential stress (Fig. 4). Yet, PDGF production is a function of circumferential stress, thus less PDGF should be

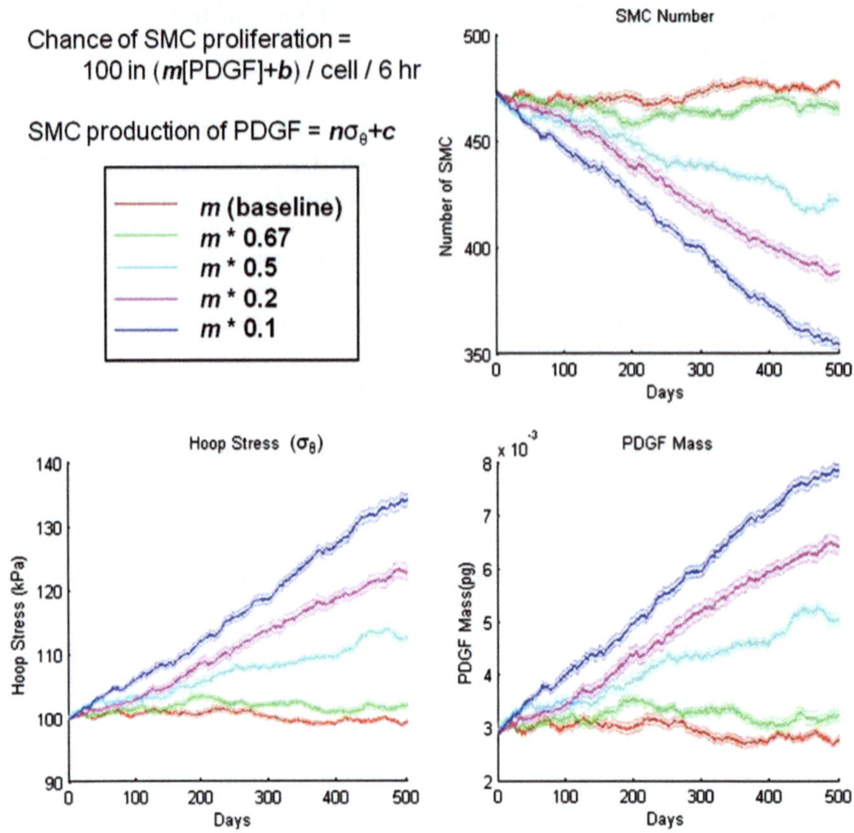

Fig. 4 Sensitivity of the number of smooth muscle cells (SMC), value of hoop stress and concentration of PDGF to changes of the *m* parameter in the chance of SMC proliferation function. Note the interdependency between PDGF, hoop stress and number of SMC. Outputs are able to obtain a new equilibrium to small changes of the *m* parameter

produced as the wall thickens (Fig. 4). Ensuring proper responses to simple perturbations, as illustrated by this example, can help identify appropriate bounds to place on free parameters.

Finally, note that multi-cell ABM rules are often based on cell level in vitro experiments, which can be well controlled but do not always have direct physiologic relevance. CMMs are typically based on either in vitro or in vivo tissue-level experiments, which can have considerable physiological relevance but often little control of the inputs sensed by the cells (e.g., humoral in addition to hemodynamic). Because each data set will be inherently limited, some advantage can be gained by ensuring that models of the same processes based on very different data should yield the same result [29]. The third step, therefore, is to enforce congruency across two (or potentially more) scales through parameter refinement. This process allows one to refine the objectively bounded model parameters in an attempt to minimize the error between common metrics between both models for simple situations. This step is imperative to synchronize both models across scales and to yield predictions that are closer to in vivo observations. This is accomplished not by compromising the strengths of either model, but by tuning each model so that it is influenced by the strengths of the other. For example, the ABM and CMM both have the ability to predict the amounts of collagen and smooth muscle under simple situations; therefore by instituting a genetic algorithm, or any preferred error minimization technique, the difference between these common metrics, at each time point, can be minimized by varying the set of parameters after each full simulation.

5 Future Directions

5.1 Potential Mulitscale Model

A compelling goal in developing multiscale models is to extend the models to both higher and lower levels of scale, with the motivation being that with every added level of scale, one gains even greater flexibility with regards to hypothesis testing, achieving biological relevance, and incorporation of disparate data sets. Starting with the multiscale CMM-ABM, a natural extension is to conjoin an intracellular signaling model (ISM) that can simulate events on a much shorter timescale and account for phenomena within the cell that ultimately impact cell behaviors and outcomes at the tissue level. We have begun to conceptualize a three-tiered multiscale model (ISM-ABM-CMM) of vascular growth and remodeling in the arterial wall (Fig. 5), and this section will briefly summarize some of the enabling components.

The relevant biology of vascular growth and remodeling during human disease occurs over years; however, biological phenomena contributing to disease progression over this timeframe occur across incremental time steps that range from seconds to weeks, depending on the spatial scale. Specifically, signaling events in the ISM have a timeframe on the order of seconds to minutes, multi-cell phenomena in the ABM have a timeframe of minutes to hours, and tissue-level continuum phenomena in the CMM have a timeframe of hours to days or months. Therefore, in the construction of the multiscale model we envision a temporal decomposition strategy to couple across scales (Box 1).

Box 1. Temporal decomposition strategy for a three-tiered multiscale model

Intracellular to multicellular time interfacing: Each time step of the ABM will represent 1 h, and multi-cell level behaviors will occur on this time-scale. For example, SMCs will proliferate over 8–24 h [44] or 8–24 time steps in the ABM, while growth factor production will occur over 12–24 h (or time steps) [44]. When the ABM calls the ISM (every 24 h for 120 simulated days), where biological events will be simulated on the order of minutes, we will institute a time-delay, or essentially "pause" the ABM, while the intracellular events are computed by the ISM.

Multicellular to tissue-level time interfacing: Each time step of the CMM will represent 1 week, and tissue-level behaviors will occur on this time-scale. For example, the strain state of the artery wall will change over the course of 1–2 weeks (depending on vessel geometry and composition alterations during this time window). Therefore, to account for these changes on a time window that is biologically realistic, the ABM will call the CMM every week (i.e., every CMM time step) in order to recompute and update tissue-level changes that occurred during this window. Thus, in the ABM, every 7 days (or 168 time steps in the ABM), the ABM module will be "paused" in order to run the CMM, whose output (computation of the mechanical state of the media) will be imported back into the ABM.

Another hurdle to achieving a unified three-tiered multiscale model is defining an appropriate level of resolution and abstraction for the relevant spatial scales such that information at one level can be mapped to higher and/or lower levels of biological scale. We have previously devised a method for coupling spatial scales between an ABM and ISM [49]. In that ABM module, we simulated individual biological cells using nine coupled agents that represented different cytoplasmic and membrane compartments within each simulated cell. In this way, we could simulate the differential behaviors of the leading versus trailing edge of a cell as it migrated across a two-dimensional substrate. Intracellular signaling events were simulated in the ISM and distributed equally to the nine agents comprising each cell, but one could envision partitioning or compartmentalizing certain reactions within spatially confined intracellular regions that would map directly to the discretized, multi-agent cells within the ABM. We have similarly devised a method for comparing spatial scales between an ABM and CMM [29]. In this case, the ABM consisted of layers of cells whereas the CMM consisted of a homogenized structural wall. Data from the ABM could thus be averaged radially and applied to the CMM at each time of interest.

We envision that a three-tiered, multiscale ISM-ABM-CMM will be united by an umbrella program implemented in Java™. This program will run the ISM, ABM, and CMM modules in parallel, while transferring information between each pair of modules. The CMM and ISM are implemented in Matlab®, and the ABM is implemented in Netlogo [63]. We have opted to use these modeling software

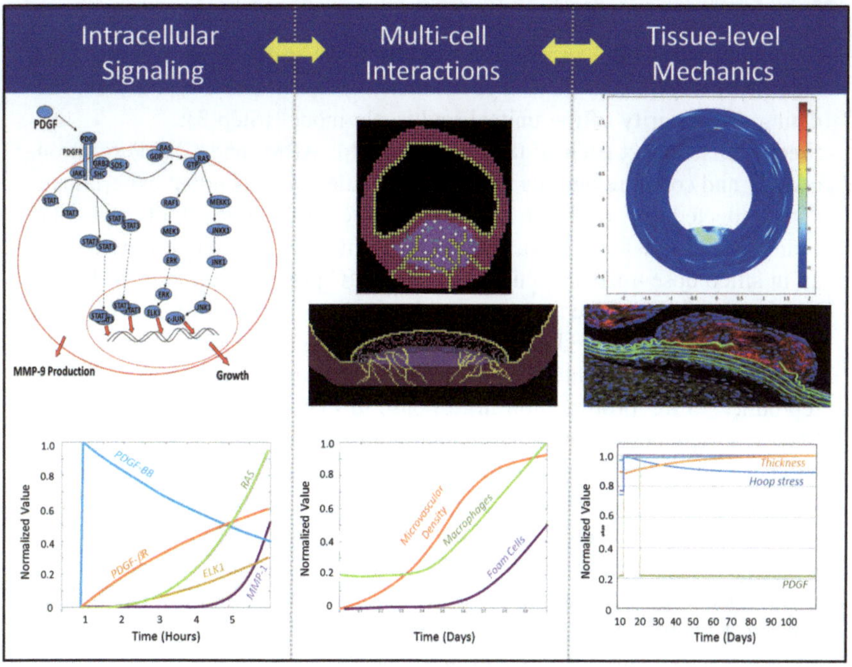

Fig. 5 Illustration of a possible approach to couple intracellular signaling models (ISM), agent based models (ABM), and constrained mixture models (CMM) to enhance computations of vascular behaviors

programs because they are well established, readily available, and easy to learn, which, importantly, will facilitate the sharing of our multiscale model. Matlab is a general-purpose simulation tool, and Netlogo, written in Java^{TM}, is freely downloadable from http://ccl.northwestern.edu/netlogo/ and is the most widely used agent-based modeling software [63]. Our group has previously published a multiscale model that unites ISM with ABM using a Java^{TM}-based umbrella program to run an ISM and an ABM in parallel [49], for the study of tissue morphogenesis during embryonic development. The umbrella program will interface the ISM, ABM, and CMM through the use of text files that can be imported and exported by both Matlab$^{®}$ and Netlogo, which will store inputs and outputs of each module and be used to communicate predictions from one level of scale to the next. Netlogo extensions are capable of running Java^{TM} code, and the Java^{TM} code will open and run Matlab.

Model validation is an important check to confirm the model is accurate and stable. Box 2 describes a possible strategy for validating a three-tiered multiscale model.

Box 2. Validation Strategy ISM-ABM-CMM

The multiscale model can be validated by first validating each of the modules (ISM, ABM, and CMM) individually (Steps 1 and 2), and then validating the integrity of the unified multiscale model (Step 3):

Step 1: Disconnect each of the modules (ISM, ABM, and CMM) from one another and compare outputs from each module to independent experimental data collected (or reported in the literature) at that level of scale. For example, to validate the ISM, one can perform a set of simulations to acquire an in silico dose–response curve relating SMC proliferation rates to PDGF-BB concentration. Then compare this predicted dose–response curve to experimentally measured proliferation rates from PDGF-BB dose–response studies. This will ensure that the internal structure of the ISM accurately reproduces SMC proliferation in response to PDGF-BB.

Step 2: Disconnect each of the modules (ISM, ABM, and CMM) from one another and compare those outputs that each module has in common with one another. This will ensure that the modules are internally consistent with one another. For example, both the CMM and the ABM will predict the thickness of an atherosclerotic plaque, but it is important to check that the predictions from both modules are congruent (and minimize the residual differences between ABM and CMM predictions), given that the parameters and "rules" governing the ABM and CMM will be derived from different sources (Fig. 5).

Step 3: Validate the unified multiscale model by comparing its outputs (i.e., outputs generated by the *integrated* ISM, ABM, and CMM) to independent experimental data. Quantitative predictions of the multiscale model at time $t > t_0$ can be compared with identical outcome metrics collected from the quantitative analyses of hypertensive patients or ApoE -/- plaques. Agreement between prediction and experiment will suggest that the multiscale model is valid for that range of parameters

The hope in pursuing this type of multiscale modeling is that one day we will be able to confidently theorize about new drug or knock-out treatments and cause-effect relationships. For example, if the angiotensin II (ANG-II) receptor type II (AT-2) is blocked, ANG-II binds to the type I receptor (AT-1) of ECs. Binding of AT-1 activates the tyrosine kinase and downstream proteins (mitogen-activated protein kinase (MAPK), Janus kinase (JNK), and signal transducer and activator of transcription (STAT)) leading to increased intracellular calcium, activation of the L-type calcium channel, and consequently arterial constriction. Activation of MAPK also stimulates fibroblast and SMC migration and proliferation via synthesis of platelet derived growth factor and tissue growth factor-β. These growth factors as well as increased aldosterone all serve to facilitate extracellular matrix production in a particular collagen, which leads to increased wall stiffening or pulse wave velocity. Stiffened arteries not only require a larger pressure to distend; flow propagation is impaired due to inadequate elastic recoil. Thus over time the increased load on the heart causes left ventricle hypertrophy and left ventricle failure. Alternatively, we can theorize about "top-down" effects. For example, how does mechanical shear force experienced by the ECs impact NO signaling and PDGF expression, which in turn affect SMCs and overall wall mechanics? Creating models to predict these types of outcomes will save time, money, and potentially lives.

Of course, the aspiration to unite ISM-ABM-CMM is met with considerable challenges, not the least of which is the requirement for additional computing power. Assuming the technical challenges can be overcome by advances in computing (e.g., parallelization, grid computing, and cloud computing), one must address the conceptual challenges in multiscale model design. The final section of this chapter will delve deeper into these challenges and suggest opportunities for innovation in multiscale modeling.

5.2 Challenges

As noted by the 1998 Bioengineering Consortium (BECON) Report of the U.S. National Institutes of Health,

> The success of reductionist and molecular approaches in modern medical science has led to an explosion of information, but progress in integrating information has lagged ... Mathematical models provide a rational approach for integrating this ocean of data, as well as providing deep insight into biological processes.

Whereas the need remains to develop more robust and faithful models at all scales (macro, micro, nano), we submit that there is a pressing need to develop approaches that integrate such models across diverse scales. Indeed, anticipating the challenges of multiscale modeling should influence the development of models at each scale for they will need to interface with the other models. Toward this end, we suggest here the following particular challenges that deserve our immediate attention.

There are several computational languages used to run numerical analysis (e.g., Matlab, Maple, Mathematica, Java Virtual Machine, FORTRAN, and C++,). Thus, a logistical challenge may arise when models, at different scales, are programmed with different languages. We proposed herein using text files as inputs/output because all our modeling platforms can read and write text files, however this process is time consuming and cumbersome. Therefore finding patches or proper interfaces between multiple numerical analysis software remains a challenge. In addition, iterative simulations may take days to complete and require considerable memory on a personal computer. Consequently large-capacity databases and fast processors/parallel systems may be required to render the computational process tractable. After a multiscale program is completed, finding ways to distill and partition model findings into digestible chunks that is are easy to disseminate and publish may be a challenge.

In addition to computational challenges, there are also many conceptual challenges. For example, events at the tissue level may depend on past events and take hours to days to occur, while at the intracellular level processes may occur in a fraction of a second. In addition to integrating across temporal scales, integrating across spatial scales (i.e., 2D versus 3D) may require further dispensation. One of the major challenges remains in simplifying the complex system due to gaps in our understanding or in an attempt to not over constrain the model. Where should one start, and how is each decision justified? Another challenge of integrating multiple discrete models is deciding what information to pass back and forth. Each model may have interdependency within itself, thus passing a concentration or stress value negates any feedback mechanisms the model had related to these parameters. Therefore, integrating models that rely on values from one another is a challenge.

Biological adaptation and variability are difficult to capture in a universal mathematical model. How biological systems change in time is what the models presented herein try to account for, but some adaptations are unpredictable. For example, natural effects (due to ageing, hormonal life cycles, ones genetic makeup, even what division cycle cells in the body are on) and external effects (due to accidents, smoking, exercise, eating habits, radiation, etc.) may alter how the general process works. Therefore, if the response of one patient or system could be very different than another, are the models unique to the patient? How general should the models be? Of course we are currently limited by our technology to measure and characterize the interactions of phenomenon of biological systems. Generally speaking, like the Heisenberg uncertainty principle, to augment our knowledge of, say, the rate of growth factor production may come at the cost of compromising physiological conditions.

Nevertheless, we feel that complex system modeling in biology is the key to developing new drugs and therapies over the next 50 years; as such there are educational needs that should be met. Having more undergraduate courses that deal with complex systems analysis in biology will equip more students with the fundamental skills. More graduate courses on the theory of modeling vascular adaptation, and biological adaptation in general, will allow for specialization and additional improvements. Having more graduate programs and/or cross-degree or dual-degree Ph.D. programs that are designed to treat high-throughput data in the context of in vivo function and quantitative modeling is needed. In addition, continued changes in academic culture that recognize the value of collaboration and teamwork on large complex systems will facilitate more advanced models. We are encouraged to hear that in April 2012, NSF and NIH jointly launched a "Core Techniques and Technologies for Advancing Big Data Science and Engineering (BIGDATA)" initiative. The need for a means to manage, analyze, visualize, and extract data from diverse, distributed data sets has been recognized. If successful, having this wealth of ordered data at our fingertips will only help to update and improve the rules and relations of multiscale modeling.

Acknowledgments This work was supported, in part, via NIH grants HL-86418 to JDH and HL-82838 to SMP.

References

1. Abbott, R.G., Forrest, S., Pienta, K.J.: Simulating the hallmarks of cancer. Artif. Life **12**, 617 (2006)
2. Alon, U.: Network motifs: theory and experimental approaches. Nat. Rev. Genet. **8**, 450 (2007)
3. An, G.: Concepts for developing a collaborative in silico model of the acute inflammatory response using agent-based modeling. J. Crit. Care **21**, 105 (2006)
4. An, G.: In silico experiments of existing and hypothetical cytokine-directed clinical trials using agent-based modeling. Crit. Care Med. **32**, 2050 (2004)
5. An, G.: Mathematical modeling in medicine: a means, not an end. Crit. Care Med. **33**, 253 (2005)
6. Baek, S., Rajagopal, K.R., Humphrey, J.D.: A theoretical model of enlarging intracranial fusiform aneurysms. J. Biomech. Eng. **128**, 142 (2006)
7. Bailey, A.M., Thorne, B.C., Peirce, S.M.: Multi-cell agent-based simulation of the microvasculature to study the dynamics of circulating inflammatory cell trafficking. Ann. Biomed. Eng. **35**, 916 (2007)
8. Bauer, A.L., Jackson, T.L., Jiang, Y.: A cell-based model exhibiting branching and anastomosis during tumor-induced angiogenesis. Biophys. J. **92**, 3105 (2007)
9. Bauer, A.L., Jackson, T.L., Jiang, Y., Rohlf, T.: Receptor cross-talk in angiogenesis: mapping environmental cues to cell phenotype using a stochastic, Boolean signaling network model. J. Theor. Biol. **264**, 838 (2010)
10. Benest, A.V., Stone, O.A., Miller, W.H., Glover, C.P., Uney, J.B., Baker, A.H., Harper, S.J., Bates, D.O.: Arteriolar genesis and angiogenesis induced by endothelial nitric oxide synthase overexpression results in a mature vasculature. Arterioscler. Thromb. Vasc. Biol. **28**, 1462 (2008)
11. Bhalla, U.S., Iyengar, R.: Emergent properties of networks of biological signaling pathways. Science **283**, 381 (1999)
12. Bian, K., Doursout, M.F., Murad, F.: Vascular system: role of nitric oxide in cardiovascular diseases. J. Clin. Hypertens (Greenwich) **10**, 304 (2008)
13. Bouvet, C., Moreau, S., Blanchette, J., de Blois, D., Moreau, P.: Sequential activation of matrix metalloproteinase 9 and transforming growth factor beta in arterial elastocalcinosis. Arterioscler. Thromb. Vasc. Biol. **28**, 856 (2008)
14. Casal, A., Sumen, C., Reddy, T.E., Alber, M.S., Lee, P.P.: Agent-based modeling of the context dependency in T cell recognition. J. Theor. Biol. **236**, 376 (2005)
15. Castro, M.M., Tanus-Santos, J.E., Gerlach, R.F.: Matrix metalloproteinases: targets for doxycycline to prevent the vascular alterations of hypertension. Pharmacol. Res. **64**, 567 (2011)
16. Chaturvedi, R., et al.: On multiscale approaches to three-dimensional modelling of morphogenesis. J. R. Soc. Interface **2**, 237 (2005)
17. Chen, N., Glazier, J.A., Izaguirre, J.A., Alber, M.S.: A parallel implementation of the cellular potts model for simulation of cell-based morphogenesis. Comput. Phys. Commun. **176**, 670 (2007)
18. Christ, G.J., Spray, D.C., el-Sabban, M., Moore, L.K., Brink, P.R.: Gap junctions in vascular tissues. Evaluating the role of intercellular communication in the modulation of vasomotor tone. Circ. Res. **79**, 631 (1996)
19. Davies, P.F.: Flow-mediated endothelial mechanotransduction. Physiol. Rev. **75**, 519 (1995)
20. DiCorleto, P.E., Bowen-Pope, D.F.: Cultured endothelial cells produce a platelet-derived growth factor-like protein. Proc. Natl. Acad. Sci. USA. **80**, 1919 (1983)
21. Drasdo, D., Forgacs, G.: Modeling the interplay of generic and genetic mechanisms in cleavage, blastulation, and gastrulation. Dev. Dyn. **219**, 182 (2000)
22. Duran, W.N., Breslin, J.W., Sanchez, F.A.: The NO cascade, eNOS location, and microvascular permeability. Cardiovasc. Res. **87**, 254 (2010)

23. Fameli, N., Kuo, K.H., van Breemen, C.: A model for the generation of localized transient [Na+] elevations in vascular smooth muscle. Biochem. Biophys. Res. Commun. **389**, 461 (2009)
24. Galkina, E., Ley, K.: Immune and inflammatory mechanisms of atherosclerosis (*). Annu. Rev. Immunol. **27**, 165 (2009)
25. Graner, F., Glazier, J.A.: Simulation of biological cell sorting using a two dimensional extended potts model. Phys. Rev. Lett. **69**, 2013–2016 (1992)
26. Grant, M.R., Mostov, K.E., Tlsty, T.D., Hunt, C.A.: Simulating properties of in vitro epithelial cell morphogenesis. PLoS Comput. Biol. **2**, e129 (2006)
27. Grimm, V., Railsback, S.F.: Individual-based modeling and ecology. Princeton University Press, Princeton (2005)
28. Hastings, N.E., Simmers, M.B., McDonald, O.G., Wamhoff, B.R., Blackman, B.R.: Atherosclerosis-prone hemodynamics differentially regulates endothelial and smooth muscle cell phenotypes and promotes pro-inflammatory priming. Am. J. Physiol. Cell Physiol. **293**, C1824 (2007)
29. Hayenga, H.N., Thorne, B.C., Peirce, S.M., Humphrey, J.D.: Ensuring congruency in multiscale modeling: towards linking agent based and continuum biomechanical models of arterial adaptation. Ann. Biomed. Eng. **39**, 2669 (2011)
30. Humphrey, J.D.: Cardiovascular solid mechanics: cells, tissues, and organs. Springer, New York (2002)
31. Humphrey, J.D.: Stress, strain, and mechanotransduction in cells. J. Biomech. Eng. **123**, 638 (2001)
32. Humphrey, J.D.: Vascular adaptation and mechanical homeostasis at tissue, cellular, and sub-cellular levels. Cell Biochem. Biophys. **50**, 53 (2008)
33. Humphrey, J.D., Rajagopal, K.R.: A constrained mixture model for arterial adaptations to a sustained step change in blood flow. Biomech. Model. Mechanobiol. **2**, 109 (2003)
34. Izaguirre, J.A., et al.: CompuCell, a multi-model framework for simulation of morphogenesis. Bioinformatics **20**, 1129 (2004)
35. Kapela, A., Bezerianos, A., Tsoukias, N.M.: A mathematical model of vasoreactivity in rat mesenteric arterioles: I. Myoendothelial communication. Microcirculation **16**, 694 (2009)
36. Kawai-Kowase, K., Owens, G.K.: Multiple repressor pathways contribute to phenotypic switching of vascular smooth muscle cells. Am. J. Physiol. Cell Physiol. **292**, C59 (2007)
37. Kohn-Luque, A., de Back, W., Starruss, J., Mattiotti, A., Deutsch, A., Perez-Pomares, J.M., Herrero, M.A.: Early embryonic vascular patterning by matrix-mediated paracrine signalling: a mathematical model study. PLoS One **6**, e24175 (2011)
38. Kuwai, T., et al.: Targeting the EGFR, VEGFR, and PDGFR on colon cancer cells and stromal cells is required for therapy. Clin. Exp. Metastasis **25**, 477 (2008)
39. Longo, D., Peirce, S.M., Skalak, T.C., Davidson, L., Marsden, M., Dzamba, B., DeSimone, D.W.: Multicellular computer simulation of morphogenesis: blastocoel roof thinning and matrix assembly in *Xenopus laevis*. Dev. Biol. **271**, 210 (2004)
40. Merks, R.M., Brodsky, S.V., Goligorksy, M.S., Newman, S.A., Glazier, J.A.: Cell elongation is key to in silico replication of in vitro vasculogenesis and subsequent remodeling. Dev. Biol. **289**, 44 (2006)
41. Nakagawa, T., Inoue, H., Sasahara, M.: Platelet-derived growth factor and renal disease. Curr. Opin. Nephrol. Hypertens **21**, 80 (2012)
42. Osol, G.: Mechanotransduction by vascular smooth muscle. J. Vasc. Res. **32**, 275 (1995)
43. Peirce, S.M., Skalak, T.C., Papin, J.A.: Coupling intracellular networks with tissue-level physiology. IBM J. Res. **50**, 1 (2006)
44. Peirce, S.M., Van Gieson, E.J., Skalak, T.C.: Multicellular simulation predicts microvascular patterning and in silico tissue assembly. Faseb. J. **18**, 731 (2004)
45. Pries, A.R., Reglin, B., Secomb, T.W.: Modeling of angioadaptation: insights for vascular development. Int. J. Dev. Biol. **55**, 399 (2011)
46. Raffetto, J.D., Khalil, R.A.: Matrix metalloproteinases and their inhibitors in vascular remodeling and vascular disease. Biochem. Pharmacol. **75**, 346 (2008)

47. Ramli, J., CalderonArtero, P., Block, R.C., Mousa, S.A.: Novel therapeutic targets for preserving a healthy endothelium: strategies for reducing the risk of vascular and cardiovascular disease. Cardiol. J. **18**, 352 (2011)
48. Ridnour, L.A., Isenberg, J.S., Espey, M.G., Thomas, D.D., Roberts, D.D., Wink, D.A.: Nitric oxide regulates angiogenesis through a functional switch involving thrombospondin-1. Proc. Natl. Acad. Sci. USA **102**, 13147 (2005)
49. Robertson, S.H. et al.: Multiscale computational analysis of *Xenopus laevis* morphogenesis reveals key insights of systems-level behavior. BMC Syst. Biol. **1**, 46 (2008)
50. Rudic, R.D., Shesely, E.G., Maeda, N., Smithies, O., Segal, S.S., Sessa, W.C.: Direct evidence for the importance of endothelium-derived nitric oxide in vascular remodeling. J. Clin. Invest. **101**, 731 (1998)
51. Schermuly, R.T., et al.: Reversal of experimental pulmonary hypertension by PDGF inhibition. J. Clin. Invest. **115**, 2811 (2005)
52. Scianna, M., Munaron, L., Preziosi, L.: A multiscale hybrid approach for vasculogenesis and related potential blocking therapies. Prog. Biophys. Mol. Biol. **106**, 450 (2011)
53. Segovia-Juarez, J.L., Ganguli, S., Kirschner, D.: Identifying control mechanisms of granuloma formation during *M. tuberculosis* infection using an agent-based model. J. Theor. Biol. **231**, 357 (2004)
54. Sessa, W.C.: eNOS at a glance. J. Cell Sci. **117**, 2427 (2004)
55. Simpson, M.J., Merrifield, A., Landman, K.A., Hughes, B.D.: Simulating invasion with cellular automata: connecting cell-scale and population-scale properties. Phys. Rev. E Stat. Nonlin. Soft Matter Phys. **76**, 021918 (2007)
56. Smith Jr., R.S., Lin, K.F., Agata, J., Chao, L., Chao, J.: Human endothelial nitric oxide synthase gene delivery promotes angiogenesis in a rat model of hindlimb ischemia. Arterioscler. Thromb. Vasc. Biol. **22**, 1279 (2002)
57. Thorne, B.C., Bailey, A.M., Peirce, S.M.: Combining experiments with multi-cell agent-based modeling to study biological tissue patterning. Br. Bioinf. **8**, 245 (2007)
58. Thorne, B.C., Hayenga, H.N., Humphrey, J.D., Peirce, S.M.: Toward a multi-scale computational model of arterial adaptation in hypertension: verification of a multi-cell agent based model. Front. Physiol. **2**, 20 (2011)
59. Tomasek, J.J., Gabbiani, G., Hinz, B., Chaponnier, C., Brown, R.A.: Myofibroblasts and mechano-regulation of connective tissue remodelling. Nat. Rev. Mol. Cell Biol. **3**, 349 (2002)
60. Tsoukias, N.M.: Nitric oxide bioavailability in the microcirculation: insights from mathematical models. Microcirculation **15**, 813 (2008)
61. Valentin, A., Cardamone, L., Baek, S., Humphrey, J.D.: Complementary vasoactivity and matrix remodelling in arterial adaptations to altered flow and pressure. J. R. Soc. Interface **6**, 293 (2009)
62. Wagenseil, J.E., Mecham, R.P.: Vascular extracellular matrix and arterial mechanics. Physiol. Rev. **89**, 957 (2009)
63. Wilensky, U.: NetLogo—http://ccl.northwestern.edu/netlogo. Center for Connected Learning and Computer-Based Modeling. Northwestern University, Evanston (1999)
64. Yamamoto, K., Ikeda, U., Shimada, K.: Role of mechanical stress in monocytes/macrophages: implications for atherosclerosis. Curr. Vasc. Pharmacol. **1**, 315 (2003)
65. Yang, J., Clark, J.W., Bryan, R.M., Robertson, C.S.: Mathematical modeling of the nitric oxide/cGMP pathway in the vascular smooth muscle cell. Am. J. Physiol. Heart Circ. Physiol. **289**, H886 (2005)
66. Yokota, T., Ma, R.C., Park, J.Y., Isshiki, K., Sotiropoulos, K.B., Rauniyar, R.K., Bornfeldt, K.E., King, G.L.: Role of protein kinase C on the expression of platelet-derived growth factor and endothelin-1 in the retina of diabetic rats and cultured retinal capillary pericytes. Diabetes **52**, 838 (2003)
67. Yoshida, T., Gan, Q., Shang, Y., Owens, G.K.: Platelet-derived growth factor-BB represses smooth muscle cell marker genes via changes in binding of MKL factors and histone deacetylases to their promoters. Am. J. Physiol. Cell Physiol. **292**, C886 (2007)

68. Yoshida, T., Owens, G.K.: Molecular determinants of vascular smooth muscle cell diversity. Circ. Res. **96**, 280 (2005)
69. Yu, J., Zhang, Y., Zhang, X., Rudic, R.D., Bauer, P.M., Altieri, D.C., Sessa, W.C.: Endothelium derived nitric oxide synthase negatively regulates the PDGF-survivin pathway during flow-dependent vascular remodeling. PLoS One **7**, e31495 (2012)
70. Zajac, M., Jones, G.L., Glazier, J.A.: Simulating convergent extension by way of anisotropic differential adhesion. J. Theor. Biol. **222**, 247 (2003)
71. Zhang, L., Strouthos, C.G., Wang, Z., Deisboeck, T.S.: Simulating brain tumor heterogeneity with a multiscale agent-based model: linking molecular signatures, phenotypes and expansion rate. Math. Comput. Model. **49**, 307 (2009)
72. Zhang, R., Wang, L., Zhang, L., Chen, J., Zhu, Z., Zhang, Z., Chopp, M.: Nitric oxide enhances angiogenesis via the synthesis of vascular endothelial growth factor and cGMP after stroke in the rat. Circ. Res. **92**, 308 (2003)

Multiscale Modeling in Vascular Disease and Tissue Engineering

Houman Zahedmanesh and Caitríona Lally

Abstract The human body, and hence the vascular system, is by its very nature a dynamic multiscale hierarchial system. This multiscale nature encompasses different length scales, from molecular and cellular levels to the tissue and organ level, as well as different physical phenomena, such as mechanical, biological and chemical processes. In arteries, vascular cells alter their growth, phenotype and extracellular matrix production in response to macro mechanical changes. These cell level events can in turn accumulate and emerge at the tissue level as pathological conditions such as atherosclerosis and intimal hyperplasia. These cardiovascular diseases evolve through adaptation of cells and tissues over days to months also demonstrating the multiscale nature of vascular diseases with respect to time. The challenge in vascular multiscale modelling is to create a framework which can incorporate the key mechanical, biological and chemical characteristics of this complex system at these various space and time scales to successfully capture the long-term behaviour of the system. Such a framework can then be used to gain additional insights with regards to pathological conditions within the vascular system and to improve the design of medical devices used to treat such pathologies. In the following chapter, a review will be presented of some relevant studies reported in literature which have used multiscale modelling approaches to elucidate the growth and remodelling mechanisms underlying vascular diseases, such as atherosclerosis, in-stent restenosis and intimal hyperplasia.

H. Zahedmanesh · C. Lally (✉)
School of Mechanical and Manufacturing Engineering,
Dublin City University, Glasnevin, Dublin 9, Ireland
e-mail: Triona.lally@dcu.ie

Stud Mechanobiol Tissue Eng Biomater (2013) 14: 241–258
DOI: 10.1007/8415_2012_159

Published Online: 10 November 2012

1 Introduction

Arterial growth, remodelling and vascular diseases are intrinsically multiscale and depend on the interactions occurring at the tissue level, cell level and the intracellular level. Consequently, multiscale computational modelling techniques can help to elucidate the mechanisms underlying the onset and progression of vascular diseases as well as vascular tissue regeneration. Mechanical perturbations at the tissue level translate to cell level mechanical signals via cell–matrix interactions. How these mechanical signals are further transduced through the cytoskeletal assembly, and other signalling pathways such as calcium channels to the cell nucleus, resulting in specific gene expressions which subsequently alter cellular behaviour, adds an additional level of complexity to these multiscale systems. Cells alter their growth, phenotype and their extracellular matrix in response to macro mechanical changes. These cell level events can then in turn accumulate and emerge at the tissue level as pathological conditions such as atherosclerosis and intimal hyperplasia.

Multiscale modelling is by its nature highly computationally expensive. With recent advances in computational capabilities, a more mechanistic approach to multiscale modelling, using discrete methodologies, has become possible which has enabled a systems approach to understanding diseases. Agent based models (ABM) or Cellular Automata (CA) models are notable examples whereby the behaviour of each individual cell can be modelled explicitly. In recent years several agent based approaches have been developed to provide a quantitative and mechanistic understanding of pathologies such as inflammation and wound healing [1, 2] atherosclerosis [3], in-stent restenosis [4–8], and intimal hyperplasia in vascular grafts [9, 10]. In agent based modelling, a population of "agents" which are autonomous individuals representing cells, are created and the rules of behaviour and interactions between the agents are defined. In the context of cell biology, agent behaviours such as migration, proliferation and differentiation can be defined for each agent using mathematical formulations which describe the migration speed, doubling time and extracellular matrix and chemokine synthesis as functions of different stimuli such as stress/strain or species concentrations. Rules of interactions between the cells such as contact inhibition and different paracrine signalling pathways can also be defined.

One important advantage of using ABM to model cell populations is their ability to better capture the discrete nature of events occurring at the cellular level compared to continuous approaches such as differential equations. In contrast, continuum methods such as the finite element method have been extensively used as a robust and reliable tool for modelling tissue level events such as mechanical interactions between the arterial wall and stents [11–15]. This has motivated development of hybrid multiscale models that take advantage of continuum methods such FEM at the tissue level and employ discrete methods such as CA or ABM to capture cell level events. Recent studies by Boyle et al. [6, 7] and Zahedmanesh et al. [58] and Zahedmanesh and Lally [10] are good examples of such multiscale modelling approaches whereby FEM was used to quantify the mechanical stimuli at the tissue level and CA and ABM were utilised to capture the cell level events.

In the following chapter, a review will be presented of some relevant studies reported in literature which have used multiscale modelling approaches to elucidate the growth and remodelling mechanisms underlying vascular diseases, such as atherosclerosis, in-stent restenosis and intimal hyperplasia.

2 Multiscale Models of Vascular Disease

2.1 *Atherosclerosis*

It has been shown that atherosclerosis usually occurs in locations where blood flow perturbations, i.e. low and oscillatory wall shear stress, take place such as at, or near, bifurcations. Damage to the endothelium can also increase penetration of low density lipoproteins (LDL) into the intima and lead to accumulation of macrophages and subsequently foam cells in the arterial wall and its chronic inflammation [16–18]. This chronic inflammation can lead to dedifferentiation of SMCs and their chemotactic migration and proliferation from the media to the intima and formation of atherosclerotic plaques. As such the events involved in the onset and progression of atherosclerosis are intrinsically multiscale chemo-mechano-biological events.

To-date models of atherosclerosis have been developed at the cell level using ABM, for example (SimAthero, [3]). Models such as SimAthero consider the biological variables that play the most important role in atherogenesis and its induced immune response, i.e., LDL, ox-LDL, chitotriosidase and the foam cells generated in the artery wall. Pappalardo et al. [3] analysed four different classes of patients to show how SimAthero could be used to analyse and predict the effects of various LDL levels in a diverse group of patients over a time scale of two years, namely (i) patients with an LDL level considered normal, where no foam cells were formed, (ii) patients with a high level of LDL with delayed drug treatment, (iii) patients with high LDL levels, treated with specific drugs aimed at reducing total LDL (statins), and (iv) patients with specific lifestyle conditions that increased the risk of LDL oxidation such as smoking. As with other ABM or multiscale models, the underlying cell rules described by the model in the virtual patients were tuned against human data, thereby providing a means of ensuring realistic behavioural outcomes. Whilst ABM of this nature can provide insights into individualised drug treatment and the underlying biology of diseases such as atherosclerosis, the inherent mechano-biological interaction driving many vascular diseases cannot be modelled by such a framework.

A multiscale model of early stage atherosclerotic plaque formation has recently been developed in order to integrate the various mechano-biological phenomena leading to fatty streak formation [19]. The different scales considered in this model are in both the spatial domain (from cellular to organism level) and the time domain (from seconds to months). At the cell level, the transport and chemical interactions of low-density lipoproteins (LDL) and other agents were modelled. This was linked to arterial thickening and mean LDL level at the organ level in a stenosed artery whilst the critical factors for atherosclerotic formation at the organ level were chosen to be mean LDL and flow induced endothelial wall shear stress. In this study, it was observed that plaque location was dependent on WSS and that plaque size, number, and growth were also dependent on mean blood LDL levels. Whilst the simulation results from these models compared favourably with some well-established biological hypotheses for atherosclerosis in terms of plaque location, number, size, and rate of formation, the most significant potential of the models is their ability to be highly individualised. Multiscale patient-specific models can take the anatomies of individual patients into account, but they can also include cell and molecular level information by considering individual baseline LDL levels and plaque growth rates.

2.2 *In-Stent Restenosis*

In-stent restenosis, or re-blockage of a vessel following stent implantation, is a vascular disease which is linked with mechanical injury at the tissue level induced by the stent which can dramatically alter the microenvironment of SMCs at the cellular level. After stent deployment, vessel injury by the stent struts leads to modulation of the medial SMCs' phenotype to a synthetic phenotype. This change of phenotype is followed by migration and proliferation of dedifferentiated medial SMCs towards the lumen and lesion formation, see Fig. 1 [20–23].

The changes in the microenvironment of cells, specifically the extracellular matrix (ECM) changes following vessel injury have been shown to regulate VSMC activation [24]. An intact and mature collagen type IV matrix has been shown to promote a quiescent and contractile phenotype, whereas its degradation leads to VSMC activation [25–28]. In addition, degradation of collagen types I and III which maintain the mechanical integrity and stability of arteries promotes a synthetic VSMC phenotype [28]. As such, basement membrane-degrading matrix metalloproteinase, i.e. MMP-2 and MMP-9, which can degrade collagen type IV [29], and fibrilar collagen types I and III [29, 30] are also strongly implicated in activation of VSMCS following arterial damage. Mechanical injury to the arterial wall has been shown to upregulate MMP-2 production [31–34], as does mechanical stretch [35, 36]. A study by Asanuma et al. [35] showed that the application of a constant mechanical stretch increased MMP expression in cultured human VSMCs. Therefore, long-term strain imposed by stents, hypertension or atherosclerosis may lead to enhanced matrix degradation by VSMCs.

Clearly therefore, during in-stent restenosis mechanical perturbations at the tissue level, due to stent implantation, lead to dramatic changes in the microenvironment of cells. Subsequently, these cell level changes initiate a cascade of event at the

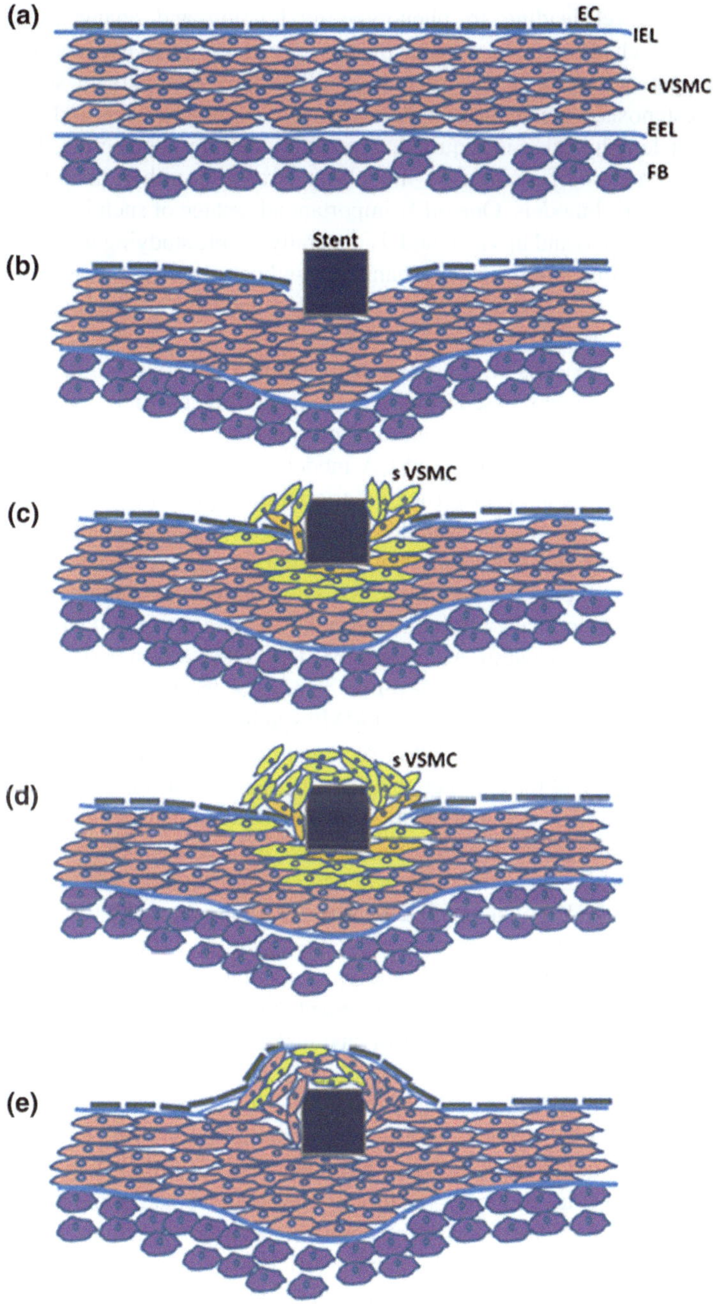

Fig. 1 Development of in-stent restenosis following stent deployment, taken from Zahedmanesh et al. [8]. **a** Normal artery, **b** de-endothelialisation and injury of the media, **c** modulation of medial VSMC phenotype and their migration and proliferation towards the lumen, **d** development of in-stent restenosis, **e** re-endothelialisation and differentiation of VSMCs back to the quiescent phenotype. *EC* endothelial cells, *IEL* internal elastic lamina, *s VSMCs* synthetic vascular smooth muscle cell, *c VSMC* contractile and quiescent vascular smooth muscle cell, *EEL* external elastic lamina

intracellular level leading to phenotype modulation and changes in cellular behaviour which ultimately emerge at the tissue level as intimal hyperplasia and loss of vessel patency. Clearly, in order to develop effective treatment strategies for in-stent restenosis, a mechanistic understanding of the mechanisms and interactions at different length scales is required. From a modelling perspective, this can be realised by means of multiscale modelling strategies rather than single-scale phenomenological models. One other important advantage of such in-silico models is that whilst in-vivo and in-vitro models generally enable studying the role of single parameters at a time, multiscale mechanistic in-silico models can provide a systems approach to the role of several different parameters and their interactions in the onset and prognosis of pathologies.

Multiscale approaches which have been adopted to model in-stent restenosis include those using agent based models to describe SMC proliferation governed by local fluid flow, structural stress and anti-proliferative drug concentration eluting from the stent [4], and lattice based CA models of SMC phenotypic modulation, proliferation and migration governed by arterial damage at the organ level [6, 7].

The authors have recently developed a mechanobiological modelling framework by coupling a FEM, that simulates strut-artery interaction, with a lattice-free ABM that simulates the key responses of VSMCs to mechanical damage [58]. In this model, the stresses induced within the arterial wall were quantified and related to a damage scale of [0–1] within each element. The ABM was superimposed on the finite element mesh whereby the level of damage at the location of each VSMC could be apraised. Damage upregulated MMP synthesis by VSMCs and resulted in the degradation of the collagen matrix surrounding each VSMC. Endothelial damage and degradation of the collagen matrix were then used as the main regulators of VSMC activation. As such, degradation of the collagen matrix and lack of nitric oxide (NO) signalling due to endothelial damage led to migration and proliferation of VSMCs and ultimately intimal growth and lesion formation, see Fig. 2.

Using this model, collagen matrix turnover following stent induced arterial injury could be evaluated quantitatively. Given that the collagen matrix is a key regulator of VSMC activation and phenotypic modulation this enabled the influence of design parameters, such as stent deployment diameter and strut geometry, on the level of in-stent restenosis to be investigated using this multiscale framework, see Fig. 3.

The model predicted that synthesis of MMP-2 reaches its maximum one week following stent deployment. As a result, the amount of collagen per VSMC drops 33 % from its initial state in the vicinity of the stent strut after 2 weeks. VSMCs were also activated due to the degradation of the collagen matrix and their numbers started to increase 2 days post stent deployment. After reaching its maximum concentration one week post-deployment, MMP concentration subsequently began to recess to normal levels, decreasing 80 % of the maximum level by the end of day 14. This was found to be consistent with the outcome of organ culture experiments conducted on damaged human saphenous veins [34], see Fig. 4.

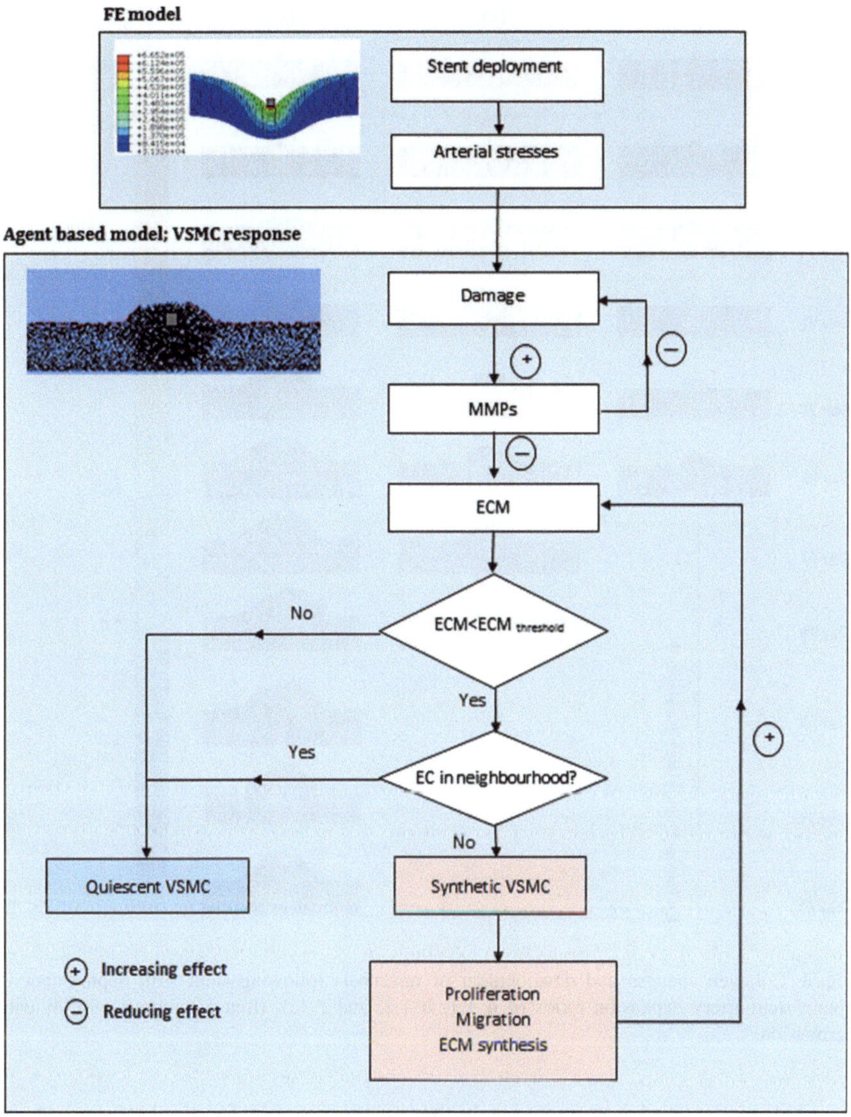

Fig. 2 Schematic of the mechanobiological model of in-stent restenosis (image courtesy of [58] with permission)

A significant increase in the level of intimal growth and VSMC number was predicted by the models when the artery-stent expansion ratio increased from 1.1 to 1.3. This was due to the nonlinear stress-stiffening response of arteries where higher expansions increased the level of arterial stresses exponentially. This prediction was verified by comparison to experimental studies which show that the stent expansion diameter is an important predictor of the level of intimal growth [37–40].

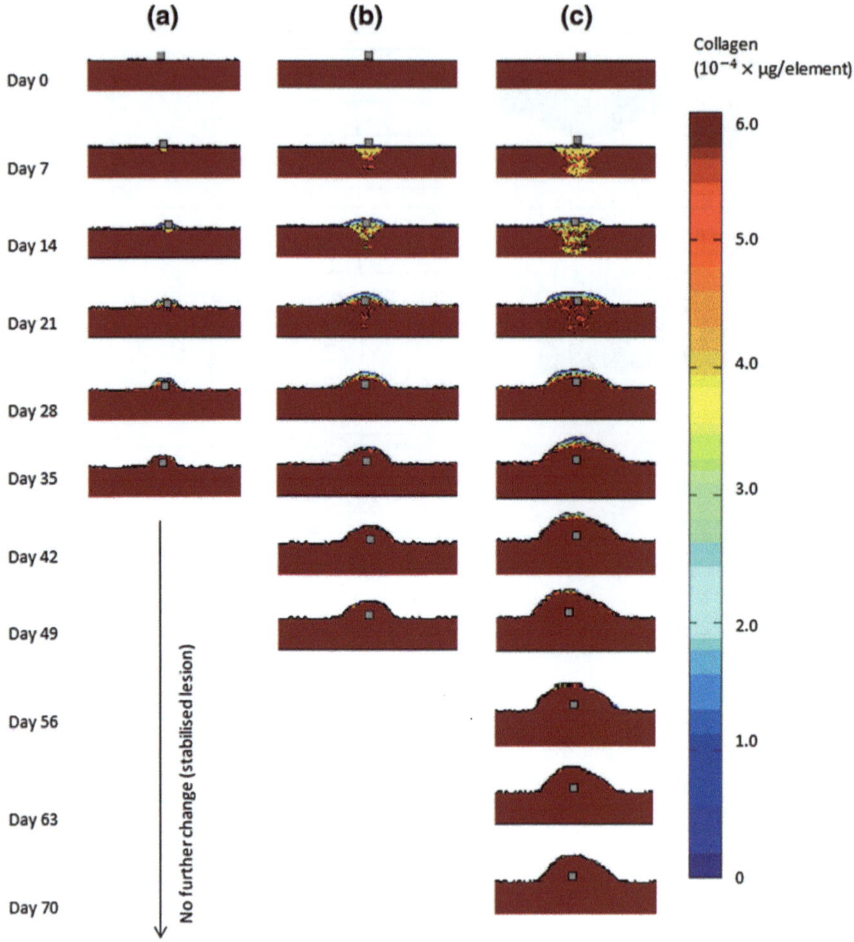

Fig. 3 Collagen changes and development of restenosis following stent strut deployment to obtain stent-artery expansion ratios of **a** 1.1, **b** 1.2, and **c** 1.3. (Image courtesy of [58] with permission)

In addition the model predicted a higher restenosis rate for thicker stent struts compared to thinner struts which is consistent with the outcome of many clinical studies [41–44].

Clearly therefore, the model could quantitatively link the mechanical and biological events occurring at different length and time scales, from the long-term mechanics of the stent-artery interaction at the tissue level to the short-term cellular events such as MMP synthesis, ECM turnover, VSMC migration, proliferation and phenotype modulation, and NO signalling between ECs and VSMCs. The model can, however, be further expanded to include the role of many more parameters at the cellular level, such as thrombocytes and vascular progenitor cells, which are

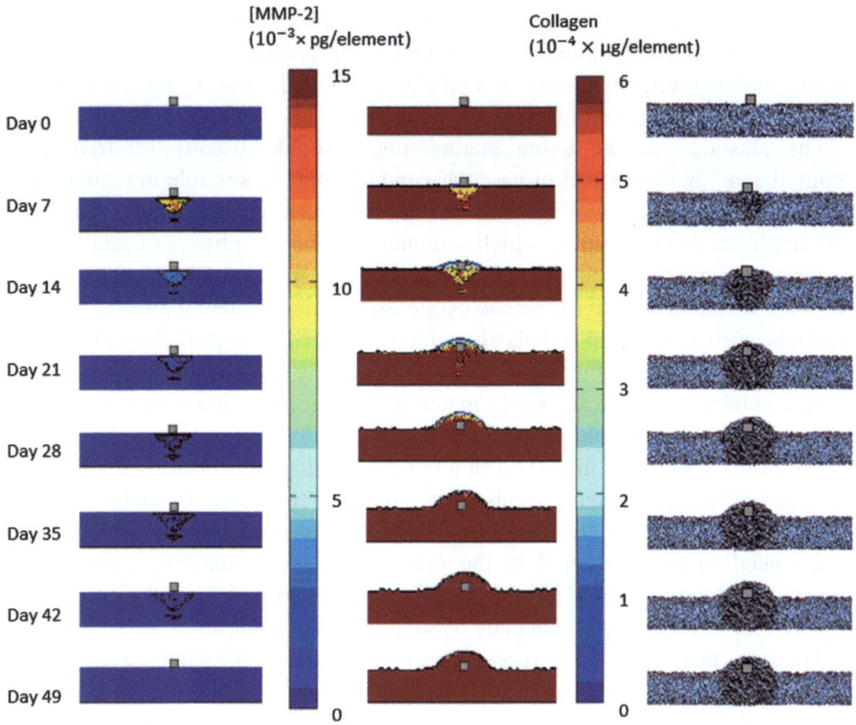

Fig. 4 Evolution of (*left*) MMP2 (*middle*) Collagen and (*right*) VSMCs, when a stent strut was deployed to obtain an expansion ratio of 1.2. (Image courtesy of [58] with permission)

recruited from the circulation system into the site of injury, as well as various individual growth factors secreted by the cells. The multiscale nature of the model can also be further extended to the intracellular level by explicit modelling of events that are initiated by cell level stimuli. As such, mechanotransduction through various signalling pathways, including Ca^{2+} and Rho signalling in VSMCs and subsequent gene expression, could potentially be included in the model.

2.3 Intimal Hyperplasia in Tissue Engineered Vascular Grafts

The need for an alternative to native blood vessels for use as bypass grafts in coronary artery bypass graft surgery has led to considerable research into tissue engineering a blood vessel substitute [45]. For the successful development of a tissue engineered blood vessel (TEBV), vascular cells and specifically VSMCs seeded on the blood vessel scaffold should ideally proliferate and populate the scaffold, synthesise an extracellular matrix similar to that of a native artery, and subsequently adopt a contractile and quiescent phenotype. Failure of VSMCs to adopt quiescent and contractile characteristics, following this initial proliferative phase whereby the cells populate and remodel the scaffold, gives rise to the most significant long-term limitation of vascular grafts and tissue engineered vessels, namely intimal hyperplasia (IH).

The elasticity of the tissue engineering scaffolds, usually referred to as "compliance" in the context of vascular conduits, plays a key role in regulation of the proliferative capacity of VSMCs and hence prevention of intimal hyperplasia. The amplitude of cyclic strain which is dictated by the compliance of scaffolds has an anti-proliferative influence on VSMCs and also increases their apoptosis rate [46–49]. In addition, physiological cyclic strain upregulates synthesis of ECM products, such as collagen and elastin, which enhances the remodelling of vascular scaffolds [48]. On the other hand, it is necessary for cells within the scaffold to receive nutrients and discard waste material in order to maintain their viability, a property governed by interstitial fluid flow which is dependent on the scaffold permeability. Interstitial fluid flow in a scaffold is also influenced by cyclic strain [50], however, and is therefore also dependant on the elasticity of the scaffold. Clearly therefore, there are numerous interlinked mechanobiological parameters which need to be considered in the design and development of viable tissue engineered blood vessels. Multiscale in-silico models provide the ideal platform to explore and optimize the mechanics of such vascular scaffolds.

Towards this goal, in a recent study the authors presented a multiscale mechanobiological framework which enabled investigation of the role of mechanical factors, such as scaffold compliance and loading regimes during cell culture, on the growth of cells within vascular scaffolds and their remodelling [10]. As previously discussed, cyclic strain and pore fluid flow are two key mechanical regulators of VSMC growth in a vascular tissue-engineered construct. As a result, cyclic strain and pore fluid flow velocity were adopted as the main regulators of VSMC growth in these models of cell and tissue growth within a TEBV. The mechanobiological modelling framework comprised two main coupled modules, (i) a module based on the finite element method (FEM) that quantified cyclic strain (ε_{cyc}) and pore fluid velocity (V_{fluid}) as the main regulators of VSMC growth in TEBVs and (ii) a biological module based on ABM that simulated migration, proliferation, apoptosis and ECM synthesis by VSMCs under the influence of the mechanical stimuli quantified in the FE module, see Fig. 5.

The FE model employed in these simulations was an axisymmetric poro-hyperelastic FE model of a tubular scaffold which could quantify cyclic strain (ε_{cyc}) and pore fluid flow velocity (V_{fluid}) in the tissue engineered construct. Pressure and pore pressure boundary conditions were applied to the luminal surface of the tissue-engineered construct while the two ends were longitudinally tethered and the pressure was set to zero at the abluminal surface.

Following remodelling by cells, the mechanical properties of the scaffolds, such as elasticity and permeability, should ideally mimic that of the native artery. Several scaffolds such as woven polyglactin 910 grafts and Poly(Trimethylene Carbonate) scaffolds have shown this response following remodelling by cells [51, 52]. Therefore, remodelling of the mechanical properties of the scaffolds was represented

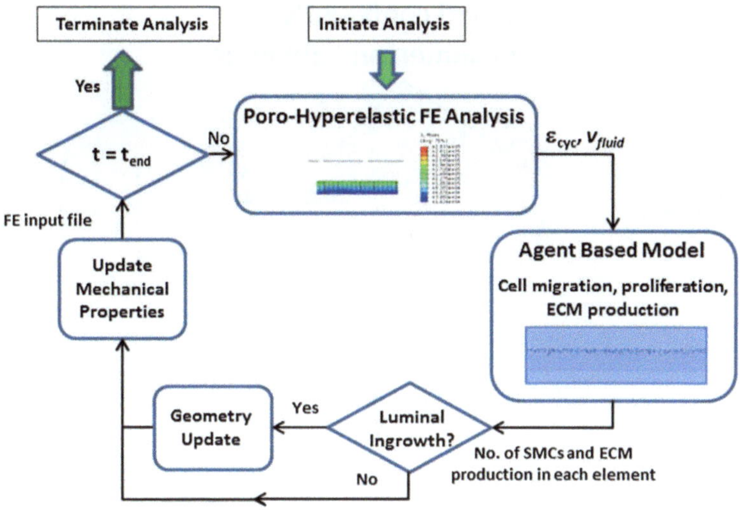

Fig. 5 Schematic of the mechanobiological modelling framework developed for simulation of the mechanoregulation of growth and remodelling in tissue engineered blood vessels. (Image courtesy of [10], with permission)

by altering the initial mechanical properties of the scaffold, i.e. stress–strain response, permeability and porosity to that of the arterial tissue in elements where the ECM content reached half of the collagen content in a native artery which is approximately 3.1×10^{-4} μg/cell [53].

Biological parameters such as the doubling time and the probability of apoptosis were defined as mathematical functions of the cyclic strain and pore fluid velocity for each VSMC. Increased cyclic strain decreased the doubling time of VSMCs and increased their probability of apoptosis whilst low pore fluid velocity also decreased the doubling time of VSMCs and increased the probability of apoptosis given that at low interstitial fluid velocities exchange of nutrients and removal of waste products within the scaffold would be impaired. On the other hand, collagen synthesis by each VSMC was increased with increases in the amplitude of cyclic strain.

Three main scenarios were then simulated using the developed mechanobiological modelling framework, (see Fig. 6); (i) VSMCs were seeded on an arterial compliant scaffold and cultured under hypotensive (50–80 mmHg), normotensive (80–120 mmHg) and hypertensive (140–200 mmHg) luminal pressure to investigate the role of loading regime for both in-vivo and in-vitro applications, (ii) VSMCs were seeded on an arterial compliant scaffold and cultured under a pulsatile luminal pressure and luminal pore pressure of 80–120 mmHg with a frequency of 1 Hz versus a non-pulsatile pressure of 100 mmHg, to study the influence of pulsatile flow in in-vitro applications and (iii) VSMCs were seeded on low compliance versus arterial compliant scaffolds and cultured under a pulsatile luminal pressure of 80–120 mmHg with a frequency of 1 Hz to study the influence of scaffold compliance for in-vivo and in-vitro applications.

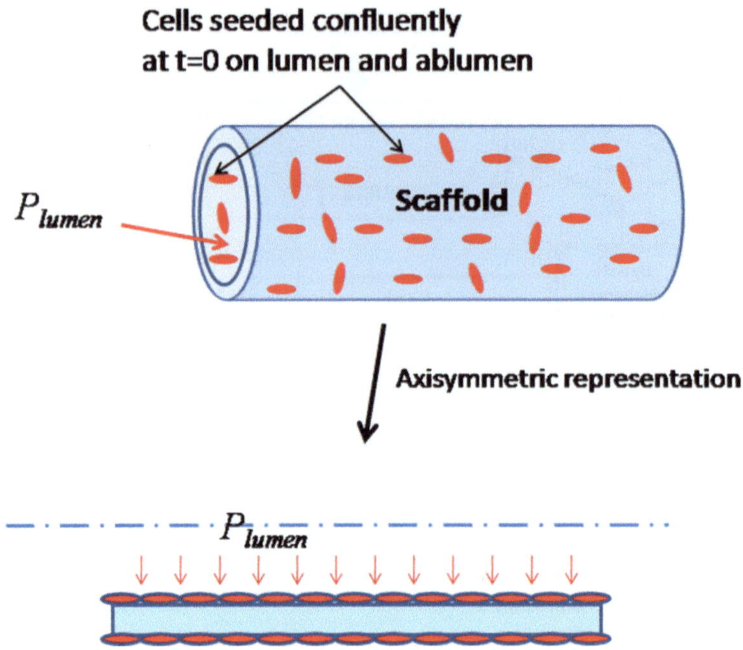

Fig. 6 Schematic of the modelled tubular scaffold (Image courtesy of [10] with permission)

The model predictions for these scenarios provided valuable insights into the multiscale mechanobiological processes involved in vascular tissue engineering and underscored the significance of mechanical factors. The study showed that any subtle decrease of cyclic strain can ultimately lead to proliferation of VSMCs towards the lumen and development of IH. Low cyclic strain contributed to more luminal ingrowth which in turn reduced the cyclic strain further and thereby expedited patency loss. Interestingly, the model predicted that under a hypertensive luminal pressure intimal growth is higher compared to normotensive and hypotensive loading regimes which corroborates findings from clinical studies which suggest that hypertension causes thickening and stiffening of arteries [54]. This response is related to the stress-stiffening mechanical response of vascular tissue which stiffens at higher pressures and therefore undergoes approximately 20 % lower cyclic strain under the hypertensive pressure regime compared to the normotenstive pressure, see Fig. 7.

The influence of loading regime during culture was also simulated using the model given that many in-vitro studies have shown that dynamic culture of VSMCs on vascular scaffolds using pulsatile flow bioreactors results in a higher number of VSMCs and improved mechanical properties compared to static culture [53, 55, 56]. The model also predicted that the amount of collagen produced in the vascular scaffold would be higher where a pulsatile flow culture was used compared to a static culture. Whilst the pulsatile flow maintained intimal growth at a considerably lower level, the model predicted enhanced ECM synthesis and remodelling of the

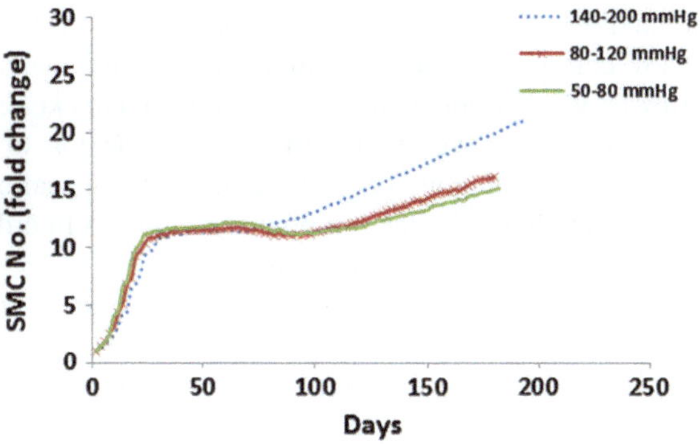

Fig. 7 The Influence of pulsatile pressure on the number of viable vascular smooth muscle cells over time in the arterial compliant scaffold, three loading regimes of hypotensive (50–80 mmHg), normotensive (80–120 mmHg) and hypertensive (140-200 mmHg) were compared. (Image courtesy of [10] with permission)

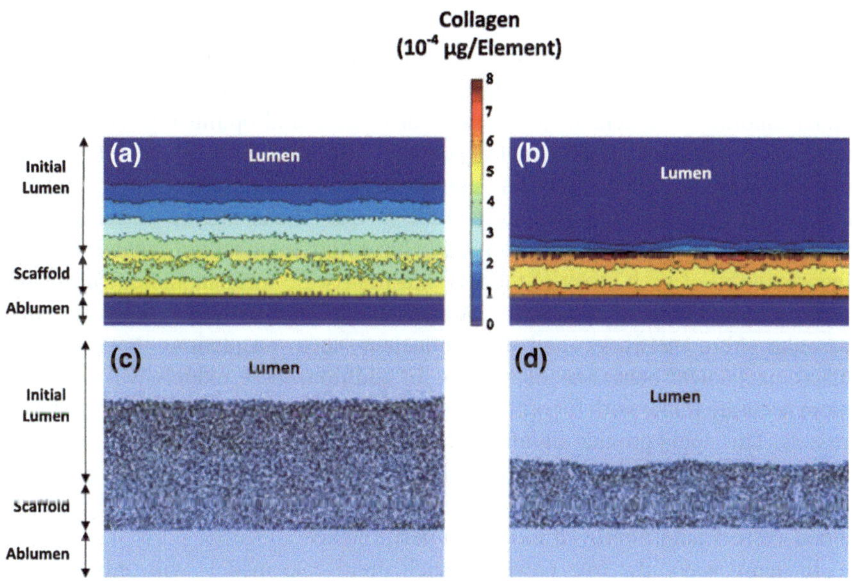

Fig. 8 ECM synthesis and viable VSMCs at day 160 **(a, c)** the TEBV was cultured under static pressure of 100 mmHg **(b, d)** the TEBV was cultured under pulsatile pressure of 80–120 mmHg. (Image courtesy of [10] with permission)

scaffold, see Fig. 8. As such the results imply that to enhance cell growth and reduce culture time, an initial non-pulsatile luminal pressure can be used until cells fully populate the scaffold, following which a pulsatile luminal pressure can enhance ECM synthesis and scaffold remodelling.

On the other hand, when the role of scaffold compliance was studied the model predicted that in low compliance scaffolds a higher level of luminal ingrowth occurs at the onset of growth which accounts for IH formation and thickening of the vessel and therefore further reduces the cyclic strain in comparison to the compliant scaffolds. Therefore, the simulations suggest that the optimal scaffold for vascular tissue engineering should have similar compliance to arteries in order to minimise IH and enhance ECM synthesis and remodelling.

Such insightful conclusions from the simulations clarify an important potential application of the multiscale mechanobiological models in vascular tissue engineering which is to serve as efficient platforms for testing hypotheses on multiscale phenomena that are not easy to test in the laboratory. The outcome of the simulations can then help to optimise in-vitro experiments and develop new tissue engineering strategies in-vivo.

3 Discussion

Whilst progress has undoubtedly been made in recent years in developing suitable multi-scale vascular mechanobiological models which can provide a basis for better understanding the mechanisms of vascular disease and optimising its treatment, there still remains considerable potential in future development of this methodology. In many respects, computational resources currently limit the complexity of established multiscale models. Arterial constitutive models are now well established which are anisotropic and have fibre remodelling capabilities [57]. Incorporation of such material models into multiscale mechanobiological models will provide a means of better exploring the cell driven adaptations of native vascular fibre networks and the influence such adaptations have on the macromechanical behaviour of arteries. In addition, these models will provide a more accurate framework for optimisation of medical devices and tissue engineered vessels. This could provide significant benefits in terms of understanding the role of resident vascular smooth muscle and stem cells in vascular repair or in the development of novel strategies in the development of vascular tissue engineered blood vessels with improved long-term performance.

In many ways, the true power of such multiscale models will only be fully realised when advances in computational capabilities enable 3D patient specific multiscale mechanobiological models to be combined with stochastic modelling approaches, such that models are capable of predicting multiple outputs for a given medical device or therapy which are dependent on patient specific anatomies, obtained from in-vivo imaging, patient history and even genetic information. In this way such multiscale models could provide preclinical data which would rival that generated from animal or even clinical trials without the associated safety, ethical or cost concerns.

Although vascular mechanobiological modeling is already helping to establish better patient selection criteria and more objective strategies for both choosing and optimising personalised therapeutic options, there is undoubtedly further scope for this predictive modeling framework in the development of the more individualised medical therapies of the future.

References

1. Mi, Q., Riviere, B., Clermont, G., Steed, DL., Vodovotz, Y. (2007) Agent-based model of inflammation and wound healing: insights into diabetic foot ulcer pathology and the role of transforming growth factor-b1. Wound Rep. Reg. **15**, 671–682
2. Li, N.Y.K., Verdolini, K., Clermont, G., Mi, Q., Rubinstein, E.N., et al.: A patient-specific in silico model of inflammation and healing tested in acute vocal fold injury. PLoS ONE **3**(7), e2789 (2008). doi:10.1371/journal.pone.0002789
3. Pappalardo, F., Cincotti, A., Motta, A., Pennisi, M.: Agent based modeling of atherosclerosis: a concrete help in personalized treatments. ICIC 2009, LNAI 5755, pp. 386–396 (2009)
4. Caiazzo, A., et al.: Towards a complex automata multiscale model of in-stent restenosis. In: Allen, G., Nabrzyski, J.,VanAlbada, G.D., Sloot, P.M.A. (eds.) Computational Science—Iccs, Part I, vol. 5544, pp. 705–714. Springer, Berlin (2009)
5. Tahir, H., Hoekstra, A.G., Lorenz, E., Lawford, P.V., Hose, D.R., Gunn, J., Evans, D.J.W.: Multi-scale simulations of the dynamics of in-stent restenosis: impact of stent deployment and design. Interface Focus **1**(3), 365–373 (2011). doi:10.1098/rsfs.2010.0024
6. Boyle, C.J., Lennon, A.B., Early, M., Kelly, D.J., Lally, C., Prendergast, P.J.: Computational simulation methodologies for mechanobiological modelling: a cell-centred approach to neointima development in stents. Philos. Trans. R. Soc. A Math. Phys. Eng. Sci. **368**, 2919–2935 (2010). doi:10.1098/rsta.2010.0071
7. Boyle, C.J., Lennon, A.B., Prendergast, P.J.: In silico prediction of the mechanobiological response of arterial tissue: application to angioplasty and stenting. J. Biomech. Eng. **133**(8), 081001 (2011). doi:10.1115/1.4004492
8. Zahedmanesh, H., Cahill, P.A., Lally, C.: Vascular stent design optimisation using numerical modelling techniques. In: Naik, G.R. (ed.) Applied Biological Engineering; Principles and Practice. InTech, ISBN 978-953-51-0412-4 (2012). doi:10.5772/37357
9. Budu-Grajdeanu, P., Schugart, R.C., Friedman, A., Valentine, C., Agarwal, A.K., Rovin, B.H.: A mathematical model of venous neointimal hyperplasia formation. Theor. Biol. Med. Model. **5**, 2 (2008). doi:10.1186/1742-4682-5-2
10. Zahedmanesh, H., Lally, C.: A multiscale mechanobiological model using agent based models, application to vascular tissue engineering. Biomech. Model. Mechanobiol. **11**, 363–377 (2012). doi:10.1007/s10237-011-0316-0
11. Migliavacca, F., Petrini, L., Colombo, M., Auricchio, F., Pietrabissa, R.: Mechanical behavior of coronary stents investigated through the finite element method. J. Biomech. **35**, 803–811 (2002)
12. Lally, C., Dolan, F., Prendergast, P.J.: Cardiovascular stent design and vessel stresses: a finite element analysis. J. Biomech. **38**, 1574–1581 (2005). doi:10.1016/j.jbiomech.2004.07.022
13. Bedoya, J., Meyer, C.A., Timmins, L.H., Moreno, M.R., Moore Jr, J.E.: Effects of stent design parameters on normal artery wall mechanics. J. Biomech. Eng. **128**(5), 757–765 (2006)
14. Zahedmanesh, H., Lally, C.: Determination of the influence of stent strut thickness using the finite element method: implications for vascular injury and in-stent restenosis. Med. Biol. Eng. Comput. **47**, 385–393 (2009). doi:10.1007/s11517-009-0432-5

15. Zahedmanesh, H., Kelly, D., Lally, C.: Simulation of a balloon expandable stent in a realistic coronary artery, determination of the optimum modelling strategy. J. Biomech. **43**, 2126–2132 (2010). doi:10.1016/j.jbiomech.2010.03.050
16. Steinberg, D.: Thematic review series: the pathogenesis of atherosclerosis: an interpretive history of the cholesterol controversy, part III: mechanistically defining the role of hyperlipidemia. J. Lipid Res. **46**, 2037–2051 (2005)
17. Vukovic, I., Arsenijevic, N., Lackovic, V., Todorovic, V.: The origin and differentiation potential of smooth muscle cells in coronary atherosclerosis. Exp. Clin. Cardiol. **11**(2), 123–128 (2006)
18. Kleemann, R., Zadelaar, S., Kooistra, T.: Cytokines and atherosclerosis: a comprehensive review of studies in mice. Cardiovasc. Res. **79**, 360–376 (2008). doi:10.1093/cvr/cvn120
19. Di Tomaso, G., Díaz-Zuccarini, V., Pichardo-Almarza, C.: A multiscale model of atherosclerotic plaque formation at its early stage. IEEE Trans. Biomed. Eng. **58**(12), 3460–3463 (2011)
20. Wieneke, H., Haude, M., Knocks, M., Gutersohn, A., von Birgelen, C., Baumgart, D., Erbel, R.: Evaluation of coronary stents in the animal model: a review. Materialwiss. Werkstofftech. **30**, 809–813 (1999). doi:10.1002/(SICI)1521-4052(199912
21. Babapulle, M.N., Eisenberg, M.J.: Coated stents for the prevention of restenosis: part I. Circulation **106**, 2859–2866 (2002). doi:10.1161/01.CIR.0000038982.49640.70
22. Welt, F.G., Rogers, C.: Inflammation and restenosis in the stent era. Arterioscler. Thromb. Vasc. Biol. **22**(11), 1769–1776 (2002). doi:10.1161/01.ATV.0000037100.44766.5B
23. Mitra, A.K., Agrawal, D.K.: In stent restenosis: bane of the stent era. J. Clin. Pathol. **59**, 232–239 (2005). doi:10.1136/jcp.2005.025742
24. Thyberg, J., Blomgren, K., Roy, J., Tran, P.K., Hedin, A.: Phenotype modulation of smooth muscle cells after arterial injury is associated with changes in the distribution of laminin and fibronectin. J. Histochem. Cytochem. **45**, 837–846 (1997). doi:10.1177/002215549704500608
25. Hirose, M., Kosugi, H., Nakazato, K., Hayashi, T.: Restoration to a quiescent and contractile phenotype from a proliferative phenotype of myofibroblasts-like human aortic smooth muscle cells by culture on type IV collagen gels. J. Biochem. **125**, 991–1000 (1999)
26. Aguilera, C.V., George, S.J., Johnson, J.L., Newby, A.C.: Relationship between type IV collagen degradation, metalloproteinase activity and smooth muscle cell migration and proliferation in cultured human saphenous vein. Cardiovasc. Res. **58**, 679–688 (2003). doi:10.1016/S0008-6363(03)00256-6
27. Monaco, S., Sparano, V., Gioia, M., Spardella, D., Di Pierro, D., Marini, S., Coletta, M.: Enzymatic processing of collagen IV by MMP-2 (gelatinase A) affects neutrophil migration and it is modulated by extracatalytic domains. Protein Sci. **15**, 2805–2815 (2006). doi:10.1110/ps.062430706
28. Adiguzel, E., Ahmad, P.J., Franco, C., Bendeck, M.P.: Collagens in the progression and complications of atherosclerosis. Vasc. Med. **14**, 73–89 (2009). doi:10.1177/1358863X08094801
29. Newby, A.C.: Matrix metalloproteinases regulate migration, proliferation, and death of vascular smooth muscle cells by degrading matrix and non-matrix substrates. Cardiovasc. Res. **69**(3), 614–624 (2006). doi:10.1016/j.cardiores.2005.08.002
30. Doronzo, G., Russo, I., Mattiello, L., Trovati, M., Anfossi, G.: Homocysteine rapidly increases matrix metalloproteinase-2 expression and activity in cultured human vascular smooth muscle cells. Thromb. Haemost. **94**(6), 1285–1293 (2005). doi:10.1160/TH05040221
31. James, T.W., Wagner, R., White, L.A., Zwolak, R.M., Brinkerhoff, C.E.: Induction of collagenase gene expression by mechanical injury in avascular smooth muscle cell derived cell line. J. Cell. Physiol. **157**, 426–437 (1993). doi:10.1002/jcp.1041570227
32. Bendeck, M.P., Zempo, N., Clowes, A.W., Galardy, R.E.: Smooth muscle cell migration and matrix metalloproteinase expression after arterial injury in the rat. Circ. Res. **75**, 539–545 (1994). doi:10.1161/01.RES.75.3.539
33. Southgate, K.M., Fisher, M., Banning, A.P., Thurston, V.J., Baker, A.H., Fabunmi, R.P., Groves, P.H., Davies, M., Newby, A.C.: Upregulation of basement-membrane-degrading

metalloproteinase secretion following balloon angioplasty of pig carotid arteries. Circ. Res. **79**, 1177–1187 (1996). doi:10.1161/01.RES.79.6.1177
34. George, S.J., Zaltsman, A.B., Newby, A.C.: Surgical preparative injury and neointima formation increase MMP-9 expression and MMP-2 activation in human saphenous vein. Cardiovasc. Res. **33**, 447–459 (1997). doi:10.1016/S0008-6363(96)00211-8
35. Asanuma, K., Magid, R., Johnson, C., Nerem, R.M., Galis, Z.: Uniaxial strain upregulates matrix-degrading enzymes produced by human vascular smooth muscle cells. Am. J. Physiol. Heart Circ. Physiol. **284**, 1778–1784 (2003). doi:10.1152/ajpheart004942002
36. Grote, K., Flach, I., Luchtefeld, M., Akin, E., Holland, S.M., Drexler, H., Schieffer, B.: Mechanical stretch enhances mRNA expression and proenzyme release of matrix metalloproteinase- 2 (MMP-2) via NAD(P)H oxidase-derived reactive oxygen species. Circ. Res. **92**, 80e–86e (2003). doi:10.1161/01.RES.0000077044.601387C
37. Kornowski, R., Hong, M.K., Tio, F.O., Bramwell, O., Wu, H., Leon, M.B.: In-stent restenosis: contributions of inflammatory responses and arterial injury to neointimal hyperplasia. J. Am. Coll. Cardiol. **31**(1), 224–230 (1998). doi:10.1016/S0735-1097(97)00450-6
38. Gunn, J., Chan, K.H., Shepherd, L., Cumberland, D.C., Crossman, D.C.: Coronary artery stretch versus deep injury in the development of in-stent neointima. Heart **88**, 401–405 (2002). doi:10.1136/heart.88.4.401
39. Lowe, H.C., Oesterle, S.N., Khachigian, L.M.: Coronary in-stent restenosis: current status and future strategies. J. Am. Coll. Cardiol. **39**, 183–193 (2002). doi:10.1016/S0735-1097(01)01742-9
40. Houbballah, R., Robaldo, A., Albadawi, H., Titus, J., LaMuraglia, G.M.: A novel model of accelerated intimal hyperplasia in the pig iliac artery. Int. J. Exp. Pathol. **92**, 422–427 (2011). doi:10.1111/j.13652613.2011.00790.x
41. Kastrati, A., Mehilli, J., Dirschinger, J., Dotzer, F., Schühlen, H., Neumann, F.J., Fleckenstein, M., Pfafferott, C., Seyfarth, M., Schömig, A.: Intracoronary stenting and angiographic results: strut thickness effect on restenosis outcome (ISARSTEREO) trial. Circulation **103**, 2816–2821 (2001). doi:10.1016/S0735-1097(03)00119-0
42. Briguori, C., Sarais, C., Pagnotta, P.: In-stent restenosis in small coronary arteries: impact of strut thickness. J. Am. Coll. Cardiol. **40**, 403–409 (2002). doi:10.1016/S0735-1097(02)01989-7
43. Pache, J., Kastrati, A., Mehilli, J., Schühlen, H., Dotzer, F., Hausleiter, J., Fleckenstein, M., Neumann, F.J., Sattelberger, U., Schmitt, C., et al.: Intracoronary stenting and angiographic results: strut thickness effect on restenosis outcome (ISARSTEREO- 2), trial. J. Am. Coll. Cardiol. **41**, 1283–1288 (2003). doi:10.1016/S0735-1097(03)00119-0
44. Morton, A.C., Crossman, D., Gunn, J.: The influence of physical stent parameters upon restenosis. Pathol. Biol. **52**, 196–205 (2004). doi:10.1016/j.patbio.2004.03.013
45. Naito, Y., Shinoka, T., Duncan, D., Hibino, N., Solomon, D., Cleary, M., Rathore, A., Fein, C., Church, S., Breuer, C.: Vascular tissue engineering: towards the next generation vascular grafts. Adv. Drug Deliv. Rev. **63**, 312–323 (2011)
46. Chapman, G.B., Durante, W., Hellums, J.D., Schafer, A.I.: Physiological cyclic stretch causes cell cycle arrest in cultured vascular smooth muscle cells. Am. J. Physiol. Heart Circ. Physiol. **278**, H748–H754 (2000)
47. Morrow, D., Sweeney, C., Birney, Y.A., Cummins, P.M., Walls, D., Redmond, E.M., Cahill, P.A.: Cyclic strain inhibits notch receptor signaling in vascular smooth muscle cells in vitro. Circ. Res. **96**, 567–575 (2005). doi:10.1161/01.RES.0000159182.98874.43
48. Kona, S., Chellamuthu, P., Xu, H., Hills, S.R., Nguyen, K.T.: Effects of cyclic strain and growth factors on vascular smooth muscle cell responses. Open Biomed. Eng. J. **3**, 28–38 (2009). doi:10.2174/1874120700903010028
49. Colombo, A., Guha, S., Mackle, J.N., Cahill, P.A., Lally, C.: Cyclic strain amplitude dictates the growth response of vascular smooth muscle cells in vitro: role in in-stent restenosis and inhibition with a sirolimus drug-eluting stent. Biomech. Model. Mechanobiol. (2012). doi:10.1007/s10237-012-0433-4

50. Buijs, J.O.D., Lu, L., Jorgensen, S.M., Dragomir-Daescu, D., Yaszemski, M.J., Ritman, E.L.: Solute transport in cyclically deformed porous tissue scaffolds with controlled pore cross-sectional geometries. Tissue Eng. Part A **15**(8), 1989–1999 (2009). doi:10.1089/ten.tea.2008.0382
51. Greisler, H.P., Joyce, K.A., Kim, D.U., Pham, S.M., Berceli, S.A., Borovetz, H.S.: Spatial and temporal changes in compliance following implantation of bioresorbable vascular grafts. J. Biomed. Mater. Res. **26**, 1449–1461 (2004). doi:10.1002/jbm.820261105
52. Song, Y., Wennink, J.W.H., Kamphuis, M.M.J., Sterk, L.M.T., Vermes, I., Poot, A.A., Feijen, J., Grijpma, D.W.: Dynamic culturing of smooth muscle cells in tubular poly (trimethylene carbonate) scaffolds for vascular tissue engineering. Tissue Eng. Part A (2010, in press). doi:10.1089/ten.tea.2009.0805
53. Hahn, M.S., Mchale, M.K., Wang, E., Schmedlen, R.H., West, J.I.: Physiologic pulsatile flow bioreactor conditioning of poly(ethyleneglycol)-based tissue engineered vascular grafts. Ann. Biomed. Eng. **35**(2), 190–200 (2007). doi:10.1007/s10439-006-9099-3
54. London, G.M., Marchais, S.J., Guerin, A.P., Pannier, B.: Arterial stiffness: pathophysiology and clinical impact. Clin. Exp. Hypertens. **26**(7–8), 689–699 (2004)
55. Jeong, S.I., Kwon, J.H., Lim, J.I., Cho, S.W., Jung, Y.M., Sung, W.J., Kim, S.H., Kim, Y.H., Lee, Y.M., Kim, B.S., Choi, C.Y., Kim, S.J.: Mechanoactive tissue engineering of vascular smooth muscle using pulsatile perfusion bioreactors and elastic PLCL scaffolds. Biomaterials **26**, 1405–1411 (2005). doi:10.1016/j.biomaterials.2004.04.036
56. Stankus, J.J., Guan, J., Fujimoto, K., Wagner, W.R.: Microintegrating smooth muscle cells in a biodegradable elastomeric fibre matrix. Biomaterials **27**(5), 735–744 (2006). doi:10.1016/j.biomaterials.2005.06.020
57. Creane, A., Maher, E., Sultan, S., Hynes, N., Kelly, D., Lally, C.: A remodelling metric for angular fibre distributions and its application to diseased carotid bifurcations. Biomech. Model. Mechanobiol. **11**(6), 869–882 (2012)
58. Zahedmanesh, H., Van Oosterwyck, H., Lally, C.: A multi-scale mechanobiological model of in-stent restenosis: deciphering the role of matrix metalloproteinase and extracellular matrix changes. Comput. Meth. Biomech. Biomed. Eng. (2012) doi:10.1080/10255842.2012.716830

Translational Research: Multi-Scale Models of the Pulmonary Circulation in Health and Disease

Alys R. Clark, Kelly S. Burrowes and Merryn H. Tawhai

Abstract The pulmonary circulation is a unique low resistance system that carries almost the entire cardiac output, and is responsible for the essential role of providing oxygenated blood to the body. As the pulmonary circulation differs from the systemic circulation in its development, structure, and function, it is often most appropriate to study the mechanisms that contribute toward pulmonary vascular disease separately from those of systemic vascular disease at the genetic, cellular, tissue and organ level. Here we review the development of multi-scale, anatomically based models of the pulmonary circulation. These models aim to describe the interaction of structural and functional aspects of the pulmonary circulation that are the most important in determining the effective uptake of oxygen to the blood. We describe how these models have been used to understand normal lung physiology and to explain outcomes in pulmonary disease. Finally, we consider the future of multi-scale modeling in the pulmonary circulation and discuss what can be learned from well-developed multi-scale models of the pulmonary airspaces that interact closely with the lung's circulatory system.

1 Introduction

Experimental or imaging studies of the pulmonary system are fraught with difficulties due to the nature of the lung's structure and function. The lung comprises trees of airways and blood vessels that are 'suspended' within an extremely

A. R. Clark (✉) · M. H. Tawhai
Auckland Bioengineering Institute, University of Auckland, Auckland, New Zealand
e-mail: alys.clark@auckland.ac.nz

K. S. Burrowes
Department of Computer Science, The University of Oxford, Oxford, UK

Stud Mechanobiol Tissue Eng Biomater (2013) 14: 259–286
DOI: 10.1007/8415_2012_152

Published Online: 15 September 2012

delicate parenchymal tissue [1]. The in vivo lung is maintained in contact with the wall of the thoracic cavity by negative pressure in the intrapleural 'space' [2]. Ex vivo experimentation with lung tissue is difficult to reconcile with in vivo behavior as when removed from the body the lung recoils to considerably smaller volumes than are encountered in breathing, and the lung changes in shape. The advent of modern imaging has provided opportunity to study the lung in situ, however this is again complicated by lung deformability and its air content. In awake humans the lung is typically functioning in an upright posture, and because the delicate tissue is readily deformable there is a substantial gradient of tissue density along the cranial-caudal axis [3]. Magnetic resonance imaging (MRI) and computed tomography (CT) are currently constrained to horizontal postures, i.e. supine or prone. The lungs are typically at a smaller volume in these postures than when upright [4], and the density gradient is in a different axis. This impacts on regional lung expansion [5], pulmonary blood volume and flow distribution [6], the relationship between ventilation ($\dot{V}$) and perfusion ($\dot{Q}$) [6], and so gas exchange function [7]. The supine or prone human lung could therefore have important differences in function to upright. Typically, MRI is preferable as an imaging modality in comparison to CT because the latter involves ionizing radiation. However, because MRI images protons in water and the lung tissue contains mostly air, its effective use in lung is restricted to tagging of blood for quantifying regional blood flow [8, 9], or imaging inhaled hyper-polarized gases [10]. That is, MRI is not yet effective at imaging lung tissue. High resolution CT provides excellent anatomical definition of the lung but its use is restricted in vulnerable populations, and it is not appropriate for most longitudinal studies. High resolution imaging data for the normal human lung is therefore typically restricted to one or two volumes, or one or both horizontal postures at a single volume. The complex interaction between airspaces, blood vessels and tissue in the lung and its movement in breathing further complicates diagnostic imaging and reconciliation of experimental data at different spatial and temporal scales.

These limitations on imaging and direct experimental measurement of lung structure and function provide strong motivation for developing mathematical models of the lung, including the pulmonary circulation, to interpret experimental results between postures, lung volumes, or different species, or at different spatial scales. A multi-scale approach to modeling the pulmonary circulation provides a means to bridge the gap between mechanisms at the genetic, cellular, tissue, vessel and organ level and to provide new insights into pulmonary function in health and disease.

Conceptual models of the pulmonary circulation have provided the basis for understanding its function since at least the 1960s. An example of an early conceptual model of the pulmonary circulation is the 'zonal model' of pulmonary blood flow [11]. This model explained how the interaction between pulmonary arterial, venous and alveolar pressure act to introduce a gravitationally dependent distribution of blood flow in the lungs. Although there is currently extensive debate on the magnitude of the zonal effect in comparison to other gravitational and structural mechanisms [12–16], the model has been significant in showing how physical laws can be combined elegantly and concisely to explain physiological function. Since that era numerous mathematical models have emerged to provide interpretation of experimental measurements of the pulmonary circulation, to summarize hypotheses, and more recently to quantify the relative effects of gravity and structure on pulmonary blood flow distribution [17].

Mathematical models of the pulmonary circulation have taken either a systems (electrical analogue) approach or a biophysical approach (based on the fluid dynamics properties of the system). Geometrically, these mathematical models have developed from the 'lumped parameter' with each of the arterial, capillary and venous systems lumped into single 'resistors' [18], through symmetric [19], fractal [20], and regular asymmetric [21] branching structures, to the current anatomically based volume-filling branching geometries [22] that represent the state-of-the art for modeling the spatially-distributed geometry of the pulmonary arterial and venous trees. Each of these model types fills a role in understanding the function of the pulmonary circulation. Following the philosophy that a mathematical model should only be as complex as to capture the essential functions under investigation (i.e. 'more is not always better'), lumped parameter and regular branching models play an important role in investigating the global response to pathologies that are diffuse (i.e. are distributed relatively evenly throughout the lung, e.g. [23, 24]). However, as they cannot capture the complex structure of the lung they cannot be used to investigate the relationship between vascular branching structure and flow distribution, or to study pathologies that present heterogeneously in the lung or within the population. To address this limitation multi-scale models have recently emerged that couple scale-specific structure and function in the macro- and micro-vessels with biophysical function in other components of the pulmonary system that influence how the circulation behaves [22, 25–32]. This type of model has been used successfully to investigate the effect of pulmonary embolism (via blood clots in the lungs) on regional and total organ gas exchange and blood pressure; pulmonary embolism has a variable impact between individuals and significant localized effects [32–34]. The evolution of these multi-scale models will be described in the following sections.

1.1 Unique Features of the Pulmonary Circulation

The pulmonary circulation is distinct from the systemic circulation in its structure and function. Notably, the lung receives almost the entire cardiac output from the right ventricle (RV). Because the height of the lung is on the order of only 30 cm, this requires far lower RV pressure (typical systolic range of 15–30 mmHg) than the LV pressure that is needed to drive blood to the systemic circulation (typical systolic range of 100–140 mmHg). The pulmonary circulation is therefore described as a 'low pressure' system, which is facilitated by its relatively low resistance.

In general, the pulmonary blood vessels have thinner walls and are less muscular than systemic blood vessels of similar size, yet localized active regulation of smooth muscle tone is an important mechanism for controlling gas exchange function in the lung [35]. In addition to active mechanisms, passive changes to the pulmonary vascular system can have a far greater effect than in the systemic circulation: pulmonary vascular resistance (PVR) and the distribution of blood can be significantly altered by changes in cardiac output, posture, and the rate or depth of breathing. Abnormalities in the pulmonary circulation can lead to impaired gas exchange efficiency or significant increases in RV afterload and impaired cardiac function. Therefore, understanding normal and abnormal function in the pulmonary circulation is essential in consideration of lung disease.

The lung provides an interface between air and blood, through which oxygen (O_2) is carried to the body and carbon dioxide (CO_2) released to the atmosphere. Air pressure varies through the breathing cycle meaning that pulmonary blood vessels (particularly the capillaries that cover the alveoli) are exposed to changes in external pressure unlike that of any other blood vessels in the body. The caliber of pulmonary vessels varies with breathing: with inflation the intra-alveolar vessels compress in height and the extra-alveolar vessels increase in diameter; with deflation the intra-alveolar vessels increase in height and the extra-alveolar vessels decrease in diameter [36]. Changes in vessel caliber with lung inflation and the pulsatility of blood flow in the lung results in a complex, yet coordinated, temporal variability in blood flow that is dependent on both the rate of breathing and heart rate. This makes the morphometry of the lung difficult to study in vitro conditions and necessitates the inclusion of air–blood interactions in modeling the pulmonary circulation.

Structurally the pulmonary circulation differs from the systemic circulation, most notably at the micro-scale. Whilst the systemic capillaries have a 3-D 'tree-like' structure, the pulmonary capillaries form a dense 2-D mesh [1]. The systemic circulation has distinct arterioles with a single smooth muscle layer that connect arteries to capillaries. There is no distinct anatomical definition of pulmonary arterioles in terms of muscularization. However, small blood vessels—from which capillary beds arise—that branch in and between the alveolar airways are sometimes termed arterioles [31, 37, 38]. As the lung carries almost the entire cardiac output via a low pressure, low resistance system, large pulmonary arteries have a relatively thin medial wall compared with systemic arteries, which makes them more distensible, and so more readily adaptable to pressure variations than the systemic arteries.

As the lung does not have distinct, muscular pre-capillary vessels it is thought that control of pulmonary blood flow, and its matching to local ventilation, is achieved by a combination of structural and gravitational influences [12–16]. This control is 'fine-tuned' via active mechanisms [39], which include a further unique attribute of the pulmonary circulation: its response to hypoxia. Whilst systemic blood vessels dilate in response to hypoxia, the pulmonary blood vessels contract when *alveolar* (air-side) oxygen is low [40, 41]. Like other blood vessels, pulmonary vessels also dilate when nitric oxide (NO) levels increase, with NO production being stimulated by increased shear stress. Therefore, a complete model of pulmonary blood flow would include important structural influences, passive influences due to blood pressure and gravity, as well as active mechanisms that translate from the cellular to the full organ scale.

2 The Macrocirculation

The pulmonary vasculature provides a direct route from the heart to the gas exchange surface. The branching structure of the pulmonary vasculature, like that of the airways, allows efficient gas exchange through its bifurcating tree-like structure that terminates in a very large alveolar-capillary surface area. After circulating through the systemic circuit to deliver oxygen and nutrients to the body tissue, blood returns to the right atrium of the heart via the vena cavae, and then to the RV. The pulmonary trunk emerges from the RV through the pulmonary valve, and is the conduit for deoxygenated blood to enter the pulmonary circuit. The left and right main pulmonary arteries enter into the hila of the lungs and divide into branches that accompany the lobar, segmental and subsegmental bronchi. Arterial branches continue to bifurcate into successively smaller daughter branches in a pattern that closely follows that of the bronchi. That is, each bronchus or bronchiole (airway) has a corresponding accompanying artery. The venous network forms a similar branching structure, but it is located in between the paired arterial-bronchial branches. A single main vein emerges from each of the lobes, and four veins feed back into the left atrium of the heart. The pulmonary vessels do, however, branch more frequently than the airways, giving rise to additional 'supernumerary vessels' that do not accompany an airway branch [42]. The functional role of these numerous additional vessels is not known. Morphometric cast-based studies have meticulously measured the geometric properties of the pulmonary arterial and venous vessels [37, 43–45] providing useful information for constructing anatomically-based models of this complex system.

2.1 Modeling the Macrocirculation: From In Vivo Imaging to In Silico Experiments

As with any simulation study, the first task in modeling the pulmonary circulation is to develop the geometric domain over which to solve a set of mathematical equations. These computational meshes can range vastly in their complexity. With respect to modeling the pulmonary macrocirculation, by far the most predominant method has been to represent the branching arterial and venous trees as symmetric structures; that is, branches within each generation have equal diameters and lengths. While this simplifies calculations tremendously (only a single pathway

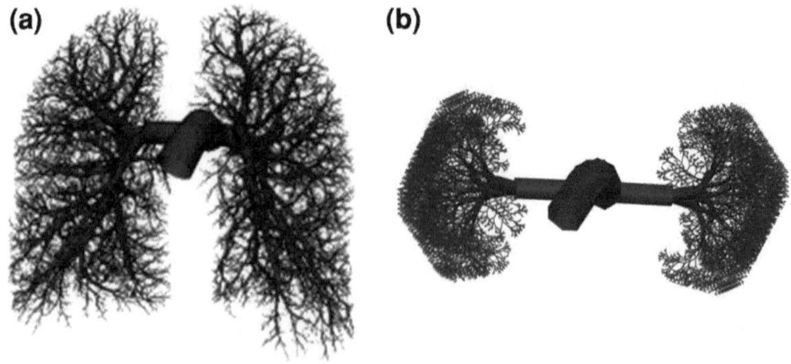

Fig. 1 **a** An anatomically based geometric model of the human arterial tree that fills the lung volume obtained from imaging data (reprinted from [28], with permission from Elsevier), and **b** a symmetric model of the arterial tree with the same average morphometric properties

needs to be considered with tens of branches, compared with tens of thousands of branches in the full lung) the significant impact that structure has on function cannot be realized. Until fairly recently, computation time limited the size of geometric models of the pulmonary circulation and airway tree. However the rapid advancement of computational power in recent decades has allowed significant steps forward in imaging technologies to describe the lung geometry, as well as in the ability to solve large systems of equations numerically over extensive and complex domains. Therefore, the structural accuracy of geometries in which models of blood flow can be solved has improved dramatically. Figure 1 shows a comparison between an anatomically based model geometry and a spatially distributed symmetric model geometry (note that symmetric models are not usually distributed spatially in this manner, nor with each pathway modeled explicitly). Each model has the same average morphometric parameters and there are an equivalent number of blood vessels in each tree. However, the anatomically based model clearly captures certain aspects of the lung structure that the symmetric model cannot.

The contribution of vascular branching structure to heterogeneity in the distribution of pulmonary perfusion and its relationship to airway structure and function was first introduced in studies that analyzed the fractal properties of blood flow distribution. The concept of the pulmonary circulation having a fractal branching structure was championed by Glenny and Robertson who demonstrated that there is a spatial correlation in local pulmonary blood flow that is independent of the size or location of the lung region considered [20, 46]. This implied an important role for the pulmonary branching networks in determining regional blood and airflow and ultimately the efficiency of the lung for gas exchange. Combined with advances in estimating regional perfusion distribution (specifically the use of inhaled or injected microspheres), fractal modeling of the pulmonary circulation pointed to inconsistencies with the traditional zonal model for perfusion. This led to a debate within the community concerning the importance of

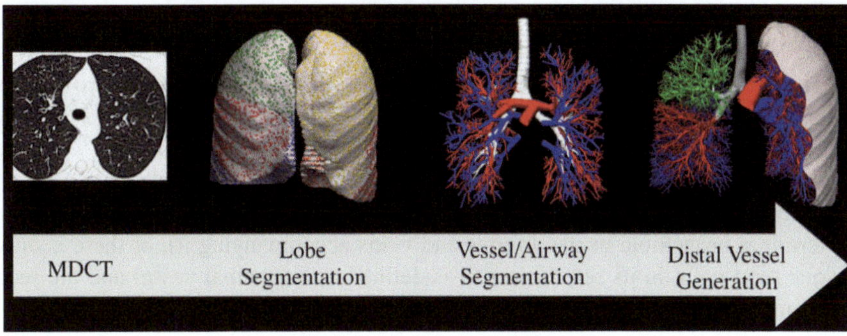

Fig. 2 The generation of an anatomically based model of the pulmonary airways and vasculature. From *left* to *right* the figure shows: a single high resolution CT image of a lung in the supine posture; rendering of volumetric MDCT of the lung and automatic placement of surface points on each lobe; initial arteries (*red*), veins (*blue*), and airways (*white*) used for model generation; volume-filling airways (*right* lung) and blood vessels (*left* lung)

branching structure in perfusion distribution and ventilation–perfusion ($\dot{V}/\dot{Q}$) matching [12–16]. In particular, the fractal model raised questions regarding the significance of gravity (e.g. West's zonal model [11]) on function [15, 16]. With this debate in mind Burrowes et al. constructed an anatomically-based finite element model of the pulmonary circulation, based on MDCT (multi-detector computed tomography) imaging which aimed to capture individual lung shape and blood vessel distribution [22].

The anatomically-based model of the pulmonary vasculature followed the spirit of previous models of the pulmonary airways developed by Tawhai et al. [47, 48]. An illustration of the generation of geometric models of the pulmonary vasculature and airways is given in Fig. 2. In each geometric model (airway or vascular), the lungs or lobes and the pulmonary trunk (central blood vessels and airways) are segmented from imaging data (in this case MDCT) using image-processing software. The segmented airways and blood vessels are represented in a finite element mesh by vectors describing their centerlines, and spatial coordinates for the location of branching points. The distal airways and their accompanying blood vessels are then generated via a volume-filling branching (VFB) algorithm [48], which generates individual representations of airways and blood vessels to a user-defined level, which is typically the level of the pulmonary acinus (a gas exchange unit). The lung or lobe volume that is segmented from the imaging data acts as a 'host volume', which is filled with a uniform grid of seed points that represent the location of the approximately 32,000 acinar units in the lung. Then, the following steps are repeated until there are no seed points remaining in the set:

1. Seed points are grouped into sets with each seed point being associated with the closest terminal (parent) vessel in the current tree.
2. The vector in the direction of the parent vessel and the coordinates of the center of mass of the seed points associated with it are used to define a splitting plane.

3. Each set of seed points is split in two (by the splitting plane) to define two new seed point sets.
4. The center of mass of the two new seed point sets is calculated and two vessels are generated along the vectors that run between the parent vessel and the new centers of mass. The length of the new vessels is defined to be 40 % of the distance between the end of the parent vessel and the center of mass.
5. If the new vessel is less than a pre-defined length (typical the length of a terminal bronchiole or the arteries and veins accompanying it), or there is only one seed point in its set, the vessel is defined as a terminal vessel and the seed point closest to that vessel is removed from the set of seed points.

Classifying each vessel in a generated tree by order allows a convenient definition of the caliber of each vessel. Morphometric studies of pulmonary vessels have shown a linear relationship between the logarithm of vessel diameter and the Strahler order of the vessel [37, 44, 45]. The Strahler ordering system defines terminal branches (in this case the terminal bronchioles of the airway tree and the vessels accompanying them) as order 1. Then, the order of the parent branch proximal to two daughter vessels is defined as the highest of the orders of the daughter branches if these orders are different, or the order of the daughter branches plus one if the order of the daughter branches is the same. Then,

$$\log D(x) = (x - N)\log R_d + \log D_N, \tag{1}$$

where $D(x)$ is the diameter of a vessel/airway of Strahler order x, N is the total number of Strahler orders in the tree, R_d is the Strahler diameter ratio (effectively the rate of decrease of vessel/airway diameter), and D_N is the diameter of the vessel/airway of the highest order in the tree. The value of D_N can be estimated from morphometric parameters or from imaging data. The geometric model developed by Burrowes et al. [22] matched well with anatomical data from cast based studies [37, 44].

By assuming that the blood vessels are rigid and adopting a steady (time-averaged) laminar flow, Burrowes et al. [22] predicted the blood flow and pressure distribution through the isolated arterial tree. Equations describing Poiseuille resistance with a gravitational dependence (dominating in the axial direction) were solved along with flow conservation equations at bifurcations. That is,

$$Q = \frac{\pi D^4}{128\mu}\left(\frac{\Delta P}{L} + \rho\, g\cos\theta\right), \tag{2a}$$

$$Q_P = Q_{D1} + Q_{D2}, \tag{2b}$$

where Q is the rate of blood flow through a vessel, D is the diameter of the vessel, μ is the viscosity of blood, ΔP is the pressure drop through the vessel, L is the length of the vessel, ρ is the density of blood, g is gravitational acceleration, θ is the angle the vessel makes with the direction of gravity, Q_P is the flow through a parent vessel, and Q_{D1} and Q_{D2} are the flows through each of the daughter vessels associated with the parent P. As flow and pressure were calculated only in an isolated arterial tree (without feedback from venous and/or capillary models) boundary conditions had to be specified at each of the ~32,000 terminal arteries. Burrowes et al. [22] assumed a constant inlet pressure (2 kPa at the main pulmonary artery) and outlet pressure (1.25 kPa at each terminal vessel) with a gravitationally dependent gradient of external (pleural) pressure of 0.25 cmH_2O per cm of lung height [49]. Regional perfusion was therefore determined via a combination of heterogeneous arterial resistance (via vascular branching and dimensional asymmetry), the hydrostatic effect of gravity acting on the weight of blood, and an induced (linear) gradient in pressure external to the blood vessels (pleural pressure).

The simple rigid tube model presented by Burrowes et al. [22] was able to predict that the asymmetry of the arterial tree could introduce a significant heterogeneity in pulmonary blood flow. The study was able to show for the first time in a computational model, a marked heterogeneity in pulmonary perfusion superimposed upon a gravitational gradient. This simple blood flow model was quickly built upon to gain physiological insight into the mechanisms determining the distribution of pulmonary perfusion [25, 26, 28]. Blood vessel elasticity was incorporated, initially using a non-linear model for vessel elasticity [28] based on experimental data for lung vessels [50], but with a functional form derived for the elasticity of systemic vessels. Later [25], a simpler functional form for vessel diameter (D) was adopted, based on the studies of Krenz and Dawson [51] who showed that the elasticity of pulmonary vessels can be assumed to be approximately linear with transmural pressure (P_{tm}) and effectively independent of vessel size and species,

$$D = D_0(1 + \alpha P_{tm}), \quad (3)$$

where $\alpha = 0.02$/mmHg is the elasticity of a vessel. Simulation studies in the new model highlighted the significant influence that vascular structure could have on pulmonary perfusion distribution [26, 28], supporting experimental studies that suggested that posture and gravitational factors have a relatively minor effect on blood flow [15, 52]. A species comparison in models of the human (biped) and sheep (quadruped) showed that differences in branching asymmetry could explain significant differences in perfusion distribution that had been observed between animal and human studies [25]. However, these models neglected two important gravitational influences on pulmonary perfusion distribution: the deformation of lung tissue under gravity (the model assumed a uniform distribution of acinar units), and the effect of alveolar inflation and air pressure on the microvasculature (only isolated arterial trees were modeled and so acinar level boundary conditions were necessarily assigned assumed pressure values).

The first limitation was addressed via concurrent development of an imaging based model of pulmonary tissue elasticity and deformation. The same lung volumes that had been segmented from imaging data and used as host volumes for airway and blood vessel trees were used to construct curvilinear finite element volume meshes of the lung using the fitting methods described by Fernandez et al. [53]. The lung-air matrix within this volume was assumed to be a compressible non-linearly elastic continuum with homogeneous and isotropic material properties [54]. Stress and strain in the tissue were related by the strain energy density function

$$W = \frac{c}{2} e^{\left(aJ_1^2 + bJ_2\right)}, \tag{4}$$

where W is the strain energy, J_1 and J_2 are the first and second invariants of the Green strain tensor, and a, b and c are constants with values set such that a uniform expansion of tissue from a theoretical stress and strain reference state (half of functional residual capacity—FRC) resulted in physiological expansion pressures of $\sim$5 cmH_2O at FRC and $\sim$25 cmH_2O at total lung capacity (TLC) [54]. Predictions of lung tissue density distribution (local expansion) in the supine posture were validated against MDCT data for the subjects that were studied. Once stress and strain have been calculated, regional elastic recoil pressure (P_e) can be estimated using

$$P_e = \frac{ce^{\gamma}}{2\lambda}(3a + b)(\lambda^2 - 1), \tag{5}$$

where λ is the isotropic stretch from reference volume to the lung volume of interest, and

$$\gamma = \frac{3}{4}(3a + b)(\lambda^2 - 1)^2. \tag{6}$$

The tethering pressure acting on a vessel or airway is then estimated locally as being equal and opposite to local P_e [29, 55]. Burrowes and Tawhai [29] used predictions of local tethering pressure and tissue deformation coupled to the model of the pulmonary macrocirculation described previously to distinguish the separate contributions of arterial vascular geometry and gravitational tissue deformation to perfusion distribution. This is an example of the strength of computational modeling in comparison to experiment: the ability to 'switch-on-and-off' features of a biological system and analyze their relative or compound effect on a phenomenon. In this case, the model of Burrowes and Tawhai showed that the deformation of lung tissue contributes significantly to observed gravitational gradients of blood flow (removal of this effect halved the predicted gradient), consistent with the concept of the lung behaving as a 'Slinky™' [56]. However it was acknowledged that the model was still lacking a description of the capillary vessels (the microcirculation), whose external pressure is air, rather than tissue tethering pressure [29]. A multi-scale approach would be needed to address this limitation of the model (see Sect. 4).

While the models described above incorporate a large arterial domain, they must take a simplified approach to the fluid dynamics of pulmonary blood flow. The use of the Poiseuille equation effectively provides a time-averaged description of blood flow, which for analyses of perfusion *distribution* is a reasonable approach [39]. However, there are conditions under which the pulsatility of blood flow may be critical, and the assumption of laminar flow in the pulmonary blood vessels may not be appropriate in the largest vessels under certain pathologies. Another class of pulmonary blood flow models has emerged in recent years, which is more inclusive of the fluid-dynamic and geometric details of the largest blood vessels, and simplified at smaller scales (beyond 5 or 6 generations). This approach has been led by the group of Taylor et al. (e.g. [57]) who solved a one-dimensional time dependent model for blood flow in the largest pulmonary vessels, assuming that the flow profile was parabolic. They used the same elasticity law given in Eq. 3 and solved the following partial differential equations representing conservation of mass and momentum

$$\frac{\partial Q}{\partial t} + \frac{4}{3}\frac{\partial}{\partial z}\left(\frac{Q^2}{s}\right) + \frac{s}{\rho}\frac{\partial p}{dz} = \frac{-8\pi\mu Q}{\rho s} + \frac{\mu}{\rho}\frac{\partial^2 Q}{\partial z^2}, \tag{7a}$$

$$\frac{\partial s}{\partial t} + \frac{\partial Q}{\partial z} = 0, \tag{7b}$$

where t is time, z is axial distance, s is vessel cross-sectional area ($s = \pi D^2/4$), and ρ is the density of blood. The flow equations were solved using a finite element scheme (a Galerkin/least squares stabilization in space and a Galerkin method in time), with a periodic flow waveform in the main pulmonary artery and a vascular impedance boundary condition at all terminal vessels (in this case the segmental vessels). Similar methods were followed by Clipp and Steele in a study of blood flow in the lamb vasculature [58], but this study included a gravitational influence depending in which 'zone' a lumped parameter group of arteries resided. This type of model provides important steps toward modeling the temporal interaction between pulmonary blood vessels and airways. However, like the steady-state models described above, it lacks a description of mechanical and fluid-dynamic features of the pulmonary microcirculation. As it is at this micro-scale that gas exchange and drug delivery will occur, and this is a scale that is hard to investigate in the in vivo environment, we now turn to discussion of how computational modeling has been used to explore function in the pulmonary capillary beds.

3 The Microcirculation

In a typical human lung, the branching airway network terminates in an astounding ~480 million alveoli, with a measured range (across six adult subjects) of 274–790 million [59]. Wrapped over the surface of each of these alveoli is a dense network of pulmonary capillaries; a total number of 280 billion has been estimated. This vast number of vessels provides a gas exchange surface area of about

80–100 m^2. The pulmonary capillary network forms a far denser mesh than that found in the systemic microcirculation. Pulmonary capillaries are, on average, shorter in length and smaller in diameter than their systemic counterpart and because of this cellular transit through the circulations differs. Measurements of pulmonary capillary morphometry have found a range of internal diameters from 1 to 15 μm, with an average of around 8 μm [1]. The density of pulmonary capillaries led in early studies to the approximation that blood flows as a sheet through interconnected posts of connective tissue [60]. Later studies, using higher resolution imaging techniques, demonstrated that pulmonary capillary blood does in fact flow through discrete tubules [61], however the sheet-flow concept is still used as a simplification in computational modeling studies (see below) and it remains a physiologically meaningful description of the pulmonary microcirculation.

The gas exchange barrier separating alveolar gas from capillary blood has a thickness of only around 0.1–0.3 μm. The passive diffusion capacity (gases diffusing down a partial pressure gradient) correlates directly with the surface area of contact and inversely with the thickness of the diffusion membrane. The structure of this tissue barrier has been optimized to enable a thin enough layer for adequate gas exchange while enabling structural integrity to be maintained over a range of intravascular pressures. However, because of the fine balance of this system the pulmonary capillaries are very sensitive to excessive increases in pulmonary blood pressure. If capillary transmural pressure exceeds a critical limit (around 24 mmHg) the blood–gas barrier may begin to rupture resulting in fluid leakage into the alveolar units. This fluid accumulation (edema) is believed to reduce the capacity for gas exchange by increasing the thickness of the diffusion barrier within the lung [62].

In large blood vessels blood can be assumed to flow as a Newtonian fluid. That is, while the viscosity of blood is shear-rate dependent (and hematocrit-dependent) the relatively large fluid velocities in the largest vessels prevent formation of cellular aggregates that would markedly change the blood viscosity, and hence viscosity can be assumed constant. In contrast, in vessels less than approximately 300 μm in diameter the non-Newtonian properties of blood can no longer be neglected. Red blood cells (RBCs) tend to preferentially migrate to the center of small vessels and travel through at a faster rate than the rest of the blood at the vessel periphery. This results in a decrease in RBC concentration and a dynamic reduction in the apparent viscosity of blood whereby apparent viscosity decreases with decreasing vessel diameter. These phenomena are known as the Fahraeus and Fahraeus–Lindqvist effects, respectively [63], and are one of the key differences in modeling macro- and micro-circulatory flow behavior. Another important aspect when discussing cellular transport within the pulmonary capillaries is the transit of neutrophils. Neutrophils are the most abundant white blood cell (WBC) in the body and play an important role in immune response. Under normal conditions the concentration of neutrophils in the lung is between 40 and 80 times higher than that found in the systemic circulation [64]. This phenomenon is known as neutrophil margination and is due to the relative size (6.8–8.3 μm [65]) and stiffness of neutrophils in comparison to the capillaries through which they transit (typical diameter 8 μm [1]) meaning that each neutrophil typically has to deform in order to pass through the pathway of capillaries from arteriole to venule [65]. Neutrophils are less deformable than RBCs and effectively block capillary pathways for the duration of the time they take to deform (which depends on the size of the neutrophil compared with the blood vessel it blocks, and the activation state of the neutrophil). The sequestering of neutrophils in the lung is thought to play a vital role in host defense, acting as the secondary line of defense by destroying any unwanted foreign material that may have penetrated the system. Models of pulmonary blood flow at the capillary level are generally designed to account for these phenomena as best as possible without losing mathematical simplicity.

3.1 Modeling the Microcirculation: Fluid Flow and Cellular Transit

There have been two general approaches taken to modeling flow through the pulmonary capillaries: the tube flow approach (typical of systemic microcirculation models) and the sheet flow approach. Figure 3 shows illustrations of a tube flow model [30] and the sheet flow model [66]. There is clearly a difference in geometrical complexity between the two. Each geometric model has its limitations and its merits, summarized below.

Tube flow models (Fig. 3) explicitly represent each individual capillary segment within a portion of the microcirculation. Flow models are set up as in any other flow problem—using equations of conservation of mass and momentum within a network of vessels. Given an initial assumed hematocrit distribution and individual capillary resistance, the pressure and flow can be calculated within the network, assuming elastic vessels. Then a rheological analysis is conducted, that incorporates the non-Newtonian properties of the blood. In this procedure the distribution of RBCs is calculated, using empirically derived models that account for the Fahraeus and Fahraeus–Lindqvist effects as well as the phase separation effect (the disproportionate distribution of RBCs and plasma at bifurcations). Conservation of RBCs and plasma is enforced at each capillary junction and the new hematocrit distribution is used to update the apparent viscosity of blood in each capillary vessel. In the pulmonary circulation this type of model includes the impact of both alveolar and intra-pleural pressures on the estimated diameter of each capillary segment. This tube flow approach has been used successfully to represent flow in both the systemic [67–69] and pulmonary microcirculation [30, 70, 71]. The use of tube flow models provides predictions of flow and cellular transit properties within each individual capillary. The discrete tubular nature of these models also enables simulation of the impact of individual capillary blockage, via neutrophils, on capillary blood flow [30, 71]. While these types of models are more realistic in structure than the sheet flow model and provide more detailed flow and cellular transit information, current computational limitations

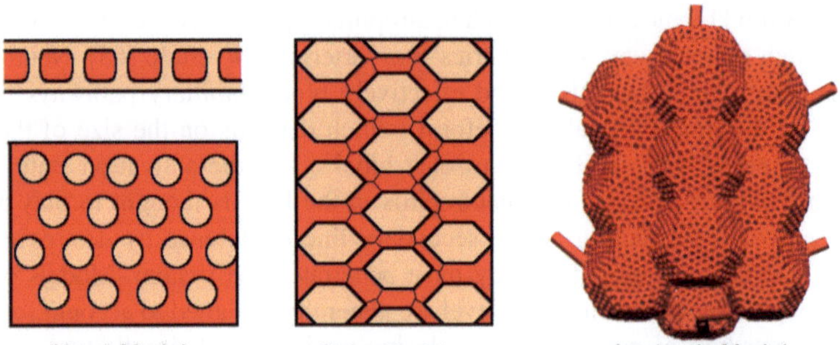

Fig. 3 A comparison between different geometric models that describe the pulmonary capillary beds. The sheet model of Fung et al. [66] assumes the capillary flow to occur between two elastic sheets connected with 'posts' of connective tissue. The tube model explicitly represents each individual capillary segment within a portion of the microcirculation. Tube models have been incorporated into anatomic models of the alveolar structure, where capillary segments cover a polyhedral representation of alveolar ducts (the anatomic model is reproduced from [30] with kind permission from Springer Science and Business Media)

mean that only a small subset of capillary tissue can be modeled. This means that a detailed tube flow model cannot be easily coupled with models of the pulmonary macrocirculation (Sect. 2.1), and flow or pressure boundary conditions must be assumed in each subset of capillary tissue. A simpler model of the capillary bed is required to couple macro- to micro-scale behaviors.

A much simpler approach to modeling the pulmonary microvasculature was proposed by Fung and Sobin [60] in 1969, and developed considerably in subsequent years [66, 72–76]. This model is termed the 'sheet flow' model as it approximates blood flow through the pulmonary capillaries covering several alveoli as a sheet of fluid flowing around "posts" of connective tissue that are bounded on either side by compliant endothelial tissue (Fig. 3). As such, the model predicts averaged flow streamlines through each inter-alveolar septum. By analyzing the fluid mechanics of flow in this structure as well as comparison with silicone models of the structure, simple equations describing blood flow in the pulmonary capillaries were derived, where flow is proportional to the fourth power of sheet height (which reflects capillary diameter). The model is able to provide good general agreement with experimental data for flow and resistance across the pulmonary capillary bed, and is able to provide a description of partial capillary bed collapse when venous blood pressures are less than alveolar air pressures [73]. Again, when the sheet flow model is considered in isolation, pressure or flow boundary conditions must be prescribed at capillary inlets, and coupling of macro- and micro-scale models must be achieved to minimize the need for these conditions.

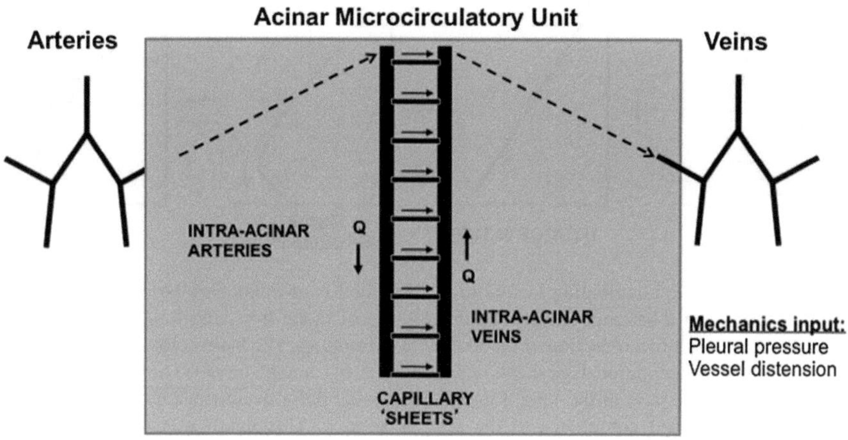

Fig. 4 A schematic of a model of pulmonary perfusion that bridges the relevant spatial scales in the lung. Arteries and veins are connected by a '*ladder-like*' structure of arterioles, venules, and capillaries that cover the alveoli in the acinar unit

4 Whole Organ Models: Bridging Spatial Scales

Macro- and micro-scale models of the pulmonary circulation can provide important insight regarding individual aspects of its function, however there is a significant limitation to these models, as boundary conditions must be prescribed at a spatial scale at which it is difficult to obtain pressure or flow measurements. A model that bridges the gap between these spatial scales is then necessary to understand how small scale phenomena emerge into whole lung function. The first steps toward an anatomically-based whole lung model were taken by Clark et al. [31], who incorporated serial and parallel micro-circulatory connections in a model of the pulmonary acinus. This model connects the arteries that accompany the respiratory bronchioles to their corresponding veins via the capillary network across several alveolar septa in what has been termed a 'ladder-like' structure (Fig. 4—gray box). There are multiple pre-capillary vessel generations in the acinus, and this model assumed that these vessels branch in a regular dichotomy, connected at each generation by capillary sheets (the sheet flow model described in Sect. 3.1). The model showed that this ladder-like structure may provide functional benefits to the lung: it predicted that the serial-parallel capillary connections provided a 17 % lower pulmonary vascular resistance than simple parallel capillary connections, when coupled to a symmetric model of the macro-vasculature.

As well as predicting a lower total pulmonary vascular resistance than previous models, the ladder-like geometrical structure was able to predict stratification of intra-acinar blood flow similar to that observed experimentally many years previously [77–80]. This stratification of blood flow may play an important role in the matching of perfusion to oxygen supply from air.

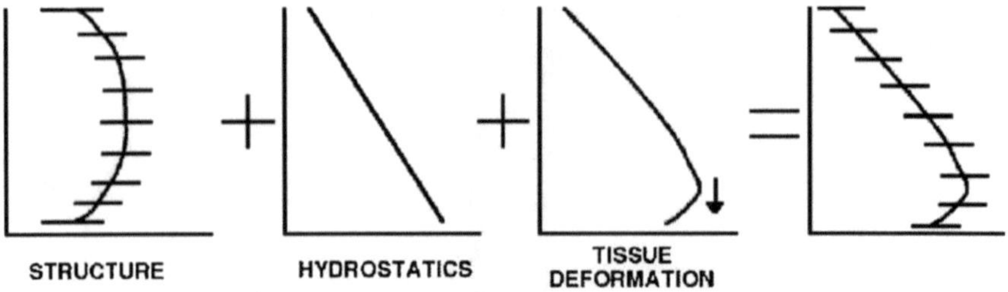

Fig. 5 The multiple contributing factors to the distribution of pulmonary blood flow. Vascular structure results in a heterogeneity in the distribution of blood flow, and long path lengths far from the heart result in a reduction of blood flow in extremities. The hydrostatic pressure gradient acts to induce a gravitational gradient in blood flow by acting directly on blood to drive it preferentially to the base of the lung. Finally, lung tissue deforms downwards, which increases the flow gradient and contributes to the size of zone 4. This figure was first published in Pulmonary Circulation [83]

The ladder-like representation of the intermediate and micro-vessels [31] has enabled bridging of the gap between arteries and veins in large-scale anatomically-based mathematical models of the pulmonary circulation [17], hence negating the constraint to prescribe boundary conditions at the pre-capillary level in both macro- and micro-scale models. This new model, which spanned the important spatial scales in the lung, is illustrated in Fig. 4. It was able—for the first time—to assess the relative contributions of the compound effects of gravity on pulmonary blood flow gradients and heterogeneity. In particular the model addressed the importance of (1) structure, (2) the hydrostatic pressure gradient in the lung, and (3) tissue deformation to blood flow distribution. Figure 5 provides a schematic of how each factor contributes to the overall distribution of perfusion. Analysis of the full model confirms that heterogeneity is introduced into the system via the asymmetric structure of the pulmonary blood vessels, as each path to and from the capillary beds has a unique resistance to flow. This means that even in zero gravity (0G) there is significant heterogeneity in the distribution of perfusion. The model predicts that heterogeneity is $\sim$47 % of its 1G value in 0G conditions. The hydrostatic pressure gradient acting directly on blood establishes a gravitational gradient in the distribution of perfusion, and introduces a height-dependent pressure gradient at the capillary level. Even without tissue deformation, this may be responsible for up to 80 % of the overall gravitational gradient in perfusion. This gradient contributes to overall heterogeneity, in the sense that variation of perfusion in the lung as a whole increases with the introduction of gravitationally varying pressures. However, local heterogeneity is a function of blood vessel branching asymmetry. Finally, the effect of gravitational deformation of lung tissue acts to alter this gravitational gradient, normally increasing its magnitude. Tissue deformation also contributes (along with structure) to an observed decrease in blood flow rates in gravitationally dependent tissue. This phenomenon is known as "zone 4" [81], and is only predicted in theoretical models that include an asymmetric geometry. The model suggested that upstream resistance, which is influenced by tissue deformation (an effective stretching of large vessels in the direction of gravity) and posture (prone and supine postures have shorter path-lengths to dependent lung tissue than upright and a smaller zone 4), is primarily responsible for the size of zone 4. This supported experimental studies that also suggested that the mechanism of perfusion reduction in zone 4 was via extra-alveolar vessel resistance [82]. The integration of structural and gravitational features to this multi-scale model of the pulmonary circulation provided important new physiological insight into the mechanisms behind the normal distribution of pulmonary perfusion that was not possible by experiment, or by previous simplified computational models. It has also opened the door for integrated studies of pulmonary function, in combination with similarly structured models of ventilation [55], as well as allowing for a subject specific approach to studying lung disease. These translational aspects of computational modeling in the pulmonary circulation are discussed in detail in Sect. 5.

5 Translational Outcomes

The ultimate goal of modeling any biological system is to provide physiological insight and translational outcomes. That is, to enable physiologists, biologists and physicians to improve knowledge and treatment of pathologies relating to the system. As the main function of the lung is the exchange of gases between alveolar air and capillary blood, any disruption to normal function in the pulmonary circulation is likely to translate to disruption to gas exchange function and hence to the ability of the body to provide oxygenated blood to tissue. A major advantage of multi-scale, anatomically based models is their ability to relate whole lung function to localized function. The state-of-the-art in this modeling approach for simulating regional gas exchange and pulmonary pathologies are introduced in this section.

5.1 Gas and Solute Exchange and Metabolism

There are many models in the literature that aim to describe gas exchange between alveolar air and capillary blood in a single 'unit' (this may be a whole lung, a single alveolus or a group of alveoli). These models have been reviewed in detail, and related to one another based on a hierarchy of modeling assumptions, by Ben-Tal [84]. In general, the relationship between the partial pressure of oxygen in alveolar air (p_{AO2}) and pulmonary capillary blood (p_{cO2}) can be expressed as an ordinary differential equation (ODE)

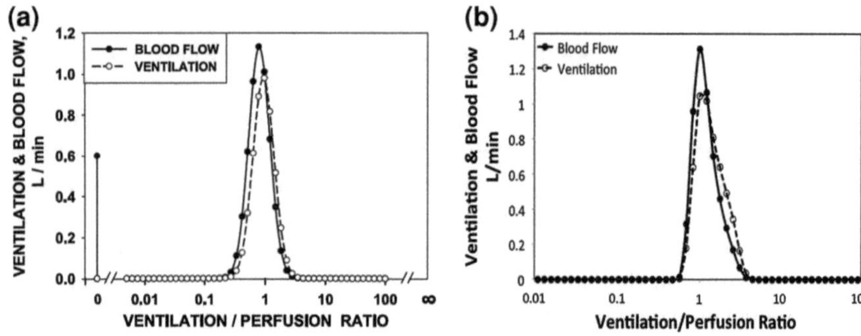

Fig. 6 **a** A typical MIGET plot obtained from a normal subject, which illustrates the distribution of ventilation and perfusion in the lung (reproduced from [88] with kind permission from Springer Science and Business Media). **b** Shows a simulated distribution of ventilation and perfusion in a normal subject obtained using the models of Clark et al. [17] and Swan et al. [55]. The two distributions compare well

$$\frac{dp_{cO_2}}{dt} = \frac{D_{O_2}}{V_c\sigma}\left(1 + \frac{4Hb}{\sigma}\frac{dS(p_{cO_2})}{dp_{cO_2}}\right)^{-1}(p_{AO_2} - p_{cO_2}), \tag{8}$$

where D_{O2} is the diffusion capacity (transfer factor) of oxygen between air and blood, V_c is capillary blood volume, σ is the solubility of oxygen in blood, Hb is the hemoglobin concentration in whole blood, and $S(p_{cO2})$ is the hemoglobin saturation function. This function can be fit to experimental data or derived from models of the chemical interactions between oxygen and hemoglobin. If one assumes equilibration between air and blood gases, the ODE given by Eq. 8 can be reduced to a system of algebraic equations which explicitly depend on capillary blood flow rates (Q_c) and predict end-expired oxygen levels. The equation for oxygen is (e.g. [85])

$$\mathrm{V}_I p_{IO_2} - V_E p_{AO_2} = Q_C k(C_{cO_2} - C_{vO_2}), \tag{9}$$

where V_I is inspired ventilation into the unit, p_{IO2} is inspired oxygen partial pressure, V_E is expired ventilation out of the unit, Q_c is the rate of capillary blood flow, C_{cO2} is the oxygen content in end-capillary blood, C_{vO2} is oxygen content entering the lungs from mixed venous blood, k is a constant factor that accounts for differences between body temperature and pressure and inspired air temperature and pressure as well as allowing consistency between the units of the left and right hand side of Eq. 9. Content (in mol l-1) is related to partial pressure by

$$C_{O_2} = \sigma p_{O_2} + 4HbS(p_{O_2}). \tag{10}$$

Critically, each of Eqs. 8 and 9 is dependent on capillary blood pressures and blood flow rates, which are both spatially and temporally variable (note that the parameter V_c in Eq. 8 is variable with Q_c as capillaries are elastic). Whilst previous modeling studies assessing the function of the pulmonary circulation had focused on blood flow alone, Burrowes et al. [34] introduced assessment of gas exchange function in a model of localized pulmonary disease by solving Eqs. 9 and 10, along with the Monod–Wyman–Changeaux model for $S(p_{cO2})$ [86]. They assumed a linear distribution of ventilation with vertical height and were able to predict baseline p_{AO2} that was consistent with physiological measurements [7]. This was expanded upon by Tawhai et al. [87] who modeled both ventilation [55] and perfusion [17] in subject specific geometries for nine subjects, and estimated gas transport using the same methodology as Burrowes et al. [34]. Figure 6 shows a normal (baseline) $\dot{V}/\dot{Q}$ distribution as predicted in a normal subject using this model, compared with a traditional MIGET plot in a normal individual [88]. The predictions of function made using anatomical geometries and physical laws by Tawhai et al. compare well to the MIGET measurements, that are essentially statistical estimates of ventilation and perfusion distributions made following global measurements of inert gas elimination from the lung. The coupling of ventilation and perfusion models to predict gas exchange function in anatomical geometries opened the door for meaningful studies of pathological pulmonary perfusion.

5.2 *Application to Pulmonary Vascular Disease*

The advent of anatomically based models of the pulmonary circulation has enabled model-based investigation of disease states that present heterogeneously in the lung (i.e. localized tissue abnormalities with relatively normal tissue in large parts of the lung) and/or in the population (i.e. diseases that affect one individual more than another for an apparently similar level of regional disease). These models have so far focused on investigation of pulmonary hypertension (PH) secondary to pulmonary embolism (PE). In PE, blood clots occlude pulmonary arteries leading to a redistribution of blood flow and pressures, which can in severe cases lead to an elevation of pulmonary arterial pressure (PAP) that can lead to RV dysfunction or failure. The obvious culprit responsible for elevated PAP is the occlusion of large pulmonary arteries, as this theoretically could increase PVR substantially, leading to RV dysfunction. However, there are significant inter-subject variations in response to PE, even when patients present with apparently similar levels of occlusion [89, 90]. The question as to what is responsible for this variable response then hinders PE diagnosis and treatment: as clot load alone is not correlated with survival [91], is it clot location (e.g. apical, basal, proximal, distal) or some humoral response that is responsible for variable outcomes? Or could inter-patient variability in response to clot simply be a result of underlying pathologies? To date, these questions have been addressed in anatomically based models of both the acinus [32] and the whole lung [33, 34, 87] in order to capture the functional consequence of arterial occlusion across all relevant spatial scales.

At the smallest spatial scale, micro-emboli (<170 μm in diameter) occluding small pulmonary arteries appear to produce a larger PAP response, for the same theoretical level of capillary bed occlusion, than their larger counterparts [92]. Small emboli are often used to induce PH and edema in animal models; computational modeling provides a means to properly interpret these experimental models by providing an understanding of the implications of micro-embolus occlusion. Under the hypothesis proposed by Read in 1969 [78] that occlusion to proximal (high flow) capillary beds may account for the relative severity of micro-emboli, Clark et al. [32] constructed an asymmetrically branching model of the blood vessels in the acinus to investigate the effect of occlusion in this structure. The idealized, symmetric ladder-like acinar structure described in Sect. 4, while useful in bridging the spatial scales in models of the full pulmonary circulation, was inappropriate for this study as it could not capture the effect of a single occlusion at the micro-scale. A multibranching geometry for pre-capillary acinar vessels was therefore constructed [93], based on morphometric data regarding acinar structure [94]. Aside from asymmetry in pre-capillary vessels, the model retained the ladder-like structure of its symmetric counterpart, with capillary connections at each level in the acinar tree. The model confirmed the hypothesis that proximal micro-emboli have a greater impact on PVR than distal emboli. However, although the blockage of normally high flow capillary beds—and so a reduction in capillary surface area—increases PVR, the model predicted a significant capacity for redistribution of blood flow through unblocked capillary beds. It suggested that a major reason for a large increase in PVR with proximal occlusions might be a result of the fact that blood must now travel through several high resistance blood vessels before reaching the pulmonary capillaries. This again highlighted the importance of a serial-parallel acinar structure being crucial to the lung's ability to carry a high volume of blood at low pressures.

Subsequent modeling studies investigated the impact of larger (mm in diameter) emboli in the anatomically-based full lung geometry of a single individual [33, 34]. Both studies simulated embolism by reducing the radius of an occluded vessel to a percentage of its original (baseline) radius, depending on the level of occlusion. Vessels were occluded based on locations determined from MDCT data in patients presenting clinically with acute PE [33], and in a probabilistic manner by occluding vessels based on their baseline blood flow rates [34]. Both studies suggested that mechanical occlusion in the absence of any response to hypoxia, vasoactive signaling, or prior pathology, was not sufficient to raise PAP to PH levels, even with 70–80 % of the capillary bed occluded. Clinically, PH is seen in some patients with far lower occlusion levels; for example McIntyre and Sasahara observed PH in some patients with <30 % estimated capillary bed occlusion in the absence of prior pulmonary disease [89, 90]. The reason for the low impact of mechanical occlusion on PAP in modeling studies is the capacity for the lung to 'recruit' capillary and large blood vessel volume in non-occluded regions, either via distension of elastic vessels (because of increased blood pressures), or in the case of the capillary bed by the direct recruitment of capillary vessels that may ordinarily be unperfused. The redistribution of blood flow post-occlusion is more

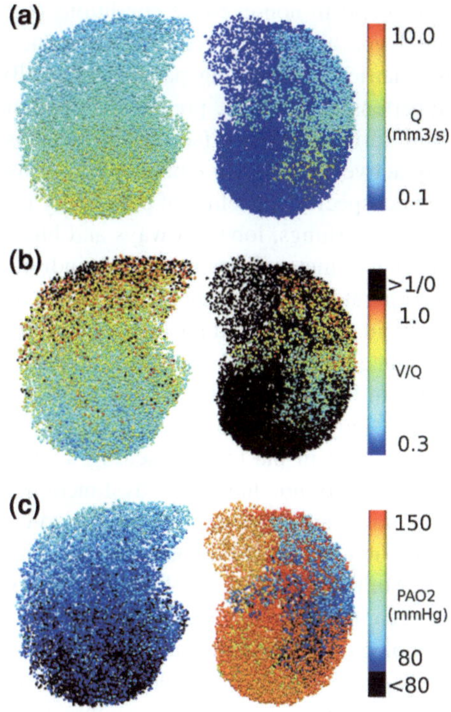

Fig. 7 Predicted acinar **a** perfusion, **b** ventilation–perfusion ($\dot{V}/\dot{Q}$) ratios and **c** alveolar oxygen partial pressures (P_{AO2}) in a subject with 40 % tissue occlusion by 7 mm pulmonary emboli. *Black regions* in **b** are areas with $\dot{V}/\dot{Q} > 1$ (more ventilation than perfusion) and in **c** are areas with $P_{AO2} < 80$ mmHg (hypoxic). This figure was first published in Pulmonary Circulation [34]

pronounced in the gravitationally non-dependent lung, which normally receives the least blood flow so has a greater capacity to carry more blood when the lung is under stress. In the absence of an underlying pathology, the small increases in PAP predicted by a model of mechanical occlusion alone [33, 34] suggests a significant role for active mechanisms in determining response to PE.

Burrowes et al. [34] incorporated the model of oxygen exchange described by Eqs. 9 and 10 to investigate the possible role of hypoxia in PE. They found that localized hypoxia was predicted in non-occluded regions following arterial occlusion (Fig. 7), as well as a significant reduction in oxygen levels in the blood that returns from the lungs to the systemic circulation. Localized hypoxia may be significant in elevating PAP in PE, with pulmonary vessels actively constricting in hypoxic conditions. Constriction of small pulmonary vessels in non-occluded regions in PE may act to restrict the ability of the lung to redistribute blood flow, and so elevate blood pressures required to maintain cardiac output. However, the redistribution of blood away from hypoxic regions may have the beneficial effect of improving $\dot{V}/\dot{Q}$ matching and so the efficiency of the diseased lung for gas exchange.

An alternative, or additional, mechanism for increased PAP in response to PE is the response to increased cardiac output. Experimental [95–97] and modeling [34, 87, 98] studies agree that PAP increases more rapidly than cardiac output in PE than under normal conditions. Tawhai et al. [87] used estimates of metabolic rate at rest and in non-exercise conditions (e.g. fidgeting or walking) [99] to estimate the level of activity required to induce hypertension in nine subject-specific geometries representing patients clinically presenting with PE. They found that hypertension could be predicted in patients whose cardiac output was consistent with upright fidgeting (so only low levels of activity or stress) in the absence of any active response to embolization. This study paved the way for the first truly subject specific modeling of pulmonary function in disease. Each of (1) anatomical geometry (lungs, lobes, airways and blood vessels), (2) height-, weight- and age-dependent metabolic rates, and (3) individualized embolus location and sizes were incorporated into the model. The patients enrolled in the study were chosen so as to avoid the confounding factor of existing lung disease, and model predictions correlated well to both accepted population norms for parameters such as PVR and PaO_2 in baseline conditions, and actual patient response. Of the patients included in the study, those with predicted global hypoxia ($P_{aO2} < 80$ mmHg) all had RV dysfunction in the clinical setting, and all the patients without predicted global hypoxia did not have RV dysfunction. Hypoxia, and/or the hypercapnia that commonly accompanies it, is known to elevate cardiac output and ventilatory rate [100] and this response may be crucial in understanding the variable outcomes in PE. Multiscale modeling of the pulmonary circulation has therefore suggested that oxygen (and carbon dioxide) transport may be crucial in understanding this pathology of the pulmonary circulation. This opens the arena for a new class of multi-scale models of the pulmonary circulation in future studies that are able to relate cell or vessel levels response to stress or oxygen tensions and emergent function at the organ level.

6 Linking to Cellular Mechanics

Although current models for the pulmonary circulation have achieved a high level of sophistication in their structure and coupling of scale-dependent function, they are currently lacking representation of important dynamic behaviors. For example, the arteries and veins are treated as passive structures that distend and recoil in response to transmural pressure only: they do not include force development in the vascular smooth muscle (VSM). Further, the studies of Burrowes et al. [22, 25, 26, 28–30, 33, 34] and Clark et al. [17, 31, 32] assumed fully developed steady-state flow to compute the *distribution* of blood, but not its *spatio-temporal fluctuations*. Multi-scale model development for the airway tree and its interaction with the parenchyma provides a roadmap for how vascular smooth muscle could be integrated with models for the pulmonary circulation [101].

Politi et al. [101] presented a multi-scale model for force development in airway smooth muscle (ASM) and its interaction with phasic force fluctuations at the airway, tissue, and organ level (illustrated in Fig. 8). The model includes molecular-scale phenomena, i.e. attachment of actin and myosin that is regulated by calcium (Ca^{2+}) dependent mechanisms, via a four-state model for the smooth muscle cross-bridge

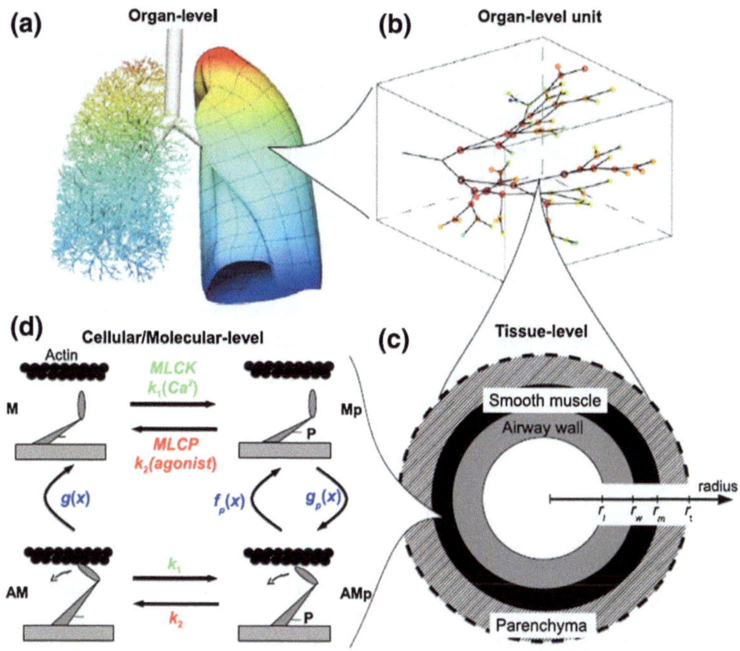

Fig. 8 An example of a multi-scale model of the pulmonary airways that incorporates: **a** an anatomically-based organ level model; **b** an organ-level tissue unit with 90 embedded airway segments with radii calculated at the tissue level; **c** a tissue level model where each airway segment is modeled as a cylinder comprising airway wall, smooth muscle cells, and a parenchymal layer; and **d** a cellular/molecular level model which describes the generation of force in smooth muscle. Reprinted from [101], with permission from Elsevier

cycle [102] (this is a modified version of the Hai-Murphy model [103]). Activation of molecular level force development is through a cellular-level Ca^{2+} model. Active force development at the cellular level acts against the passive load of the airway wall, and a constantly oscillating load from the breathing lung parenchyma. The parenchyma is modeled as in Tawhai et al. [54], and the forces that tether the parenchyma to the airway wall are calculated as spatially- and temporally-varying elastic recoil pressures plus additional tethering forces that develop locally as the airway contracts, following the relationship for tethering force proposed by [104]. Donovan et al. [105] extended the molecular level model to include 'cross-linkers' that represent the passive properties of the ASM. The critical component of the multi-scale model is its connection of the spatial scales, which is defined at the individual airway level. That is, at each airway at each time point the active tension developed by the cellular and molecular models is explicitly balanced against force development in the surrounding tissues.

The basic mechanisms for interaction of VSM with its surrounding force environment are the same as for ASM, however the VSM responds differently to agonist stimulation [102]. Reaction to agonist (such as methacholine) is relevant in the study of airway reactivity, as dose-response to methacholine challenge is a standard test for quantifying asthma severity. However in VSM the relevant stimuli are the partial pressures of O_2 and CO_2, and nitric oxide (NO) release via endothelial sensing of shear stress. The multi-scale model for the airways is focused on explaining pathology (airway hyper-responsiveness, a hallmark of asthma) whereas understanding the dynamics of VSM in the pulmonary circulation is necessary simply to explain the normal function of the lung. That is, PO_2-mediated vasoconstriction and NO-mediated vasodilation operate simultaneously in the normal lung, with their balance (along with other signaling pathways) determining the local VSM tension. Transitioning the organ- and vessel-level models for the pulmonary circulation to studies of e.g. pulmonary hypertension, where wall remodeling may be a significant component, will require a multi-scale approach that spans to cellular signaling.

References

1. Weibel, E.R.: Morphometry of the Human Lung. Springer, Berlin (1963)
2. Levitzky, M.G.: Pulmonary Physiology, 7th edn. The McGraw-Hill Companies, Inc., New York (2007)
3. Grassino, A.E., Anthonisen, N.R.: Chest wall distortion and regional lung volume distribution in erect humans. J. Appl. Physiol. **39**(6), 1004–1007 (1975)
4. Whitfield, A., Waterhouse, J., Arnott, W.M.: The total lung volume and its subdivisions. II. The effect of posture. Brit J Soc Med **4**, 86–97 (1950)
5. Hoffman, E.A., Sinak, L.J., Riman, E.L.: Effect of body position on regional lung expansion: A computer tomographic approach. Physiologist **26**(4), A-69 (1983)
6. Amis, T., Jones, H., Hughes, J.: Effect of posture on inter-regional distribution of pulmonary perfusion and VA/Q ratios in man. Respir. Physiol. **56**, 169–182 (1984)
7. West, J.B.: Regional differences in gas exchange in the lung of erect man. J. Appl. Physiol. **17**(6), 893–898 (1962)
8. Hopkins, S.R., Henderson, A.C., Levin, D.L., Yamada, K., Arai, T., Buxton, R.B., Prisk, G.K.: Vertical gradients in regional lung density and perfusion in the supine human lung: the slinky effect. J. Appl. Physiol. **103**(1), 240–248 (2007)
9. Prisk, G.K., Yamada, K., Henderson, A.C., Arai, T.J., Levin, D.L., Buxton, R.B., Hopkins, S.R.: Pulmonary perfusion in the prone and supine postures in the normal human lung. J. Appl. Physiol. **103**, 883–894 (2007)
10. Albert, M.S., Cates, G.D., Driehuys, B., Happer, W., Saam, B., Springer Jr., C.S., Wishnia, A.: Biological magnetic resonance imaging using laser-polarized 129Xe. Nature **370**(6486), 199–201 (1994)
11. West, J.B., Dollery, C.T., Naimark, A.: Distribution of blood flow in isolated lung; relation to vascular and alveolar pressures. J. Appl. Physiol. **19**, 713–724 (1964)
12. Hughes, M., West, J.B.: Point: Gravity is the major factor determining the distribution of blood flow in the human lung. J. Appl. Physiol. **104**(5), 1531–1533 (2008)
13. West, J.: Importance of gravity in determining the distribution of pulmonary blood flow. J. Appl. Physiol. **93**(5), 1888–1889 (2002)
14. Glenny, R.W.: Counterpoint: gavity is not the major factor determining the distribution of blood flow in the healthy human lung. J. Appl. Physiol. **104**(5), 1533–1535 (2008)
15. Glenny, R.W., Bernard, S., Robertson, H.T., Hlastala, M.P.: Gravity is an important but secondary determinant of regional pulmonary blood flow in upright primates. J. Appl. Physiol. **86**(2), 623–632 (1999)

16. Glenny, R.W., Lamm, W.J.E., Albert, R.K., Robertson, H.T.: Gravity is a minor determinant of pulmonary blood flow distribution. J. Appl. Physiol. **71**, 620–629 (1991)
17. Clark, A.R., Tawhai, M.H., Burrowes, K.S.: The interdependent contributions of gravitational and structural features to the distribution of pulmonary perfusion in a multi-scale model of the pulmonary circulation. J. Appl. Physiol. **110**, 943–945 (2011)
18. Rideout, V., Katra, J.: Computer simulation of the pulmonary circulation. Simulation **12**, 239–245 (1969)
19. Parker, J.C., Cave, C.B., Ardell, J.L., Hamm, C.R., Williams, S.G.: Vascular tree structure affects lung blood flow heterogeneity simulated in three dimensions. J. Appl. Physiol. **83**(4), 1370–1382 (1997)
20. Glenny, R.W., Robertson, H.T.: Fractal modeling of pulmonary blood flow heterogeneity. J. Appl. Physiol. **70**(3), 1024–1030 (1991)
21. Bshouty, Z., Younes, M.: Distensibility and pressure-flow relationship of the pulmonary circulation. II. Multibranched model. J. Appl. Physiol. **68**(4), 1514–1527 (1990)
22. Burrowes, K.S., Hunter, P.J., Tawhai, M.H.: Anatomically-based finite element models of the human pulmonary arterial and venous trees including supernumerary vessels. J. Appl. Physiol. **99**, 731–738 (2005)
23. Marshall, B., Marshall, C.: A model for hypoxic constriction of the pulmonary circulation. J. Appl. Physiol. **64**(1), 68–77 (1988)
24. Nelin, L.D., Krenz, G.S., Rickaby, D.A., Linehan, J.H., Dawson, C.A.: A distensible vessel model applied to hypoxic pulmonary vasoconstriction in the neonatal pig. J. Appl. Physiol. **74**(5), 2049–2056 (1993)
25. Burrowes, K.S., Hoffman, E.A., Tawhai, M.H.: Species-specific pulmonary arterial asymmetry determines species differences in regional pulmonary perfusion. Ann. Biomed. Eng. **37**(12), 2497–2509 (2009)
26. Burrowes, K.S., Hunter, P.J., Tawhai, M.H.: Investigation of the relative effects of vascular branching structure and gravity on pulmonary arterial blood flow heterogeneity via an image-based computational model. Acad. Radiol. **12**(11), 1464–1474 (2005)
27. Burrowes, K.S., Swan, A.J., Warren, N.J., Tawhai, M.H.: Towards a virtual lung: multi-scale, multi-physics modelling of the pulmonary system. Philos. Trans. R. Soc. A **366**(1879), 3247–3263 (2008)
28. Burrowes, K.S., Tawhai, M.H.: Computational predictions of pulmonary blood flow gradients: gravity versus structure. Respir. Physiol. Neurobiol. **154**(3), 515–523 (2006)
29. Burrowes, K.S., Tawhai, M.H.: Coupling of lung tissue tethering force to fluid dynamics in the pulmonary circulation. Int. J. Numer. Methods. Biomed. Eng. **26**, 862–875 (2010)
30. Burrowes, K.S., Tawhai, M.H., Hunter, P.J.: Modeling RBC and neutrophil distribution through an anatomically based pulmonary capillary network. Ann. Biomed. Eng. **32**(4), 585–595 (2004)
31. Clark, A.R., Burrowes, K.S., Tawhai, M.H.: Contribution of serial and parallel micro-perfusion to spatial variability in pulmonary inter- and intra-acinar blood flow. J. Appl. Physiol. **108**(5), 1116–1126 (2010)
32. Clark, A.R., Burrowes, K.S., Tawhai, M.H.: The impact of micro-embolism size on haemodynamic changes in the pulmonary micro-circulation. Respir. Physiol. Neurobiol. **175**, 365–374 (2011)
33. Burrowes, K.S., Clark, A.R., Marcinkowski, A., Wilsher, M.L., Milne, D.G., Tawhai, M.H.: Pulmonary embolism: predicting disease severity. Philos. Trans. R. Soc. A **369**(1954), 4145–4148 (2011)
34. Burrowes, K.S., Clark, A.R., Tawhai, M.H.: Blood flow redistribution and ventilation–perfusion mismatch during embolic pulmonary occlusion. Pulm. Circ. **1**(3), 365–376 (2011)
35. MacLean, M., Herve, P., Eddahibi, S., Adnot, S.: 5-hydroxytryptamine and the pulmonary circulation: receptors, transporters and relevance to pulmonary arterial hypertension. Br. J. Pharmacol. **131**(2), 161–168 (2000)
36. Howell, J.B.L., Permutt, S., Proctor, D.F., Riley, R.L.: Effect of inflation of the lung on different parts of pulmonary vascular bed. J. Appl. Physiol. **16**(1), 71–76 (1961)

37. Horsfield, K.: Morphometry of the small pulmonary arteries in man. Circ. Res. **42**, 537–593 (1978)
38. Pump, K.K.: The circulation in the peripheral parts of the human lung. Chest **49**(2), 119–129
39. Clough, A.V., Audi, S.H., Molthen, R.C., Krenz, G.S.: Lung circulation modeling: status and prospects. Proc. IEEE **94**(4), 753–768 (2006)
40. Hillier, S.C., Graham, J.A., Hanger, C.C., Godbey, P.S., Glenny, R.W., Wagner Jr., W.W.: Hypoxic vasoconstriction in pulmonary arterioles and venules. J. Appl. Physiol. **82**(4), 1084–1090 (1997)
41. Marshall, B.E., Marshall, C.: Continuity of response to hypoxic pulmonary vasoconstriction. J. Appl. Physiol. Respir. Environ. Exerc. Physiol. **49**, 189–196 (1980)
42. Elliot, F.M., Reid L.: Some new facts about the pulmonary artery and its branching pattern. Clin Radiol **16**, 193–198 (1965)
43. Huang, W., Yen, R.T., McLaurine, M., Bledsoe, G.: Morphometry of the human pulmonary vasculature. J. Appl. Physiol. **81**(5), 2123–2133 (1996)
44. Horsfield, K., Gordon, W.I.: Morphometry of pulmonary veins in man. Lung **159**, 211–218 (1981)
45. Singhal, S., Henderson, R., Horsfield, K., Harding, K., Cumming, G.: Morphometry of the human pulmonary arterial tree. Circ. Res. **33**(2), 190–197 (1973)
46. Glenny, R.W., Robertson, T.J.: Fractal properties of pulmonary blood flow: charaterization of spatial heterogeneity. J. Appl. Physiol. **69**(2), 532–545 (1990)
47. Tawhai, M.H., Hunter, P.J., Tschirren, J., Reinhardt, J.M., McLennan, G., Hoffman, E.A.: CT-based geometry analysis and finite element models of the human and ovine bronchial tree. J. Appl. Physiol. **97**(6), 2310–2321 (2004)
48. Tawhai, M.H., Pullan, A.J., Hunter, P.J.: Generation of an anatomically based three-dimensional model of the conducting airways. Ann. Biomed. Eng. **28**(7), 793–802 (2000)
49. West, J.B.: Respiratory Physiology—The Essentials. Williams and Wilkins, Baltimore (1995)
50. Yen, M.: Elastic properties of pulmonary blood vessels. In: Respiratory Physiology: An Analytical Approach, pp. 553–560. Marcel Dekker, Inc. (1989)
51. Krenz, G.S., Dawson, C.A.: Flow and pressure distributions in vascular networks consisting of distensible vessels. Am. J. Physiol. Heart Circ Physiol **284**(6), H2192–H2203 (2003)
52. Glenny, R.W., Lamm, W.J.E., Bernard, S.L., An, D., Chornuk, M., Pool, S., Wagner Jr., W.W., Hlastala, M.P., Rovertson, H.T.: Physiology of a microgravity environment, selected contribution: redistribution of pulmonary perfusion during weightlessness and increased gravity. J. Appl. Physiol. **89**(3), 1239–1248 (2000)
53. Fernandez, J.W., Mithraratne, P., Thrupp, S.F., Tawhai, M.H., Hunter, P.J.: Anatomically based geometric modelling of the musculo-skeletal system and other organs. Biomech. Model. Mechanobiol. **2**(3), 139–155 (2004)
54. Tawhai, M., Nash, N., Lin, C., Hoffman, E.: Supine and prone differences in regional lung density and pleural pressure gradients in the human lung with constant shape. J. Appl. Physiol. **107**(3), 912–920 (2009)
55. Swan, A.J., Clark, A.R., Tawhai, M.H.: A computational model of the topographic distribution of ventilation in healthy human lungs. J. Theor. Biol. **300**, 222–231 (2012)
56. Hopkins, S.R., Henderson, A.C., Levin, D.L., Yamada, K., Arai, T., Buxton, R.B., Prisk, G.K.: Vertical gradients in regional lung density and perfusion in the supine human lung: the Slinky effect. J. Appl. Physiol. **103**(1), 240–248 (2007)
57. Spilker, R.L., Feinstein, J.A., Parker, D.W., Reddy, V.M., Taylor, C.A.: Morphometry-based impedance boundary conditions for patient-specific modeling of blood flow in pulmonary arteries. Ann. Biomed. Eng. **35**(4), 546–559 (2007)
58. Clipp, R., Steele, B.N.: Impedance boundary conditions for the pulmonary vasculature including the effects of geometry, compliance, and respiration. IEEE Trans. Biomed. Eng. **56**(3), 862–870 (2009)
59. Ochs, M., Nyengaard, J.R., Jung, A., Knudsen, L., Voigt, M., Wahlers, T., Richter, J., Gundersen, H.J.: The number of alveoli in the human lung. Am. J. Respir. Crit. Care Med. **169**(1), 120–124 (2004)

60. Fung, Y.C., Sobin, S.S.: Theory of sheet flow in lung alveoli. J. Appl. Physiol. **26**, 472–488 (1969)
61. Guntheroth, W.G., Luchtel, D.L., Kawabori, I.: Pulmonary microcirculation: tubules rather than sheet or post. J. Appl. Physiol. **53**(2), 510–515 (1982)
62. Maina, J.N., West, J.B.: Thin and strong! The bioengineering dilema in the structural and functional design of the blood gas barrier. Physiol. Rev. **85**, 811–844 (2005)
63. Fahraeus, R., Lindqvist T.: The viscosity of the blood in narrow capillary tubes. J. Appl. Physiol. **96**, 562–568 (1931)
64. Hogg, J.: Neutrophil kinetics and lung injury. Physiol. Rev. **67**(4), 1249–1295 (1987)
65. Doerschuk, C.: Neutrophil rheology and transit through capillaries and sinusoids. Am. J. Respir. Crit. Care Med. **159**, 1693–1999 (1999)
66. Fung, Y.C., Sobin, S.S.: Elasticity of the pulmonary alveolar sheet. Circ. Res. **30**(4), 451–469 (1972)
67. Pries, A.R., Secomb, T.W.: Microcirculatory network structures and models. Ann. Biomed. Eng. **28**, 916–921 (2000)
68. Pries, A.R., Secomb, T.W., Gaehtgens, P., Gross, J.F.: Blood flow in microvascular networks. Experiments and simulation. Circ. Res. **67**(4), 826–834 (1990)
69. Fenton, B., Wilson, D., Cokelet, G.: Analysis of the effect of measured white blood cell entrance time on hemodynamics in a computer model of a mircovascular bed. Pflugers Arch. **403**, 396–401 (1985)
70. Dhadwal, A., Wiggs, B., Doerschuk, C., Kamm, R.: Effects of anatomic variability on blood flow and pressure gradients in the pulmonary circulation. J. Appl. Physiol. **83**(5), 1711–1720 (1997)
71. Huang, Y., Doerschuk, C.M., Kamm, R.D.: Computational modeling of RBC and neutrophil transit through the pulmonary capillaries. J. Appl. Physiol. **90**(2), 545–564 (2001)
72. Fung, Y.C., Sobin, S.S.: Pulmonary alveolar blood flow. Circ. Res. **30**(4), 470–490 (1972)
73. Fung, Y.C., Yen, R.T.: A new theory of pulmonary blood flow in zone 2 condition. J. Appl. Physiol. **60**(5), 1638–1650 (1986)
74. Sobin, S.S., Fung, Y.C., Tremer, H.M., Rosenquist, T.H.: Elasticity of the pulmonary microvascular sheet in the cat. Circ. Res. **30**(4), 440–450 (1972)
75. Sobin, S.S., Tremer, H.M., Fung, Y.C.: Morphometric basis of the sheet-flow concenpt of the alveolar microcirculation in the cat. Circ. Res. **26**(3), 397–414 (1970)
76. Sobin, S.S., Tremer, H.M., Lindal, R.G., Fung, Y.C.: Distensibility of human pulmonary capillary blood vessels in the interalveolar septa. Fed. Proc. **38**, 990 (1979)
77. Read, J.: Redistribution of stratified pulmonary blood flow during exercise. J. Appl. Physiol. **27**(3), 374–377 (1969)
78. Read, J.: Stratified pulmonary blood flow: some consequences in emphysema and pulmonary embolism. Br. Med. J. **2**, 44–46 (1969)
79. Wagner, P., McRae, J., Read, J.: Stratified distribution of blood flow in secondary lobule of the rat lung. J. Appl. Physiol. **22**(6), 1115–1123 (1967)
80. West, J.B., Maloney, J.E., Castle, B.L.: Effect of stratified inequality of blood flow on gas exchange in liquid-filled lungs. J. Appl. Physiol. **32**(3), 357–361 (1972)
81. Hughes, J.M., Glazier, J.B., Maloney, J.E., West, J.B.: Effect of lung volume on the distribution of pulmonary blood flow in man. Respir. Physiol. **4**(1), 58–72 (1968)
82. Hopkins, S.R., Arai, T.J., Henderson, A.C., Levin, D.L., Buxton, R.B., Prisk, G.K.: Lung volume does not alter the distribution of pulmonary perfusion in dependent lung in supine humans. J. Physiol. **588**(Pt 23), 4759–4768 (2010)
83. Tawhai, M.H., Clark, A.R., Burrowes, K.S.: Computational models of the pulmonary circulation: insights and the move towards clinically directed studies. Pulm. Circu. **1**(2), 224–238 (2011)
84. Ben-Tal, A.: Simplified models for gas exchange in the human lungs. J. Theor. Biol. **238**, 474–495 (2006)
85. Kapitan, K., Hempleman, S.: Computer simulation of mammalian gas exchange. Comput. Methods Biol. Med. **16**(2), 91–101 (1986)

86. Monod, J., Wyman, J., Changeaux, J.: On the nature of allosteric transitions: a plausible model. J. Mol. Biol. **12**, 88–112 (1965)
87. Tawhai, M., Clark, A., Wilsher, M., Millne, D., Subramaniam, K., Burrowes, K.: Spatial redistribution of perfusion and gas exchange in patient specific models of pulmonary embolism. In: 2012 IEEE International Symposium on Biomedical Imaging. Barcelona, Spain
88. Wagner, P.D.: The multiple inert gas elimination technique (MIGET). Intensive Care Med. **34**(6), 994–1001 (2008)
89. McIntyre, K., Sasahara, A.: Hemodynamic alterations related to extent of lung scan perfusion defect in pulmonary embolism. J. Nucl. Med. **4**, 166–170 (1971)
90. McIntyre, K., Sasahara, A.: The hemodynamic response to pulmonary embolism in patients without prior cardiopulmonary disease. Am. J. Cardiol. **28**(3), 288–294 (1971)
91. Ghaye, B., Ghuysen, A., Bruyere, P.J., D'Orio, V., Dondelinger, R.F.: Can CT pulmonary angiography allow assessment of severity and prognosis in patients presenting with pulmonary embolism? What the radiologist needs to know. Radiographics **26**(1), 23–39; discussion 39–40 (2006) (discussion 39–40)
92. Malik, A.: Pulmonary microembolism. Physiol. Rev. **63**, 1115–1207 (1983)
93. Tawhai, M.H., Hunter, P.J.: Characterising respiratory airway gas mixing using a lumped parameter model of the pulmonary acinus. Respir. Physiol. **127**, 241–248 (2001)
94. Haefeli-Bleuer, B., Weibel, E.R.: Morphometry of the human pulmonary acinus. Anat. Rec. **220**, 401–414 (1988)
95. Delcroix, M., Mélot, C., Lejeune, P., Leeman, M., Naeije, R.: Effects of vasodilators on gas exchange in acute canine embolic pulmonary hypertension. Anesthesiology **72**, 77–84 (1990)
96. Delcroix, M., Mélot, C., Vachiery, J.-L., Lejeune, P., Leeman, M., Vanderhoeft, P., Naeije, R.: Effects of embolus size on hemodynamics and gas exchange in canine embolic pulmonary hypertension. J. Appl. Physiol. **69**(6), 2254–2261 (1990)
97. Hasinoff, I., Ducas, J., Schick, U., Prewitt, R.: Pulmonary vascular pressure-flow characteristics in canine pulmonary embolism. J. Appl. Physiol. **68**(2), 462–467 (1990)
98. Mélot, C., Delcroix, M., Closset, J., Vanderhoeft, P., Lejeune, P., Leeman, M., Naeije, R.: Starling resistor vs. distensible vessel models for embolic pulmonary hypertension. Am. J. Physiol. Heart Circ. Physiol. **268**(2), H817–H827 (1995)
99. Levine, J.A., Schleusner, S.J., Jensen, M.D.: Energy expenditure of nonexercise activity. Am. J. Clin. Nutr. **72**, 1451–1454 (2000)
100. Nishimura, M., Kiyamoto, K., Suzuki, A., Yamamoto, H., Tsuji, M., Kishi, F., Kawakami, Y.: Ventilatory and heart rate responses to hypoxia and hypercapnia in patients with diabetes mellitus. Thorax **44**, 215–257 (1989)
101. Politi, A.Z., Donovan, G.M., Tawhai, M.H., Sanderson, M.J., Lauzon, A., Bates, J.H.T., Sneyd, J.: A multiscale, spatially distributed model of asthmatic airway hyper-responsiveness. J. Theor. Biol. **266**, 614–624 (2010)
102. Wang, I., Politi, A.Z., Tania, N., Bai, Y., Sanderson, M.J., Sneyd, J.: A mathematical model of airway and pulmonary arteriole smooth muscle. Biophys. J. **94**(6), 2053–2064 (2008)
103. Hai, C., Murphy, R.: Cross-bridge phosphorylation and regulation of latch state in smooth muscle. Am. J. Physiol. **254**, C99–C106 (1988)
104. Lai-Fook, S.J., Hyatt, R.E.: Effect of parenchyma and length changes on vessel pressure-diameter behavior in pig lungs. J. Appl. Physiol. **47**(4), 666–669 (1979)
105. Donovan, G., Bullimore, S., Elvin, A., Tawhai, M., Bates, J., Lauzon, A., Sneyd, J.: A continuous-binding cross-linker model for passive airway smooth muscle. Biophys. J. **99**(10), 3164–3171 (2010)

Part III
Wounds and Infection

A Multilevel Finite Element Approach to Study Pressure Ulcer Aetiology

Cees W. J. Oomens

Abstract A pressure ulcer is a form of tissue degeneration as a result of sustained mechanical loading. In the last 3 decades a lot of research has been done to understand the aetiology of pressure ulcers. It has become clear that the initial signs of tissue damage are found at the cell scale. That is where the damage process starts that eventually leads to severe wounds. In order to define damage thresholds or to understand what cells "feel" it is necessary to have information on the mechanical status of cells at a scale in the order of micrometres. How the external loading, that is gravitational body forces and reaction forces at supporting surfaces on patients in a bed or a wheel chair, is transferred to a local mechanical state within tissues depends on tissue morphology, mechanical properties and other boundary conditions and requires an analysis at the scale of the order of centimetres to a meter. This cannot be done in one single analysis covering the entire range of scales. This chapter summarizes some work that our group has done in the last 10 years on multi-scale modelling of soft tissues that was aimed at understanding some of the phenomena that play a role in pressure ulcer development. The work has shown the potential of multi-scale modelling to gain insight in the very complex interactions at cell level. It was shown that the heterogeneity in the microstructure has a profound impact on the way cells deform as well as the mechanical property changes of the cell after they become damaged.

C. W. J. Oomens (✉)
Biomedical Engineering Department, Eindhoven University of Technology, Eindhoven, The Netherlands
e-mail: c.w.j.oomens@tue.nl

Stud Mechanobiol Tissue Eng Biomater (2013) 14: 289–298

DOI: 10.1007/8415_2012_158

Published Online: 6 November 2012

1 Introduction

A pressure ulcer is a localized injury to the skin and/or underlying tissues, usually over a bony prominence that results from pressure and shear [1]. In more technical terms it is a form of tissue degeneration as a result of sustained mechanical loading. In the last 3 decades a lot of research has been done to understand the aetiology of pressure ulcers. It has become clear that the initial signs of tissue damage are found at the cell scale [2, 3]. The damage process starts with the disruption of single cells and if continued may eventually lead to severe, chronic wounds. Bouten et al. [3] proposed a hierarchical approach: "To investigate the differential response of the various tissue functional units to mechanical loading and their relative contributions to the aetiology of pressure ulcers, a hybrid methodology involving a combination of computer and experimental studies must be involved." This has led to a series of experimental model systems ranging from single cell studies [2], to studies with tissue engineered bio-artificial muscle [4–6] and animal studies with brown Norway rats [7, 8]. Each of the model systems represented a different length scale and hierarchical position in the system that was studied. Of course, each system had benefits and drawbacks, but all contributed to an overall understanding on the effect of ischaemia and deformation on the development of tissue damage. And although each model system was very useful within its own system boundaries, for some questions to be answered it is necessary to connect systems and bridge the different scales.

In order to define damage thresholds or to understand what cells "feel" it is required to know the mechanical status of cells at a scale in the order of micrometres. However, if we want to know how the external loading, that is gravitational body forces and reaction forces at supporting surfaces on patients in a bed or a wheel chair, is transferred to a local mechanical state within tissues, we require an analysis at the scale of the order of centimetres to a metre, because it all depends on tissue morphology, mechanical properties and boundary conditions. This cannot be done in one single analysis covering the entire range of scales, but requires a multi-scale analysis using different models at different length scales and methods to connect the models in a correct way.

In this chapter, I will summarize some of the work that was done in our department in this area. I assume the reader is familiar with continuum mechanics and the finite element method and will only briefly touch the rather complex mathematics involved in the coupling of the different length scales. For this the reader is referred to the original articles.

In the next section the concept of the multi-scale or multi-level modelling will be explained. After that, two studies will be discussed. The first by Breuls et al. [4] was aimed at the role of physical and geometrical heterogeneity of the microstructure on the damage development in cells. The second study by Nagel et al. [9] was focussed on the effect that a change in mechanical stiffness of single cells, after they are being damaged, has on the load distribution over the surrounding cells.

2 Multi-level Modelling: The Concept

Multi-level modelling of continuum mechanics problems can be done in various ways. The method that is described here, was originally developed to study the behaviour of composite polymer materials and metals [10, 11].

In our application a tissue is considered to consist of cells embedded in matrix material. The microscopic length scale of the cells (μm) is orders of magnitude smaller than the macroscopic dimensions (cm). Because of this, two models have to be developed: one at the micro-level and one at the macro-level. The problem is how to couple these models, or which boundary conditions have to be applied to the micro-model. Furthermore, it is not trivial how the average mechanical state of the micro-model can be transferred back to the macro model. The latter is called the *homogenization* of the mechanical state of the micro-model.

There are several approaches for this coupling. One is, to use *periodic boundary conditions*. When an external load is applied, the stress and strain fields in the microstructure will show large gradients due to the microstructural heterogeneity. However, due to the differences in scale, the microstructural deformation field around a macroscopic point will be approximately the same as the deformation field around neighbouring points. The repetitive deformations justify the assumption of local periodicity, meaning that the microstructure can be thought of as repeating itself near a macroscopic point (illustrated in Fig. 1). However, the microstructure itself may differ from one macroscopic point to another. The repetitive microstructural deformations suggest that macroscopic stresses and strains around a certain macroscopic point can be found by averaging micro-structural stresses and strains, in a small representative area of the microstructure attributed to that point.

Now consider a two-dimensional representative volume element (RVE). The periodic boundary condition implies:

1. The shapes of two opposite edges remain identical.
2. The stress vectors on opposite edges of the RVE are opposite to satisfy stress continuity.

In the implementation of numerical codes this is implemented as tying the displacements of nodes from one edge to corresponding nodes of the opposite edge. The RVE deformations, orientation and overall dimensions are determined by the positions of the three vertex points 1, 2 and 4 and the tying conditions. Next to this the condition of opposite stress vectors on opposite boundaries have to be satisfied. By averaging the deformations and the stresses in the RVE a coupling can be made between the microscopic RVE and the macroscopic model.

In a finite element implementation of the theory in each integration point of the elements of the macroscopic model a microscopic RVE is defined, which again is a finite element model. The macroscopic deformations $\boldsymbol{F}_{\mathrm{macro}}$ are transferred to the corresponding microscopic FE models, by prescribing the corner nodes and applying periodic boundary conditions. The RVE problems are solved and in a full

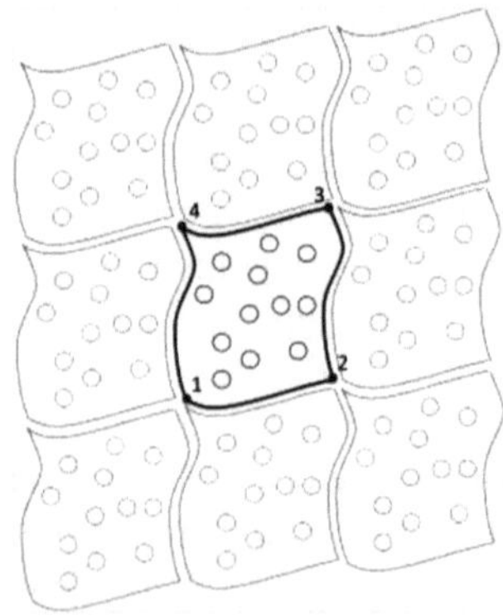

Fig. 1 Two-dimensional RVE illustrating periodic boundary conditions, which can be thought of as surrounded by identically shaped RVE's

multi-scale analysis the macroscopic stresses $\boldsymbol{\sigma}_{\mathrm{macro}}$ and tangent stiffness matrices ${}^4\boldsymbol{S}_{\mathrm{macro}}$ are computed from the microscopic models and supplied to the macroscopic integration points. In this case only constitutive equations for the structures in the RVE have to be available. It will be clear that this method is very expensive, because of computer time and memory. On the other hand the method is very suitable for implementation on parallel computers. From the following sections it will be clear that it is not always necessary to perform a full multi-scale analysis in the described way.

3 The Role of Heterogeneity in the Stress/Strain State of Individual Cells

Breuls [4] used the multi-level method to study the effect of microstructure on the load transfer of a macroscopically loaded tissue engineered muscle to the deformation of single cells. The model described an in vitro experiment that he performed on Bioartificial Muscles (BAM's). These tissue engineered BAM's are small strips consisting of a collage/matrigel extracellular matrix (ECM) and oriented myotubes. This in vitro model system was used in several investigations to study how cells behave under different loading conditions, by indenting the strips with a spherical indenter and then follow cell death with vital staining's for apoptosis and necrosis on a confocal microscope [5, 12, 13].

Breuls wanted to study the influence of the microstructure on what the cell "feels" when the entire construct is loaded. Because the cells in the constructs formed long myotubes, mostly lying in the direction of the long axis of the construct a two-dimensional plane strain model of the cross section seemed appropriate. Micromodel A in Fig. 2 represented the BAM cross section. For comparison the same analysis was performed on a cross section that was more similar to real muscle tissue (Micromodel B). Because of symmetry only one quarter of the cross-section was used for the macroscopic mesh. The mesh consisted of 14 quadratic, rectangular elements with 4 integration points each. In each

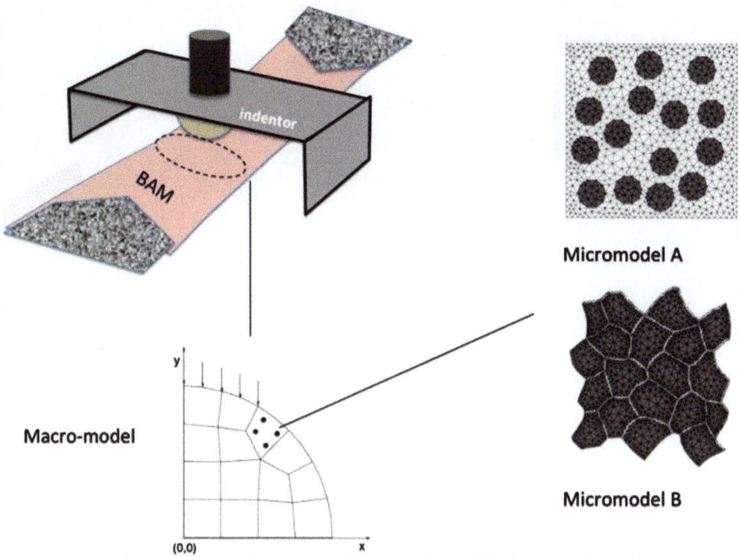

Fig. 2 Construction of a multi-level model of a bio-artificial muscle (BAM). In every integration point of the macroscopic mesh an RVE is defined representing the microstructure. Micro model A represents a typical microstructure expected in a BAM. Micro model B represents the cross section of a mature muscle (adapted from Breuls [12])

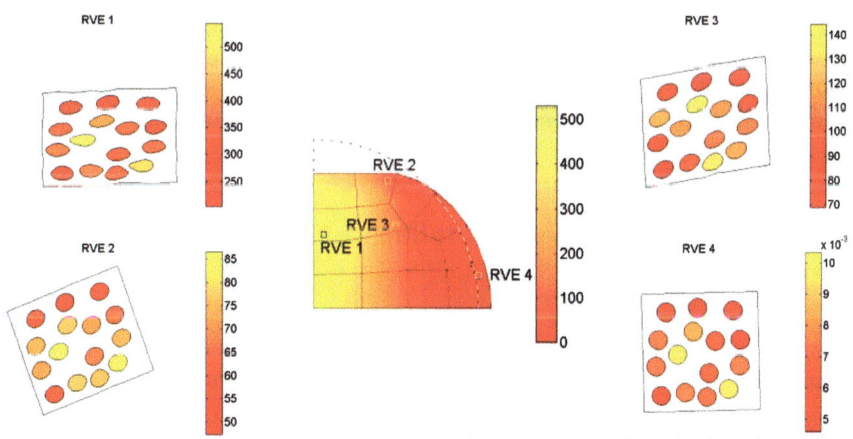

Fig. 3 Results of a simulation with microstructure A. The deformed macroscopic geometry (*centre*) is shown, surrounded with four selected RVEs. The color bar next to the macroscopic mesh represents the macroscopic SED (in N/m^2). The color scale bars next to the RVEs represent the SED, averaged over the cell area (Adapted from: Breuls [12])

integration point, the same RVE was defined. Both cells as well as ECM were modelled as elastic materials, using a Neo-Hookean material model. At $x = 0$ and $y = 0$, displacements were suppressed in y- and x-direction, respectively. To

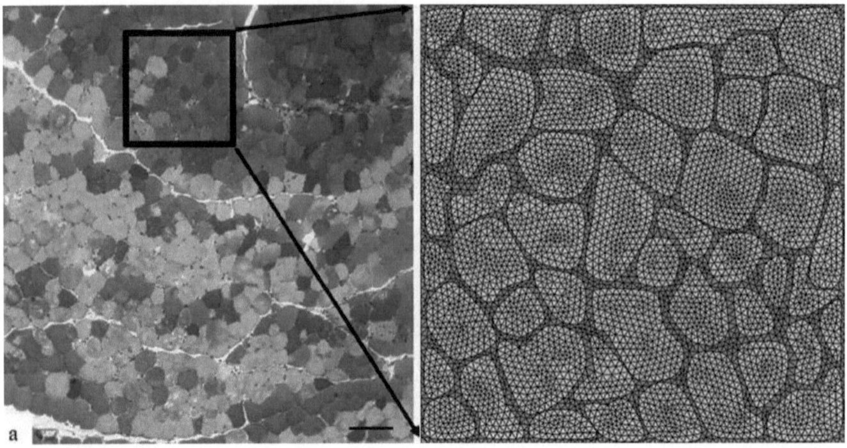

Fig. 4 Histological image and RVE mesh that was derived from it (with permission from [9])

simulate the transverse compression a constant load was applied by prescribing displacements in the *y*-direction of nodal points at the top-edge of the macroscopic mesh (See Fig. 2).

Figure 3 shows a result from a simulation with the micromodel A. It was clear that locally in a RVE there is a large variation in the deformation that individual cells experience. The macroscopic strain energy density at the level of RVE 1 is in the order of 500 N/m^2, but the averaged strain energy density for individual cells ranges from 200 to 550 N/m^2. Taking this into account, it is not so strange that the initial signs of tissue damage in skeletal muscle are always some individual cells with a disrupted cytoskeleton, surrounded by perfectly healthy cells.

4 How Do Stiffness Changes Influence the Stress/Strain State of Single Cells?

Nagel [9] studied the effect of cellular stiffening on the damage evolution in deep tissue injury. Starting point for his work was the observation by Ganz and Gefen [7, 14] that rat muscles showed a 1.6–3.3 fold increase in stiffness after being damaged (in vitro as well as in vivo). Such a rise in stiffness may lead to a redistribution of stress and strain that might lead to a further accumulation of damage. Based on these observations Nagel et al. spawned the hypothesis, that a change in material properties of necrotic tissue during mechanical loading leads to a significant redistribution of strains, so that previously unaffected regions exceed the damage threshold. They did not apply a full multi-level analysis but used a two step approach. First the effect of cellular stiffening was investigated at the level of an RVE. On the micro-level a finite element mesh was created to act as an RVE.

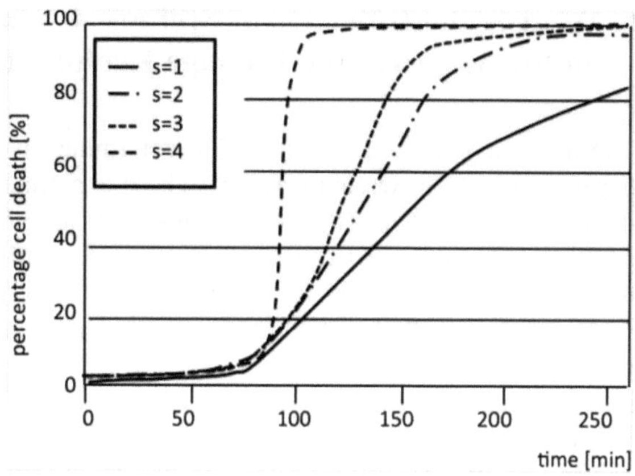

Fig. 5 The percentage of dead cells as a function of time for a shear strain of 0.8 and different stiffening factors (s = the stiffness of a cell after cell death divided by the original stiffness), Adapted from [9]

This mesh was derived from a histological image of a muscle cross-section of a rat Tibialis Anterior muscle. Figure 4 shows the image and the mesh that was created from it.

It was a 2-dimensional plane stress mesh using quadratic triangular elements. The damage law was derived from in vitro experiments that were described by Gefen et al. [13] and allowed the description of the typical sigmoid strain threshold versus time curve like observed in these experiments. The damage law was implemented and different amounts of stiffening were applied to cells that died due to excessive prolonged deformation to study the effect of this stiffening on the strain distribution inside the RVE, but also to find the average stiffness change of the RVE as a whole.

When applying this damage evolution law it appeared that indeed the percentage of damaged cells at a certain time increased with increasing strain. An applied macroscopic shear strain to the RVE of 0.6 led to 100 % cell death after 4 h. The time at which cells started to die was also slightly time dependent.

In this context it was important to know what happened when cells changed mechanical properties after cell death. A stiffening of cells after cell death did not have an impact on the amount of initial cell damage, nor on the onset time of further damage progression. However, it did influence the final number of cells that died after a certain amount of time and it also had an effect on the rate of cell damage progression. Of course the level of strain (and thus the number of dying cells) has a big influence on how much a stiffness change of the cells effects the progression of cell death. This is illustrated in Fig. 5:

By performing a sensitivity analysis with different levels of macroscopic strain and stiffness increase of a single cell after cell death, it was possible to construct for an RVE a stiffness versus percentage of death cells curve. This was used in a macroscopic analysis to study the effect of cell stiffening on the damaged area a macroscopic model.

5 What is the Influence of Stiffness Changes of Single Cells on the Macroscopic Damage Evolution?

The macroscopic model was used to determine the damage evolution in animal experiments on Brown Norway rats. These experiments are extensively described in Stekelenburg et al. [8] and Loerakker et al. [15]. Brown Norway rats were anesthetized and placed in supine position in an MR compatible loading device with their left hind limbs fixed. The Tibialis Anterior was mechanically loaded with an indenter filled with fluid to make it visible on MR images. The indentation was applied slowly, but kept constant for a period of 2 h and then the load was relieved. Because an MR compatible system was used it was possible to make high resolution images before and after indentation. These images were used to create dedicated finite element models of each individual rat and each loading condition. T2-weighted imaging was used to measure the location and magnitude of tissue changes including oedema and tissue damage.

The procedure for the damage evolution in the macro model worked as follows. In a first step the full indentation as observed in the experiments was applied to the Finite Element Model. This resulted in an initial, non-uniform strain state in the loaded. When a strain in an integration point was less than a threshold γ_l there was no damage. Cells start to be damage when this lower threshold is exceeded. Also an upper threshold γ_h was defined, based on the RVE studies. Above this threshold 100 % cell damage was assumed. Between the thresholds γ_l and γ_h the percentage n of cell death was linearly interpolated:

$$n[\%] = \begin{cases} 0 & \text{if} \quad \hat{E}_s < \gamma_l, \\ \frac{\hat{E}_s - \gamma_l}{\gamma_h - \gamma_l} 100 & \text{if} \quad \gamma_l \leq \hat{E}_s \leq \gamma_h, \\ 100 & \text{if} \quad \hat{E}_s > \gamma_h, \end{cases} \tag{1}$$

Based on the sensitivity analysis with the RVE's the amount of stiffening around an integration point was calculated with a quadratic function of n. With the original stiffness c_{10}, the current stiffness of the damaged muscle c_{10}^d is defined as:

$$c_{10}^d = f c_{10} \tag{2}$$

with:

$$f = an^2 + bn + 1 \tag{3}$$

After checking, how many cells were damaged, locally the new stiffness c_{10}^d and the simulation was repeated until convergence was reached.

Depending on the stiffening factor used for the individual cells the damaged area changed. For stiffening factors of 2, 3 and 4 fold for the cells the damaged area in the macroscopic model increased with factors of 1.2, 1.45 and 1.65 respectively.

6 Conclusion

This chapter gave a summary of two examples of multi-scale work, used in pressure ulcer aetiology studies. The objective was to show in which cases a multi-scale analysis is useful or indispensible. The first was aimed at studying how heterogeneity in the cell shape and location in a muscle influences the damage evolution. The second example was focussed on how stiffening of cells, after they are damaged, influences the strain distribution in a muscle and further progress in damage development.

Neither of the simulations involved a full multi-scale analysis, with a coupling from both macro to micro and back again. In both cases it was not really necessary to make a full-scale coupling and in this way simulations were much more cost effective. However, a full-scale coupling is possible in the way as described in the first sections. Because each RVE is based on the same FEM mesh many calculations in integration points can be run parallel and in Eindhoven we typically run this kind of simulations on a cluster with 64 or 128 CPU's.

References

1. EPUAP/NPUAP—Pressure ulcer prevention & treatment, clinical practice guideline. http://www.npuap.org/ (2009)
2. Bouten, C.V.C., Knight, M.M., Lee, D.A., Bader, D.L.: Compressive deformation and damage of muscle cell sub-populations in a model system. Ann. Biomed. Eng. **29**, 153–163 (2001)
3. Bouten, C.V.C., Oomens, C.W.J., Baaijens, F.P.T., Bader, D.L.: The aetiology of pressure sores: skin deep or muscle bound? Arch. Phys. Med. Rehab. **84**, 616–619 (2003)
4. Breuls, R.G.M., Sengers, B.G., Oomens, C.W.J., Bouten, C.V.C., Baaijens, F.P.T.: Predicting local cell deformations in engineered tissue constructs: A multilevel finite element approach. J. Biomech. Eng. **124**, 198–207 (2002)
5. Gawlitta, D., Oomens, C.W.J., Bader, D.L., Baaijens, F.P.T., Bouten, C.V.C.: Temporal differences in the influence of ischemic factors and deformation on the metabolism of engineered skeletal muscle. J. Appl. Physiol. **103**, 464–473 (2007)
6. Gawlitta, D., Li, W., Oomens, C.W.J., Baaijens, F.P.T., Bader, D.L., Bouten, C.V.C.: The relative contributions of compression and hypoxia to development of muscle tissue damage: An in vitro study. Ann. Biomed. Eng. **35**, 273–284 (2007)
7. Linder-Ganz, E., Gefen, A.: Mechanical compression induced pressure sores in rat hind limb: muscle stiffness, histology and computational models. J. Appl. Physiol. **96**, 2034–2049 (2004)
8. Stekelenburg, A., Strijkers, G.J., Parusel, H., Bader, D.L., Nicolay, K., Oomens, C.W.J.: The role of ischemia and deformation in the onset of compression induced. J. Appl. Physiol. **102**, 2002–2011 (2007)
9. Nagel, T., Loerakker, S., Oomens, C.W.J.: A theoretical model to study the effects of cellular stiffening on the damage evolution in deep tissue injury. Comp. Meth. Biomech. Biomed. Engng. **12**, 585–598 (2009)
10. Kouznetsova, V., Brekelmans, W.A.M., Baaijens, F.P.T.: An approach to micro-macro modeling of heterogeneous materials. Comput. Mech. **27**, 37–48 (2001)
11. Smit, R.J.M., Brekelmans, W.A.M., Meijer, H.E.H.: Prediction of the mechanical behavior of nonlinear heterogeneous systems by multi-level finite element modeling. Comp. Meth. Appl. Mech. Engng. **155**, 181–192 (1998)

12. Breuls, R.G.M., Bouten, C.V.C., Oomens, C.W.J., Bader, D.L., Baaijens, F.P.T.: Compression induced cell damage in engineered muscle tissue: an in vitro model to study pressure ulcer aetiology. Ann. Biomed. Eng. **31**, 1357–1365 (2003)
13. Gefen, A.G., van Nierop, B.J., Bader, D.L., Oomens, C.W.J.: Strain-time cell death threshold for skeletal muscle in a tissue-engineered model system for deep tissue injury. J. Biomech. **41**, 2003–2012 (2008)
14. Gefen, A., Gefen, N., Linder-Ganz, E., Margulies, S.S.: In vivo muscle stiffening under bone compression promotes deep pressure sores. J. Biomech. Engng **127**, 512–524 (2005)
15. Loerakker, S., Manders, E., Strijkers, G.J., Nicolay, K., Baaijens, F.P.T., Bader, D.L., Oomens, C.W.J.: The effects of deformation, ischaemia and reperfusion on the development of muscle damage during prolonged loading. J. Appl. Phys. **111**, 1168–1177 (2011)

Discrete and Continuum Multiscale Behaviour in Bacterial Communication

Sara Jabbari and John R. King

Abstract The interacting effects operating on subcellular (gene regulatory processes), cellular (interactions between neighbouring cells) and population (signalling molecule transport) scales are exemplified and explored through simple multiscale models. Specific attention is focused on how the upregulation (or downregulation) of small numbers of discrete cells can influence the behaviour of the population as a whole, by investigating toy models for positive autoregulation and by the simulation of a much more detailed model for quorum sensing within a Gram-positive population of bacteria. The implications for delays associated with gene expression are also investigated in a spatio-temporal context through the analysis of blow-up behaviour, as a mathematical symptom of upregulation through positive feedback, in some model reaction-diffusion delay equations.

1 Bacterial Communication

The belief that bacteria within a colony operate independently of one another was abandoned in the 1970s with the discovery of cell-density dependent behaviour in the marine bacterium *Vibrio fischeri* [16]. It was found that these bacteria emit light only when present at sufficiently high concentrations, most likely either to obtain camouflage from predators (by mimicking moonlight on the water's surface) or to attract mates; at lower bacterial concentrations, the amount of light

S. Jabbari
University of Birmingham, Edgbaston, Birmingham B15 2TT, England
e-mail: s.jabbari@bham.ac.uk

J. R. King (✉)
Centre for Mathematical Medicine and Biology, University of Nottingham,
University Park, Nottingham NG7 2RD, UK
e-mail: john.king@nottingham.ac.uk

Stud Mechanobiol Tissue Eng Biomater (2013) 14: 299–320
DOI: 10.1007/8415_2012_155

Published Online: 17 October 2012

emitted would not merit the energy expended on generating the bioluminescence, meaning that this phenotype should be induced only at high population density. This understanding between the cells is mediated by extracellular signal molecules and the process has been termed 'quorum sensing' (the behaviour of the bacteria alters once a quorum has been attained). Cell communication is now known to be prevalent throughout the bacterial kingdom [5, 28].

Though signalling pathways vary between species (most notably between Gram-positive and Gram-negative bacteria) the phenomenon generally consists of the production of diffusible signal molecules and their excretion into the environment, allowing detection by other cells. The level of signal molecules in the environment is a reflection of the population size or density, and the number of signal molecules detected by an individual cell will determine the regulatory gene cascade which is triggered and ultimately the behaviour of the cell. A sufficiently well-mixed population should therefore act in a synchronised manner.

Since its discovery in the form of bioluminescence, quorum sensing has been found to regulate a spectrum of bacterial behaviours, many of which, such as biofilm formation, virulence, sporulation and swarming motility [5, 28], are medically significant. Biofilm formation in particular causes huge problems on medical implant devices such as catheters. Modelling quorum sensing in biofilms poses additional complications since, unlike the relatively homogeneous populations that occur in other contexts, biofilms are highly heterogeneous, comprising layers of mixed populations of cells with quorum-sensing signal production (and therefore cell phenotype) varying throughout the biofilm. Spatial variation is something on which we focus in the models presented later, albeit in contexts much simpler than those arising in biofilms.

The goal of quorum sensing is often to coordinate a particular behaviour which would be futile were the whole population not engaged. For instance, it is believed that *Staphylococcus aureus* switches on the production of virulence factors, thereby attacking the host, only when it is present in sufficient numbers to overwhelm the resulting immune response [25]. Thus the population should act in a concerted fashion. In other contexts, however, quorum sensing may not be regulating the response of a large population. Since signal molecule build-up can arise as a result of a confined environment, rather than a large population, quorum sensing can also serve to detect environmental conditions (this is sometimes referred to as diffusion sensing [23]). For instance, the same quorum-sensing system in *S. aureus* (i.e. the *agr* operon) is thought also to trigger endosome escape [22] (only one or two cells are contained within a host cell), whereby signal aggregation determines the escape time.

We have also recently used a mathematical model of quorum sensing by *Clostridium acetobutylicum* (which uses a modified *agr* operon) [10] to illustrate that, if the quorum-sensing system is composed of fewer feedback loops (thus softening the response), it is possible that it can be used to anticipate hostile environments and, rather than coordinating behaviour at the whole population level, send only a portion of cells to a sporulation fate (making them capable of surviving these conditions). By maintaining a number in a vegetative state, the cells should still be able to profit from any potential favourable change in the environment, thus facilitating a "hedge-betting" approach. Thus it should be possible that quorum sensing is also used to control behaviour at a level somewhere between the whole population and the single cell.

Quorum sensing can therefore serve to regulate at a variety of levels. Multiscale modelling is consequently required to describe quorum sensing accurately: subcellular gene regulation informs behaviour of an individual cell which influences increasing numbers of cells, potentially through to the whole population phenotype. At each of these levels, signal is relayed back to the intracellular processes, linking each level together. Unlike the majority of previous mathematical models of quorum sensing, we shall focus on heterogeneity of the phenotype within a population of cells, rather than the population as a whole becoming fully quorate, investigating effects associated with both discrete and continuous models. Many quorum-sensing systems have been shown mathematically to be bistable, enabling switch-like behaviour to arise between phenotypes (i.e. between downregulated and upregulated behaviour). This bi- (or indeed multi-) stability is dependent upon the number and nature of feedback loops within the system [26] and quorum-sensing-induced transcription being significantly faster than the basal rate [7]. For the case study proposed in Sect. 4 of this chapter, we consider a system with the appropriate conditions to be bistable.

More broadly, our focus here is on the spatio-temporal modelling of autoinductive (positive-feedback) processes such as those central to quorum sensing and, specifically, on investigating effects that go beyond those that can be captured by the classical [i.e. partial-differential equation (PDE)] models for such phenomena. In keeping with this focus, we investigate in Sects. 2 and 3 two model problems associated with the most explicit mathematical abstraction of a positive feedback process leading to upregulation, namely blow-up behaviour, investigating firstly the effects of delays associated with gene expression and secondly the effects of discreteness on cell–cell communication. The latter is complemented by an exploration in Sect. 4 of the influence of discreteness on wave propagation and pinning in a more mechanistic, and hence significantly more complicated, discrete model for a specific quorum-sensing system.

The ostensibly simple model problems we explore first seek to illustrate the implications of a number of key effects: positive feedback (reflected by the source terms in the systems studied below), nonlinearity (describing effects associated with cooperativity in gene regulation: we focus on the simplest (quadratic) possibility, as applicable when, for example, regulation is governed by protein dimerisation), discreteness (associated with the population being made up of individuals or distinct compartments: it should be stressed that, while the discrete models below take the form of finite-difference approximations to PDEs, here it is the discrete problem that is to be regarded as the 'true' model and its continuum limit as the approximation, rather than vice versa) and delays (which are inevitably present in autoregulation for a variety of reasons). Our analysis will apply ideas from (multiscale) matched asymptotic expansions and similarity methods, as well as numerical simulations.

The remainder of the chapter can be summarised as follows. Sects. 2 and 3 analyse respectively the effects of delay and discreteness on blow-up behaviour (reflecting upregulation due to a positive feedback loop) in minimal models, in each case investigating both linear and nonlinear (cooperative) feedback. Sect. 4 explores related effects by further characterising a preexisting [7, 9] model for quorum sensing governed by the so-called *agr* operon. Finally, Sect. 5 briefly discusses some of the implications of the results.

2 Spatio-Temporal Blow-Up Phenomena: The Effects of Delays

The context for the current analysis is provided by two very widely studied PDE models. Firstly, the simplest (linear) reaction-diffusion model[1] takes the form

$$\frac{\partial u}{\partial t}(x,t) = \frac{\partial^2 u}{\partial x^2}(x,t) + u(x,t) \quad -\infty < x < +\infty, \tag{1}$$

wherein the linear source term embodies the simplest class of positive-feedback mechanism, leading (for finite-mass initial data) to infinite-time blow up[2] in the form

$$\begin{aligned} u &\sim \frac{M}{2(\pi t)^{\frac{1}{2}}} e^{t-x^2/4t} \quad \text{as} \quad t \to +\infty, \quad x = O(t^{\frac{1}{2}}) \\ u &\sim \frac{1}{2(\pi t)^{\frac{1}{2}}} \Omega\left(\frac{x}{t}\right) e^{t-x^2/4t} \quad \text{as} \quad t \to +\infty, \quad x = O(t), \end{aligned} \tag{2}$$

for some constant M and arbitrary function Ω that satisfies $\Omega(0) = M$. Secondly,

$$\frac{\partial u}{\partial t}(x,t) = \frac{\partial^2 u}{\partial x^2}(x,t) + u^2(x,t) \quad -\infty < x < +\infty \tag{3}$$

is an archetypal model for nonlinear positive feedback and one that has spawned an extensive literature concerning its finite time blow up, an effect already illustrated by its spatial homogeneous solution

$$u(t) = \frac{1}{t_c - t} \tag{4}$$

for constant $t_c > 0$; the diffusion term in (3) turns out to have a rather limited influence in mitigating such blow up, the competition between autoinduction

[1] We for the most part adopt dimensionless forms containing the minimal numbers of parameters in this section and the next.

[2] Throughout we associate such blow up with upregulation (e.g. quorum sensing).

driving a local increase in u and diffusion acting to smear out such a peak being an intrinsic feature of such models.

Here we incorporate delays in feedback, typically associated in the autoregulation of gene expression with the time needed for transcription and translation of further copies of the relevant transcription factor but also attributable to intracellular transport, for example. Thus in generalising (1) we investigate

$$\frac{\partial u}{\partial t} = \frac{\partial^2 u}{\partial x^2}(x,t) + u(x,t-T) \quad -\infty < x < +\infty, \tag{5}$$

where $T > 0$ corresponds to the delay time and we lump the entire signalling and genetic machinery into the single variable $u(x,t)$, which could be thought of as the concentration of a signalling molecule within a population of cells.

In the spatially homogeneous case, solutions to (5) can be written as a superposition over modes of the form

$$u(t) = A(\lambda)e^{\lambda t}, \quad \lambda = e^{-\lambda T} \tag{6}$$

for constants A, the large time behaviour being dominated by the real root for λ, whereby

$$\begin{aligned} \lambda &\sim 1 - T \quad \text{as} \quad T \to 0, \\ \lambda &\sim \frac{1}{T}\ln T \quad \text{as} \quad T \to +\infty, \end{aligned} \tag{7}$$

i.e. the longer the delay, T, the lower the growth rate; it is this real root that we denote by λ in what follows. The large-time behaviour in the spatially structured case (5) then follows readily: setting

$$u(x,t) = e^{\lambda t}v(x,t)$$

and treating v to be slowly varying in t (an assumption whose validity is readily confirmed *a posteriori*) implies

$$(1 + \lambda T)\frac{\partial v}{\partial t} \sim \frac{\partial^2 v}{\partial x^2}$$

and hence

$$u \sim \frac{(1+\lambda T)^{\frac{1}{2}}M}{2(\pi t)^{\frac{1}{2}}} e^{\lambda t - (1+\lambda T)x^2/4t} \quad \text{as} \quad t \to +\infty, \quad x = O(t^{\frac{1}{2}}), \tag{8}$$

so that the delay has the effect of reducing the effective diffusivity as well as the growth rate. The behaviour as $t \to +\infty$ with $x = O(t)$ is also instructive and can be characterised by applying the Liouville–Green (JWKB) method in the form

$$u \sim a(x,t)e^{-f(x,t)} \tag{9}$$

to give

$$\frac{\partial f}{\partial t}+\left(\frac{\partial f}{\partial x}\right)^2+e^{T\partial f/\partial t}=0,$$

the large-time behaviour to which takes the form

$$f=tF(\eta),\quad \eta=x/t$$

whereby (since $F(\eta)$ satisfies a first-order ODE of Clairaut's form) F is given parametrically in terms of $P\equiv dF/d\eta$ by

$$F-\eta P+e^{T(F-\eta P)}+P^2=0,\quad -\eta-T\eta e^{T(F-\eta P)}+2P=0,$$

so that

$$\begin{aligned}F&\sim-\lambda+(1+\lambda T)\eta^2/4 &&\text{as}\quad \eta\to 0,\\ F&=\eta^2/4+\text{exponentially small terms} &&\text{as}\quad \eta\to+\infty,\end{aligned}$$

the former being consistent with (8) [i.e., in the terminology of matched asymptotic expansions, matching with (8)] and the latter implying that the purely diffusive effects dominate the far field, contrasting with the zero-delay case (2) in which the feedback term continues to generate an exponentially growing contribution.

We now turn to the more complicated nonlinear case

$$\frac{\partial u}{\partial t}=\frac{\partial^2 u}{\partial x^2}(x,t)+u^2(x,t-T)\quad -\infty<x<+\infty$$

(in this case we could rescale T to 1 but we forego the opportunity). That this case is both significantly more involved and qualitatively distinct from the zero-delay version (3) is already apparent from the spatially uniform solution, which we denote $U(t)$: applying the Liouville–Green method in the form

$$U(t)=e^{\Phi(t)}\quad as\quad t\to+\infty$$

gives

$$\Phi(t)\sim 2\Phi(t-T),\quad \Phi\sim\Phi_o e^{t\ln 2/T}$$

for some constant Φ_0 that depends on the initial data (implying strong sensitivity to the initial state); constructing correction terms yields

$$U(t)\sim\frac{4\ln 2}{T}\Phi_0 e^{t\ln 2/T}e^{\Phi_0 e^{t\ln 2/T}}\quad \text{as}\quad t\to+\infty,\tag{10}$$

implying that blow up is in infinite, rather than in finite, time, albeit at a much faster-than-exponential rate. In the spatially structured case we set

$$u(x,t) = U(t)v(x,t)$$

to give

$$\frac{\partial v}{\partial t}(x,t) = \frac{\partial^2 v}{\partial x^2}(x,t) + \frac{\dot{U}(t)}{U(t)}(v^2(x,t-T) - v(x,t)).$$

We conjecture, then, that the large time behaviour takes the quasi-steady form

$$\mathrm{v}(x,t) \sim V(\zeta), \quad \zeta = \left(\frac{\dot{U}(t)}{U(t)}\right)^{\frac{1}{2}} x \quad \text{as} \quad t \to +\infty \quad \text{with} \quad \zeta = O(1) \tag{11}$$

whereby $V(\zeta)$ satisfies the novel nonlinear functional differential equation

$$\frac{d^2V}{d\zeta^2}(\zeta) + V^2\left(\frac{\zeta}{\sqrt{2}}\right) - V(\zeta) = 0, \quad V = O(e^{-|\zeta|}) \quad \text{as} \quad |\zeta| \to \infty, \tag{12}$$

where we have made use of (10), which implies in (11) that $x = O(e^{-t\ln 2/2T})$, i.e. that u is largest in an exponential narrow spike, and that the neglected terms are exponentially smaller than those included. The dominant balance for larger x is simply the heat equation

$$\frac{\partial u}{\partial t}(x,t) \sim \frac{\partial^2 u}{\partial x^2}(x,t); \tag{13}$$

u remains hyperexponentially large out to $x = O(e^{t\ln 2/2T})$, a caustic of the resulting rays associated with the Liouville–Green approximation to (12) being present in this exponentially large range of x. Confirmation of the applicability of this postulated scenario of course requires in particular the existence of a non-trivial solution of the boundary-value problem (12), which represents an open problem; stability is also important: if a solution to (12) exists, it will be unstable to modes associated with translations of t (or equivalently of Φ_0) and of x, but should not be to any others if it is to represent the generic large-time behaviour.

3 Spatio-Temporal Blow-Up Phenomena: The Effects of Discreteness

The analog of the linear problem (1) now reads

$$\frac{du_i}{dt}(t) = \upsilon^2(u_{i+1}(t) - 2u_i(t) + u_{i-1}(t)) + u_i(t), \quad -\infty < i < +\infty \tag{14}$$

wherein the integers i are to be viewed as labelling distinct cells, the simplest interpretation being that the signalling molecules (concentration $u_i(t)$) can be transported directly across the membranes of adjacent cells, the constant υ^2 being a measure of the rate of transport relative to that of production. Setting $x = i/\upsilon$, the large-time behaviour for $x = O(t^{\frac{1}{2}})$ takes the form in (2) [note that (1) corresponds to the large υ limit of (14)] and the only novelty here concerns the regime $x = O(t)$, wherein the Liouville–Green [as in (9)] implies

$$\frac{\partial f}{\partial t} + 4\upsilon^2 \sinh^2\left(\frac{\partial f}{\partial x}/2\upsilon\right) + 1 = 0,$$

so that

$$F - \eta P + 2\upsilon^2(\cosh(P/\upsilon) - 1) + 1 = 0, \quad -\eta + 2\upsilon \sinh(P/\upsilon) = 0$$

and hence

$$F = \upsilon\eta \ln\left(\left(1 + \frac{\eta^2}{4\upsilon^2}\right)^{\frac{1}{2}} + \frac{\eta}{2\upsilon}\right) - 2\upsilon^2\left(\left(1 + \frac{\eta^2}{4\upsilon^2}\right)^{\frac{1}{2}} - 1\right) - 1,$$

so that

$$F \sim -1 + \eta^2/4 \quad \text{as} \quad \eta \to 0, \quad F \sim \upsilon\eta \ln(\eta/\upsilon) - \upsilon\eta - 1 + 2\upsilon^2 \quad as \quad \eta \to +\infty,$$

so the transfer by a finite distance between neighbouring cells that is associated with discreteness leads to slower decay in the far field than occurs in the continuous case. More significant for what follows is that blow up occurs at a similar rate over the range $x = O(t^{\frac{1}{2}})$, i.e. diffusion suffices to drive many cells into an upregulated state at a similar time.

The situation with the nonlinear case

$$\frac{du_i}{dt}(t) = \upsilon^2(u_{i+1}(t) - 2u_i(t) + u_{i-1}(t)) + u_i^2(t) \quad -\infty < i < +\infty \tag{15}$$

(where we could choose to scale υ out) is markedly different: blow up occurs in finite time and does so generically over three cells only [in the case of quadratic nonlinearity adopted in (15)]—the central one of these ($i = 0$, say) has

$$u_0(t) \sim 1/(t_c - t) \quad \text{as} \quad t \to t_c^-, \tag{16}$$

as in (4), with transport having a negligible effect, while blow-up in the neighbouring cells is driven by this one in the form

$$u_1, u_{-1} \sim \upsilon^2 \ln(1/(t_c - t)) \quad \text{as} \quad t \to t_c^-,$$

exhibiting less dramatic blow up than (16). Thus upregulation occurs within a small number of cells only; in practice the nonlinearity will saturate at high concentrations, leading to upregulation propagating out from this initial 'quorate' subpopulation.

The limit $\upsilon \to \infty$ is both physically meaningful and of mathematical interest (specifically, in the associated continuum limit, blow up can occur at an arbitrary real value of x, whereas in the discrete version the blow up point needs to be at an integer i: the switch from continuous to discrete translation invariance in going from the PDE to the differential-difference system has a number of important implications for the proper understanding of such discrete problems, including for the pinning phenomena noted later). We adopt the modified definition $x = (i - i_c)/\upsilon$, where $i_c \in \mathbb{R}$ is to be chosen such that blow up in the continuum limit occurs at $x = 0$. Setting

$$u_i(t;\upsilon) \sim u(x,t) + \frac{1}{\upsilon^2}\mathrm{v}(x,t) \quad \text{as} \quad \upsilon \to +\infty \tag{17}$$

implies

$$\frac{\partial u}{\partial t} = \frac{\partial^2 u}{\partial x^2} + u^2, \quad \frac{\partial v}{\partial t} = \frac{\partial^2 \mathrm{v}}{\partial x^2} + 2uv + \frac{1}{12}\frac{\partial^4 u}{\partial x^4}, \tag{18}$$

wherein the fourth derivative term is the sole remnant of discreteness thus far. The blow-up behaviour of the first of (18) is well known and motivates the intermediate-asymptotic variables

$$u = \frac{1}{(t_c - t)} f(\xi,\tau), \quad \mathrm{v} = \frac{1}{(t_c - t)^2} g(\xi,\tau), \quad \xi = \frac{x}{(t_c - t)^{\frac{1}{2}}}, \quad \tau = -\ln(t_c - t).$$

Introducing

$$\kappa(\tau) = -\frac{\partial^2 f}{\partial \xi^2}\bigg|_{\xi=0}$$

with $\kappa \to 0$ as $\tau \to +\infty$ to be determined, it can be shown that

$$f \sim 1 + \kappa(\tau)\left(1 - \frac{1}{2}\xi^2\right) + \kappa^2(\tau)\left(-5 + \frac{1}{4}\xi^4\right) \quad \text{as} \quad \kappa \to 0 \tag{19}$$

with

$$\frac{d\kappa}{d\tau} = -4\kappa^2, \quad \kappa = 1/4(\tau + \tau_c) \tag{20}$$

arising as a solvability condition (cf., [3, 6], for example). Since $g(\xi,\tau)$ satisfies a linear problem, we may superpose the 'complementary function' contribution

$$g \sim -T_c(1 + \kappa(\tau)(1 - \xi^2)), \tag{21}$$

corresponding to a shift of T_c/υ^2 in the blow up time t_c, with the constant T_c depending on the initial data, and a 'particular integral'

$$g \sim \kappa(\tau)\beta_1 + \kappa^2(\tau)\left(\beta_2 + \frac{1}{8}\xi^2\right), \tag{22}$$

where $\beta_1 = -1/8$ follows from the solvability condition on the $\kappa^2(\tau)$ term (and calculation of the constant β_2 would similarly require a solvability condition at higher order in the expansion).

It is clear from (20) that the expansion (19) is non-uniform at $\rho \equiv \xi/\tau^{\frac{1}{2}} = O(1)$ and on this scaling we have

$$\frac{1}{2}\rho\frac{\partial f}{\partial \rho} \sim f^2 - f, \quad f \sim 1/(1+\rho^2/8),$$
$$\frac{1}{2}\rho\frac{\partial g}{\partial \rho} \sim 2(f-1)g, \quad g \sim -(T_c + 1/(32\tau))/(1+\rho^2/8)^2,$$

where the first term in the numerator of g matches with (21) and the second with (22). The fourth derivative term associated with discreteness first causes a non-uniformity in the expansion in the $\xi = O(1)$ region, doing so for $T = O(1)$ where we set

$$t = t_c + T/\upsilon^2, \quad u_i = \upsilon^2 U_i,$$

discreteness coming fully into play here: at leading order the full balance

$$\frac{dU_i}{dT} = U_{i+1} - 2U_i + U_{i-1} + U_i^2$$

applies, but the required leading-order solution is fortunately spatially uniform, i.e. we may set

$$U_i \sim \frac{1}{T_c - T} + \frac{1}{\ln \upsilon}\frac{W_i(T)}{(T_c - T)^2} \quad \text{as} \quad \upsilon \to +\infty \tag{23}$$

to give

$$\frac{dW_i}{dT} = W_{i+1} - 2W_i + W_{i-1},$$

the required solution to which is

$$W_i = \frac{1}{8}\left(T_c - T - \frac{1}{2}(i - i_c)^2\right) - \frac{1}{64}, \tag{24}$$

the final contribution being required to match into g and the others into f.

Because (23) and (24) happen to satisfy both the discrete problem and its continuous limit, the $T = O(1)$ timescale turns out to be innocuous; the one on which significant behaviour occurs corresponds to a nonuniformity arising for $\rho = O(1)$ and has scalings [as also implied by (23)]

$$t = t_c + \frac{1}{\upsilon^2} T_c + \frac{1}{\upsilon^2 \ln \upsilon} T^\ddagger, \quad u_i = \upsilon^2 \ln \upsilon U_i^\ddagger$$

giving

$$\frac{dU_i^\ddagger}{dT^\ddagger} \sim U_i^{\ddagger 2}, \quad U_i^\ddagger \sim 1 \Big/ \left(\frac{1}{64} - T^\ddagger + \frac{(i - i_c)^2}{16} \right) \tag{25}$$

the diffusion term being logarithmically negligible; the extent of the analytical progress that is possible with this problem is thus striking.

The conclusions of this $\upsilon \to +\infty$ analysis are of limited practical significance, but are instructive in terms of the general issues associated with discrete to continuous limits. Firstly, the blow up time in (25) $T^\ddagger = T_c^\ddagger$ depends on i_c with $T_c^\ddagger \in \left[\frac{1}{64}, \frac{1}{32} \right]$, the lower bound arising when the blow up point of the continuous limit coincides with a cell location (i.e. with integer i) and the upper bound when it is 'half way between' cells (integer $i + 1/2$), i.e. (and not surprisingly) upregulation is fastest when the peak of the initial profile (when specified throughout $\mathbb{R}$) can be regarded as coinciding with a cell. Secondly, the signal molecule distribution at the blow up time is given as $\upsilon \to +\infty$ with $i - i_c = O(1)$ by

$$u_i \sim \upsilon^2 \ln \upsilon \frac{16}{((i - i_c)^2 - \min(i - i_c)^2)}$$

for integer i; this represents single point blow up, but for the two neighbouring cells there are additional contributions to $U_i^\ddagger$ of the form $-\ln (T_c^\ddagger - T^\ddagger)/\ln \upsilon$, the associated further non-uniformity manifesting itself in the three-point blow up described above.

We now turn to a more complicated differential-difference formulation that was developed (see [9]) specifically to investigate quorum-sensing behaviour in *S. aureus*, again focusing our discussion of phenomena that are explicitly associated with discreteness and are multiscale in the sense that they reflect both subcellular genetic regulation and population-level transport of signalling molecules.

4 A Quorum-Sensing Case-Study: The *agr* Operon

4.1 Background

We now move away from the generic and highly simplified representations of processes such as quorum sensing given in the previous sections and investigate a specific quorum sensing system using distinct equations to represent each relevant component. The result is a spatially structured model of a population of bacteria utilising an *agr* operon. This quorum-sensing system is found in a number of

Gram-positive bacteria. First identified in the pathogen *S. aureus* [17], homologues of the *agr* operon have since been found to also affect virulence factor production in *Listeria monocytogenes* [24], *Enterococcus faecalis* (the *fsr* system) [14] and *Clostridium botulinum* [2] and *perfringens* [1, 13] amongst others [29]. This system, however, is not restricted solely to pathogenic bacteria: it has also been found to regulate granulose formation and sporulation in *C. acetobutylicum* [27] and cell morphology in *Lactobacillus plantarum* [4] for instance. The prevalence of this system and the fact that it controls such a wide range of cell phenotypes in different species render it particularly important to understand. Furthermore, since the signal molecules can be secreted from the cells, cross-talk between strains and species has been found to occur (for examples relating to *S. aureus* see e.g. [11, 19] or [20]). Thus modelling of the *agr* operon traverses a number of scales—from subcellular gene regulation, through cell phenotype, to population behaviour and the repercussions upon multiple populations. We present here a model of a single species, but examples involving cross-talk between species and strains can be found in [8] and [9] and between two quorum-sensing systems within one strain in [12] or [21] for example (though the last two studies consider quorum-sensing in Gram-negative bacteria, so the systems studied are not *agr* homologues).

4.2 Model Formulation

Known *agr* operons have varying degrees of feedback contained within them: some have every element of the operon up-regulated in response to signal molecules, others only the elements controlling signal synthesis or those governing signal detection and response. In order to ensure bistability[3] in our system, we consider an *agr* system in which every element is induced in response to increased levels of quorum-sensing signal molecule, i.e. that first identified in *S. aureus*.

The operon consists of two genes (*agrB* and *agrD*) that interact to produce the signal molecule (termed AIP: autoinducing peptide) and two genes (*agrA* and *agrC*) which produce proteins to detect and respond to the AIP (these form a two-component system); see Fig. 1. Ultimately, sufficiently high levels of signal molecule induce high levels of phosphorylated AgrA (AgrA $\sim$ P), which increase transcription of both the *agr* genes and the downstream target genes (e.g. those controlling virulence factor production). We have previously derived a spatially structured model of the *agr* operon [9] which, under suitable simplifying assumptions, could always achieve upregulation should there be no interference with the operon. In order to investigate bistability, we cannot employ all of these simplifications and detail must be added back into the model. The full (ordinary

[3] Note that we are thus concerned in this section with the interactions between two stable states, whereas the focus in the previous sections was on how the solution is driven away from an unstable state (namely the trivial one).

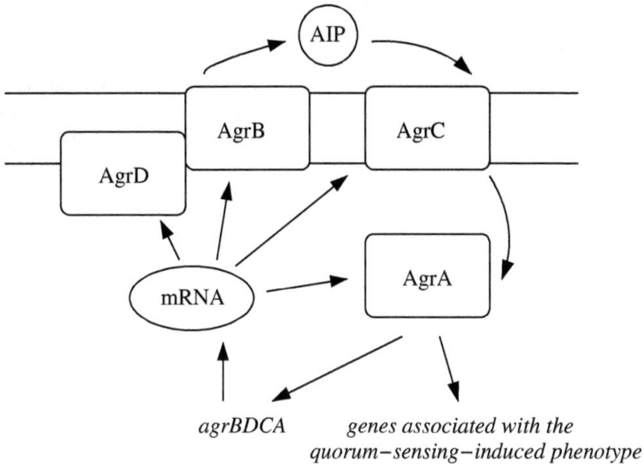

Fig. 1 Schematic representation of the *agr* system. *Arrows* illustrate the positive feedback loop. AgrB and AgrC are transmembrane proteins, while AgrD is anchored to the cell membrane

differential equation) model can be found in [7] and the simplifications which have been made to arrive at the model in this chapter can be identified from [9]. For conciseness, these details are omitted here. While we provide only the nondimensional model, the full dimensional model is easily extrapolated from this and, again, details can be found in [7]. The spatially structured model we consider is accordingly given by:

$$\frac{\partial M_i}{\partial \tau} = \frac{1}{\epsilon} \nu P_i - M_i + 1, \tag{26}$$

$$\frac{\partial A_i}{\partial \tau} = \lambda(M_i - A_i) - \eta A_i R_i^* + \epsilon \mu \eta A_P, \tag{27}$$

$$\frac{\partial B_i}{\partial \tau} = \alpha(M_i - B_i), \tag{28}$$

$$\frac{\partial S_i}{\partial \tau} = \lambda(B_i - S_i) - \epsilon k_s T_i S_i, \tag{29}$$

$$\frac{\partial T_i}{\partial \tau} = \lambda(B_i - T_i), \tag{30}$$

$$\frac{\partial a_i}{\partial \tau} = \frac{1}{\epsilon^2} \frac{k\beta\phi}{\eta} T_i S_i - \frac{1}{\epsilon} \beta R_i a_i + \frac{1}{\epsilon} \beta \gamma R_i^* - \lambda_a a_i + D \frac{(a_{i+1} - 2a_i + a_{i-1})}{h^2}, \tag{31}$$

$$\frac{\partial R_i}{\partial \tau} = \lambda(B_i - R_i) - \epsilon \frac{\eta}{\phi} R_i a_i + \epsilon \frac{\eta\gamma}{\phi} R_i^*, \tag{32}$$

Table 1 Variables representing elements of the *agr* operon. In addition, P represents *agr* activity levels (see text)

Variable	*agr* element
M	mRNA
A	AgrA (response regulator)
B	AgrB (processes the signal molecule)
S	Transmembrane AgrD (signal precursor)
T	Transmembrane AgrB
a	AIP (signal molecule)
R	Transmembrane AgrC (receptor)
R^*	Phosphorylated AgrC (AIP-bound)
A_P	Phosphorylated AgrA (activated)

$$\frac{\partial R_i^*}{\partial \tau} = R_i a_i - (\lambda + \gamma)R_i^*, \tag{33}$$

$$\frac{\partial A_{Pi}}{\partial \tau} = A_i R_i^* - (\lambda + \epsilon\mu)A_{Pi}, \tag{34}$$

$$\frac{\partial P_i}{\partial \tau} = A_{Pi}(1 - P_i) - uP_i, \tag{35}$$

in $0<x<1$, where $i = 0,\ldots,p$. All variables except P_i represent subcellular processes in a population of cells on a one-dimensional interval with the region divided into p equally spaced (of dimensionless size $h = 1/p$) compartments; we treat each compartment as containing a subpopulation of identically behaving cells. In the current context it is preferable to view i as labelling distinct subpopulations rather than individual cells; since it is not our intention here to explore the detailed biological implications of the model results, the distinction is in any case somewhat moot. In (31), D/h^2 captures the role played by υ^2 in Sect. 3, governing the rate of transport between compartments. $P_i(x,\tau)$ defines the quorum-sensing activity level of the population of cells in compartment i ($P \ll 1$ is quorum-sensing inactive, while $P \sim 1$ is fully active); see Table 1 for definitions of the other variables and Table 2 for the rates represented by the nondimensional parameters.

Focusing on signal accumulation, we assume there is no flux across the boundaries (i.e. signal molecules cannot be lost to, or gained from, the surrounding environment) yielding the conditions:

$$a_{-1} = a_1 \quad \text{and} \quad a_{p+1} = a_{p-1} \tag{36}$$

(see [9]). For each simulation we vary the initial conditions of the system, with different sections of cells beginning in an inactive (non-quorum-sensing) state, and others in an active (quorum-sensing) state. Hence we monitor diffusion of the signal molecules across the region, envision the resulting cell communication and investigate the ability of these cells either to retain their initial state or to pull the

Table 2 Interpretations of the nondimensional parameters

Nondimensional parameter	Interpretation
α	AgrB becoming transmembrane
β	AIP binding to receptors
η	Receptor loss through AIP binding
γ	Spontaneous separation of AIP and receptors
λ	Natural protein degradation
λ_a	Natural AIP degradation
μ	Housekeeping dephosphorylation of AgrA
ϕ	Activation of AgrA
D	Diffusion coefficient of AIP
k	AIP production
k_S	AgrD loss through AIP production
u	Unbinding of active AgrA from the DNA binding site
v	Ratio of activated *agr* transcription to basal transcription
ϵ	Ratio of basal to QS-induced transcription

other cells into the same state. To represent an already active cell, we use the following (dimensionless) initial conditions:

$$\begin{aligned} &M_i(x^*,0) = B_i(x^*,0) = S_i(x^*,0) = T_i(x^*,0) = 10, \\ &A_i(x^*,0) = R_i(x^*,0) = A_{Pi}(x^*,0) = 5, \quad a_i(x^*,0) = 100, \\ &R_i^*(x^*,0) = 30, \quad P_i(x^*,0) = 0.9, \end{aligned} \tag{37}$$

where x^* is the region(s) containing these cells. These initial conditions loosely characterise the up-regulated steady state of a cell in the spatially homogeneous case (this varies with diffusion rate). For inactive cells, we simply assume zero initial conditions for all variables in the relevant compartment(s).

4.3 Numerical Solutions

The discrete system is solved in Matlab v7.14 using the ODE15s solver. In the numerical solutions displayed, the interval is broken into $p = 50$ compartments (chosen for ease of illustration). To verify that the results are qualitatively reproduceable in the continuum limit, the equations have also been solved with larger numbers of discretisations.

To investigate multi-stability, we first consider the well-mixed non-spatial version of (26)–(35). We find that a key parameter, λ, can generate bistability in the system, see Fig. 2. λ captures the degradation rate of all intracellular proteins in the *agr* system. Low values of this parameter result in the preservation of the *agr* machinery to the extent that the response regulator (A_P) will always build-up, reach a critical level and trigger activation ($P \rightarrow 1$). Increasing λ renders this less and less likely to occur, eventually resulting in guaranteed downregulation. In

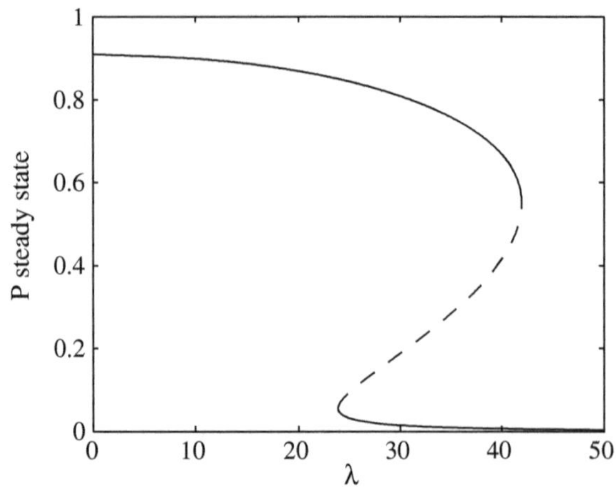

Fig. 2 Steady-state diagram for P when the model is well-mixed (e.g. when $D \to \infty$ and there is no spatial variation in the initial condition). Stable steady states are given by *solid lines* and unstable by the *dashed line*. Protein degradation rate is given by λ. A range of λ exists where the system is bistable: the cells will either reach an active or inactive state, depending upon the initial conditions of all variables. Outside this region, either activation (low λ) or inactivation (high λ) is guaranteed

between, there exists an interval of λ in which the cells can achieve either state, depending upon the initial conditions. For the following numerical simulations we consider two values to examine what can happen as you near either extreme of this region. All other parameters (except D which is varied for each simulation and stated in the corresponding caption) are taken to be unity and scaled with the small parameter $\epsilon = 0.1$ [as written in (26)–(35)] where necessary, as per [7].

4.4 $\lambda = 25$: Slower Protein Degradation Renders the Achievement of Quorum Sensing Activity Easier

We begin by considering $\lambda = 25$. On examination of Fig. 2, one can see that this falls to the lower end of the bistable region, meaning that the cells are more likely to be drawn towards an active state in the well-mixed scenario (this being initial condition dependent). In Fig. 3, several solutions are illustrated with varying initial conditions and diffusion rates. We depict the steady-state value of P across the full region, i.e. the final level of quorum sensing activity. For each subfigure, the number of compartments, n, initially selected to be quorum sensing active increases as we move through the plots (the exact number is given above each graph).

In Fig. 3a, the rate of signal molecule diffusion is so low that the whole interval of cells simply retain their initial states: insufficient "communication" occurs between them and they act as single entities. Increasing D in Fig. 3b (so that the signal molecules will spread out more readily) yields a different pattern which enables the visualisation of cell communication: the central cells are able to drag

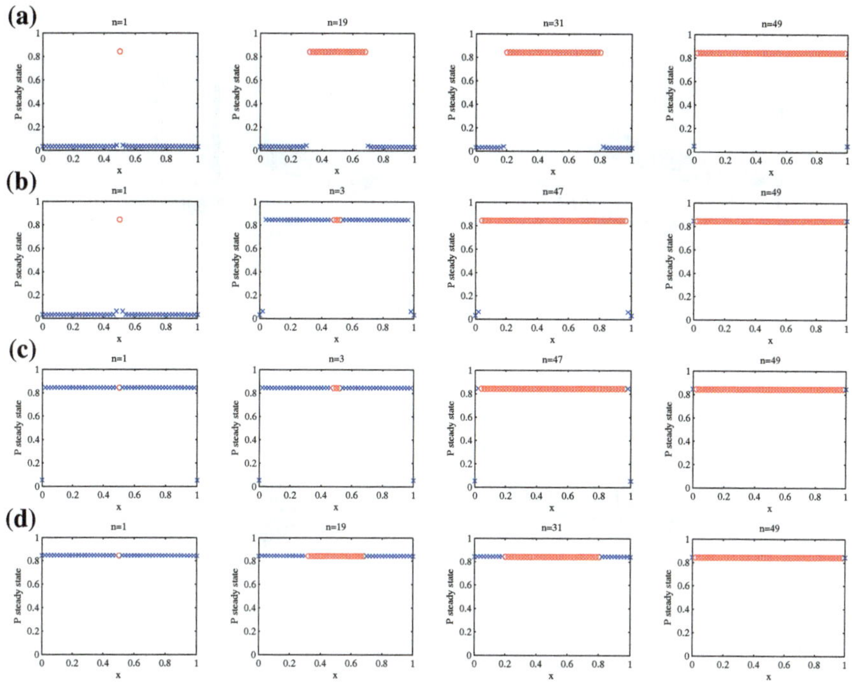

Fig. 3 The steady state of P, representing the final quorum sensing activity levels across the interval, with $\lambda = 25$. Higher values of $P \in [0, 1]$ represent compartments of cells which have been induced into an active state by the quorum sensing machinery and signal accumulation. n central compartments begin in an active state (*red circles*), as given by (37). $51 - n$ outer compartments have zero initial conditions for all variables (*blue crosses*). In **a** $D = 4 \times 10^{-5}$, **b** $D = 8.12 \times 10^{-5}$, **c** $D = 8.2 \times 10^{-5}$ and **d** $D = 8.4 \times 10^{-4}$. Thus as we move down the plots, the diffusion rate increases and as we move across, the number of compartments initially starting active increases. Complementary time-dependent solutions are illustrated in Fig. 4

the surrounding cells into an up-regulated state. Interestingly, for $3 \leq n \leq 45$, all compartments except two at either extremity become active, with the remaining four receiving insufficient signal for them to become active: the very outer compartments receive signal from only one direction; that these are therefore not up-regulated has a knock-on effect on the adjacent cells. The population is effectively divided into active and inactive populations at steady state because the subcellular processes are unable to even out sufficiently across the interval for the entire population to upregulate (i.e. to achieve quorum).

Successively increasing D first leads to only the very outer compartments remaining down-regulated and then to all the cells becoming quorum-sensing active, see Fig. 3c and d. Thus while the machinery is the same at the subcellular level, the phenotype of particular cells can be strikingly different, the effect being associated with the well-known phenomenon of wave pinning (failure of signal propagation) that occurs in such discrete systems (e.g. [15, 18]) but not their corresponding continuous limits.

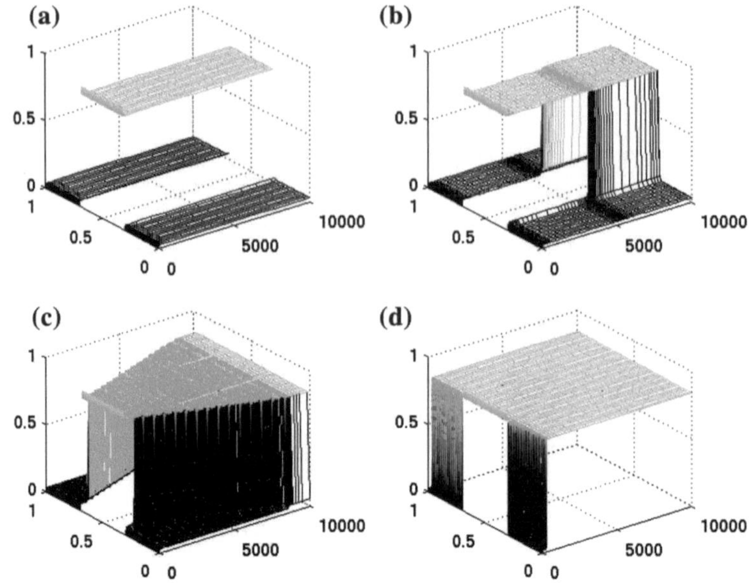

Fig. 4 Time dependent solutions when $\lambda = 25$ (these correspond with the steady state plots in Fig. 3), identical initial conditions for each are employed ($n = 19$ central compartments begin in an active state) and with various diffusion rates: **a** $D = 4 \times 10^{-5}$, **b** $D = 8.12 \times 10^{-5}$, **c** $D = 8.2 \times 10^{-5}$ and **d** $D = 8.4 \times 10^{-4}$. Moving through the plots, it is easy to see the effect of increasing the diffusion rate: more and more outside compartments are drawn into the up-regulated state by the central (initially active) compartments

In Fig. 4 some complementary time-dependent solutions are given for each value of D employed in Fig. 3. The time taken for the cells to evolve to their steady state varies significantly with diffusion rate and initial conditions.

4.5 $\lambda = 40$: Faster Protein Degradation Requires Greater Signal Accumulation for Quorum to be Attained

An alternative scenario is that protein degradation, compared to other reactions within the cell, is sufficiently high that the quorum-sensing machinery must work much harder to make one or more cells active, making the down-regulated state in some sense more stable than the active one (i.e. the converse case to that described in Sect. 4.4); enhancing protein degradation could provide a mechanism for preventing virulence, so scenarios such as that described here are of interest in analysing possible treatments for bacterial infections. Again, a number (n) of central compartments is chosen to begin in an active state, but this time the cells are more likely to be drawn into an inactive state by loss of signal molecules to neighbouring inactive cells.

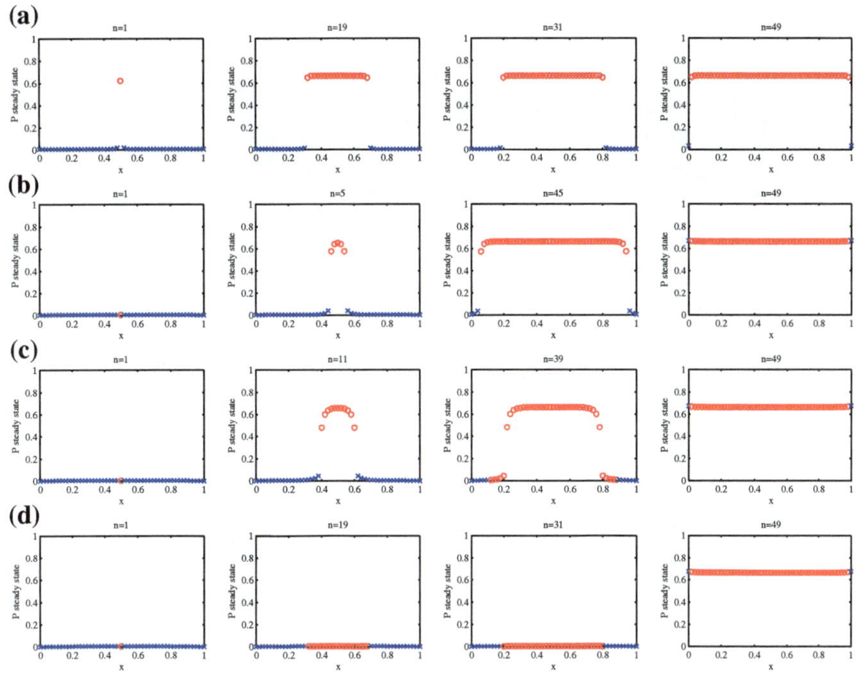

Fig. 5 The steady state of P, representing the final quorum sensing activity levels across the interval, with $\lambda = 40$. Compared with Fig. 3, this higher value of λ renders more signal accumulation required to achieve upregulation. n central compartments begin in an active state (*red circles*), as given by (37). $51 - n$ outer compartments have zero initial conditions for all variables (*blue crosses*). In **a** $D = 1 \times 10^{-3}$, **b** $D = 8 \times 10^{-3}$, **c** $D = 2.2 \times 10^{-2}$ and **d** $D = 2.3 \times 10^{-2}$. Thus as we move down the plots, the diffusion rate increases and as we move across, the number of compartments initially starting active increases. Complementary time-dependent solutions are illustrated in Fig. 6

Contrasting Fig. 5a and b we see that low diffusion (cutting off communication with adjacent cells) or sufficiently large n (making the initially active population larger) is required to retain (localised) upregulation. See also the time-dependent solutions in Fig. 6. These solutions are clean-cut reflections of the essence of quorum sensing: a small population in its own right can be active in the correct environment, but if this population migrates to a more open environment containing inactive cells, the associated signal dilution could imply that the subcellular mechanisms will force the smaller population to react to its neighbours and transition to near-identical behaviour.

For values of D within a certain range, we obtain some particularly interesting behaviour, however, in which the distinction between the central and outer cells can become much more blurred than in Sect. 4.4. In Fig. 5c it is evident that some of the central cells can lose their quorum-sensing activity, with those left in the centre effectively forming a new smaller population of active cells. Increasing the diffusion rate beyond this range means that (unless all but two compartments begin active) the

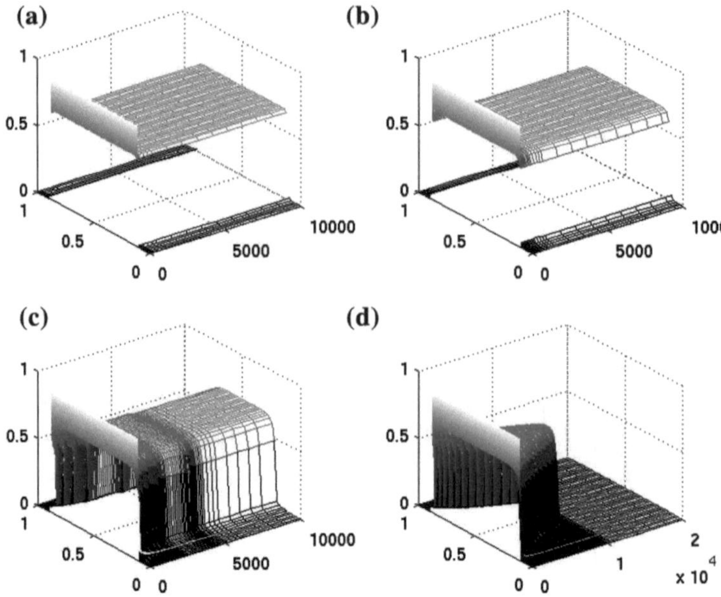

Fig. 6 Time dependent solutions when $\lambda = 40$ (so these correspond with the steady state plots in Fig. 5), identical initial conditions for each are used ($n = 39$ central compartments begin in an active state) and with various diffusion rates: **a** $D = 1 \times 10^{-3}$, **b** $D = 8 \times 10^{-3}$, **c** $D = 2.2 \times 10^{-2}$ and **d** $D = 2.3 \times 10^{-2}$. As the diffusion rate is increased, the signal molecules spread out further and, since greater signal accumulation is required than in Fig. 4 because the protein degradation rate is higher, when this arises the cells are all dragged into an inactive state

signal will diffuse out too quickly and any activity is lost right across the full interval, see Fig. 5d (in medical terms, this might represent a successful treatment).

5 Discussion

We have sought in the analysis above to demonstrate some of the explicitly multiscale behaviour that can arise in even the simplest spatio-temporal models describing gene-regulatory and signalling processes: we emphasise that capturing true biological complexity requires much more detailed, and hence involved, models, but contend that the phenomena which arise in our model problems are both of mathematical interest in their own right and instructive to gaining insight into the qualitative properties that are shared by more realistic models arising in integrative systems biology, but which may be obscured by the complexity of such models. Given the breadth of the field, we have chosen to focus on two particular classes of effect that are absent from the PDE models most commonly explored in classical mathematical biology, namely delay effects and, most extensively, the implications of the discreteness that is intrinsically associated with populations being made up of individual cells.[4] We conclude by revisiting some aspects of these phenomena and noting some obvious omissions.

Delay effects (Sect. 2) have a significant impact on the time dependence of the initial stages of the behaviour of an autoinductive process. Discreteness (Sect. 3) by contrast has little influence on the rate of upregulation but can affect significantly the extent to which initial quorum-sensing behaviour (say) is localised. Wave propagation and pinning (Sect. 4) make explicit how discreteness can block propagation of a switch in phenotype and hence whether or not an entire population may upregulate or downregulate in response to the state of a subpopulation.[5] It scarcely need be said that numerous potentially important effects have been ignored, even at the level of generic modelling (combinations of discreteness and delays, stochastic behaviour etc.): applications to systems biology of such multiscale spatio-temporal systems can be expected to continue to raise a wide variety of novel mathematical challenges.

Acknowledgments SJ thanks the MRC for funding in the form of a Special Training in Biomedical Informatics Fellowship and both authors the BBSRC for support under the SysMO and SysMO2 initiatives (COSMIC and COSMIC 2).

References

1. Cheung, J.K., Keyburn, A.L., Carter, G.P., Lanckriet, A.L., An Immerseel, F., Moore, R.J., Rood, J.I: The VirSR two-component signal transduction system regulates NetB toxin production in *Clostridium perfringens*. Infect. Immun. **78**, 3064–3072 (2010)
2. Cooksley, C.M., Davis, I.J., Winzer, K., Chan, W.C., Peck, M.W., Minton, N.P.: Regulation of neurotoxin production and sporulation by a putative agrBD signalling system in proteolytic *Clostridium botulinum*. Appl. Environ. Microb. **76**, 4448–4460 (2010)
3. Dold, J.W.: Analysis of the early stage of thermal runaway. Q. J. Mech. Appl. Math. **38**, 361–387 (1985)
4. Fujii, T., Ingham, C., Nakayama, J., Beerthuyzen, M., Kunuki, R., Molenaar, D., Sturme, M., Vaughan, E., Kleerebezem, M., de Vos, W.: Two homologous Agr-like quorum-sensing systems cooperatively control adherence, cell morphology, and cell viability properties in *Lactobacillus plantarum* WCFS1. J. Bacteriol. **190**, 7655–7665 (2008)
5. Henke, J.M., Bassler, B.L.: Bacterial social engagements. Trends Cell Biol. **14**, 648–656 (2004)
6. Hocking, L.M., Stewartson, K., Stuart, J.T., Brown, S.N.: A nonlinear instability burst in plane parallel flow. J. Fluid Mech. **51**, 705–735 (1972)
7. Jabbari, S., King, J.R., Koerber, A.J., Williams, P.: Mathematical modelling of the *agr* operon in *Staphylococcus aureus*. J. Math. Biol. **61**, 17–54 (2010)

[4] The quorum-sensing phenomena described above allow bacterial populations to function in some regards as multicellular organisms, and the types of qualitative phenomena that we have sought to classify are of course also relevant to plants, fungi and animals.

[5] Discreteness means that it is only meaningful for an entire cell or compartment, rather than an infinitesimally small part of one, to become upregulated, an effect that cannot of course be captured by continuum models.

8. Jabbari, S., King, J.R., Williams, P.: Cross-strain quorum sensing inhibition by *Staphylococcus aureus*. Part 1: a spatially homogeneous model. Bull. Math. Biol. **74**, 1292–1325 (2012)
9. Jabbari, S., King, J.R., Williams, P.: Cross-strain quorum sensing inhibition by *Staphylococcus aureus*. Part 2: a spatially inhomogeneous model. Bull. Math. Biol. **74**, 1326–1363 (2012)
10. Jabbari, S., Steiner, E., Heap, J.T., Winzer, K., Minton, N.P., King, J.R.: The putative influence of the *agr* operon upon survival mechanisms used by *Clostridium acetobutylicum*. Submitted to Math. Biosci. (2012)
11. Ji, G., Beavis, R., Novick, R.P.: Bacterial interference caused by autoinducing peptide variants. Science **276**, 2027–2030 (1997)
12. Kuttler, C., Hense, B.A.: Interplay of two quorum sensing regulation systems of *Vibrio fischeri*. J. Theor. Biol. **251**, 167–180 (2008)
13. Ma, M., Vidal, J., Saputo, J., McClane, B.A., Uzal, F.: The VirS/VirR two-component system regulates the anaerobic cytotoxicity, intestinal pathogenicity, and enterotoxemic lethality of *Clostridium perfringens* Type C isolate CN3685. mBio **2**, e00338-10 (2011)
14. Nakayama, J., Chen, S., Oyama, N., Nishiguchi, K., Azab, E.A., Tanaka, E., Kariyama, R., Sonomoto, K.: Revised model for Enterococcus faecalis fsr quorum-sensing system: the small open reading frame fsrD encodes the gelatinase biosynthesis-activating pheromone propeptide corresponding to staphylococcal AgrD. J. Bacteriol. **188**, 8321–8326 (2006)
15. Muratov, C.B., Shvartsman, S.Y.: Signal propagation and failure in discrete autocrine relays. Phys. Rev. Lett. **93**, 118101 (2004)
16. Nealson, K.H., Hastings, J.W.: Bacterial bioluminescence: its control and ecological significance. Microbiol. Rev. **43**, 496–518 (1979)
17. Novick, R.P.: Autoinduction and signal transduction in the regulation of staphylococcal virulence. Mol. Microbiol. **48**, 1429–1449 (2003)
18. Owen, M.R.: Waves and propagation failure in discrete space models with nonlinear coupling and feedback. Phys. D Nonlinear Phenom. **173**, 59–76 (2002)
19. Otto, M., Süßmuth, R., Vuong, C., Jung, G., Götz, F.: Inhibition of virulence factor expression in *Staphylococcus aureus* by the *Staphylococcus epidermidis agr* pheromone and derivatives. FEBS Lett. **450**, 257–262 (1999)
20. Otto, M., Echner, H., Voelter, W., Götz, F.: Pheromone cross-inhibition between *Staphylococcus aureus* and *Staphylococcus epidermidis*. Infect. Immun. **69**, 1957–1960 (2001)
21. Pèrez, P.D., Weiss, J.T., Hagen, S.J.: Noise and crosstalk in two quorum-sensing inputs of *Vibrio fischeri*. BMC Syst. Biol. **5**, 153 (2011)
22. Qazi, S.N.A., Counil, E., Morrissey, J., Rees, C.E.D., Cockayne, A., Winzer, K., Chan, W.C., Williams, P., Hill, P.J.: agr expression precedes escape of internalized *Staphylococcus aureus* from the host endosome. Infect. Immun. **69**, 7074–7082 (2001)
23. Redfield, R.J.: Is quorum sensing a side effect of diffusion sensing?. Trends Microbiol. **10**, 365–370 (2002)
24. Riedel, C.U., Monk, I.R., Casey, P.G., Waidmann, M.S., Gahan, C.G.M., Hill, C.: AgrD-dependent quorum sensing affects biofilm formation, invasion, virulence and global gene expression profiles in *Listeria monocytogenes*. Mol. Microbiol. **71**, 1177–1189 (2009)
25. Salmond, G.P.C., Bycroft, B.W., Stewart, G.S.A.B., Williams, P.: The bacterial 'enigma': cracking the code of cell–cell communication. Mol. Microbiol. **16**, 615–624 (1995)
26. Smits, W.K., Kuipers, O.P., Veening, J.-W.: Phenotypic variation in bacteria: the role of feedback regulation. Nat. Rev. Microbiol. **4**, 259–271 (2006)
27. Steiner, E., Scott, J., Minton, N.P., Winzer, K.: An agr quorum sensing system that regulates granulose formation and sporulation in *Clostridium acetobutylicum*. Appl. Environ. Microbiol. **78**, 1113–1122 (2012)
28. Waters, C.M., Bassler, B.L.: Quorum sensing: cell-to-cell communication in bacteria. Annu. Rev. Cell. Dev. Biol. **21**, 319–346 (2005)
29. Wuster, A., Babu, M.M.: Conservation and evolutionary dynamics of the *agr* cell-to-cell communication system across firmicutes. J. Bacteriol. **190**, 743–746 (2008)

Wound Healing: Multi-Scale Modeling

Fred J. Vermolen and Amit Gefen

Abstract This chapter is meant as an overview of our already published work that we carry out on modeling wound healing on the cellular, colony and tissue scale, though we detail the description of some stochastic principles that appear in our models. The relation between the scales is described in terms of the underlying biological and mathematical concepts. We also present the implications and applicability of the mathematical models studied.

1 Introduction

Wound healing is a very complicated process with the following partly overlapping phases: inflammation—proliferation—remodeling. During the post-bleeding inflammatory phase macrophages and white blood cells (leukocytes) enter the wound site to clear up invading harmful agents and bacteria through the broken network of capillaries. If a patient suffers from diabetes, then the capillary walls are suffering from an increased stiffness by which they can break down, and extend less due to a decreased flexibility, and thereby transport less blood containing oxygen and indispensable nutrients. Co-agulation of blood occurs to shut-off the wound. This is followed by angiogenesis, to restore the capillary network, dermal regeneration, which involves contraction due to traction forces exerted by

F. J. Vermolen (✉)
Delft Institute of Applied Mathematics, Delft University of Technology,
Mekelweg 4, 2628 Delft, The Netherlands
e-mail: f.j.vermolen@tudelft.nl

A. Gefen
Department of Biomedical Engineering, Tel Aviv University, Tel Aviv, Israel
e-mail: gefen@eng.tau.ac.il

Stud Mechanobiol Tissue Eng Biomater (2013) 14: 321–345
DOI: 10.1007/8415_2012_156

Published Online: 25 November 2012

(myo-)fibroblasts, as well as wound closure by the keratinocytes that form the basis of the epidermis (epithelium).

Many in-vitro experimental and clinical in-vivo studies have been carried out to scrutinize the biological mechanisms that take place during the very complex process of wound healing. Unfortunately, still many of the underlying biology is still unclear despite the long lasting research in wound healing. In order to improve and to prevent wounds, such as pressure ulcers or diabetic ulcers, it is important to quantify the influence of the related partial processes taking place during the healing of wounds. This quantification can be done using statistical analyses on raw data using for instance genetic algorithms or other forms of artificial intelligence such as neural networks. Since much data lack detailed quantitative aspects, this holds for in-vivo data in particular, mathematical modeling is also a very helpful tool for the quest of the interrelations between the parameters involved. The challenge is either to build a complicated mathematical model that contains as many of the biological parameters as possible, or to construct simple models that contain a minimum number of parameters such that only those parameters and processes that have the largest impact on the healing kinetics are taken into account. The first class of models will involve many biological parameters that need to be determined using complicated inverse modeling or any other type of regression analysis, in which the valid question arises whether the set of parameters determined is the actual solution or that one should take another combination of the parameters involved which reproduces the experiments (almost) equally well. In other words, the question of uniqueness arises in a natural setting. This concern is overcome by the construction of a simplified formalism of a certain (partial) biological process occurring in wound healing. In this paper, we will highlight the latter class of mathematical models: simplified models for partial processes occurring during wound healing. We will look at models designed for various scales and attempt to describe the relations between these models in terms of the underlying biology and mathematics.

Since wound healing involves basic biological processes like cell migration as a result of chemico-mechanical stimuli and random walk, cell proliferation and growth, cell differentiation, cell death, secretion and signaling of growth factors, we will incorporate many of these processes in a different way into the models at the various scales considered. To apply these processes, one basically considers the following mathematical approaches:

- Continuum-based partial differential equations involving transport (random walk, chemo-tenso taxis) and mechanical balances (visco-elasticity) on a tissue scale;
- Cellular scale involving discrete lattice models like the cellular Potts model, cellular automata models (involving a minimization of a virtual energy with a Monte-Carlo like scheme), or the continuous semi-stochastic approach by Vermolen and Gefen [1] and Byrne and Drasdo [2];

- Phenomenological models where the wound healing is modeled as a moving boundary problem where the interface moves as a result of a growth factor and local curvature.

The first approach involves very complicated models where many badly known biological input-parameters are needed. A big advantage is the fact that these models take relatively many biological parameters and subprocesses into account and that large tissue areas and large wounds can be modeled. This class of models can be applied to real-like in-vivo wounds of the order of centimeters or even larger. The domain of computation needs to be discretized to obtain a finite-element (or any other discretization) discretization in order to approximate the solution of the resulting boundary value problems formulated in terms of partial differential equations and its initial/boundary conditions. The parameter space and limited availability of appropriate values is a serious drawback of this class of models. An example concerns the availability of diffusion coefficients (i.e. random walk) or chemotactic coefficients or proliferation coefficients, see [3–6, 7–10, 11, 12, 13, 14, 15, 16] to mention a few of them. The second class of models only takes few parameters into account, but stays close to biology if one models in-vitro experiments. An extension to in-vivo cases is not straightforward since one typically will need to consider a large domain of computation and thereby making the number of cells or lattice points to be considered extraordinarily large. However, information from experiments concerning cell motility as a function of the acidity for instance, can be incorporated in a relatively straightforward manner. Examples are the studies presented in [1, 2, 17, 18, 19, 20]. The third model class takes few parameters as well, however, there is not much biology involved. An advantage of this class of models is, if the model has been set up in a clever way, that the small number of parameters involved can be adjusted such that experimental cases can be modeled in both in-vitro and in-vivo situations. See for the instance [21, 9, 22, 23].

In the manuscript, we will describe these classes of models and discuss their applicability. We will mainly focus on a recently developed continuity-based model from the second class on cellular level. Of course the models from the cellular automata-class, such as the cellular Potts model, can be positioned in the same kind of models. This continuous-based model mimics the migration of a collection of cells on a planar substrate, where we also take into account a bacterially infected zone where an increased acidity, resulting from the competition of cells and bacteria on oxygen and nutrients, impairs cellular mobility without the use of a predefined computational lattice. We will show some examples of simulations. In this model cell motion is a partly stochastic process. Cell death and cell division are modeled as stochastic processes. The original formulation of the model can be found in Vermolen and Gefen [1]. Furthermore, we will show some results from a newly constructed cell deformation model under the influence of chemotaxis. Finally, we address how the results from a small scale model can be used as input for a large scale model.

2 Mathematical Models: From Cell Scale to Tissue Scale

In this section, we consider some mathematical models at various scales where we introduce the models first to make the present manuscript complete. In the next chapter, we will describe the link between modeling at various scales in terms of the underlying biology and mathematics. In this chapter, we will mainly focus on cell migration, proliferation and death.

2.1 *The Cell Scale*

In this class of model, we consider the deformation during migration of individual cells. The cells are assumed to migrate as a result of a chemical signal. We bear in mind that the mathematical description of the influencing signal is generic and can easily be adapted and used to model cell deformation and growth as a result of a mechanical signal.

2.1.1 Random Walk: From Bacteria or Cells to Probability

The model can be applied to bacterial sources where individual bacteria make the surrounding tissue more acid by the effective production of biotic lactates as a result of the competition between the bacteria and cells for the available nutrients and oxygen, which make white blood cells move towards the infectious bacteria, or it can be applied, for instance, to the migration of fibroblasts or keratinocytes, among others, towards the wound region due to the signaling agents released by platelets that are in the coagulated area of the wound. In the case of modeling individual randomly moving bacteria, we use a random walk model with a stochastic differential equation based on Ito-processes. The model may also incorporate the bacteria in an upscaled way so that only bacterial densities are considered. First, we consider the individual random walk of bacteria. Then, in three dimensions, the equation of motion does not contain any deterministic drift, hence for the motion of the bacterium, we obtain

$$dX(t) = \sigma dW(t), \qquad dY(t) = \sigma dW(t), \qquad dZ(t) = \sigma dW(t), \qquad \text{for } t > 0, \tag{1}$$

subject to the prescribed initial bacterial condition $(X(0), Y(0), Z(0)) = (X_0, Y_0, Z_0)$, where the co-ordinate positions are independent. Here $W(t)$ is a Wiener process, or Brownian Motion such that the position of the bacterium is distributed normally with mean coordinates (X_0, Y_0, Z_0) and variance of $\sigma^2 t$ for each coordinate direction. Formally, the Wiener process satisfies the following requirements:

- $W(0) = 0$;
- The increments, $W(t_{k+1}) - W(t_k)$ and $W(t_k) - W(t_{k-1})$ are independent for any $0 \leq t_{k-1} < t_k < t_{k+1}$;
- For $0 \leq s \leq t$, the increment $W(t) - W(s)$ has the Gaussian distribution with mean 0 and variance $t - s$, i.e. $W(t) - W(s) \sim N(0, t - s)$.

Further, $W(t)$ is 'stochastically continuous' ($\lim_{t \to s} P(|W(t) - W(s)| > \epsilon) = 0$), where P stands for the probability. The formal analytic solution,

$$X(t) = X_0 + \sigma W(t), \qquad Y(t) = Y_0 + \sigma W(t), \qquad Z(t) = Z_0 + \sigma W(t), \quad \text{for } t > 0, \tag{2}$$

can be given, however the differential form is more useful in this study from a practical point of view. The equations are classically numerically solved using the Euler–Maruyama Method, given by

$$\hat{\mathbf{X}}_{i+1} = \hat{\mathbf{X}}_i + \sigma \Delta \mathbf{W}_{i+1}, \qquad \hat{\mathbf{X}}_0 = (X_0, Y_0, Z_0). \tag{3}$$

Here each component of $\Delta\mathbf{W}$ is a normally distributed stochastic parameter with zero mean and variance Δt, denoted by $N(0, \Delta t)$, and it can be proved that [24] each component v satisfies $\Delta W_{i+1}^v = W_{i+1}^v - W_i^v \sim N(0,1)\sqrt{\Delta t}$, in other words, a Gaussian distribution with zero mean and a variance of Δt. We show a run of the solution of the stochastic differential equations with one bacterium initially located at $(0,0,0)$ with mobility $\sigma = 2.6833 \cdot 10^{-5}$ m$/\sqrt{s}$. This value was chosen from [25] and corresponds to the classical bacillum in Fig. 1. Figure 1 shows the trajectory of the bacterium over time in three dimensions. Since Fig. 1 only gives one specific run, the trajectory itself is a stochastic parameter and hence for many purposes the probability density function is of more importance. To this extent, since $dW(t) \sim N(0,1)\sqrt{dt}$ and $W(t) \sim N(0,1)\sqrt{t}$, the probability density for the position of the bacterium at time t for each coordinate direction satisfies

$$f_v(t, v) = \frac{1}{\sqrt{2\pi\sigma^2 t}} \exp\left(-\frac{(v - v_0)^2}{2\sigma^2 t}\right), \qquad v \in (X, Y, Z). \tag{4}$$

Since the Brownian motion in each coordinate direction is an independent stochastic event, the multi-variate probability density is given by

$$f(x, y, z) = \frac{1}{(2\pi\sigma^2 t)^{\frac{3}{2}}} \exp\left(-\frac{(\mathbf{x} - \mathbf{X}_0)^2}{2\sigma^2 t}\right), \tag{5}$$

which solves the initial value problem in $\mathbb{R}^3$

$$\frac{\partial f}{\partial t} - \frac{\sigma^2}{2} \Delta f = 0, \qquad f(0, (x, y, z)) = \delta(\mathbf{x} - \mathbf{X}_0). \tag{6}$$

Here $\delta(\mathbf{x})$ represents the Dirac Delta Distribution in three dimensions, with characteristics

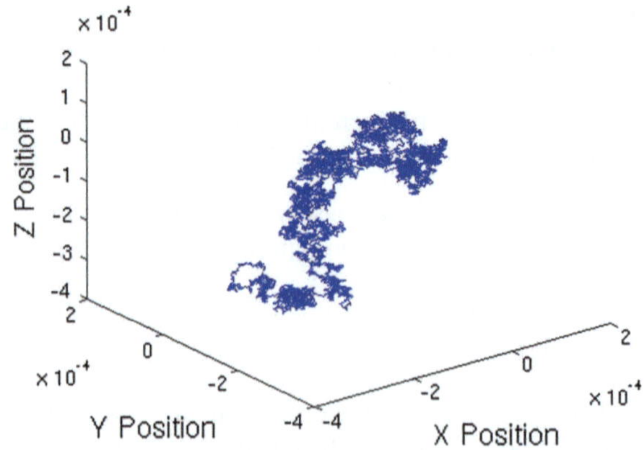

Fig. 1 The trajectory of a bacterium originally located at $(0, 0, 0)$ and moving according to Brownian motion with $\sigma = 2.6833 \cdot 10^{-5}$ m/$\sqrt{\text{s}}$, which corresponds to a bacillum at 37 °C

$$\delta(\mathbf{x}) = 0, \text{ for all } (x, y, z) \neq (0, 0, 0), \qquad \int_{\Omega \ni (0,0,0)} \delta(x, y, z) d\Omega = 1, \tag{7}$$

where Ω is subset of $\mathbb{R}^3$ with nonzero measure. For $\frac{\sigma^2}{2}$, which represents the diffusivity, we used $\frac{\sigma^2}{2} = 3.6 \cdot 10^{-10}$ m^2/s (bacillum at 37 °C). Note that if D represents the diffusivity of a species, then $\sigma = \sqrt{2D}$. Equation (5) represents a fundamental solution to the three-dimensional diffusion equation (in an unbounded domain), and it represents the probability density that the bacterium is localized at position (x, y, z) at time t. Note that Eq. (5) is very helpful in deriving the relation between the stochastic differential equation of Langevin type with zero drift, see Eq. (1) and the diffusion equation (12). The probability that a region Ω contains the bacterium at time t is then given by

$$P(t, \Omega) = \int_{\Omega} f(t, (x, y, z)) d\Omega, \tag{8}$$

and note that $\int_{\mathbb{R}^3} f(t, (x, y, z)) d\Omega = 1$ for $t \geq 0$. We remark that if drift is incorporated through $\mu = (\mu_x, \mu_y, \mu_z)$, then Eq. (1) becomes

$$dX(t) = \mu_x dt + \sigma dW(t), \quad dY(t) = \mu_y dt + \sigma dW(t), \quad dZ(t) = \mu_z dt + \sigma dW(t), \tag{9}$$

for $t > 0$, with exact solution, if μ_x, μ_y, μ_z and σ are constant,

$$X(t) = X_0 + \mu_x t + \sigma W(t), \quad Y(t) = Y_0 + \mu_y t + \sigma W(t), \quad Z(t) = Z_0 + \mu_z t + \sigma W(t), \tag{10}$$

It is easy to check the validity of the exact solution using Ito's calculus. The drift term could possibly result from chemotaxis or fluid flow and induces a temporarily shifting mean in the probability density, hence Eq. (5) is altered into

$$f(x,y,z) = \frac{1}{(2\pi\sigma^2 t)^{\frac{3}{2}}} \exp\left(-\frac{(\mathbf{x}-\mu t)^2}{2\sigma^2 t} \right). \tag{11}$$

It can be shown by the use of some elementary algebra that this function solves the Fokker–Planck equation

$$\frac{\partial f}{\partial t} + \nabla \cdot (\mu f) - \frac{\sigma^2}{2} \Delta f = 0, \qquad f(0,(x,y,z)) = \delta(\mathbf{x} - \mathbf{X}_0). \tag{12}$$

The above concepts are very standard and were, for the case of unbiased random walk, originally derived by Einstein to study Brownian motion of a particle. Note that we modeled the bacteria as point-sources so far. The extension to multiple bacteria, say n, is somewhat straightforward upon approximating the bacterial motion of each bacterium as independent stochastic processes. The probability follows from the binomial distribution that is used to compute the probability of k successes out of n trials where the probability of success is given by p. Since then the probability that a certain region, say Ω possesses $k \leq n$ bacteria is determined through

$$p(t,\Omega;k) = \binom{n}{k} (P(t,\Omega))^k (1 - P(t,\Omega))^{n-k}. \tag{13}$$

Hence the probability that this region Ω contains at least one bacteria is given by

$$p(t,\Omega;k \geq 1) = 1 - (1 - P(t,\Omega))^n \approx nP(t,\Omega), \tag{14}$$

where the last approximation is only accurate for $P(t,\Omega) \ll 1$. This approximation enables us to approximate the probability density function for n particles by $nf(t,(x,y,z))$ at those positions away from the initial bacterial positions. Note also that for $t > 0$ the probability density function(s) becomes finite at each position and that we can take the limit $\text{meas}(\Omega) \to 0$, to get an arbitrarily small probability as the volume considered tends to zero. Hence the approach can be extended to solving f in the case of a multi-bacterial environment under the application of the superposition principle for linear diffusion equations. These concepts can be used to model the bacterial density using the same partial differential equations. One can also evaluate a convolution over the domain of computation to get the bacterial density in case of a (piecewise) continuous initial bacterial distribution.

2.1.2 A Cell Deformation Model

In the literature, many models for cell deformation exist [26, 17], to mention a few of them. As far as we know, one of the major issues is that most of these models are based on the solution of partial differential equations, and often the level-set method is used to compute the position of the cell boundary, see for instance [26]. The level-set method requires the solution of a set of partial differential equations, such as the level-set function itself, whose zero level curve typically coincides with the cell boundary, the extension of the boundary velocity and a tedious re-initialization procedure which can be done by the fast-marching method based on the shortest-path optimization procedure, or via the solution of another nonlinear partial differential equation. Despite the enormous flexibility of the level-set method in terms of the ability to track interfaces also in cases where topological changes take place, the method is very expensive. Therefore, we choose to present a simpler method, which has been published only very recently in Vermolen and Gefen [17]. This model is based on the sensitivity of cells to a chemical and can therefore be applied to simulate cell migration and deformation as a result of chemotaxis. To this extent, the cell boundary, either in 2D or in 3D, is divided into gridnodes, which have the ability to move according to the gradient of the concentration of a certain chemical. This chemical could be a source of nutrition, oxygen, a growth factor or a poisonous chemical. Further, these points are connected to their neighbors and to the nucleus via springs, see Fig. 2 for a schematic representation. In this way, surface tension of the cell membrane and the connection between the membrane and nucleus via the ligaments in the cytoplasm are dealt with. First, we consider the modeling of the chemical sources and subsequently we consider the equations of motion of the points on the cell boundary.

To approximate the concentration of the chemical that gives raise to chemotaxis, we will use an approach based on Fundamental solutions of the diffusion equations in unbounded domains such as Eq. (5). For the release of the chemical

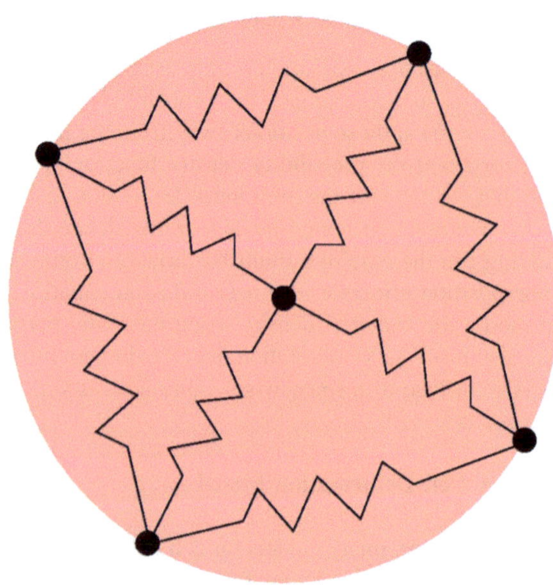

Fig. 2 A schematic of the distribution of springs that forms the backbone of the cell skeleton in the model

agent that attracts cells, we assume all sources to be very small compared to cell areas and therefore, we approximate these sources as *point-sources*. Here, we will assume that a point source is able to move, for instance via Brownian motion, but that the diffusion-field that surrounds it, sets in either instantaneously or gradually builds up in time. These point sources can correspond to either bacteria or to points on the cell boundary of other cells. First we consider the *instantaneous* diffusion field. For this purpose, we consider a point-source in 2D that moves according to a trajectory $(X(t), Y(t))$ moving under (biased) Brownian motion for instance [see Eq. (1) or (9)], for which we have for $(x, y) \in \mathbb{R}^2$ and $t > 0$

$$-D\Delta c = \gamma(t)\delta(x - X(t), y - Y(t)). \tag{15}$$

Here c denotes the concentration of the chemical, which diffuses with a diffusion coefficient D. Further, γ denotes the strength of the point-source, which may change in time as a result of being present or not being present, and $\delta(.)$ denotes the Dirac Delta Function. The solution to this differential equation is given by

$$c(t, (x, y)) = -\frac{\gamma(t)}{2\pi D}\ln((x - X(t))^2 + (y - Y(t))^2), \tag{16}$$

in $\mathbb{R}^2$, which can be found in textbooks like for instance [27]. For the 3-D case, we report that the Green's Function is given by

$$c(t, (x, y, z)) = \frac{1}{4\pi D||\mathbf{x} - \mathbf{X}(t)||}, \tag{17}$$

In the case of multiple, say n, sources, with intensities γ_j and positions $(X_j(t), Y_j(t))$, linearity of the diffusion equation allows us to use the superposition principle, to obtain

$$c(t, (x, y)) = -\sum_{j=1}^{n}\frac{\gamma_j(t)}{2\pi D}\ln(||\mathbf{x} - \mathbf{X}_j(t)||^2). \tag{18}$$

For a continuously distributed source-function $Q(t, (x, y))$ that is non-zero in $\Omega \subset \mathbb{R}^2$, we get the following convolution-based solution

$$c(t, (x, y)) = -\tfrac{1}{2\pi D}\int_{\Omega} Q(t, (\overline{x}, \overline{y}))\ln(||\mathbf{x} - \overline{\mathbf{x}}||^2)d\overline{\Omega}, \tag{19}$$

where the above integral is evaluated over $(\overline{x}, \overline{y})$. The 3D case can be treated analogously.

For the case of a transient diffusion field, we proceed analogously to the steady-state case with the application of delta-functions to deal with the point sources, then we arrive at

$$\tfrac{\partial c}{\partial t} - D\Delta c = \gamma(t)\delta(x - X(t), y - Y(t)), \quad \text{for } (x, y) \in \mathbb{R}^2, t > 0, \tag{20}$$

in which we assume that $(X(t), \mathbf{Y}(t))$ represents a certain trajectory. Initially, the concentration is assumed to be zero, and then the following solution for the 2-dimensional case, as a Green's Function, is derived according to the principles outlined in Evans [27]:

$$c(t,(x,y)) = \int_0^t \frac{\gamma(s)}{4\pi D(t-s)} \exp\left(-\frac{||\mathbf{x}-\mathbf{X}(s)||^2}{4D(t-s)}\right) ds, \tag{21}$$

for the two-dimensional case and

$$c(t,(x,y,z)) = \int_0^t \frac{\gamma(s)}{(4\pi D(t-s))^{3/2}} \exp\left(-\frac{||\mathbf{x}-\mathbf{X}(s)||^2}{4D(t-s)}\right) ds, \tag{22}$$

for the 3-D case, see also Eq. (5). Using this Green's Function, any solution with sources having a compact support, but non-zero measure can be constructed, or any initial condition, by the application of superposition arguments that result into a convolution. For completeness, we give the result for n discrete point sources at the points $(X_j(t), Y_j(t))$ and strength $\gamma_j(t)$ for $j \in \{1, \ldots, n\}$:

$$c(t,(x,y)) = \sum_{j=1}^{n} \int_0^t \frac{\gamma_j(s)}{4\pi D(t-s)} \exp\left(-\frac{||\mathbf{x}-\mathbf{X}_j(s)||^2}{4D(t-s)}\right) ds, \tag{23}$$

as well as for a 'continuous' source that lives in $\Omega \subset \mathbb{R}^2$, we get the following solution by the use of convolution

$$c(t,(x,y)) = \int_0^t \int_\Omega \frac{Q(s,(\overline{x},\overline{y}))}{4\pi D(t-s)} \exp\left(-\frac{||\mathbf{x}-\overline{\mathbf{x}}||^2}{4D(t-s)}\right) d\overline{\Omega} ds. \tag{24}$$

The treatment in $\mathbb{R}^3$ is fully analogous.

Next we consider the dynamics of the points on the cell boundary. Inertia is neglected in the present formalism. The computational domain may be given by a flat two-dimensional substrate, where we consider projections of cells or by a three-dimensional domain where cells move through extracellular matrix or a gel-like medium. We divide the circumference of the cell into N points. On each point, the cell detects a chemical signal and each point moves according to the concentration gradient that is constructed by a (sequence of) fundamental solutions. Further the direction of motion, as well as the velocity of the points are determined by the degree of deformation. To this extent, we use the following phenomenological law for the motion of the gridpoints on the cell boundary

$$\mathbf{v}_i = \beta \nabla c(t, \mathbf{x}_i) + \alpha(\mathbf{x}_c(t) + \hat{\mathbf{x}}_i - \mathbf{x}_i(t)), \text{ for } i \in \{1, \cdots, N\}, \tag{25}$$

where β denotes a mobility parameter of the cell boundary. This parameter is a measure for the deformation rate of the cell and also represents a measure of the sensitivity of the cell boundary to the concentration gradient. This β-term models chemotaxis. Further, the α-term models the 'desire' of the cell to attain its equilibrium shape and size. The parameter $\alpha > 0$ is a measure of the stiffness of the cell. The position of the point i, which depends on time, is denoted by $\mathbf{x}_i(t)$. The velocity of the point is denoted by $\mathbf{v}_i$. Further, the position of the cell center is given by $\mathbf{x}_c(t)$ and the initial positions of the boundary nodes minus the initial position of the cell center are denoted by $\hat{\mathbf{x}}_i$. The above equation guarantees that the velocity is directed towards the largest increase of the concentration, and that its magnitude depends on the magnitude of the concentration gradient. Note that in the case of repulsion, for instance due to a poison, the sign of the β-term should be reversed.

The movement of the cell boundary makes the cell deform and change its position. Furthermore, the cell area or volume changes as well. Since the cell consists of both fluids and solid polymeric matter, the cell is classically modeled as a visco-elastic medium. This means that the volume of the cell is not necessarily conserved. It is possible to inhibit volumetric changes by enlarging the α-parameter if the volume of the cell increasingly differs from the initial cell volume. The model is described in more detail in Vermolen and Gefen [17]. An example of a three-dimensional computation of the model is shown in Fig. 3. The input-data were the same as in Vermolen and Gefen [17], see Table 1.

In this figure it can be seen how a cell deforms and migrates to engulf the bacteria. Once the bacteria have been neutralized, the cell deforms back to its equilibrium shape. In [17], the model is extended to multiple cells where each cell secretes an agent that attracts the other cells. The model is based on the assumption that a cell registers the difference between the present concentration profile and the concentration profile from its own secretion. A repulsive force between gridnodes on different cells is introduced to prevent the cells from overlapping. The phenomenological relation of the repulsive force is inspired by the Lennard–Jones potential from electromagnetics. Since the medium through in which the cells deform is nonhomogeneous and anisotropic, a stochastic component is added to the equation via a Wiener process. This makes Eq. (26) stochastic:

$$d\mathbf{x}_i = \beta\nabla c(t,\mathbf{x}_i)dt + \alpha(\mathbf{x}_c(t) + \hat{\mathbf{x}}_\mathbf{i} - \mathbf{x}_i(t))dt + \sigma d\mathbf{W}(t), \quad \text{for } i \in \{1,\cdots,N\}, \quad (26)$$

where $\mathbf{W} = (W_x, W_y, W_z)$ is a vector with Wiener processes W_x, W_y and W_z and σ is a measure for the uncertainty (standard deviation) induced by nonhomogeneities of the medium. The first two terms are deterministic and hence represent classical drift. Some computed results with a stochastic contribution can be found in Vermolen and Gefen [17]. In this manuscript we only show a deterministic run in Fig. 3. In Fig. 4, we plot the times of engulfing a bacterium versus the cell stiffness and mobility. It can be seen that an increase of cell stiffness and/or a decrease of cell mobility delay the engulfment of bacteria. This computation can be used to quantify the influence of cell stiffening and motility decrease due to certain diseases. This simulation models the effectiveness of the immune response as a function of the properties of the immunity cells like white blood cells.

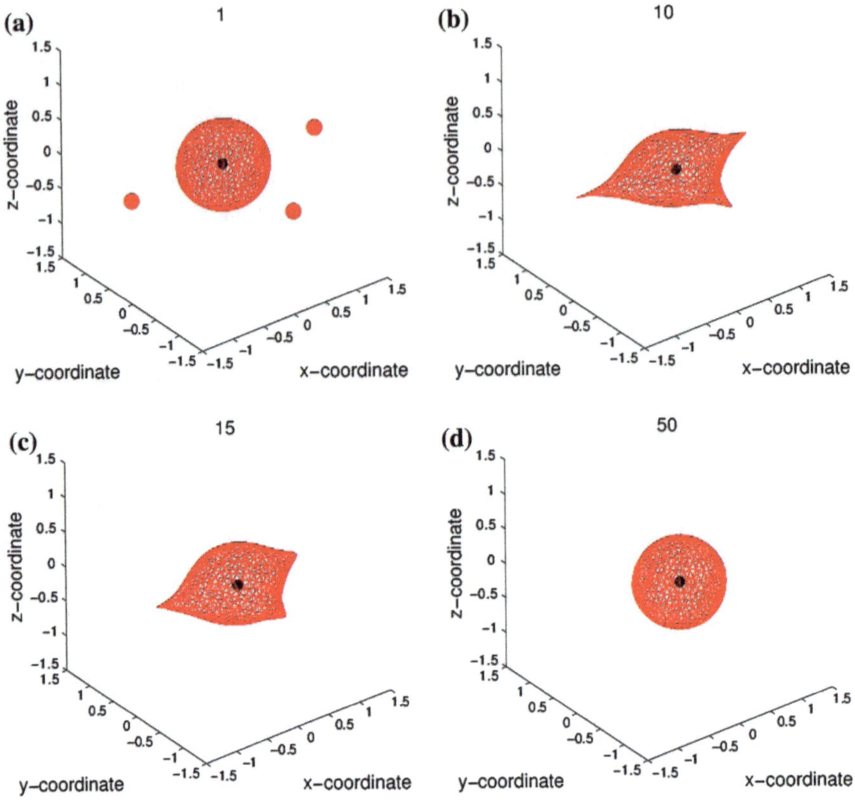

Fig. 3 Snapshots at consecutive times of a cell engulfing bacteria

Table 1 Values for the various parameters used

Parameter	Value	Unit
β	5	$mm^4/h/mol$
γ	1	$mol/mm^3/h$
ϵ	0.01	mm^6/h
D	$\frac{1}{4\pi}$	mm^2/h
α	1.5	1/h

2.2 The Colony Scale

To make the present manuscript as self-contained as possible, we repeat some of the equations presented in Vermolen and Gefen [1]. Presently, we are also extending the model to simulate infected cell colonies. To this extent, we consider a flat substrate on which cells are allowed to move. The projection of the cells onto the substrate is assumed to be circular. Upon moving, each cell exerts a traction

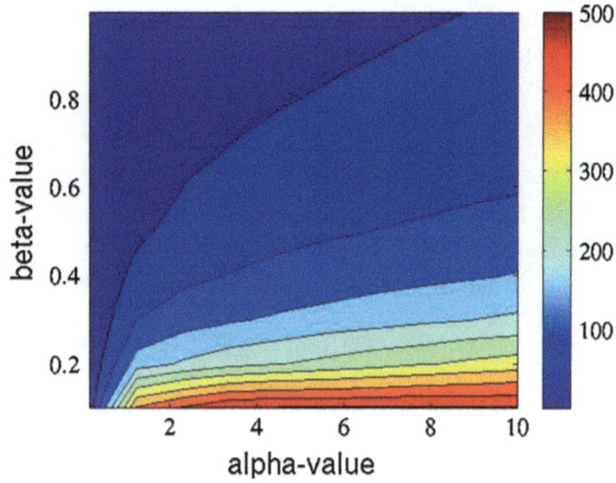

Fig. 4 The engulfment time versus cell stiffness, α, and cell mobility, β

force onto the substrate. This force generates a strain field around the cell that is sensed by the other cells if the strain energy density exceeds a certain threshold. In this way cells that are distant from each other sense each other's presence and hence these cells communicate with one-another, if they are not too far away from each other. This inter-cell communication over substrates through mechanical forces and sensing was experimentally observed by Byrne and Drasdo [2]. In this model, the deformation of the cells is not modeled, and cells are treated as circles with a constant radius R and hence only the coordinate positions of the cell centres need to be computed and stored. In the present manuscript, we will disregard the randomness in the motion of the cells. In [1] a random contribution to cell movement is introduced via a uniform probability distribution. One could improve this formulation through a standard normal distribution so that the stochastic component is built up by a Wiener process. Besides movement of cells, we also incorporate the basic biological processes like cell division and death. Despite the fact that cell division and death can be predicted more-or-less if the entire history of a cell is known, these two fundamental processes are modeled as stochastic processes. The reason is that the history of the cells is not known and that the circumstances, although modeled as idealized, are not known well. In the model, we consider $n(t)$ cells. Due to cell division and death the integer n depends on time. These cells are able to divide or die with respective probabilities p and q per unit of time. Further, the viable cells pull the substrate with a force $\hat{F}$, hence for cell i, we have

$$F_i = \begin{cases} \hat{F}, & \text{if the cell is viable} \\ 0, & \text{else} \end{cases} \tag{27}$$

The cells sense the strain energy density and the direction of the largest increase of this parameter, i.e. the gradient. Given a cell radius R and a Young's Modulus of the substrate E_s, the strain energy density as a result of cell i pulling on the substrate is computed by

$$M_i^0 = \frac{1}{2}\sigma\epsilon = \frac{1}{2}E_s(\mathbf{r}_i)\epsilon^2 = \frac{F_i^2}{2E_s(\mathbf{r}_i)\pi^2R^4}, \tag{28}$$

where the last step can be evaluated using some standard Hookean relations from mechanics, see [1]. In [1], we show by the use of finite-element simulations, that the strain energy density away from the cell can be approximated by

$$M_i(\mathbf{r}) = M_i^0 \exp\{-\lambda_i \tfrac{||\mathbf{r}-\mathbf{r}_i||}{R}\}, \qquad \text{for } \mathbf{r} \in \Omega, i \in \{1,\ldots,n\}, \tag{29}$$

where $\mathbf{r}_i$ represents the location of cell i, projected on Ω, further λ_i is a measure of how much the signal is attenuated, where we have

$$\lambda_i = \frac{E_s(\mathbf{r}_i)}{E_i}. \tag{30}$$

Here E_i represents the Young's Modulus of the cell. Using the additivity-property of the strain energy density, the strain energy density for a superposition of cells is given by

$$M(\mathbf{r}) = \sum_{j=1}^{n} M_j(\mathbf{r}) = \sum_{j=1}^{n} M_j^0 \exp\{-\lambda_j \tfrac{||\mathbf{r}-\mathbf{r}_j||}{R}\}. \tag{31}$$

Let $\mathbf{r}_i(t)$ denote the position of the cell center i at time t, then using this expression, the mechanical stimulus sensed by the ith cell is computed via

$$\begin{aligned} M(\mathbf{r}_i) &= \sum_{j=1}^{n} M_j(\mathbf{r}_i) = \sum_{j=1}^{n} M_j^0 \exp\{-\lambda_j \frac{||\mathbf{r}_i-\mathbf{r}_j||}{R}\} \\ &= M_i^0 + \sum_{j=1_{j\neq i}}^{n} M_j^0 \exp\{-\lambda_j \frac{||\mathbf{r}_i-\mathbf{r}_j||}{R}\}, \qquad \text{for all } i \in \{1,\ldots,n\}. \end{aligned} \tag{32}$$

Note that the cell's own contribution is also incorporated in this formula. Next, we repeat some of the formulas in Vermolen and Gefen [1] for the determination of the direction in which the cell is moving. The direction is determined by all the vectors connecting the other cells felt by the considered cell. The weight factors are given by the strength of the signal, in this case the strain energy density, experienced by the cell. This implies the following (deterministic) direction:

$$\mathbf{z}_i = \sum_{j=1_{j\neq i}}^{n} M_j(\mathbf{r}_i(t)) \frac{\mathbf{r}_j-\mathbf{r}_i}{||\mathbf{r}_j-\mathbf{r}_i||}, \text{ for all } i \in \{1,\ldots,n\}, \tag{33}$$

where a contributing term is mapped onto zero if $||\mathbf{r}_i - \mathbf{r}_j|| = 0$. The unit vector follows from the normalization:

$$\hat{\mathbf{z}}_i = \frac{\mathbf{z}_i}{||\mathbf{z}_i||}. \tag{34}$$

The velocity vector is constructed by the multiplication of this direction vector by the signal strength that is sensed by the considered cell to obtain:

$$\mathbf{r}_i'(t) = \alpha_i M(\mathbf{r}_i)\hat{\mathbf{z}}_i. \tag{35}$$

Here the cell velocity is modeled as instantaneous. In [18], we present a modification to incorporate the inertia effects into this formulation. The above equation can also be enriched with a stochastic contribution as is done in Vermolen and Gefen [1] using a uniform distribution or using Brownian Motion as in Eq. (9). Further, α_i is a velocity parameter with a dimension $\left[\frac{m^2 s}{kg}\right] = \left[\frac{m^3}{Ns}\right]$, determined by

$$\alpha_i = \left(\frac{F_i}{\hat{F}}\right)^2 \beta_i \frac{R^3}{f}. \tag{36}$$

Note that $\hat{F}$ is a property of the specific phenotype of the cell. The coefficient β_i, with unit s^{-1}, accounts for the mobility of the viable cell over the substrate surface. In [18], we incorporate the concentration of an infectious agent that typically results from bacteria. The cell-substrate friction effectively represents the averaged contribution of focal adhesions along the entire base of the cell without considering each localized connections of integrins. In [1], it is shown that α_i is determined by

$$\alpha_i = \frac{\beta_i R^3}{\mu \hat{F}^2} F_i, \tag{37}$$

where μ (=0.2 following [28]) denotes the cell friction coefficient. In [18], inertia is taken into account. Since this effect is known to be small, this effect is omitted in the present manuscript.

The cells will push each other away if they are too close together. This will give rise to repelling contact forces once these cells impinge elastically. The contact forces are due to the linear deformation of the cell bodies. In the present manuscript, the principles outlined in Gefen [28] are used. In [1], the derivation of the *invagination force* based on the work in Gefen [28] is given. In this manuscript, the final result for the strain energy density is given, which reads as

$$M^{ij} = \frac{6}{15} \frac{\sqrt{R^* E^* h^{\frac{5}{2}}}}{\pi R^3}, \tag{38}$$

where $h = \max(2R - ||\mathbf{r}_{ij}||, 0)$ is the indentation of cell i into cell j and vice versa. The final result for the total strain energy density function becomes

$$\hat{M}_i(\mathbf{r}) = M_i(\mathbf{r}) - M^{ij}, \tag{39}$$

Table 2 Input data

Quantity name	Symbol	Value	Unit
Substrate elasticity	E_s	5	kPa
Cell elasticity	E_c	0.5	kPa
Cell radius	R	4	μm
Cell traction force	$\hat{F}$	1	μN
Cell death probability	p	0.001	–
Cell division probability	q	0.005	–
Probability of velocity perturbation	p_{mp}	0	–
Cell mobility coefficient	β_i	$0.167 \cdot 10^{-3}$	s^{-1}
Initial relaxation parameter	κ	1000	s^{-1}
Friction coefficient	μ	0.2	–

where $\hat{M}_i$ and M^{ij} respectively denote the total strain energy density and the contribution to the strain energy density from the elastic interaction between neighboring cells. This quantity should be regarded as some energy relative to a certain energy level or as a potential in order to allow it to have negative values. For more details, we refer to [1]. The data that we use here can be found in Table 2.

We show some snapshots of a cell-colony simulation for 'wound healing' in Fig. 5. The red dots denote the cells that are moving towards each other as a result of mechanical pulling and their mechanical sensing. The snapshots at consecutive times show how the 'wound' closes. Further, we show the 'wound area' versus time in Fig. 6. It can be seen that first the 'wound' expands a bit and subsequently the 'wound' contracts. The curve shows a bit of noise that originates mainly from the randomness in cell division and cell death. The overall curve looks like a sigmoid relation. This is confirmed by in-vitro experiments on cell colonies. If one likes to model angiogenesis and its relation to wound healing, one could use a cellular automata model for instance and combine this model with the presently described model.

2.3 The Tissue Scale

In order to be able to perform simulations over larger volumes and areas of tissues, individual cells are no longer tracked. Instead, cell densities are considered. In other words, the number of cells per unit volume or area is considered. This approach gives a system of partial differential equations (PDEs) where densities of several cell types are considered. In [29], among many other studies, a PDE-model for cutaneous wound healing is considered in terms of tracking the densities of fibroblasts, endothelial cells (to model angiogenesis), and keratinocytes are considered. The right-hand side in the above PDE contains a logistic growth term to account for an increase of cell density towards an equilibrium (i.e. the undamaged state). Furthermore, the levels of oxygen, VEGF, and extra-cellular matrix are

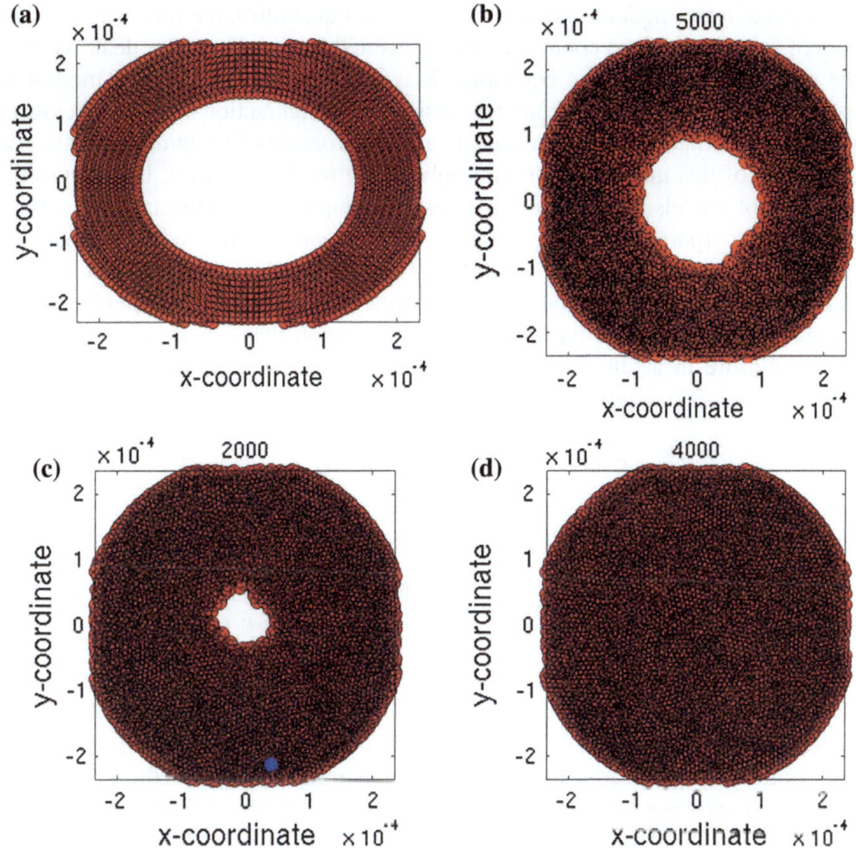

Fig. 5 Four snapshots of a closing gap: **a** the initial state, **b** after 4 h, **c** after 15 h, and **d** after 32 h

Fig. 6 The gap area as a function of time

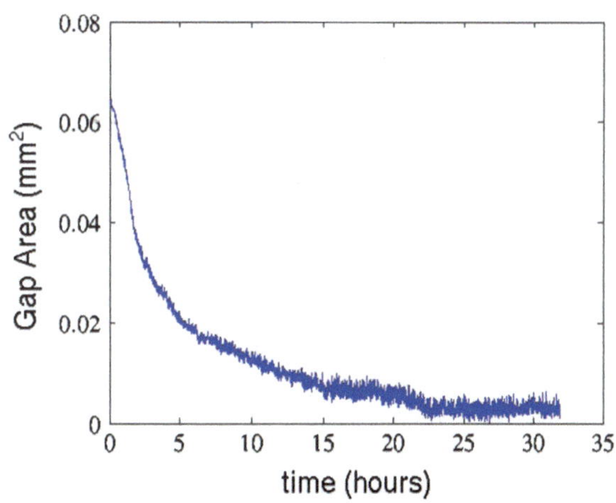

considered in the model. Next to these biological quantities, the local stress-strain pattern as a result of the contractile forces exerted by fibroblasts are dealt with by the solution of visco-elastic equations (Maxwell-model). This class of models is useful to model processes like angiogenesis, wound contraction and wound closure and this class has also be extended to model processes like tumor growth. An advantage of this model class is its applicability to in-vivo cases. Unfortunately, this class of models that contains a system of complicated nonlinear PDEs suffers from the incorporation of huge number of parameters which often are hard to measure. In this manuscript, we will not present a detailed model for wound healing, however, we will give a very simplified flavor of this model class by the consideration of a single partial differential equation that can be used to model wound closure in an in-vitro setting. To this extent, we consider the relatively simple Fisher–Kolmogorov equation, which reads as

$$\frac{\partial u}{\partial t} - D\Delta u = ku(1 - \frac{u}{u_0}), \qquad t > 0, (x, y, z) \in \Omega, \tag{40}$$

subject to some initial condition, that reads as

$$u(0, (x, y, z)) = u_0(x, y, z) = \begin{cases} u_0, & (x, y, z) \notin \Omega_w, \\ 0, & (x, y, z) \in \Omega \setminus \Omega_w. \end{cases} \tag{41}$$

Here u denotes the cell density, u_0 denotes the undamaged equilibrium cell density and Ω_w denotes the area of the initial wound. Furthermore, in the PDE cell motion is determined through random walk only and the right-hand side models growth of the cell population towards an equilibrium. A snapshot at 12.5 h of the cell density (number of cells per unit volume or area, normalized to unity) is shown in Fig. 7. We used an initial circular 'wound' of radius 1 mm, $D = 10^{-4}\,\text{mm}^2/\text{h}$, and $k = 0.7\,\text{h}^{-1}$. If one also encounters chemotaxis, then the cells move according to the concentration gradient of a generic chemical. Each cell moves according to the aforementioned concentration gradient, hence in the case of u cells per unit area or volume, the amount and direction of cell movement are determined by the concentration gradient of the chemical multiplied by the cell density u, which gives

$$\frac{\partial u}{\partial t} - D\Delta u + \nabla \cdot (u\beta\nabla c) = ku(1 - \frac{u}{u_0}), \qquad t > 0, (x, y, z) \in \Omega, \tag{42}$$

where β and c, respectively, denote the sensitivity and motility of the cells as a result of chemotaxis, and the concentration of the chemical that gives rise to chemotaxis. Here also a profile of the chemoattractant needs to be determined, which already increases the parameter space considerably. Note that the above equations corresponds to a drift term that is given by $\beta\nabla c$ in the stochastic counterpart. This equation models mobility of cells towards the concentration gradient of c, whereas reversing the sign would model mobility of cells away from the concentration gradient. In the case of a bounded domain of computation, then

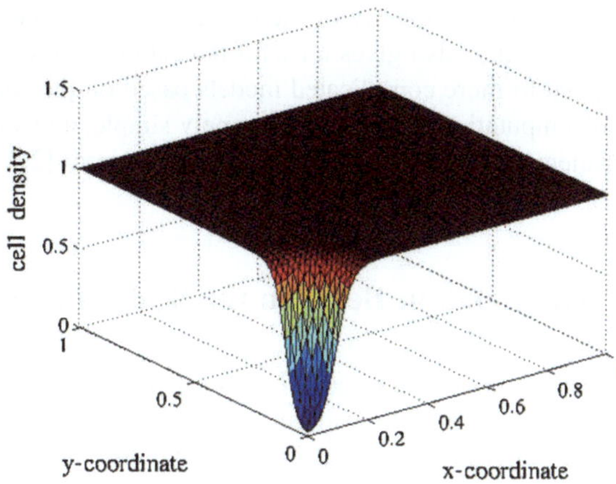

Fig. 7 A snapshot of the solution to the Fisher–Kolmogorov equation as an elementary model for wound healing

one needs to formulate a boundary condition, such as setting the cell density equal to the equilibrium undamaged cell density or by imposing a no-flux condition. The Fisher–Kolmogorov equation admits traveling wave solutions, see [30] for instance. Furthermore, in a bounded domain, one often solves this PDE by the use of discretization techniques, such as the finite-element method. The wound boundary, which moves in time, is classically tracked as a level-curve of the cell density. The choice of the value for the level-curve is somewhat arbitrary, however, it gives a good qualitative picture of the kinetics of wound healing predicted by this simplified version of the PDE-continuum based models. Using the principles that were outlined Sect. 2.1.1, one can also regard the cell density as a measure for the likelihood to encounter a cell at a certain position and time. If cells are considered as point masses, then the principles outlined in Sect. 2.1.1 are helpful. However, if cells get compressed, then one should incorporate the cell volumes or areas. In the case of 'supersaturated' cell colonies, the cells are compressed and thereby their sizes are small. On the contrary, the 'subsaturated' cell colonies, in general contain elongated cells and thereby the sizes are larger. In both cases the same portion of the computational domain can be considered and hence the likelihood to find a cell at a certain position and time could thereby be equal. Hence the probability measures nonlinearly with the cell density in case of relatively large cells which cannot be treated by the use of point sources. Therefore, it was argued by Vermolen et al. [21] that the likelihood proceeds nonlinearly and that the curve of the likelihood versus the cell density exhibits a concave-downward relationship. An alternative point of view is to use the solution u to represent the quality of the tissue at a certain position, where $u = u_0$ represents the undamaged quality which could be equal to unity. The quality of the overall tissue could be quantified by

$$Q(\Omega) = \frac{\int_{\Omega} u d\Omega}{u_0 Area(\Omega)}. \tag{43}$$

In this way, the quality of the tissue can be evaluated over time within a certain region, which also gives a useful indication of how the wound heals. This can be applied to more complicated models based on the solution of PDEs. We note that the computational model is extremely simple, and we refer for more complicated mathematical models to Vermolen and Javierre [29].

3 The Relation Between the Scales Involved

The three model classes give insights in their own way in their own scales. The cell-level models tell us how the shape of the cell actually behaves during the process of migration, which plays a major role in wound closure, wound contraction, but also in processes like tumor growth, as well as in the immune system. The measure of how much a cell can deform depends on its stiffness. Diseases that impair the cell stiffness (by for instance making the cell stiffer), will influence how the cell deforms when it is moving. In the case of the immune system, more elastic cells need to migrate over an only smaller distance to engulf a bacterium or any other harmful agent, see Fig. 4. This means that the bacteria or agents are neutralized within less time, and that less energy is consumed if the cell is flexible. In future studies, we will analyze the energy consumption of the immune system. Thereby, it can be concluded that the cell stiffness influences the efficiency of the immune system, next to the known parameters like the blood vessel stiffness and the number of white blood cells in the sense that if the cells are stiffer for some reason then the immunity response becomes less efficient. This holds for the immune system but also for the cells (for instance fibroblasts) that converge during processes like wound healing. The reason is that if cells converge to each other and if cells are relatively stiff, then it will take more time until cells are in physical contact. During the early stages of wound healing, flexible cells will be elongated as they move towards each other. Hence at the earliest stages at which the wound is closed, the cells are elongated if they are flexible. In the course of time as more cells have appeared due to cell division, the cells will get their cobble-stone shape. An example of a micrograph with different cell shapes in a cell colony is shown in Fig. 8.

If the cells are very stiff, then wound closure will be retarded since the cells are not able to elongate and hence the state of wound closure with elongated cells, which is the first stage of the wound being fully or partly closed, does not exist. Hence, the cell deformation model is very helpful in predicting the macroscopic closure rate of the wound. This issue will be investigated quantitatively in future.

The results from the cell colony model describe the nature how large numbers of cells converge, divide and die during processes like wound healing or tumor growth. In these models cell velocities, as well as cell division and death rates are used. These quantities are relatively easy to measure and thereby a good estimate of the wound healing kinetics can be obtained. Furthermore, the biological nature of wound healing is evaluated by monitoring the shape of the wound edge, and a

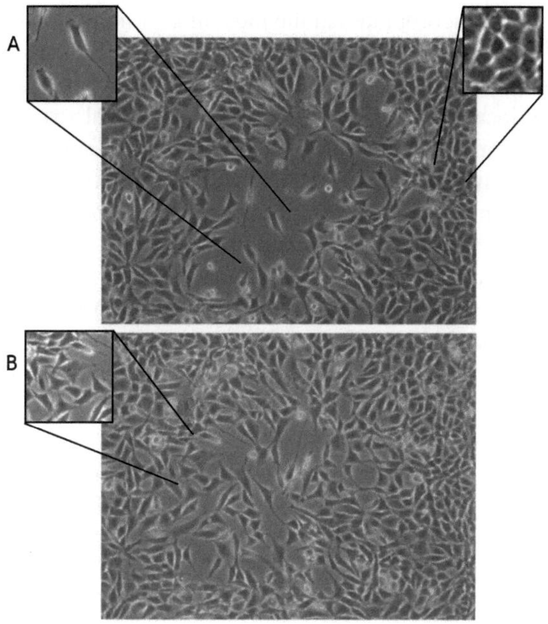

Fig. 8 Two time-sequence micrographs, taken 4 h apart, which demonstrate shape changes in NIH3T3 fibroblasts which cover a local damage site: **A** oval elongated shape of migrating cells. **B** Multi-polar cell shapes when cells are resting and well-spread at a sub-confluent density; note the 4-poles cell in the center of the magnified frame. **C** Polygonal cell shapes in dense confluent sites. The scale-bar represents 100 μm

close link to in-vitro experiments with one-layer cell colonies can be established easily. Note that in the cell deformation model we considered a chemical signal driving cell edge mobility. Without any complication, it is mentioned that the mathematical nature of this signal is very generic and hence that this signal can also represented by a strain energy density as in the cell colony model. This generic nature of course also applies for the cell colony model where we could use a chemical signal using the same mathematical principles. Both the cell-scale and colony-scale models can model processes like chemotaxis and tensotaxis.

The transfer of the information from the cell colony model and hence also from the cellular model, in a certain sense is done by considering the theory outlined in Sect. 2.1.1 in case of random walk with a well-defined drift component. It can be seen that the cell diffusivity is related to the average cell velocity. This, however, only holds for the treatment of cells as so-called point sources that move independently. Incorporating the cell areas or volumes will make the treatment more challenging. However, if the cells or bacteria are sufficiently small compared to the mean distance they travel over a certain amount of time, then the approach in Sect. 2.1.1 is quite reasonable. The incorporation of chemotaxis, or analogously tensotaxis, the amount of biomass per unit area or volume that moves over a certain distance within a unit of time is proportional to the number of cells per unit volume or area, that is the cell density, times the concentration gradient scaled by a factor of proportionality. This is how the classical linear version of the Keller–Segell model for chemotaxis results. The continuum-based scale allows the use of larger tissue areas and hence allows to consider realistic in-vivo wound sizes. The transfer from colony models to PDE models, also strongly depends on the modes that are incorporated in the modeling. If migration of cells is not only random, nor determined by any chemical signals, then, the tensotaxis could be modeled, which results into a completely different model where the diffusivity-tensor depends on the local strains. We note that the total local strain-tensor at any point within the domain of computation is determined by taking summing over the contributions of all individual cells. In a PDE-setting, one has to evaluate the cell density, which is the number of cells per unit of area or volume, and the forces that they exhibit. In our colony model, the strain energy density is evaluated as a result of adding the contributions of all individual cells that are present in the colony. To upscale this tensotaxis is by all means less trivial to carry out then upscaling processes like chemotaxis or random walk, and hence this is a challenge for future research. Though, it is tempting to upscale the models and many processes, though in some cases it is very difficult to incorporate all the information from the models. An example is the size of the cells in the colony models. The cell size certainly has an influence on the modeling outcomes in the sense that if large cells disappear then a relatively large gap arises. This enlargement of the cell radius makes the profile more prone to noise.

4 Modeling Several Processes in Wound Healing

The models that we considered here are very generic of nature and until here the presentation has mainly focused on wound closure or gap closure. Apart from gap closure, infections are very common to occur in clinical or real-world wounds. To this extent, we also plan to model bacterial infections, in which bacteria compete with the basis cells, such as fibroblasts or keratinocytes, on nutrients and oxygen, and thereby increase the tension of biotic lactates, which increase the acidity. The colony model has been extended with bacteria that move around and divide randomly. Bacterial motion is purely modeled as a Wiener process, by the use of equation (Sect. 2.1.1) with a certain division and death probability and release rate of biotic lactates. This chemical release is modeled by the use of Green's Fundamental solutions to the diffusion equation, and therewith in fact, the concentration of lactates as a result of a score of bacteria is determined using a superposition argument. A pilot study has been carried out in Vermolen and Gefen [18]. In this work, it is assumed that the cellular mobility decreases with increasing concentration of biotic lactates. A final conclusion of this work is that the decrease of motility causes gaps not to close anymore, hence the initial wound does no longer close entirely and that 'micro-gaps', which result from local cell death, and which would normally be occupied by newly appearing daughter cells from cell division, are no longer filled up due to decreased cellular motility. Hence the decrease of cellular motility leaves the gap open (for a long time) and can also be held responsible for the decrease of the quality of tissue. In principle white blood cells clear up contaminants and bacteria, and therefore we are working on a colony-model that also clears up the bacteria by introducing white blood cells into the model.

We extended the simulations that we showed here for the three-dimensional cells that deform and migrate to the case of white blood cells (leukocytes) that leave a small blood vessel to head for an infection to neutralize the bacteria present. This modeling is currently done by the use of colony models and cell deformation models. Here a translation to the use of PDEs for continuum models is also to be made. We are also in the process of doing this and the results will appear in future papers. A final stage is the remodeling stage where the tissue remodels to transform from a scarred state into the fully undamaged state. To simulate this remodeling process, which is important in the context of hypertrophic scar formation as a result of burns, both the cellular and PDE-based models will be very useful since fibroblasts having several properties due to various chemico-mechanical environments will be taken into account.

Until now, we described the modeling of several biological processes: cell division, cell migration (due to random walk, tensotaxis or chemotaxis). Immobile processes like maturation towards cell division or cell differentiation can be modeled in cell colonies like stochastic processes. In fact, if the entire history path in terms of the chemical and mechanical environment is known then the time at which the cell differentiates or divides is determined. This advocates for a deterministic approach for cell division or differentiation processes. An example of such a model can be found in the age-structured model by de Vries et al. [31] for the computation of age-distributions in population dynamics or Prokharau et al. [32] for the modeling of cell differentiation with a maturation space variable (which corresponds to complete differentiation whenever this variable is one and to a fully undifferentiated state whenever the value zero holds). The latter model also contains biological processes like cell migration and cell division. This modeling class is based on solving an advection equation for the cell density per unit of maturation and can be extended to the incorporation of the physical space to model cell migration. In real world situations, the entire history path of the cells is not exactly known or even hardly known. To this extent, the hypothesis of deterministic modeling is violated and one has to rely on stochastic processes. In the current paper we limited ourselves to modeling cell division as a purely random process. Probably it is more accurate to model cell differentiation by means of both a stochastic and deterministic component.

5 Conclusions

We presented a review of our ongoing work in simulation of wound healing on various scales. All scales give their own bits of information: The cell-based model for cell deformation can be used to analyze the shape changes a cell experiences under the influence of an attracting or repulsing chemical or under the influence of a local strain pattern. The cell-colony models can be used to look at the dynamics of large number of cells upon using just a few, well accessible parameters as a function of the local chemical condition of the substrate or tissue. Further, the third scale is based on the continuum-hypothesis and is hence based on (systems of) PDEs. These PDEs can be solved using discretization techniques such as finite-element techniques or discontinuous Galerkin methods combined with limiters when the equations are chemotaxis-dominated (or mathematically speaking, predominantly hyperbolic). The paper describes the relations between the various scales involved in terms of stochastic and deterministic relations.

References

1. Vermolen, F.J., Gefen, A.: A semi-stochastic cell-based formalism to model the dynamics of migration of cells in colonies. Biomech. Model. Mechanobiol. **11**(1–2), 183–195 (2012)
2. Byrne, H., Drasdo, D.: Individual-based and continuum models of growthing cell populations: a comparison. J. Math. Biol. **58**, 657–687 (2009)
3. Sherratt, J.A., Murray, J.D.: Mathematical analysis of a basic model for epidermal wound healing. J. Math. Biol. **29**, 389–404 (1991)
4. Filion, J., Popel, A.P.: A reaction diffusion model of basic fibroblast growth factor interactions with cell surface receptors. Ann. Biomed.Eng. **32**(5), 645–663 (2004)
5. Maggelakis, S.A.: A mathematical model for tissue replacement during epidermal wound healing. Appl. Math. Modell. **27**(3), 189–196 (2003)
6. Gaffney, E.A., Pugh, K., Maini, P.K.: Investigating a simple model for cutaneous wound healing angiogenesis. J. Math. Biol. **45**(4), 337–374 (2002)
7. Murray, J.D.: Mathematical Biology II: Spatial Models and Biomedical Applications. Springer, New York (2004)
8. Maggelakis, S.A.: Modeling the role of angiogenesis in epidermal wound healing. Discr.Cont. Syst. **4**, 267–273 (2004)
9. Adam, J.A.: A simplified model of wound healing (with particular reference to the critical size defect). Math. Comput. Model. **30**, 23–32 (1999)
10. Vermolen, F.J., Adam, J.A.: A finite element model for epidermal wound healing. In:Computational Science, ICCS 2007. Springer, Heidelberg, pp. 70–77
11. Olsen, L., Sherratt, J.A., Maini, P.K.: A mechanochemical model for adult dermal wound closure and the permanence of the contracted tissue displacement role. J. Theor. Biol. **177**,113–128 (1995)
12. Alarcon, T., Byrne, H., Maini, P., Panovska, J.: Mathematical modeling of angiogenesis and vascular adaptation. In: Paton, R., McNamara, L. (eds.) Studies in Multidisciplinary, vol. 3, pp. 369–387 (2006)
13. Vermolen F.J.: A simplified finite element model for tissue regeneration with angiogenesis. ASCE J. Eng. Mech. 135(5), 450–460 (2009)
14. Javierre, E., Vermolen, F.J., Vuik, C, van der Zwaag, S.: A mathematical approach to epidermal wound closure: model Analysis and Computer Simulations. J. Math. Biol.doi: 10.1007/s00285-008-0242-7; http://www.springerlink.com/content/w4j663334 5j7228k/fulltext.pdf (2008)
15. Tranquillo, R.T., Murray, J.D.: Continuum model of fibroblast-driven wound contraction inflammation-mediation. J. Theor. Biol. **158**(2), 135–172 (1992)
16. Pettet, G.J., Byrne, H.M., McElwain, D.L.S., Norbury, J.: A model of wound healing angiogenesis in soft tissue. Math. Biosci. **136**, 35–63 (1996)

17. Vermolen, F.J., Gefen, A.: A phenomenological model for chemico-mechanically induced cell shape changes during migration and cell–cell contacts. Biomech. Model. Mechanobiol. (2012, in press)
18. Vermolen, F.J., Gefen, A.: A semi-stochastic cell-based model for in-vitro infected 'wound healing through motility reduction. J. Theor. Biol. (Submitted, 2012), doi:10.1016/j.jtbi.2012.11.007
19. Groh, A., Louis, A.K.: Stochastic modeling of biased cell migration and collagen matrix modification. J. Math. Biol. **61**, 617–647 (2010)
20. CJones, G.W., Chapman, S.J.: Modeling growth in biological materials. SIAM Rev. **54**(1), 52–118 (2012)
21. Vermolen, F.J., Gefen, A., Dunlop, J.W.C.: In vitro 'wound' healing: experimentally based phenomenological modeling. Adv. Eng. Mater. **14**(3), B76–B88 (2012)
22. Vermolen, F.J., van Baaren, E., Adam, J.A.: A simplified model for growth factor induced healing of circular wounds. Math. Comput. Model. **44**, 887–898 (2006)
23. Vermolen, F.J., Javierre, E.: On the construction of analytic solutions for a diffusion-reaction equation with a discontinuous switch mechanism. J. Comput. Appl. Math. doi:10.1016/j.cam.2009.05.022 (2009)
24. Steele, J.M.: Stochastic Calculus and Financial Applications. Springer, New York (2001)
25. Kim, Y.-Ch.: Diffusivity of bacteria. Korean J. Chem. Eng. **13**(3), 282–287 (1996)
26. Neilson, M.P., MacKenzie, J.A., Webb, S.D., Insall, R.H.: Modeling cell movement and chemotaxis using pseudopod-based feedback. SIAM J. Sci. Comput. **33**(3), 1035–1057 (2011)
27. Evans, L.C.: Partial Differential Equations, Graduate Studies in Mathematics, vol 19, 2nd edn. Americal Mathematical Society, Providence **49**, 22–25 (1998).
28. Gefen, A.: Effects of virus size and cell stiffness on forces, work and pressures driving membrane invagination in a receptor-mediated endocytosis. J. Biomech. Eng. (ASME) **132**(8), 4501–4505 (2010).
29. Vermolen, F.J., Javierre, E.: A finite-element model for healing of cutaneous wounds combining contraction, angiogenesis and closure. J. Math. Biol. doi:10.1007/s00285-011-0487-4
30. Sachdev, P.L.: Nonlinear Diffusive Waves. Cambridge University Press, Cambridge (1987)
31. de Vries, G., Hillen, Th., Lewis, M., Müller, J., Schönflsch, B.: A Course in Mathematical Biology: Quantitative Modeling with Mathematical and Computational Methods. SIAM, Philadelphia (2006)
32. Prokharau, P.A., Vermolen, F.J., Garcia-Aznar, J.M.: Model for direct bone apposition on pre- existing surfaces during peri-implant osseointegration. J. Theor. Biol. **304**, 131–142 (2012)

Part IV
Cancer Research

Modeling Multiscale Necrotic and Calcified Tissue Biomechanics in Cancer Patients: Application to Ductal Carcinoma In Situ (DCIS)

Paul Macklin, Shannon Mumenthaler and John Lowengrub

Abstract Tissue necrosis and calcification significantly affect cancer progression and clinical treatment decisions. Necrosis and calcification are inherently multiscale processes, operating at molecular to tissue scales with time scales ranging from hours to months. This chapter details key insights we have gained through mechanistic continuum and discrete multiscale models, including the first modeling of necrotic cell swelling, lysis, and calcification. Among our key findings: necrotic volume loss contributes to steady tumor sizes but can destabilize tumor morphology; steady necrotic fractions can emerge even during unstable growth; necrotic volume loss is responsible for linear ductal carcinoma in situ (DCIS) growth; fast necrotic cell swelling creates mechanical tears at the perinecrotic boundary; multiscale interactions give rise to an age-structured, stratified necrotic core; and mechanistic, patient-calibrated DCIS modeling allows us to assess our working biological assumptions and better interpret pathology and mammography. We finish by outlining our integrative computational oncology approach to developing computational tools that we hope will one day assist clinicians and patients in their treatment decisions.

P. Macklin (✉) · S. Mumenthaler
Keck School of Medicine, Center for Applied Molecular Medicine,
University of Southern California, Los Angeles, CA, USA
e-mail: Paul.Macklin@usc.edu
URL: http://MathCancer.org

J. Lowengrub
Departments of Mathematics, Chemical Engineering and Materials Science,
and Biomedical Engineering, University of California, Irvine, CA, USA
URL: http://math.uci.edu/~lowengrb

Stud Mechanobiol Tissue Eng Biomater (2013) 14: 349–380 349
DOI: 10.1007/8415_2012_150

Published Online: 6 October 2012

1 Introduction

At its most basic level, cancer is a disease of uncontrolled cell proliferation: cancer cells, either through mutations or epigenetic alterations, overexpress oncogenes and underexpress tumor suppressor genes (TSGs). Consequently, the cells enter into and progress through the cell cycle more often than they should and disregard apoptotic signals, resulting in a net increase in proliferation and aberrant tissue growth. (See recent cancer biology tutorials for modelers and physical scientists, such as [50, 52, 53]). Accordingly, cell proliferation and apoptosis, along with genetic mutations and epigenetic alterations in genes controlling these processes, have been major foci of both basic cancer research and modeling. Most cancer therapies attempt to manipulate these processes either by cytostatic (suppressing entry to or progression through the cell cycle) or cytotoxic (inducing apoptosis: programmed cell death) mechanisms. For example, chemotherapy agents such as doxorubicin are considered to be cytotoxic [10]; therapies that target hormone-addicted cells (e.g., tamoxifen in estrogen-driven breast cancer) are considered to be cytostatic [74].

Key biological and clinical terms	
Basement membrane (BM)	(≈100 nm) thick plasto-viscoelastic membrane separating epithelial and stromal tissues
Extracellular matrix (ECM)	Fibrous supportive scaffolding in stromal tissue
Oncogene	A growth-promoting gene
Tumor suppressor gene (TSG)	A growth-inhibiting gene
Apoptosis	Well-regulated, programmed cell death
Anoikis	Apoptosis due to loss of attachment to the BM
Necrosis	Disordered cell death
Oncosis	Cell death at the start of (or preceding) necrosis, marked by rapid cell swelling
Adenosine triphosphate (ATP)	The immediate product of aerobic cell metabolism, and the "currency" of cell energy
(Apoptotic) caspase	Proteases responsible for degrading intracellular proteins during apoptosis
In situ carcinoma	Cancer contained by an intact BM
Ductal carcinoma in situ (DCIS)	An in situ precursor to invasive ductal breast cancer
Comedonecrosis	Necrotic tissue filling the lumen of a gland, most typically with intraductal breast cancers
Invasive ductal carcinoma (IDC)	An invasive breast cancer derived from ductal cells
Van Nuys Prognostic Index (VNPI)	A system for evaluating DCIS and guiding treatment

Necrosis—the disorderly death of cells due to rapid injury or energy depletion—has seen less attention in basic cancer research and computational modeling. Indeed, cancer apoptosis publications outnumber cancer necrosis in PubMed by over three to one after excluding tumor necrosis factor (TNF) citations that are more directly related to apoptotic signaling than necrosis. Many prominent mathematical models do not incorporate necrosis (e.g., [8]), while others generically model cell death while failing to differentiate between apoptosis and necrosis. For example, the recent ductal carcinoma in situ (DCIS) model in [83] provided an excellent model of cell death due to energy depletion, but the work did not differentiate this death process (necrosis) from death due to detachment from the basement membrane (anoikis). As we shall see below, apoptosis and necrosis take widely divergent courses, particularly in cases of DCIS that exhibit *comedo*necrosis (necrosis filling the lumen of a gland).

Those models that do include necrosis have often modeled it as an instantaneous or fast time scale process by immediately removing necrotic cells from the simulations (e.g., [2]). Others have modeled necrosis as simple volume loss terms in continuum models (e.g., [12, 89, 90]), or as inert, persistent debris in discrete models (e.g., [21, 71]). While these are more true to the generally longer time scale of necrosis, they still fail to account for the multiscale processes involved and their potential biomechanical impact on tumor progression. None of these or other prior works have examined calcification of necrotic debris.

And yet necrosis plays a prominent, essential role in many carcinomas. A 1 mm tumor spheroid with a typical 100 μm viable rim is over 50 % necrotic by volume. Cell death in such a significant fraction drastically alters mass transport throughout a tumor and can lead to steady size dynamics as proliferative cell flux out of the viable rim balances with fluid flux released by degrading necrotic cells [13, 50, 52]. See Fig. 1 (left). Necrosis has a proven prognostic value in breast cancer, particularly ductal carcinoma in situ (DCIS) [72, 92]: presence or absence of comedonecrosis is a prominent part of the Van Nuys Prognostic Index (VNPI) [84]. Moreover, DCIS is primarily detected as subtle patterns of calcified necrotic tissue in mammograms [27, 29, 82]. See Fig. 1 (right). 90 % of all cases of nonpalpable DCIS are detected and diagnosed on the basis of microcalcifications alone [69]. Prominent tissue necrosis is also observed in other cancer types and can similarly be an important prognostic indicator [76], such as in glioblastoma multiforme [1, 70] and colorectal cancer [77]. Secretions by necrotic cells may promote inflammation in neighboring "normal" tissue (tumor-associated stroma) [9, 24, 31], thereby promoting progression from in situ to invasive carcinoma [26, 37, 79].

In this chapter, we shall explore recent efforts by our modeling groups to shed light on the impact of necrotic tissue biomechanics on tumor progression through increasingly sophisticated computational modeling. After a brief introduction in Sect. 2 to the biological background of apoptosis, necrosis, and calcification, we examine our earliest continuum-scale modeling of necrotic tumor growth [51, 58–62] in Sect. 3. Continuum conservation laws describe the biomechanics, while smaller scales are integrated as constitutive relations. The work gave early and extensive insights on the impact of necrotic core biomechanics on tumor growth, characteristic features, sizes, morphology, and stability.

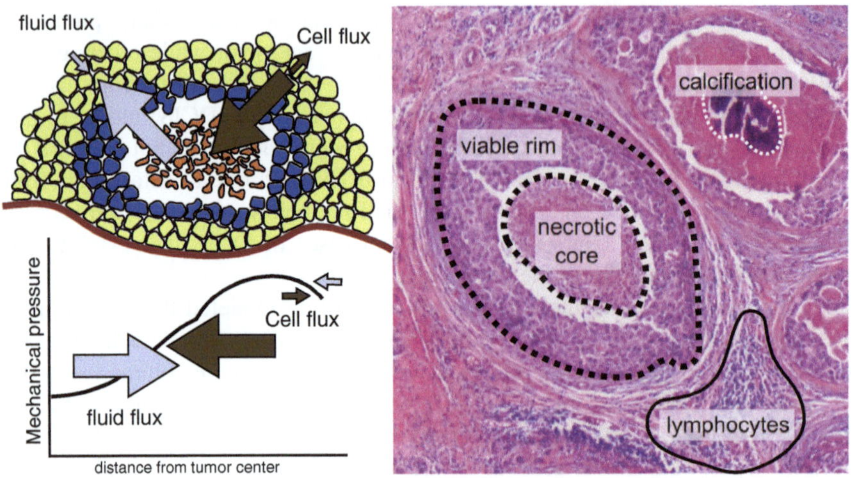

Fig. 1 *Left*: Proliferation in the viable rim (*yellow cells*) generates a cell flux (*dark gray arrows*) that can balance with fluid flux (*pale blue arrows*) created by lysing cells in the necrotic core (brown cell debris), resulting in steady tumor sizes. Adapted with permission from [52]. *Right*: Typical ductal carcinoma in situ (DCIS) duct cross-sections showing the outer viable rim, inner necrotic core, calcifications, and an inflammatory response. Adapted from [56] with permission

In Sect. 4, we present a multiscale agent-based cell model [56] by Macklin and colleagues and apply it to ductal carcinoma in situ. The model is the first to incorporate the vast range of time scales of necrosis and calcification; tissue-scale biomechanics emerge from interactions among time-varying forces, adhesion characteristics, and individual cell volumes. This work, which included the first patient-specific calibration to pathology, gave new mechanistic insights on the impact of multiscale necrotic and calcified tissue biomechanics on features observed in patient pathology and mammography. We conclude by discussing the next steps in multiscale modeling of necrotic and calcified tissues, and we outline our vision for the future of clinically-focused *integrative computational oncology*. It is our belief that integrative modeling will increasingly push the envelope to advance the state-of-the-art across biology, engineering, mathematics, computing, and the clinical sciences.

2 Biological Background

2.1 Basic Biology of Apoptosis

Apoptosis is a tightly-regulated, energy-consuming process [25, 36] that begins when intrinsic or extrinsic signals activate *initiator caspases* (e.g., Caspase-9) in the cytoplasm [25, 38]. This is generally regulated in one of two ways. In the first,

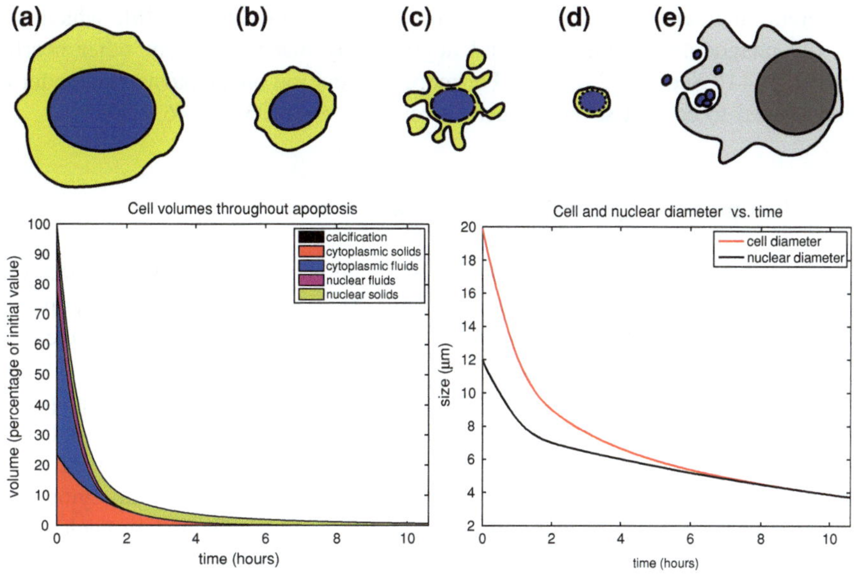

Fig. 2 Apoptosis schematic. *Top*: **a**, **b** While pro-apoptotic signals work to activate initiator caspases and then effector caspases to degrade subcellular structures and DNA, the cell rapidly shrinks by removing fluid. **c** The cell sheds its cytoplasm as membrane-encapsulated blebs. **d**, **e** Chromatin is condensed. DNA is fragmented, encapsulated into apoptotic bodies, and phagocytosed by nearby cells. *Bottom*: Preliminary simulation [64] of apoptotic cell volume composition (*left*) and nuclear/total diameters (*right*). Figures provided courtesy of [64]

mitochondrial membranes are permeabilized and release cytochrome c and other proteins into the cytoplasm to activate the initiator caspases. In the second, pro-apoptotic signals directly activate the initiator caspases independently of the mitochondria [38]. Mitochondria-regulated apoptosis disrupts ATP (energy) production by decreasing the mitochondrial membrane potential. The cell's remaining ATP store is depleted by energy-intensive processes throughout apoptosis [80]. See [65, 80] for greater detail on early regulation of apoptosis. While we do not describe them here, there are also caspase-independent apoptosis mechanisms [25, 38].

After apoptosis is initiated, various ion pumps on the cell's surface quickly remove water from the cell, resulting in significant volume loss [6, 14, 65, 67]. See Fig. 2 (top: a, b) and Fig. 2 (bottom). Indeed, cell shrinkage and separation from neighboring cells are some of the first visible signs of apoptosis in histopathology. The initiator caspases cleave and activate *effector* caspases (e.g., Caspase-3), which degrade cellular proteins [25, 38]. The cytoplasm collects in bulbous "blebs" that are shed from the cell. See Fig. 2 (top: c). These blebs surround cell protein fragments with intact membrane, and thus typically do not trigger inflammation [25, 44, 65].

In the nucleus, the chromatin condenses and is henceforth degraded by endogeneous endonucleases into short fragments of DNA [Fig. 2 (top: d)]. Protein cross-linking (e.g., by transglutaminase [32]) helps to bundle these fragments into coherent, membrane-encapsulated *apoptotic bodies* [4, 25, 44, 46], which are finally phagocytosed (ingested) and degraded by macrophages or other nearby cells [25, 44]. See Fig. 2 (top: e). For more information on apoptosis, the reader is encouraged to consult several excellent reviews (e.g., [25, 44, 46, 65, 87]).

2.1.1 Estimates of Apoptosis Time Scales

In [55–57], Macklin et al. analyzed the experimental biology and clinical literature to estimate the main apoptosis time scales. The overall duration of apoptosis was estimated at 8–9 h, with an approximately 2 h lag until detectable cleaved Caspase-3 activity, and an additional hour of lag prior to detection by TUNEL assay. Mumenthaler et al. are now directly measuring these and other apoptosis time scales with in vitro experiments on MCF-7 and related breast cancer cell lines [68]. In preliminary results, we observed cell water loss to be very fast: most water is lost within the first hour of apoptosis. We also observed that the cytoplasm blebbs and loses much of its volume within 3 h, leaving a degrading nucleus for the remainder of apoptosis. These preliminary observations are consistent with other experiments (e.g., [30, 88]), which estimated apoptosis to last 8–9 h [30], and measured rapid 60 % volume losses early in apoptosis [88].

2.2 Basic Biology of Necrosis and Calcification

In contrast to apoptosis, necrosis is a relatively energy-independent process, spanning a variety of time and spatial scales [46, 66]. In the context of cancer biology, necrosis is most frequently the result of cellular energy depletion, rather than a "planned" event [6]. Thus, while apoptotic cells generally appear sporadically as isolated, shrunken cells, necrotic tumor cells are found in large contiguous regions (i.e., necrotic cores) where oxygen and glucose are too low to sustain cell survival [44, 87]. Necrosis also differs from apoptosis in that it triggers an inflammatory response, due to the dysregulated release of intracellular proteins into the microenvironment [6, 46]. Indeed, inflammatory responses can readily be seen in pathology images near necrotic tumors; see Fig. 1 (right) for one such example.

In the early stages of necrosis (more properly called *oncosis* [46, 65]), energy depletion causes the cell's ion pumps to shut down, resulting in rapid swelling by osmosis. This swelling has traditionally been a key feature differentiating necrotic/on-co-tic cell death from apoptotic cell death in pathology and in vitro biology [44, 46, 65]. The cell swells to several times its original volume, lyses (splits open), and slowly leaks fluids and other protein contents into the surrounding microenvironment [46]. See Fig. 3 (top: a–c and bottom: left). Disintegrating lysosomes can release enzymes that help to further degrade the cell [6]. As in apoptosis, the nucleus displays some (irregular) chromatin condensation and

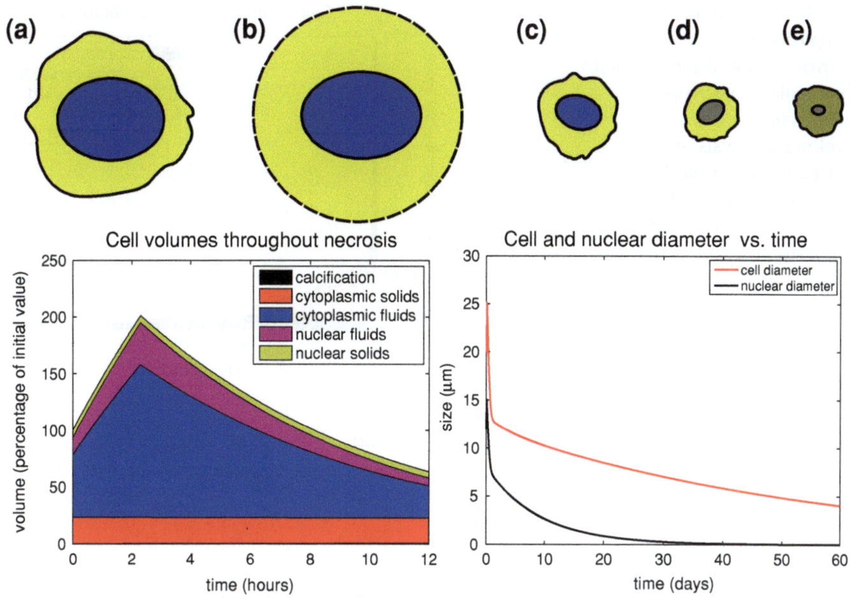

Fig. 3 Necrosis schematic. *Top*: **a**, **b** The necrotic cell swells until its membrane splits and leaks its fluid content **c**. Its remaining nuclear solid content degrades **d**, and the remaining cytoplasm degrades and is sometimes calcified **e**. *Bottom*: Preliminary simulation [64] of early necrotic cell volume composition (*left*) and nuclear/total diameters (*right*). Figures courtesy of [64]

shrinkage (pyknosis). However, the DNA is not cleaved into regularly-sized fragments, nor is it encapsulated into apoptotic bodies. Instead, it remains and degrades over time, eventually rupturing and dissipating into the remaining cytoplasm. See Fig. 3 (top: d and bottom: right). In many tumors, necrotic tissue is removed by infiltrating macrophages; see the mathematical modeling of this process (and corresponding references) in [73]. We note that this brief overview of early-to-mid necrosis is a simplification, and the lines between apoptosis and necrosis can be blurred. For example, apoptotic bodies that are not cleared can become necrotic [44, 65], and while necrosis is seemingly "passive", it involves numerous significant biochemical processes [7, 46]. Excellent reviews of necrosis can be found in [6, 7, 46, 66, 87].

2.2.1 Dystrophic Calcification

In DCIS, the necrotic core is separated from immune cells (and the stroma) by an intact basement membrane, preventing the removal of necrotic material. Instead, it remains and continues to degrade. In this and other cancers where the necrotic material is not cleared but rather persists for long periods of time, the necrotic core can undergo *dystrophic calcification* [40, 47]. In this process, calcium ions interact with remaining phospholipids in the necrotic cell (the membrane, vesicles, etc.) to

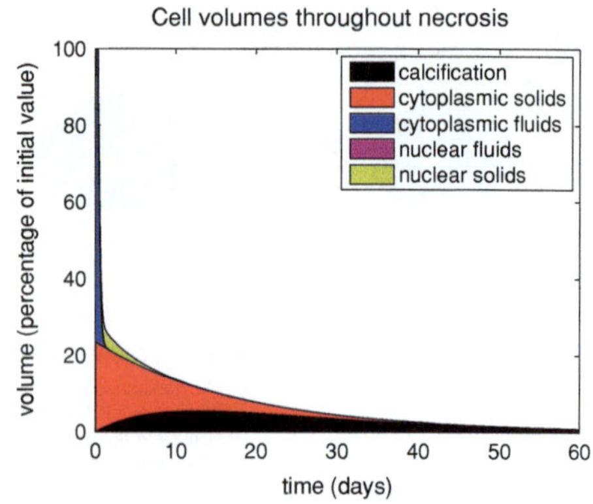

Fig. 4 Cell volume composition during later necrosis: Viewed on the long time scale of necrosis and calcification, early cell swelling and lysis are a fast perturbation on the longer-term trends of solid content calcification and degradation. Figure provided courtesy of [64]

nucleate and grow calcium phosphate crystals [47]. See Fig. 3 (top: e), the example in Fig. 1 (right), and Fig. 4. How this process transpires in vivo is still poorly understood.

2.2.2 Estimates of Necrosis and Calcification Time Scales

In [55–57], Macklin et al. estimated the various time scales of necrosis and calcification, many of which have not been experimentally measured for carcinoma. We estimated initial cell swelling and lysis to occur on the order of 2–6 h. Based upon experimental reports on aortic calcification [34, 42, 48] and our previous computations [55–57], we estimate calcification to take on the order of 15–20 days [56]. Based upon our insights from [56] (See Sect. 4.3) and the existence of necrotic tissue with intense eosin staining (a sign of cytoplasm with significant water loss and little calcification) and compact, partly-degraded nuclei, we estimate that water loss occurs more quickly than pyknosis, and that pyknosis is a faster process than calcification. We have recently hypothesized and found good evidence that calcifications degrade at a very long time scale (on the order of two to three months) [56]. See Sect. 4.5. Thus, necrosis and calcification have processes that operate on time scales ranging from hours to months.

3 Early Continuum Modeling Results: Impact of Necrotic Core Mechanics on Tumor Progression, Morphology, and Stability

Following earlier tumor growth models that included necrotic cores [12, 89, 90] and an earlier non-necrotic free boundary formulation of tumor growth [18], Macklin and Lowengrub developed a model of non-symmetric avascular tumor growth in heterogeneous tissues which included necrosis [51, 58–62]. We modeled the tumor as an incompressible fluid (with constant cell density) moving in a porous medium—the ECM. We used a sharp interface approach, where $\Omega(t)$ denoted the moving tumor volume with boundary $\Sigma(t)$; we denoted the surrounding host tissue by Ω_{H}. In [60], we set Ω_{H} to enclose Ω in an $L \approx 100 - 200\,\mu$m ring of tissue:

$$\Omega \cup \Omega_{\mathrm{H}} = \{\mathbf{x} : |\mathbf{x} - \mathbf{x}_{\mathrm{center}}(t)| \leq R(t) + L\}, \tag{1}$$

where

$$R(t) = \max\{|\mathbf{x} - \mathbf{x}_{\mathrm{center}}(t)|, \mathbf{x} \in \Omega(t)\}, \tag{2}$$

and where $\mathbf{x}_{\mathrm{center}}$ is the center of mass of $\Omega(t)$. We scaled space by L (the nutrient diffusion length scale) and time by a mechanical relaxation time scale $\lambda_{\mathrm{R}}^{-1}$. The time is rescaled in all plots to correspond to the cell mitosis time scale $\lambda_{\mathrm{M}}^{-1} \approx 24$ h. See [51, 58–60] for more details.

We introduced a single nondimensional "nutrient" σ which was required for cell survival and drove growth. The nutrient was released by the host vasculature at $\partial(\Omega \cup \Omega_{\mathrm{H}})$, diffused through the non-vascularized nearby host tissue Ω_{H} to the tumor, and was then consumed by tumor cells in Ω. Following [18] and as described in [58], we make the quasi-steady assumption: nutrient transport and consumption occur on much faster time scales than cell proliferation and tissue deformation, and so on the time scale of simulation, $\partial\sigma/\partial t \approx 0$. Thus, σ satisfies

$$\begin{aligned} 0 &= \nabla \cdot (D_{\mathrm{H}} \nabla \sigma) && \mathbf{x} \in \Omega_{\mathrm{H}} \\ 0 &= \nabla \cdot (D_{\mathrm{T}} \nabla \sigma) - \sigma && \mathbf{x} \in \Omega \end{aligned} \tag{3}$$

subject to boundary and matching conditions

$$\begin{gathered} [\sigma]_{\Sigma} = 0 \qquad [D\nabla\sigma \cdot \mathbf{n}]_{\Sigma} = 0 \\ \sigma(\mathbf{x})\big|_{\partial(\Omega_{\mathrm{H}} \cup \Omega)} = 1, \end{gathered} \tag{4}$$

where for any $\mathbf{x} \in \Sigma$, the jump function $[f(\mathbf{x})]_{\Sigma}$ is defined as

$$[f(\mathbf{x})]_{\Sigma} = \lim_{\Omega \ni \mathbf{y} \to \mathbf{x}} f(\mathbf{y}) - \lim_{\Omega_{\mathrm{H}} \ni \mathbf{y} \to \mathbf{x}} f(\mathbf{y}). \tag{5}$$

In [60], $D_{\mathrm{T}} = 1$ as a result of nondimensionalization. The nutrient is used to implicitly define viable and necrotic regions (Ω_{V} and Ω_{N}, respectively) of the tumor:

$$\begin{aligned} \Omega_{\mathrm{V}} &= \{\mathbf{x} \in \Omega \text{ such that } \sigma(\mathbf{x}) \geq \sigma_{\mathrm{N}}\} \\ \Omega_{\mathrm{N}} &= \{\mathbf{x} \in \Omega \text{ such that } \sigma(\mathbf{x}) < \sigma_{\mathrm{N}}\}, \end{aligned} \tag{6}$$

where σ_{N} is the necrotic threshold value of σ. Note that $\Omega = \Omega_{\mathrm{V}} \cup \Omega_{\mathrm{N}}$.

Within the tumor's viable rim, cells were assumed to proliferate at a rate proportional to σ and apoptose at a constant background rate. In Ω_{N}, the model degraded necrotic debris and released volume, acting as a biomechanical stress relief. We assumed the host tissue was in homeostasis (proliferation and apoptosis were in balance), but cells and tissue could be displaced by forces generated by the tumor.

The tissue moved with velocity $\mathbf{u}$ in response to forces generated by cell proliferation and death. Under the incompressibility and constant cell density assumptions, the local rate of volume change is given by $\nabla \cdot \mathbf{u}$. In dimensionless form,

$$\begin{array}{ll} \nabla \cdot \mathbf{u} = 0 & \mathbf{x} \in \Omega_{\mathrm{H}} \\ \nabla \cdot \mathbf{u} = G(\sigma - A) & \mathbf{x} \in \Omega_{\mathrm{V}} \\ \nabla \cdot \mathbf{u} = -GG_N & \mathbf{x} \in \Omega_{\mathrm{N}}, \end{array} \tag{7}$$

where G, A, and G_{N} are dimensionless parameters characterizing the rates of cell proliferation, apoptosis, and necrotic tissue volume loss relative to the time scale $\lambda_{\mathrm{R}}^{-1}$. See [60] for greater detail on the nondimensionalization and these parameters.

We introduced a dimensionless proliferation-generated mechanical pressure p as a simplified model of tissue stress, and assumed a Darcy flow response: cells can respond to the pressure by overcoming cell–cell and cell-ECM adhesion and moving through the porous medium (the ECM) supporting the cells. Moreover, the ECM itself can deform in response to p. Hence, $\mathbf{u} = -\mu \nabla p$, where μ is the tissue mobility (its ability to respond to pressure gradients). Assuming constant cell–cell adhesive forces and cell density throughout Ω_{V}, cell–cell adhesion can be modeled as a surface tension proportional to the curvature κ along $\Sigma(t)$. Thus, as in [18],

$$\begin{array}{lll} -\nabla \cdot (\mu_{\mathrm{H}} \nabla p) = 0 & & \mathbf{x} \in \Omega_{\mathrm{H}} \\ -\nabla \cdot (\mu_{\mathrm{T}} \nabla p) = \left\{ \begin{array}{l} G(\sigma - A) \\ -GG_{\mathrm{N}} \end{array} \right. & \begin{array}{l} \\ \\ \end{array} & \begin{array}{l} \mathbf{x} \in \Omega_{\mathrm{V}} \\ \mathbf{x} \in \Omega_{\mathrm{N}} \end{array} \end{array} \tag{8}$$

subject to boundary and matching conditions

$$\begin{array}{c} [p]_{\Sigma} = \kappa \, [\mu \nabla p \cdot \mathbf{n}]_{\Sigma} = 0 \\ p(\mathbf{x})\big|_{\partial(\Omega_{\mathrm{H}} \cup \Omega)} = 0. \end{array} \tag{9}$$

In [60], $\mu_{\mathrm{T}} = 1$ as result of nondimensionalization.

We implicitly tracked the moving boundary position using the level set method: an auxiliary distance function ϕ satisfies $\phi < 0$ in Ω, $\phi > 0$ in Ω_{H}, $\phi = 0$ on Σ, the outward normal vector is given by $\mathbf{n} = \nabla \phi$, and $\kappa = \nabla \cdot \mathbf{n}$. The outward normal velocity of $\Sigma(t)$ is obtained by evaluating $\mathbf{u} \cdot \mathbf{n} = -\lim_{\Omega \ni \mathbf{y} \to \mathbf{x}} \mu_{\mathrm{T}} \nabla p(\mathbf{y}) \cdot \mathbf{n}$ for any $\mathbf{x} \in \Sigma$. The motion of Σ then becomes an advection equation for ϕ [51, 58–62]. We solved Eqs. (3–4) and (8–9) using a second-order accurate ghost fluid method [51, 58–62]. We let $D = D_{\mathrm{H}}/D_{\mathrm{T}}$ and $\mu = \mu_{\mathrm{H}}/\mu_{\mathrm{T}}$ denote the relative oxygenation and mechanical compliance of the surrounding host tissue, respectively.

3.1 Impact of Necrotic Core Biomechanics: Key Results

As in earlier tumor spheroid models [12, 89, 90] and early non-symmetric necrotic tumor simulations in [93], our theoretical and numerical analyses [51] showed that even with $A = 0$, volume creation in the proliferative rim could balance with

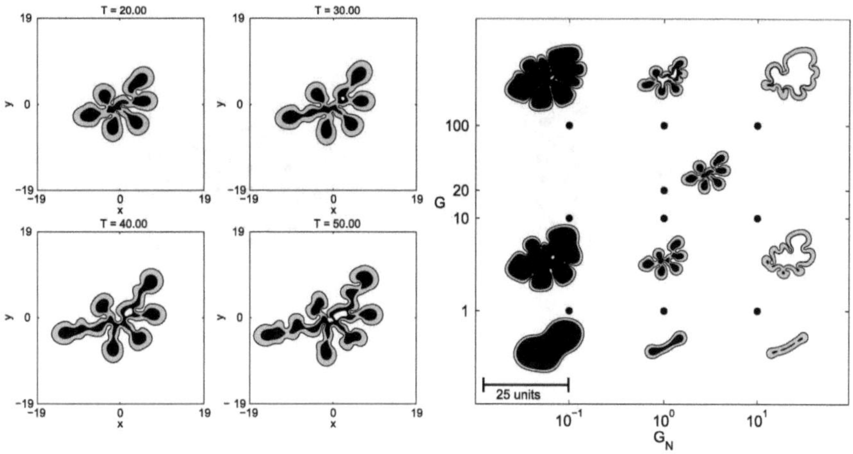

Fig. 5 *Left*: Growth of a necrotic avascular tumor into well-perfused, mechanically-stiff tissue. The invasive "fingers" and necrotic regions acquire relatively fixed, characteristic sizes. *Right*: Impact of the rate of cell proliferation (G) and necrotic volume loss (G_N) on invasive fingering growth. G acts primarily as a time scale (tumor morphologies are the same but evolve more quickly with increased G), whereas larger values of G_N can destabilize the morphology (seen here as changing rounded protrusions into invasive fingers). *Legend*: Viable (*gray*) and necrotic tissue (*black*) grow in host tissue (*white*). Figures adapted with permission from [60]

volume loss in the necrotic core, leading tumor spheroids to grow to a steady size. Our work had an additional insight: even during growth (and overall morphological instability), tumor proliferation and necrotic volume loss could equilibrate *locally*, leading to (1) a near-constant necrotic volume fraction, and (2) the emergence of characteristic feature sizes and shapes. For example, a tumor growing into well-perfused ($D > 1$), mechanically-stiff ($\mu \leq 1$) tissue develops invasive fingers with a characteristic width. See Fig. 5 (left) for such an example.

The qualitative tumor behavior (classified as fragmenting, fingering, or hollow/compact growth) was primarily dependent upon the microenvironmental parameters D and μ. However, the quantitative behavior—viable rim thick-ness, necrotic volume fraction, overall growth rate, etc.—was strongly dependent upon tumor cell characteristics, particularly the necrosis parameters σ_N and G_N. The viable rim size was determined by the balance of nutrient penetration into the host tissue (D), apoptosis (A), and the tumor cells' resistance to hypoxia (σ_N). The size of the necrotic core was primarily determined by the rate of volume loss in necrotic tissue (G_N). See Fig. 5 (right), where we show how the tumor varied with G_N for several values of G. A key finding was that while moderate rates of necrotic volume loss indeed contribute to the emergence of a steady state size for the spherical case, *fast* necrotic volume loss (large G_N) can destabilize the tumor morphology.

This work revealed a few outstanding problems with continuum necrosis models of the time. First, defining the necrotic region implicitly through σ as in Eq. (6) could cause unexpected behavior for complex tumor morphologies.

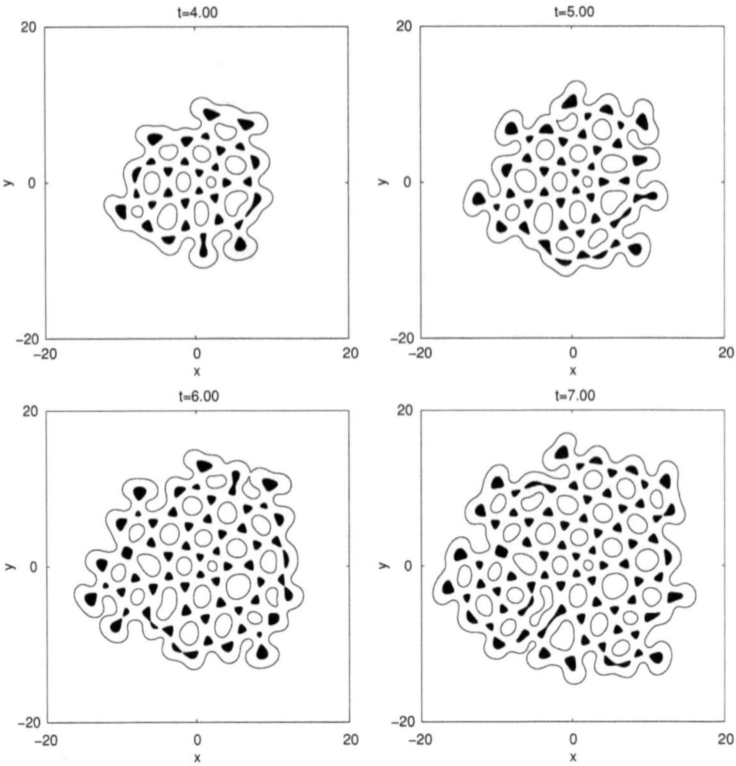

Fig. 6 A simulation of repeated encapsulation of host tissue by a growing tumor [58]. *Legend*: *White* viable tumor. *Black* necrotic tumor. Figure adapted with permission from [58]

Particularly unstable tumors could experience frequent connection and disconnection of invasive fingers or bulbs [51, 58, 60]. (See Fig. 6 for an example of repeated encapsulation of host tissue). Connection or reconnection of invasive fingers or bulbs lead to rapid depletion of nutrient in the newly encapsulated host and tumor tissue, leading to a jump in necrosis. Subsequent disconnection would rapidly reperfuse the encapsulated regions, leading to the condition where $\sigma > \sigma_N$ in previously necrotic tissue. This necrotic tissue would "come back to life"—an impossibility. We solved these problems by introducing an additional level set function ϕ_N to separately track the necrotic core boundary [61, 62].

Second, because the continuum model linked together many biophysical effects into very few parameters (much to the benefit of mathematical analysis!), it was difficult to directly calibrate the model to experimental measurements. Model calibration required force-fitting the parameters to match experimental growth rate measurements, and then tuning the remaining parameters to match the simulated morphologies (as informed by parameter space investigations) to clinical or other observations (e.g., as in [33]). While this makes data-driven simulations possible, it can hinder the acceptance of mathematical modeling in the biological and clinical communities, who are concerned that complex models with too many free parameters can be tuned to any desired behavior without necessarily being biologically correct. Upscaling mechanistic cell-scale models can solve such problems, as in [23].

Lastly, even if the necrotic biomechanical properties can be rigorously estimated, continuum models like this one would need further refinement to incorporate both the slow and fast dynamics known to play a role in necrosis. In the next section, we will next describe a mechanistic, patient-calibrated agent-based model developed by Macklin and collaborators in [56] to examine these and other issues.

4 Recent Agent-Based Modeling Results: Impact of Necrotic Core Biomechanics on DCIS

Agent-based modeling affords us the opportunity to examine the multiscalarity of necrosis and calcification by implementing both fast and slow time scale processes in individual cells and investigating the emergent whole-tumor biomechanics and clinical progression. We present recent work by Macklin et al. in simulating DCIS for individual patients [54–57]. The work discussed below includes the most detailed model of cell necrosis to date, and the first model of calcification. It also includes the first patient-specific calibration method to use clinically-accessible pathology from a single time point, as might be available in a standard biopsy.

4.1 Model Overview

In [54–57], Macklin et al. developed a patient-calibrated, lattice-free agent-based cell model and applied it to DCIS. Each virtual cell (an agent) has a position $\mathbf{x}$, velocity $\mathbf{v}$, and time-dependent physical properties. In particular, each cell has a volume $V(t)$ and nuclear volume $V_{\mathrm{N}}(t)$, which can readily be expressed as equivalent spherical cell and nuclear radii $R(t)$ and $R_{\mathrm{N}}(t)$, respectively. The cell also has a maximum adhesion interaction distance $R_{\mathrm{A}} > R(t)$, which models both the cell's deformability and uncertainty in its morphology [56]. See Fig. 7 (left).

The cell's velocity (and hence position) is governed by the balance of forces acting upon it: cell–cell adhesion ($\mathbf{F}_{\mathrm{cca}}$) and "repulsion" (resistance to deformation: $\mathbf{F}_{\mathrm{ccr}}$), cell-BM adhesion and repulsion ($\mathbf{F}_{\mathrm{cba}}$ and $\mathbf{F}_{\mathrm{cbr}}$), fluid drag ($-\nu\mathbf{v}$), cell–ECM adhesion ($\mathbf{F}_{\mathrm{cma}} = -c_{\mathrm{cma}}E$, where E is the local ECM density), and the net locomotive (motile) force $\mathbf{F}_{\mathrm{loc}}$. These forces are balanced by Newton's second law (conservation of linear momentum). As in [22, 35, 75], we use the "inertialess" assumption of fast force equilibration to explicitly express the velocity of cell i:

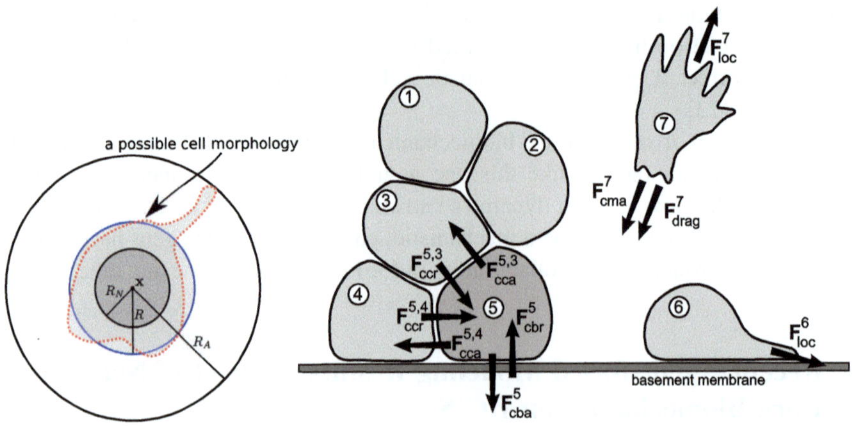

Fig. 7 *Left*: Cell position **x**, maximum adhesion interaction distance R_{A}, volume V (*light gray area*), nuclear volume V_{N} (*dark gray area*), and equivalent radii R and R_{N}. *Right*: Key forces in the model, labeled for cell 5. Figures reprinted with permission from [56]

$$\mathbf{v}_i = \frac{1}{\underbrace{\nu + c^i_{\mathrm{cma}} E}_{\text{cell-medium interactions}}} \left(\overbrace{\sum_{\substack{j=1 \\ j \neq i}}^{N(t)} \left(\mathbf{F}^{ij}_{\mathrm{cca}} + \mathbf{F}^{ij}_{\mathrm{ccr}} \right)}^{\text{cell-cell interactions}} + \overbrace{\mathbf{F}^i_{\mathrm{cba}} + \mathbf{F}^i_{\mathrm{cbr}}}^{\text{cell-BM interactions}} + \mathbf{F}^i_{\mathrm{loc}} \right), \tag{10}$$

where $N(t)$ is the number of simulated cells/agents at time t. For this discussion, we set $E \equiv 0$ and $\mathbf{F}_{\mathrm{loc}} = \mathbf{0}$ to model nonmotile cells contained in a lumen without ECM. See [56] for the specific forms of the forces, which were modeled using potential functions with finite interaction distances, consistent with the maximum adhesion interaction distance R_{A}. These forces are labeled for cell 5 in Fig. 7 (right).

Each cell has a phenotypic state $\mathcal{S}(t) \in \{\mathcal{A}, \mathcal{P}, \mathcal{Q}, \mathcal{N}\}$, where $\mathcal{A}$ cells are apoptosing, $\mathcal{P}$ cells are proliferating (in non-G_0), $\mathcal{Q}$ cells are quiescent (in G_0), and $\mathcal{N}$ cells are necrotic. Transitions between phenotypic states are governed by microenvironment and signaling-dependent stochastic processes. For example, quiescent cells enter the cell cycle with a (nondimensionalized) O_2-dependent probability:

$$\begin{aligned} \mathrm{Prob}(\mathcal{S}(t+\Delta t) = \mathcal{P} | \mathcal{S}(t) = \mathcal{Q}) &= 1 - \exp\left(- \int_t^{t+\Delta t} \overline{\alpha}_{\mathrm{QP}} \frac{\mathrm{O}_2(s) - \mathrm{O}_{2,\,\mathrm{hypoxic}}}{1 - \mathrm{O}_{2,\,\mathrm{hypoxic}}} \, \mathrm{d}s \right) \\ &\approx \overline{\alpha}_{\mathrm{QP}} \left(\frac{\mathrm{O}_2(t) - \mathrm{O}_{2,\,\mathrm{hypoxic}}}{1 - \mathrm{O}_{2,\,\mathrm{hypoxic}}} \right) \Delta t, \end{aligned} \tag{11}$$

where $\overline{\alpha}_{\mathrm{QP}}$ is the normoxic $\mathcal{Q} \to \mathcal{P}$ transition rate (when $\mathrm{O}_2 = 1$), and $\mathrm{O}_{2,\mathrm{hypoxic}}$ is the hypoxic oxygen threshold. The $\mathcal{Q} \to \mathcal{A}$ transition is similar but does not depend upon O_2. Cells become irreversibly necrotic ($\mathcal{S} = \mathcal{N}$) when $O_2 < O_{2,\text{hypoxia}}$. The proliferative and apoptotic states have fixed durations τ_P and τ_A. Cell volume and other key properties are controlled by a "sub-model" for each phenotypic state. Proliferating cells in $\mathcal{P}$ divide in half after progressing through S, G_2, and M; their two daughters spend G_1 growing (linearly) to their mature volumes and then return to $\mathcal{Q}$. Apoptotic cells are removed from the simulation after τ_A. We do not impose contact inhibition (a common feature for cellular automata models: reduced $\mathcal{Q} \to \mathcal{P}$ transitions for cells when surrounded by neighbor cells); this is because patient pathology for Ki-67 (a proliferation marker) frequently shows proliferating cells that are completely surrounded by other cells. As we shall see, a properly-calibrated mechanistic model can predict quantitatively-reasonable DCIS growth without need for contact inhibition. See [56] for full details on the proliferative and apoptotic sub-models.

4.1.1 Necrosis Sub-model

Let τ denote the elapsed time spent in the necrotic state. Define τ_{NL} to be the length of time for the cell to swell, lyse, and lose its water content, τ_{NS} the time for all surface receptors to become functionally inactive, and τ_{C}, the time for calcification to occur. We assume that $\tau_{\text{NL}} < \tau_{\text{NS}} < \tau_{\text{C}}$.

We assume a constant rate of calcification, reaching a radiologically-detectable level at $\tau = \tau_{\text{C}}$. If C is the nondimensional degree of calcification (scaled by the detection threshold), then $C(\tau) = \tau/\tau_{\text{C}}$ for $0 \leq \tau \leq \tau_{\text{C}}$, and $C(\tau) = 1$ otherwise. (We do not track further calcification after τ_{C}). We model the degradation of any surface receptor S (scaled by the non-necrotic expression level) by exponential decay with rate constant $\log 100/\tau_{\text{NS}}$, so that $S(\tau_{\text{NS}}) = 0.01S(0)$. We set $S(\tau) = 0$ for $\tau > \tau_{\text{NS}}$.

To model the necrotic cell's volume change, let f_{NS} be the maximum percentage increase in the cell's volume (just prior to lysis), and let V_0 be the cell's volume at the onset of necrosis. Then we model:

$$V(\tau) = \begin{cases} V_0\left(1 + f_{\text{NS}} \frac{\tau}{\tau_{\text{NL}}}\right) & \text{if } 0 \leq \tau < \tau_{\text{NL}} \\ V_{\text{N}} & \text{if } \tau_{\text{NL}} < \tau. \end{cases} \tag{12}$$

To model uncertainty in the cell morphology during lysis, we randomly perturb its location $\mathbf{x}$ such that its new radius $R(\tau_{\text{NL}})$ is contained within its swelled radius $R(0)(1 + f_{\text{NS}})^{\frac{1}{3}}$.

4.1.2 Other Model Details and Numerical Implementation

As we described in [56], microenvironmental quantities are modeled with reaction-diffusion equations throughout the computational domain. Uptake terms (e.g., for O_2) are created by a coarse-graining technique: first construct a high-resolution uptake term that resolves each cell's volume, then average it across a lower-resolution mesh (mesh size $\sim 1/10$ the appropriate diffusion length scale) before solving the reaction-diffusion equation. We apply Dirichlet conditions on the BM, and use Neumann conditions wherever the lumen intersects the computational boundary.

We represent the basement membrane using a signed distance function d satisfying $d > 0$ in the lumen, $d < 0$ in the stroma, $d = 0$ on the basement membrane, and $|\nabla d| \equiv 1$. We introduce an auxiliary data structure to reduce the overall computational cost from $\mathcal{O}\left(N(t)^2\right)$ to $\mathcal{O}(N(t))$, where $N(t)$ is the number of simulation objects at time t [56]. We implemented the model in cross-platform, object-oriented C++; we currently plan to open source the simulation framework in the next year. Towards that end, we introduced MultiCellXML, a new XML-based standard for sharing multicell agent simulation data. The supplementary material for [56] include sample DCIS simulation datasets (in MultiCellXML 1.0 format) and open source postprocessing and visualization code. Please see http://MathCancer.org/JTB_DCIS_2012/.

4.1.3 Calibration to Individual Patients, and Key Necrosis Parameter Values

In [56], we introduced the first calibration method to use individual patient pathology from a single time point, based upon processing several DCIS-affected ducts for the patient, as described in [23]. The proliferative index (PI: the percentage of Ki-67 positive cells in the viable rim) and apoptotic index (AI: the percentage of cleaved Caspase-3 positive cells in the viable rim) were combined with estimates of the proliferative time scale ($\tau_P = 18$ h) and apoptotic time scale ($\tau_A = 8.6$ h) and a population dynamic argument to calibrate the $\mathcal{A} \leftarrow \mathcal{Q} \leftrightarrow \mathcal{P}$ phenotypic transitions in the model. The cell density and experimental reports on cell mechanical response to deformation (see the references in [56]) were used to calibrate the mechanical parameters of the model. We calibrated oxygen transport by solving steady-state reaction-diffusion equations in a simplified cylindrical duct geometry and matching to the patient's measured viable rim thickness. In [56], we applied the calibration to a single anonymized DCIS patient with high-grade solid-type DCIS with comedonecrosis; we show the simulation (in a 1.5 mm, 2-D longitudinal section of duct) after 45 days of growth in this patient in Fig. 8. We recently combined this calibration method with an upscaling/coarse-graining argument to derive patient-specific predictions of surgical excision volumes in [23].

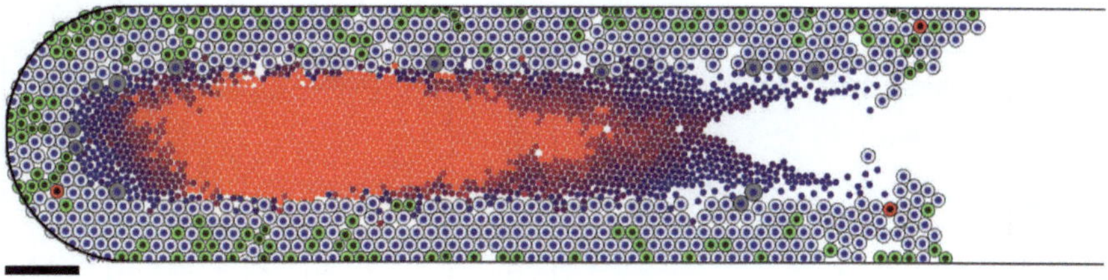

Fig. 8 Patient-calibrated DCIS simulation: After calibrating to a patient's pathology data as in [56], we simulate 45 days of DCIS growth. *Legend*: *Viable rim:* The *black curve* denotes the basement membrane. The *small blue circles* are cell nuclei, quiescent cells ($\mathcal{Q}$) are *pale blue*, proliferating cells ($\mathcal{P}$) are *green*, and apoptosing cells ($\mathcal{A}$) are *red*. *Necrotic core:* Necrotic cells ($\mathcal{N}$) are *grey* until they lyse; their solid fraction remains as debris (*dark circles* in center of duct). The shade of *red* indicates the level of calcification; *bright red* debris are clinically-detectable microcalcifications ($\mathcal{N}$ with $t > \tau_C$). *Bar*: 100 μm. Adapted with permission from [56]

4.2 DCIS Growth is Linear; Mammography and Pathology Sizes are Linearly Correlated; Origins in Necrotic Cell Water Loss

In [56], we post-processed the simulation in one-hour increments to determine the mean proliferative index, apoptotic index, viable rim thickness, and density as functions of time, as well as the farthest viable cell position ($x_V(t)$: the virtual pathology size) and the farthest calcified cell position ($x_C(t)$: the virtual mammography size). Open source postprocessing code is given at MathCancer.org. We plot x_V (solid blue curve) and x_C (dashed red curve) in Fig. 9 (left). After early transient dynamics, a linear (constant-rate) growth pattern emerges. The tumor advances at approximately 10.2 mm/year (obtained by the linear least-squares fit of x_V), whereas the calcification grows at 9.15 mm/year (linear least-squares fit of x_C). Due to these linear growth rates, the tumor's mammography and pathology sizes were predicted to be linearly correlated, with a linear least-squares correlation:

$$\text{pathology size} \approx 0.4203\,\text{mm} + 1.117\,\text{mammography size}, \qquad (13)$$

where all measurements are in mm. See the blue points in Fig. 9 (right).

These predictions are qualitatively and quantitatively consistent with clinical estimates of DCIS growth. Linear DCIS growth has been reported in a clinical study correlating changes in mammographic size with time between mammograms [15]. Another clinical study on microcalcifications reported that high-grade DCIS grows at 7.1 mm per year (along an axis to the nipple) [86]. They also analyzed the data in [15], deriving 13 and 6.8 mm/year mean and median growth rates, respectively. According to our relationship in Eq. (13), these correspond to pathology growth rates on the order of 7.6–14.5 mm/year. Hence, both our mammography and pathology growth rate predictions are quantitatively consistent with the clinical literature. [78] compared the maximum calcification diameter in

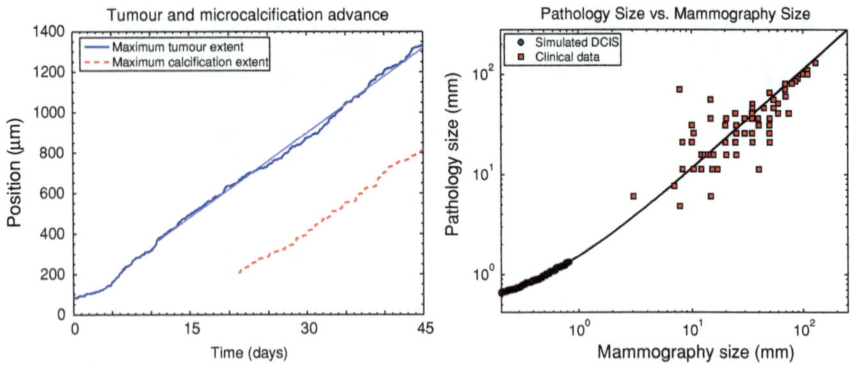

Fig. 9 *Left*: Over long times, the DCIS advances linearly at approximately 1 cm/year (*top curve*); the calcification also grows linearly. *Right*: The simulation (*blue circles*) predicts a linear correlation between the DCIS mammography and pathology sizes. When extrapolated over two orders of magnitude, the predicted correlation shows good agreement with clinical reports (*red squares*). Figures reproduced with permission from [56]

mammograms (corresponds to x_C) with the measured pathologic tumor size (corresponds to x_V) in 87 patients, finding a significant linear correlation between these measurements. When we extrapolate our linear relationship in Eq. (13) over two orders of magnitude (from the 1 mm scale to the 1 and 10 cm scales, approximating 1–10 years of growth), our extrapolated mammography-pathology correlation (the curve) shows an excellent quantitative agreement with these 87 data points (red squares) in Fig. 9 (right). This is a surprising and interesting result, which suggests that absent major alterations in signaling or microenvironmental factors, a patient's long-time growth dynamics may be established very early in progression.

These clinical phenomena can be understood as emergent from the underlying biophysics of the viable rim and necrotic core. Due to oxygen transport limitations, cell proliferation is confined to an approximately 80 μm viable rim. As the tumor grows, a steady pattern of flux emerges: proliferating cells towards the tumor leading edge are directed primarily towards empty space ahead of the tumor. Farther back, it is more mechanically favorable for mitosing cells to push their neighbors towards the duct center (against fewer cells) than along the duct (against more cells). Viable cells get pushed into hypoxic regions of the lumen, where they become necrotic and accumulate to fill the duct. This results in a linear growth pattern, as forward-directed proliferative cell flux is constrained to the leading edge of the tumor.

Necrotic cell lysis sustains this process. Whenever a necrotic cell lyses, its former volume is converted to a small core of cellular debris and a large pocket of (released) fluid, which is easily occupied by other cells. Thus, the earlier flux dynamic is maintained: proliferating cells on the outer edge of the duct push interior cells towards the necrotic core, diverting much of the overall cell flux inwards rather than towards the tumor leading edge. Hence, necrotic cell lysis acts as a mechanical stress relief, analogously to the mechanical pressure sink terms used in [58–61].

This can be further confirmed by altering the necrosis model. In [55, 57], we used a more gradual model of necrotic cell volume loss, where fluid "leakage" was spread over $\tau_C = 15$ days. The tumor advance accelerated as the viable rim grew, consistent with exponential growth. In those simulations, the rate of biomechanical stress relief in the necrotic core was too slow, causing more of the proliferative cell flux to be directed along the duct, preventing sustained linear growth. When we set $\tau_{NL} = \tau_C = 15$ days, we observed accelerating, exponential-like growth (blue curve after initial transient dynamics) [56]. See Fig. 10 (left). Generally, we found that all simulations exhibited exponential-like growth for approximately τ_{NL} time after the first instance of necrosis. For sufficiently small τ_{NL} (under 1 day), the brief exponential growth phase could not be detected. This mechanism suggested to us that because the lumen/necrotic core acts as a "reservoir" of mechanical stress relief to absorb proliferative cell flux, DCIS growth should be fastest in small ducts, and slowest in larger ducts. In simulations, we found this to be supported [56]. See Fig. 10 (right). We found an inverse relationship between duct radius R_{duct} and the DCIS growth rate x'_V (the red curve in Fig. 10 (right)):

$$x'_V \approx 20.52 + e^{6.085 - 0.02584 R_{\text{duct}}} \ \mu\text{m/day}. \tag{14}$$

Notice that as $R_{\text{duct}} \to \infty$, we find a minimum growth rate of 7.5 mm/year, or a mammography growth rate (by Eq. 13) of 6.7 mm/year. Cases with slower growth would need to be attributed to reduced oxygen or altered cell signaling.

4.3 Proliferative Cell Flux and Multiscale Necrosis Lead to a Stratified, Age-Structured Necrotic Core

Thus far, we have focused upon the gross macroscopic behavior of DCIS: the emergent growth rate and the relationship between mammography and pathology. We now turn our attention to the finer microstructure of the tumor. In Fig. 11 (top), we highlight several characteristic cross-sections of our DCIS simulation at 45 days.

In *Slice a*, there is a viable rim of thickness comparable to the remainder of the tumor, but with little visible evidence of necrosis. Biologically, this section of the tumor is no different than portions with necrosis (i.e., hypoxia is significant). This raises the possibility that in cases where too few ducts are sampled, a pathologist may fail to observe comedonecrosis, potentially (and incorrectly) changing the patient's Van Nuys Prognostic Index score [84] and affecting treatment decisions. This could be particularly true in cases where $\langle \text{PI} \rangle / \tau_P \approx \langle \text{AI} \rangle / \tau_A$, as little net cell flux from the viable rim to the necrotic core would be expected [56].

Farther from the tumor leading edge in *Slice b*, a ring of necrotic debris surrounds a hollow duct lumen. In cross sections like this, there has not yet been sufficient tumor cell flux from the viable rim to completely fill the lumen with necrotic debris. Farther

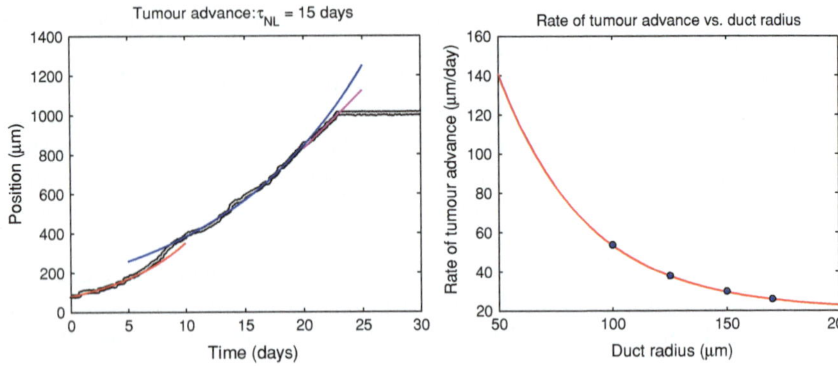

Fig. 10 *Left*: Delaying cell lysis leads to a completely filled necrotic core, which redirects proliferative cell flux along the duct. This results in exponential-like growth (*blue curve*). *Right*: Larger ducts have a greater "reservoir" available to absorb proliferative cell flux through necrotic cell lysis, leading to slower growth than in smaller ducts. Figures reproduced from [56] with permission

still from the leading edge in *Slice c*, there has been sufficient cell flux to fill the lumen with necrotic material; an outermost band of intact necrotic nuclei encircles a central region of mostly degraded nuclei (modeled here simply as partly calcified). Farther from the leading edge in *Slice d*, a thin outermost band of relatively intact necrotic nuclei surrounds an inner band of mostly degraded necrotic material and an inner core of microcalcification. In *Slice e*, the microcalcification is larger, and the outermost band of intact necrotic nuclei is largely gone. The necrotic core is increasingly calcified with distance from the tumor leading edge.

Overall, the model predicts an age-ordered necrotic core microstructure, with oldest material in the center surrounded by increasingly newer, less-degraded, and less-calcified material. Indeed, all these cross-sections can be found in our patient. See the hematoxylin and eosin (H&E) stained section in Fig. 11 (bottom). Slice b corresponds to *Duct 1*, where a ring of relatively intact necrotic debris (red arrows) surrounds an as-yet unfilled lumen. Slice c corresponds to *Duct 2*, where the entire lumen has been filled necrotic debris, which is more intact at its outer edge (red arrow), and increasingly degraded in its center (green arrow). Slice d corresponds to *Duct 3*, where a thinner ring of mostly intact nuclei (red arrows) surrounds an intermediate layer of mostly degraded debris (green arrows) and a central core of microcalcifications (white arrows). (Note that Duct 3 is likely the intersection of two or more ducts near a branch point.) Slice e corresponds to *Duct 4*, where a ring of degraded necrotic debris (green arrow) surrounds a larger calcification (white arrow). The *inset* shows a different duct from the patient that is similar to slice a.

This stratified structure arises from the overall flux of cells from the viable rim into the necrotic core, working in concert with the multiple time scales during

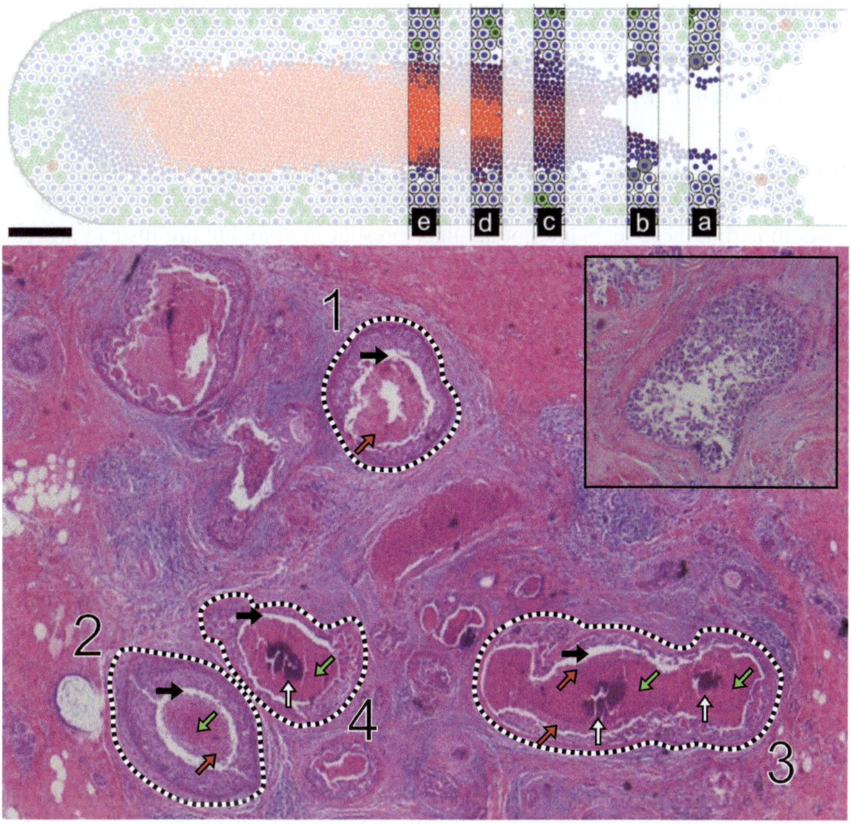

Fig. 11 *Top*: Patient-calibrated DCIS simulation from Fig. 8 with selected cross-sections highlighted to emphasize the emergent necrotic core microstructure. Near the leading edge (slice a), little necrotic debris has accumulated in the lumen. Farther back, relatively intact necrotic debris forms a ring near the necrotic boundary (slice b). Farther still, the lumen is completely filled with necrotic debris, with increasing degradation towards the center (slice c). Farther back, the oldest material is calcified, surrounded by relatively degraded debris (slice d). Calcification increases with distance from the leading edge (slice e). Reproduced with permission from [56]. *Bottom*: All the predicted necrotic core microstructures are observed in the patient's hematoxylin and eosin (H&E) pathology. *Red arrows* show necrotic debris with relatively intact nuclei. *Green arrows* show relatively degraded necrotic debris. *White vertical arrows* show calcification. *Black arrows* show the mechanical tear at the perinecrotic boundary. Simulated slice b predicts the microstructure seen in duct 1. Simulated slice c corresponds to duct 2. Simulated slice d corresponds to duct 3. Simulated slice 3 corresponds to duct 4. *Inset:* A duct similar to slice a. Pathology images adapted with permission from [56]

necrosis. If any of these scales were changed or removed, the microstructure would be altered. Indeed, better accounting for the time scales of nuclear degradation and fluid loss would likely improve the quantitative match to the patient pathology [56, 64].

4.4 Fast Time Scale Necrotic Cell Lysis and Volume Loss are Responsible for Mechanical "Tears" at Perinecrotic Boundary

One notable feature of nearly every DCIS pathology section is a "tear" at the perinecrotic boundary. See the black arrows in Fig. 11 (bottom). The conventional wisdom is that these tears are not actually present in vivo, but are instead artifacts that arise from tissue dehydration during sample preparation.

In [57], we implemented a preliminary necrosis sub-model where fluid volume was lost through the membrane gradually throughout necrosis, at a rate proportional to surface area and the remaining fluid fraction:

$$\frac{dV}{d\tau} = -\left(\frac{2}{\tau_{\mathrm{N}}}\log 100\right)\left(4\pi R^2\right)\left(\frac{V - V_{\mathrm{S}}}{V}\right), \tag{15}$$

where $0 < \tau < \tau_{\mathrm{C}} = \tau_{\mathrm{N}}$ is the elapsed time since entering the necrotic state, V_{S} is the cell's solid fraction, and the coefficient was chosen to make this nonlinear ODE satisfy $V(\tau_{\mathrm{N}}) \approx V_{\mathrm{S}}$. Fast cell swelling and lysis were neglected. The simulation, plotted at 30 days in Fig. 12, did not predict a tear at the perinecrotic boundary. We therefore hypothesized that if the perinecrotic tear is not an artifact, it must be caused by a fast time scale process. In [56], based upon a more thorough review of necrosis biology (see Sect. 2.2), we implemented the current model with rapid necrotic cell swelling followed by rapid volume loss. These simulations did recapitulate the perinecrotic tear. See the tumor leading edge in Fig. 8.

The mechanistic model is based upon the balance of actual forces with biophysically sound parameter values, is calibrated to actual patient data, and successfully makes quantitative, validated predictions on DCIS progression. In light of this care we put into the biological and clinical accuracy of the model, we conclude that mechanical separation of the viable rim and necrotic core at the perinecrotic boundary, although exacerbated by tissue dehydration, is in fact a real phenomenon, rather than a simple artifact. Based upon this new insight, we now interpret tears and cracks in pathology sections as indicators of a tissue's local biomechanical strength.

4.5 Evidence of Calcification Degradation at a Very Long Time Scale

Our simulations (Fig. 8) predict a linear/casting-type calcification, where the calcification forms a long, solid "plug" in the center of the duct. See Fig. 13 for a mammographic image of casting-type microcalcifications. Other calcification morphologies (e.g., fine pleomorphic) are not predicted by the biophysical assumptions of our model. While casting-type calcifications correlate with comedonecrosis [85], they are only present in approximately 30–50 % of DCIS

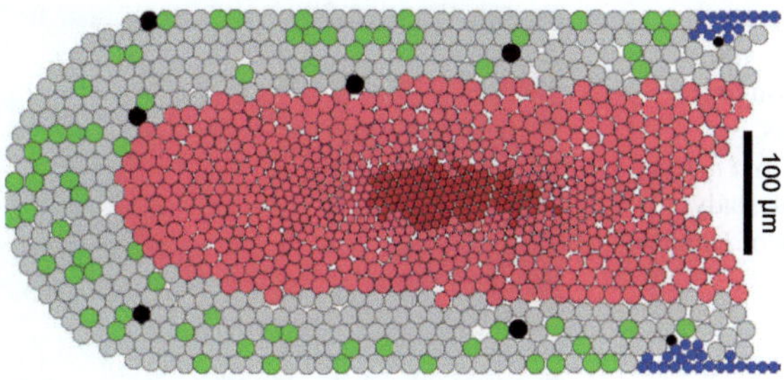

Fig. 12 Early DCIS simulations [57] neglected fast necrotic cell swelling and implemented a gradual volume loss over 15 days. The simulations could not reproduce the tear at the perinecrotic boundary. Necrotic cell lysis was too slow to sustain linear growth. Adapted with permission from [57]

[28, 39]. Moreover, casting-type calcifications can be absent from small, high-grade DCIS, while present in larger, low-grade DCIS [28]. Additional biophysics (e.g., secretions, heterogeneous adhesion mechanics, or degradation over long time scales) are required to model the broader spectrum of observed calcifications in DCIS. Our H&E images (Fig. 11) support this idea. The central regions of many calcifications—which we have shown are associated with the "oldest" necrotic material—demonstrate significant cracks that suggest extensive degradation and weak cohesion.

Phospholipids—such as those from subcellular structures that likely form a "backbone" for the formation of microcalcifications—degrade with half-lives on the order of 80 [3]–300 h [45] in non-pathologic tissue. Given this time scale, we would expect necrotic tissues and their associated microcalcifications to degrade over the course of a few months. This may partly explain rare cases of spontaneous resolution of calcifications in mammograms, where calcifications become smaller or occult without alternative explanations [81]: in slow-growing DCIS (e.g., with both high PI and AI, as observed in high-grade DCIS [11]), calcifications may be degraded more quickly than they are replaced by new necrotic material.

5 Discussion and Looking Forward

As we have seen, tissue necrosis and calcification are truly multiscale processes. Early tissue-scale modeling [51, 58–62] (Sect. 3) provided key insights on the role of tissue necrosis in steady tumor spheroid sizes, and its potentially destabilizing role when volume loss is rapid. Notably, these models can sufficiently predict the impact of the necrotic core on the long-time volume and morphology of a tumor, allowing quantitative predictions of progression. However, continuum modeling has thus far focused on the slower time scale processes of fluid loss and solid degradation; reformulation would be required to incorporate fast time scale processes like swelling and lysis. This is an interesting shortcoming, given that these are *key* features used to differentiate necrosis (and oncosis) from apoptosis in pathology.

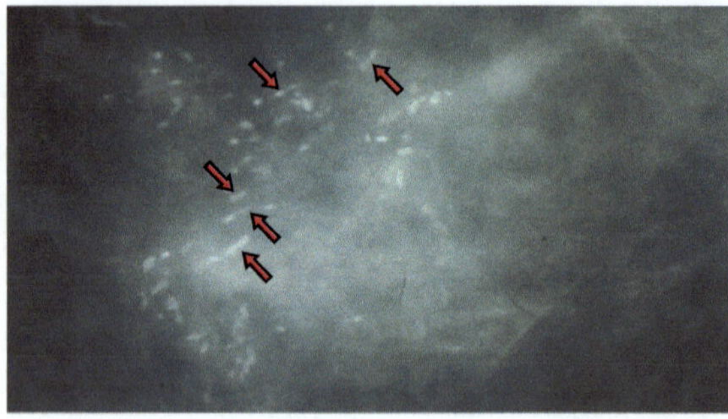

Fig. 13 Mammogram of a DCIS patient with characteristic casting-type microcalcifications, labeled here with *red arrows*. Image courtesy of Andy Evans, University of Dundee/ NHS Tayside

Models that consider the full spread of time scales in necrosis and calcification can produce a rich spectrum of behaviors that match observations in pathology [55–57] (Sect. 4). As hypothesized in [57] and investigated in [56], fast cell swelling and lysis—so fundamentally characteristic of early necrosis—are responsible for the tears ("artifacts") at the perinecrotic boundary that we consistently see in pathology. From a continuum point of view, these are rapid perturbations that create persistent and sharp discontinuities in the cell and necrotic debris distributions.

The simulated tumor microstructure—a viable rim (with greatest proliferation at the outermost edge) surrounding a stratified, age-structured necrotic core—arises from the multiscalarity of tissue necrosis and calcification. In the necrotic core, the structure mirrors tissue age due to the steady flux from the viable rim into the necrotic core: the newest, least degraded material surrounds increasingly degraded debris, with central calcifications in the oldest tissues [56]. All these features are consistently observed in patient pathology. Our work revealed a long-time deterioration of calcifications that may explain key features in mammography.

5.1 Next-Generation Hybrid Multiscale Modeling

Improved multiscale and hybrid mathematics and computational techniques are necessary for further advances. In the agent-based model, each necrotic cell agent must remain in memory on the order of simulated months; by later times, necrotic agents outnumber viable agents by three to one or more. And yet the vast majority of these objects are engaged in the slow time scale processes of calcification and solid degradation—processes that are well-suited to continuum modeling!

Lowengrub and colleagues are now developing a sophisticated continuum model of necrotic cell calcification in DCIS [17]. We apply a phase field approach [91] to model the tumor as a mixture of fluid, extracellular matrix, and cells. The model can separately track the necrotic and calcified cell fractions. We also include a sophisticated model of the basement membrane, which can deform in response to mechanical stresses introduced by the growing tumor [16]. Preliminary results recapitulate the gross features observed in DCIS pathology: a viable rim of appropriate thickness surrounding a necrotic core with a calcified center [17]. See Fig. 14.

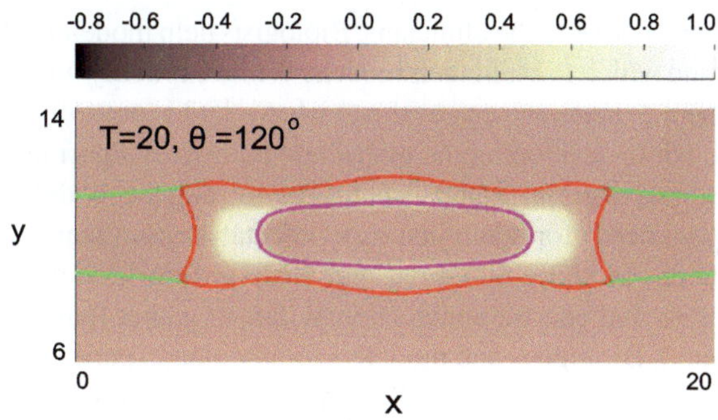

Fig. 14 Preliminary continuum simulation of solid-type DCIS with comedonecrosis and calcifications [17]. *Legend*: *Green curve*: deformed basement membrane. *Red curve*: viable tumor boundary. *Magenta curve*: calcified necrotic debris. *White shading*: non-calcified necrotic tissue

We plan to integrate these discrete and continuum approaches in a hybrid model, as outlined in [43, 50]. A key issue is determining the rate constants for the continuum model. As water loss in necrotic cells does not occur at a fixed rate, it may be best to simulate using the discrete model until most fluid has been lost, then "convert mass" to the continuum model for the slower time scale processes. A more detailed analysis of the full agent-based model could yield the correct average per-volume rate of volume loss in the necrotic tissues, similarly to the upscaling approach we developed in [23, 54]. Other approaches may include introduction of an age structuring variable, as is often used today in mathematical ecology (e.g., [5, 41, 49]).

5.2 A Vision for Quantitative, Integrative Computational Oncology

An integrative modeling approach—where clinicians, modelers, and biologists work in close-knit teams throughout the modeling process—is necessary to push computational oncology towards clinical application. Conversely, just as the space race in the 1950s and 1960s fueled advances throughout engineering, physics, and mathematics, efforts to push the envelope in patient-specific modeling are advancing the state-of-the-art in mathematical modeling, computational algorithms, experimental methods, and clinical practice. Moreover, quantitatively and explicitly stating our working biological hypotheses gives us the opportunity to rigorously and systematically test and refine what can best be described as *current cancer biology orthodoxy*. We close this chapter by outlining our vision of clinically-oriented integrative computational oncology, and its possible impact beyond the clinic.

Model Design Clinicians and modelers jointly identify important unanswered clinical questions. This helps modelers avoid investigating unnecessary tangents while bringing fresh perspective to the clinicians. Biologists help modelers identify working hypotheses around which to build their models. While explicitly writing these out and "translating" them to code, we can evaluate what is and is not truly known in cancer biology. Lastly, while developing the model and numerical algorithms, assessing the expected clinical and experimental data helps in choosing the modeling approach; the model may expose needs for additional experimental measurements.

Data generation, model calibration, and early testing Modelers and clinicians jointly plan studies and choose which clinical data to gather (pathology, radiology, case histories, etc.). Biologists and modelers jointly plan experiments to supplement the clinical data and inform the model's constitutive relations. These data are integrated into the model with the help of statisticians, image processing specialists, and others. Early simulations help test and refine the data, model, and calibration.

Simulation, analysis, validation, and feedbacks The calibration procedure is applied to simulate cancer in individual patients. The simulation data are postprocessed, yielding quantitative predictions that we validate for each patient. This quantitative focus allows us to assess and improve our underlying biological hypotheses. If the predictions are accurate, trials may be planned to assess the model's ability to assist individual treatment decisions. The modelers, clinicians, and biologists jointly identify future refinements and experiments. They also jointly select new modeling foci as suggested by both clinical needs and model-derived insights.

5.2.1 Application of Integrative Modeling to Breast Cancer

This approach guides our work on breast cancer. We have built a team that now includes oncologists, pathologists, radiologists, biologists and modelers [63], and we are continuing to recruit complementary expertise (e.g., in analytical pathology, tissue bioengineering, etc.). We have jointly identified that patient-specific predictions of progression from in situ to invasive carcinoma would be of immense clinical value, and would naturally build upon our increasingly accurate in situ models. To that end, we are developing key modeling technologies, such as improved BM and ECM mechanics [20] and multiscale matrix metalloproteinase transport-reaction kinetics [19]. Early modeling results will help guide future experimental design.

Given the critical role of tissue necrosis in DCIS progression, we are developing next-generation models of intracellular fluid transport, solid synthesis, and dystrophic calcification to more accurately describe individual cell volume and composition changes during these processes [64], based upon in vitro measurements we are currently gathering [68]. By this approach, it should soon be possible to accurately simulate common pathology stains based upon each cell agent's composition. This, in turn, should make possible new and innovative quantitative comparisons to patient pathology, better refinement of the otherwise nigh-unmeasurable necrosis time scales, and ultimately more accurate predictions of clinical progression.

The interested reader can find up-to-date information on these efforts (including frequent news postings, animations, tutorials, simulation data, and software) at MathCancer.org. We also encourage the interested reader to visit the newly-established *Consortium for Integrative Computational Oncology* at the University of Southern California, where we are developing this approach with a focus on building community and training the next generation of interdisciplinary cancer scientists.

5.2.2 Broader Implications and Spillover Benefits

The quest for quantitative accuracy in patient-specific modeling drives advances in mechanistic modeling. Quantitative testing allows us to choose among competing models, where multiple models may be qualitatively compelling, but fewer are quantitatively reasonable. To the extent that rigorously-calibrated models can successfully make quantitative predictions in individual patients, we gain new confidence in the underlying models. Because the models are built to be universal (cancer cells are just cells with different phenotypic parameter values), these advances will be of use across computational biology. Likewise, efficient numerical simulation of these increasingly sophisticated models is driving advances in applied parallel computing and hybrid and multiscale modeling. Any derived algorithms will be of benefit across applied mathematics and engineering.

If we should reach the point where we can integrate in vitro measurements with clinical data to accurately predict cancer progression and therapy response in individual patients, the implications are vast: new insights from wetlab biology could be immediately evaluated for potential impact in individual patients in combination with current therapies, offering accelerated discovery and clinical translation. Ultimately, it is our goal that this approach will help bridge the gap between theoretical modeling, wetlab biology, and clinical practice to develop and deliver patient-calibrated predictive tools. We believe that such tools will one day help clinicians and their patients to make optimal, personalized treatment decisions that incorporate both accepted clinical practice and cutting-edge research results.

Acknowledgments PM and SM thank the National Institutes of Health for the Physical Sciences Oncology Center grant 5U54CA143907 for Multi-scale Complex Systems Transdisciplinary Analysis of Response to Therapy–MC-START. PM thanks the USC James H. Zumberge Research and Innovation Fund (2012 Large Interdisciplinary Award) for support through the new Consortium for Integrative Computational Oncology (CICO), and the USC Undergraduates Research Associate Program (URAP) for student support. JL gratefully acknowledges partial support from the National Institutes of Health, National Cancer Institute, for funding through grants 1RC2CA148493-01, P50GM76516 for a Center of Excellence in Systems Biology at the University of California, Irvine, and P30CA062203 for the Chao Comprehensive Cancer Center at the University of California, Irvine. JL also acknowledges support from the National Science Foundation, Division of Mathematical Sciences.

PM thanks David Agus (USC Center for Applied Molecular Medicine); Andrew Evans, Jordan Lee, Colin Purdie, and Alastair Thompson (U. of Dundee/NHS Tayside); and Paul Newton (USC Department of Aerospace and Mechanical Engineering) for enlightening discussions. PM thanks Andrew Evans for Fig. 13. The authors thank Ying Chen (U. California at Irvine) for Fig. 14.

References

1. Altman, D.A., Atkinson, D.S., Brat, D.J.: Glioblastoma multiforme. Radiographics **27**(3), 883–888 (2007)
2. Anderson, A.R.A: A hybrid mathematical model of solid tumour invasion: the importance of cell adhesion. Math. Med. Biol. **22**(2), 163–186 (2005)
3. Ayre, K.J., Hulbert, A.J: Dietary fatty acid profile influences the composition of skeletal muscle phospholipids in rats. J. Nutr. **126**(3), 653–662 (1996)
4. Bacsó, Z., Everson, R.B., Eliason, J.F: The DNA of annexin V-binding apoptotic cells is highly fragmented. Cancer Res. **60**(16), 4623 (2000)
5. Bai, Z.: A periodic age-structured epidemic model with a wide class of incidence rates. J. Math. Anal. Appl. **393**(2), 367–376 (2012)
6. Barros, L.F., Hermosilla, T., Castro, J.: Necrotic volume increase and the early physiology of necrosis. Comp. Biochem. Physiol. A. Mol. Integr. Physiol. **130**(3), 401–409 (2001)
7. Barros, L.F., Kanaseki, T., Sabriov, R., Morishima, S., Castro, J., Bittner, C.X., Maeno, E., Ando-Akatsuka, Y., Okada, Y.: Apoptotic and necrotic blebs in epithelial cells display similar neck diameters but different kinase dependency. Cell Death Diff. **10**(6), 687–697 (2003)
8. Basanta, D., Strand, D.W., Lukner, R.B., Franco, O.E., Cliffel, D.E., Ayala, G.E., Hayward, S.W., Anderson, A.R.A: The role oftransforming growth factor-β-mediated tumor-stroma interactions in prostate cancer progression: An integrative approach. Cancer Res. **69**(17), 7111–7120 (2009)
9. Basu, S., Binder, R.J., Suto, R., Anderson, K.M., Srivastava, P.K: Necrotic but not apoptotic cell death releases heat shock proteins, which deliver a partial maturation signal to dendritic cells and activate the NF-κB pathway. Int. Immunol. **12**(11), 1539–1546 (2000)
10. Blagosklonny, M.V., Robey, R., Bates, S., Fojo, T.: Pretreatment with DNA-damaging agents permits selective killing of checkpoint-deficient cells by microtubule-active drugs. J. Clin. Investig. **105**(4), 533–539 (2000)
11. Buerger, H., Mommers, E.C., Littmann, R., Diallo, R., Brinkschmidt, C., Poremba, C., Dockhorn-Dworniczak, B., van Diest, P.J., Böcker, W.B: Correlation of morphologic and cytogenetic parameters of genetic instability with chromosomal alterations in in situ carcinomas of the breast. Am. J. Clin. Pathol. **114**, 854–859 (2000)
12. Byrne, H., Chaplain, M.A.J: Necrosis and apoptosis: Distinct cell loss mechanisms in a mathematical model of avascular tumour growth. J. Theor. Med. **1**(3), 223–235 (1998)
13. Byrne, H., Preziosi, L.: Modelling solid tumour growth using the theory of mixtures. Math. Med. Biol. **20**(4), 341–366 (2003)
14. Cantoni, O., Guidarelli, A., Palomba, L., Fiorani, M.: U937 cell necrosis mediated by peroxynitrite is not caused by depletion of ATP and is prevented by arachidonate via an ATP-dependent mechanism. Mol. Pharm. **67**(5), 1399–1405 (2005)
15. Carlson, K.L., Helvie, M.A., Roubidoux, M.A., Kleer, C.G., Oberman, H.A., Wilson, T.E., Pollack, E.W., Rochester, A.B: Relationship between mammographic screening intervals and size and histology of ductal carcinoma in situ. Am. J. Roentenol. **172**(2), 313–317 (1999)
16. Chen, Y.: Modeling tumor growth in complex, dynamic geometries. Ph.D. dissertation, University of California, Irvine Department of Mathematics (2012)
17. Chen, Y., Lowengrub, J.S.: Tumor growth in complex, evolving geometries: A diffuse domain approach (2012, in preparation)
18. Cristini, V., Lowengrub, J.S., Nie, Q.: Nonlinear simulation of tumor growth. J. Math. Biol. **46**(3), 191–224 (2003)
19. D'Antonio, G., Macklin, P.: A multiscale hybrid discrete-continuum model of matrix metalloproteinase transport and basement membrane-extracellular matrix degradation (2013, in preparation)
20. D'Antonio, G., Macklin, P., Preziosi, L.: An agent-based model for elasto-plastic mechanical interactions between cells, basement membrane and extracellular matrix. Math. Biosci. Eng. http://MathCancer.org/Publications.php#dantonio12_mbe (2012)

21. Drasdo, D., Höhme, S.: A single-scale-based model of tumor growth *in vitro*: monolayers and spheroids. Phys. Biol. **2**(3), 133–147 (2005)
22. Drasdo, D., Kree, R., McCaskill, J.S: Monte-carlo approach to tissue cell populations. Phys. Rev. E **52**(6), 6635–6657 (1995)
23. Edgerton, M.E., Chuang, Y.L., Macklin, P., Yang, W., Bearer, E.L., Cristini, V.: A novel, patient-specific mathematical pathology approach for assessment of surgical volume: Application to ductal carcinoma in situ of the breast. Anal. Cell. Pathol. **34**(5), 247–263 (2011)
24. Eichbaum, C., Meyer, A.S., Wang, N., Bischofs, E., Steinborn, A., Bruckner, T., Brodt, P., Sohn, C., Eichbaum, M.H.R: Breast cancer cell-derived cytokines, macrophages and cell adhesion: Implications for metastasis. Anticancer Res. **31**(10), 3219–27 (2011)
25. Elmore, S.: Apoptosis: A review of programmed cell death. Toxicol. Pathol. **35**(4), 495–516 (2007)
26. Erez, N., Truitt, M., Olson, P., Hanahan, D.: Cancer-associated fibroblasts are activated in incipient neoplasia to orchestrate tumor-promoting inflammation in an NF-κB-dependent manner. Cancer Cell **17**(2), 135–147 (2010)
27. Evans, A.: The diagnosis and management of pre-invasive breast disease: Radiological diagnosis. Breast Cancer Res. **5**(5), 250–253 (2003)
28. Evans, A., Clements, K., Maxwell, A., Bishop, H., Handby, A., Lawrence, G., Pinder, S.E: Lesion size is a major determinant of the mammographic features of ductal carcinoma in situ: Findings from the Sloane project. Radiology **53**(3), 181–4 (2010)
29. Evans, A., Pinder, S., Wilson, R., Sibbering, M., Poller, D., Elston, C., Ellis, I.: Ductal carcinoma in situ of the breast: Correlation between mammographic and pathologic findings. Am. J. Roentgen. **162**(6), 1307–1311 (1994)
30. Fall, C.P., Bennett, J.P: Characterization and time course of MPP+-induced apoptosis in human SH-SY5Y neuroblastoma cells. J. Neurosci. Res. **55**(5), 620–628 (1999)
31. Festjens, N., Berghe, T.V., Vandenabeele, P.: Necrosis, a well-orchestrated form of cell demise: Signalling cascades, important mediators and concomitant immune response. Biochim. Biophys. Acta Bioenergetics **1757**(9–10), 1371–1387 (2006)
32. Fesus, L.: Transglutaminase-catalyzed protein cross-linking in the molecular program of apoptosis and its relationship to neuronal processes. Cell. Mol. Neurobiol. **18**, 683–694 (1998)
33. Frieboes, H.B., Zheng, X., Sun, C.H., Tromberg, B., Gatenby, R., Cristini, V.: An integrated computational/experimental model of tumor invasion. Cancer Res. **66**(3), 1597–1604 (2006)
34. Gadeau, A.P., Chaulet, H., Daret, D., Kockx, M., Daniel-Lamaziè, J.M., Desgranges, C.: Time course of osteopontin, osteocalcin, and osteonectin accumulation and calcification after acute vessel wall injury. J. Histochem. Cytochem. **49**(1), 79–86 (2001)
35. Galle, J., Loeffler, M., Drasdo, D.: Modeling the effect of deregulated proliferation and apoptosis on the growth dynamics of epithelial cell populations in vitro. Biophys. J. **88**(1), 62–75 (2005)
36. Garland, J.M., Halestrap, A.: Energy metabolism during apoptosis. J. Biol. Chem. **272**(8), 4680–4688 (1997)
37. Grivennikov, S.I., Greten, F.R., Karin, M.: Immunity, inflammation, and cancer. Cell **140**(6), 883–899 (2010)
38. Hengartner, M.O: The biochemistry of apoptosis. Nature **407**(6805), 770–776 (2000)
39. Hofvind, S., Iversen, B.F., Eriksen, L., Styr, B.M.S., Kjellevold, K., Kurz, K.D: Mammographic morphology and distribution of calcifications in ductal carcinoma in situ diagnosed in organized screening. Acta Radiologica **52**(5), 481–487 (2011)
40. Huether, S., McCance, K.: Understanding Pathology, 5 edn. chap. 3. Mosby, St. Louis, MO USA (2011)
41. Iannelli, M., Ripoll, J.: Two-sex age structured dynamics in a fixed sex-ratio population. Nonlinear Anal. Real World Appl. **13**(6), 2562–2577 (2012)
42. Jian, B., Narula, N., Li, Q.Y., Mohler, E.R. III, Levy, R.J: Progression of aortic valve stenosis: TGF-β1 is present in calcified aortic valve cusps and promotes aortic valve interstitial cell calcification via apoptosis. Ann. Thorac. Surg. **75**(2), 457–465 (2003)

43. Jin F., Chuang Y.L., Cristini V., Lowengrub J.S.: Hybrid continuum-discrte tumor models. In: Cristini, V., Lowengrub, J.S. (eds.) Multiscale Modeling of Cancer, chap. 8, pp. 153–182. Cambridge University Press, Cambridge, UK (2010)
44. Kerr, J.F.R., Winterford, C.M., Harmon, B.V: Apoptosis. its significance in cancer and cancer therapy. Cancer **73**(8), 2013–2026 (1994)
45. Krause, R.F., Beamer, K.C: Lipid content and phospholipid metabolism of subcellular fractions from testes of control and retinol-deficient rats. J. Nutr. **104**(5), 629–637 (1974)
46. Krysko, D.V., Berghe, T.V., D'Herde, K., Vandenabeele, P.: Apoptosis and necrosis: Detection, discrimination and phagocytosis. Methods **44**(3), 205–221 (2008)
47. Kumar, V., Abbas, A.K., Aster, J.C., Fausto, N.: Pathologic Basis of Disease, 8 edn. chap. 1. Saunders Elsevier, Philadelphia, PA USA (2009)
48. Lee, J.S., Basalyga, D.M., Simionescu, A., Isenburg, J.C., Sinionescu, D.T., Vyavahare, N.R: Elastin calcification in the rate subdermal model is accompanied by up-regulation of degradative and osteogenic cellular responses. Am. J. Pathol. **168**(2), 490–498 (2006)
49. Levin, S.A., Goodyear, C.P: Analysis of an age-structured fishery model. J. Math. Biol. **9**, 245–274 (1980)
50. Lowengrub, J.S., Frieboes, H.B., Jin, F., Chuang, Y.L., Li, X., Macklin, P., Wise, S.M., Cristini, V.: Nonlinear modeling of cancer: Bridging the gap between cells and tumors. Nonlinearity **23**(1), R1–R91 (2010)
51. Macklin, P.: Numerical simulation of tumor growth and chemotherapy. M.S. thesis, University of Minnesota School of Mathematics (2003)
52. Macklin, P.: Biological background. In: Cristini, V., Lowengrub, J.S. Multiscale Modeling of Cancer, chap. 2, pp. 8–24. Cambridge University Press, Cambridge, UK (2010)
53. Macklin, P.: Basic ductal carcinoma in situ (DCIS) pathobiology for modelers. http://MathCancer.org/Resources.php#DCIS_biology_tutorial (online tutorial). (2012)
54. Macklin, P., Edgerton, M.E., Cristini, V.: Agent-based cell modeling: application to breast cancer. In: Cristini, V., Lowengrub, J.S. Multiscale Modeling of Cancer, chap. 10, pp. 216–244. Cambridge University Press, Cambridge, UK (2010)
55. Macklin, P., Edgerton, M.E., Lowengrub, J., Cristini, V.: Discrete cell modeling. In: Cristini, V., Lowengrub, J.S. Multiscale Modeling of Cancer, chap. 6, pp. 92–126. Cambridge University Press, Cambridge, UK (2010)
56. Macklin, P., Edgerton, M.E., Thompson, A.M., Cristini, V.: Patient-calibrated agent-based modelling of ductal carcinoma in situ (DCIS): From microscopic measurements to macroscopic predictions of clinical progression. J. Theor. Biol. **301**, 122–140 (2012)
57. Macklin, P., Kim, J., Tomaiuolo, G., Edgerton, M.E., Cristini, V.: Agent-based modeling of ductal carcinoma in situ: Application to patient-specific breast cancer modeling. In: Pham T. (ed.) Computational Biology: Issues and Applications in Oncology, chap. 4, pp. 77–112. Springer, NewYork (2009)
58. Macklin, P., Lowengrub, J.S: Evolving interfaces via gradients of geometry-dependent interior poisson problems: application to tumor growth. J. Comput. Phys. **203**(1), 191–220 (2005)
59. Macklin, P., Lowengrub, J.S: An improved geometry-aware curvature discretization for level set methods: application to tumor growth. J. Comput. Phys. **215**(2), 392–401 (2006)
60. Macklin, P., Lowengrub, J.S: Nonlinear simulation of the effect of microenvironment on tumor growth. J. Theor. Biol. **245**(4), 677–704 (2007)
61. Macklin, P., Lowengrub, J.S: A new ghost cell/level set method for moving boundary problems: Application to tumor growth. J. Sci. Comp. **35**(2–3), 266–299 (2008)
62. Macklin, P., McDougall, S., Anderson, A.R.A., Chaplain, M.A.J., Cristini, V., Lowengrub, J.: Multiscale modeling and nonlinear simulation of vascular tumour growth. J. Math. Biol. **58**(4–5), 765–798 (2009)
63. Macklin, P., Mumenthaler, S., Jordan, L.B., Purdie, C.A., Evans, A.J., Thompson, A.M.: Integration of pathology, radiology, and in vitro data in patient-calibrated cancer simulations: Recent advances and future outlook for ductal carcinoma in situ (DCIS). To be presented at: Annual Meeting of the Society for Mathematical Biology (SMB). http://www.mathcancer.org/publications/macklin12_SMB.pdf (2012)

64. Macklin, P. et al.: Agent-based cell modeling with improved subcellular fluid and solid transport: comparison with ductal carcinoma in situ pathology (2013, in preparation)
65. Majno, G., Joris, I.: Apoptosis, oncosis, and necrosis. an overview of cell death. Am. J. Pathol. **146**(1), 3–15 (1995)
66. Majno, G., Joris, I.: Cells, Tissues, and Disease: Principles of General Pathology, 2nd edn. Oxford University Press, New York (2004)
67. McCarthy, J.V., Cotter, T.G: Cell shrinkage and apoptosis: a role for potassium and sodium ion efflux. Cell Death Diff. **4**(8), 756–770 (1997)
68. Mumenthaler, S., Macklin, P., et al.: Time-course volume and water fraction measurements in cycling and apoptotic breast cancer cell lines (2013, in preparation)
69. Muttarak, M., Kongmebhol, P., Sukhamwang, N.: Breast calcifications: which are malignant? Singapore Med. J. **50**(9), 907–913 (2009)
70. Noch, E., Khalili, K.: Molecular mechanisms of necrosis in glioblastoma: The role of glutamate excitotoxicity. Cancer Biol. Ther. **8**(19), 1791–1797 (2009)
71. Norton, K.A., Wininger, M., Bhanot, G., Ganesan, S., Barnard, N., Shinbrot, T.: A 2D mechanistic model of breast ductal carcinoma in situ (DCIS) morphology and progression. J. Theor. Biol. **263**(4), 393–406 (2010)
72. Ottesen, G.L., Christensen, I.J., Larsen, J.K., Larsen, J., Baldetorp, B., Linden, T., Hansen, B., Andersen, J.: Carcinoma in situ of the breast: correlation of histopathology to immunohistochemical markers and DNA ploidy. Breast Cancer Res. Treat. **60**(3), 219–226 (2000)
73. Owen, M.R., Byrne, H.M., Lewis, C.E: Mathematical modelling of the use of macrophages as vehicles for drug delivery to hypoxic tumour sites. J. Theor. Biol. **226**(4), 377–391 (2004)
74. Pearson O.H., Manni A., Arafah B.M.: Antiestrogen treatment of breast cancer: An overview. Cancer Res. 42(8 Suppl.):3424s–3428s (1982)
75. Ramis-Conde, I., Drasdo, D., Anderson, A.R.A., Chaplain, M.A.J: Modeling the influence of the e-cadherin-beta-catenin pathway in cancer cell invasion: A multiscale approach. Biophys. J. **95**(1), 155–165 (2008)
76. Richards, C.H., Mohammed, Z., Qayyum, T., Horgan, P.G., McMillan, D.C: The prognostic value of histological tumor necrosis in solid organ malignant disease: a systematic review. Future Oncol. **7**(10), 1223–1235 (2011)
77. Richards, C.H., Roxburgh, C.S.D., Anderson, J.H., McKee, R.F., Foulis, A.K., Horgan, P.G., McMillan, D.C: Prognostic value of tumour necrosis and host inflammatory responses in colorectal cancer. Brit. J. Surg. **99**(2), 287–294 (2012)
78. de Roos, M.A.J., Pijnappel, R.M., Post, W.J., de Vries, J., Baas, P.C., Groote, L.D: Correlation between imaging and pathology in ductal carcinoma in situ of the breast. World J. Surg. Onco. **2**(1), 4 (2004)
79. Rüegg, C.: Leukocytes, inflammation, and angiogenesis in cancer: fatal attractions. J. Leukoc. Biol. **80**(4), 682–684 (2006)
80. Scarlett, J.L., Sheard, P.W., Hughes, G., Ledgerwood, E.C., Ku, H.K., Murphy, M.P: Changes in mitochondrial membrane potential during staurosporine-induced apoptosis in Jurkat cells. FEBS Lett. **475**(3), 267–272 (2000)
81. Seymour, H.R., Cooke, J., Given-Wilson, R.M: The significance of spontaneous resolution of breast calcification. Brit. J. Radiol. **72**(853), 3–8 (1999)
82. Sickles, E.A: Mammographic features of 300 consecutive nonpalpable breast cancers. Am. J. Roentgen. **146**(4), 661–663 (1986)
83. Silva, A.S., Gatenby, R.A., Gillies, R.J., Yunes, J.A: A quantitative theoretical model for the development of malignancy in ductal carcinoma in situ. J. Theor. Biol. **262**(4), 601–613 (2010)
84. Silverstein, M.J., Lagios, M.D., Craig, P.H., Waisrnan, J.R., Lewinsky, B.S., Colburn, W.J., Poller, D.N: A prognostic index for ductal carcinoma in situ of the breast. Cancer **77**(11), 2267–74 (1996)
85. Stomper, P.C., Connolly, J.L., Meyer, J.E., Harris, J.R: Clinically occult ductal carcinoma in situ detected with mammography: analysis of 100 cases with radiologic-pathologic correlation. Radiology **172**(1), 235–241 (1989)

86. Thomson J.Z., Evans, A.J., Pinder, S.E., Burrell, H.C., Wilson, A.R.M., Ellis, I.O.: Growth pattern of ductal carcinoma in situ (DCIS): a retrospective analysis based on mammographic findings. Br. J. Cancer 85(2):225–227 (2001)
87. Trump, B.E., Berezesky, I.K., Chang, S.H., Phelps, P.C: The pathways of cell death: Oncosis, apoptosis, and necrosis. Toxicol. Pathol. **25**(1), 82–88 (1997)
88. Walsh, G.M: Mechanisms of human eosinophil survival and apoptosis. Clin. Exp. Allergy **27**(5), 482–487 (1997)
89. Ward, J.P., King, J.R: Mathematical modelling of avascular tumour growth. IMA J. Math. Appl. Med. Biol. **14**(1), 36–69 (1997)
90. Ward, J.P., King, J.R: Mathematical modelling of avascular-tumour growth II: modelling growth saturation. Math. Med. Biol. **16**(2), 171–211 (1999)
91. Wise, S.M., Lowengrub, J.S., Frieboes, H.B., Cristini, V.: Three-dimensional multispecies nonlinear tumor growth–I. model and numerical method. J. Theor. Biol. **253**(3), 524–543 (2008)
92. Yagata H., Harigaya K., Suzuki M., Nagashima T., Hashimoto H., Ishii G., Miyazaki M., Nakajima N., Mikata, A (2003) Comedonecrosis is an unfavorable marker in node-negative invasive breast carcinoma. Pathol. Int. **53**, 8):501–506
93. Zheng, X., Wise, S.M., Cristini, V.: Nonlinear simulation of tumor necrosis, neo-vascularization and tissue invasion via an adaptive finite-element/level set method. Bull. Math. Biol. **67**(2), 211–259 (2005)

Integration of Molecular Signaling into Multiscale Modeling of Cancer

Zhihui Wang and Vittorio Cristini

Abstract Multiscale modeling has now been well-accepted as a powerful tool to quantitatively represent, simulate, understand, and predict cancer progression and development across multiple biological scales. In this chapter, we focus on a specific type of multiscale cancer models where molecular signaling profiles are explicitly linked to the determination of cellular phenotypic changes. These models are particularly suitable for exploring the relationship between signaling dynamics within each individual cancer cell and the emergent cancer behavior on the multicellular level. We also discuss current challenges and future directions of this molecular signaling-incorporated multiscale cancer modeling approach.

Abbreviations

ABM	Agent-based model or agent-based modeling
EGF	Epidermal growth factor
EGFR	EGF receptor
EMT	Epithelial–mesenchymal transition
ERK	Extracellular signal-regulated kinase
NSCLC	Non-small cell lung cancer
ODE	Ordinary differential equation
PDE	Partial differential equation
PLCγ	Phopholipase Cγ

Z. Wang (✉) · V. Cristini
Department of Pathology, University of New Mexico,
Albuquerque, NM 87131, USA
e-mail: zwang@salud.unm.edu

V. Cristini
Department of Chemical and Biomedical Engineering,
University of New Mexico, Albuquerque, NM 87131, USA

Stud Mechanobiol Tissue Eng Biomater (2013) 14: 381–394
DOI: 10.1007/8415_2012_151

Published Online: 15 September 2012

TGFβ	Transforming growth factor β
2D	Two-dimensional
3D	Three-dimensional
TEM	Transendothelial migration

1 Introduction

Cancer growth is a multistage complex process originating from molecular and genetic cell abnormalities [1, 2]. Data-driven mathematical/computational modeling has recently gained recognition for its potential to integrate the large volume of experimental data currently available, to simulate and analyze the behavior of complex biological systems, including cancer, and to optimize and predict clinical therapies and outcome [3–5]. Because cancer growth indeed spans multiple spatial and temporal biological scales (from genes and proteins to individual biological cells, and tissues, up to the entire organism) [6, 7], modeling of cancer across different biological scales, i.e., multiscale cancer modeling, that accounts for intracellular signaling dynamics, individual cell properties, and multicellular tumor growth environment is potentially more appropriate to predict cancer progression and development and generate experimental intervention strategies. Focusing on only one scale, as does the vast majority of current cancer models [7], simply neglects the correlative dependence and interplay between different scales. However, since a multiscale cancer model has to quantify parameters on, and relationships between biological processes that occur at different scales, the complexity of model development is significantly increased.

There are three main types of modeling approach currently employed in the cancer modeling community at large: continuum, discrete, and hybrid, and readers are referred to [8, 9] for a detailed discussion on each modeling approach. Briefly, continuum models benefit from the knowledge gained in fundamental physical principles [7], and are capable of capturing larger-scale volumetric tumor growth dynamics [10]. However, it is very difficult to use continuum models to explore heterogeneity in both the tumor and its surrounding microenvironment [11]. Discrete models can address these shortcomings, since they can work on the scale of individual cells or a cluster of cells [12]. Additionally, they can easily incorporate biological rules generated from biomedical data. However, a major drawback of discrete models is their compute intense nature due to the detail that each cell is modeled in, which often limits the model to a relatively small number of cells. For these reasons, hybrid modeling, i.e., the integration of both continuum and discrete descriptions, currently appears to be a more appealing approach in the cancer modeling field [6, 12].

Agent-based modeling (ABM) is a discrete-based hybrid modeling approach [13]. In an ABM, agents (often representing individual cells) interact or communicate with other immediate agents and their common microenvironment according to a set of pre-defined biologically inspired computational rules. Tightly coupling and iteratively refining these roles at all stages of model development with in vitro or in vivo experiments is important [14]. An ABM produces emergent behavior that arises from the behavior and interactions evolving at a cellular level. To date, the ABM has been used to simulate a variety of cancer aspects, such as somatic evolution in tumorigenesis [15, 16], the growth dynamics of multicellular tumor spheroids [17, 18], and cancer cell invasion [19, 20]. In this chapter, we focus on the introduction of the design and development of a specific type of ABM, i.e., molecular signaling incorporated multiscale ABM. This type of ABM is able to address the role of diversity in cell populations and also within each individual cell, and thus have the capacity to explore the relationship between molecular signaling properties and upper level cancer behavior.

2 Modeling Approach

Cancer is a complex disease involving a series of irreversible genomic changes that affect intrinsic cellular programs [2]. Wet-lab cancer researchers probably do not appreciate a model which misses correlations of molecular-level alterations with cancer cell properties, because it is essentially the aberration of signaling pathways that contributes to the initiation and progression of cancer [21]. Cancer is also a context-dependent disease [22], implicating that its progression behaviors depend on the microenvironment where the activities of cancer cells take place. More precisely, cancer cells bi-directionally communicate with their microenvironment, not only responding to various external cues but also impacting their surroundings, e.g., by producing various signals and degrading the neighboring tissue through proteases [23]. ABM is of particular interest to cancer modelers because, as we will introduce below, it allows researchers to explore how cancer growth and invasion properties (due to cell proliferation and migration) emerge as a result of individual dynamics, including cell–cell and cell-environment interactions and intracellular signaling of individual cells.

In the following, the design concept and development of the most recent molecular signaling-incorporated multiscale ABMs will be discussed. Special focus is given to the demonstration of how these two scales are explicitly linked by what algorithm to determine cell phenotypic transitions upon what molecular changes.

2.1 EGFR Signaling and Cellular Phenotypic Transition

A set of molecular-multicellular ABMs within brain tumors and non-small cell lung cancer (NSCLC) have been developed. These models (as reviewed in [4, 12, 24]) effectively ground and foster future models incorporating multiple scales from the molecular up to the cellular level and beyond.

2.1.1 EGFR Signaling

The epidermal growth factor receptor (EGFR) is mutated and overexpressed in many cancers, including brain cancer and lung cancer [25]. Epidermal growth factor (EGF) binds EGFR and promotes dimerization and subsequent autophosphorylation, resulting in the downstream activation of a number of key cell decision-making proteins such as phospholipase Cγ (PLCγ), extracellular signal-regulated kinase (ERK), and many others [26]. A number of EGFR-related pathway kinetic models have been developed [27–29], but regardless of differences in their complexity and scale (i.e., the number of molecular entities or molecular events), all of these models use mathematical kinetic equations to describe molecular interactions. The change in concentration of a certain protein pathway component over time is determined based on the following ordinary differential equation form:

$$\frac{d(X_i)}{dt} = \sum v_{\text{Production}} - \sum v_{\text{Consumption}}, \tag{1}$$

where X_i represents one of the pathway components; the change in concentration of X_i is the result of the reaction rates producing X_i minus the reaction rates consuming it. If the initial concentrations of pathway components or reaction rate constants are not yet available in the literature, their values either have to be investigated experimentally or are fitted to published time-dependent quantitative (or sometimes even qualitative) observations.

2.1.2 Microenvironment

A two-dimensional (2D) environment made up of a discrete lattice or a three-dimensional (3D) environment composed of a discrete cube are constructed to investigate tumor growth dynamics. Each grid point is occupied by a single cell or is empty. Heterogeneous environments are attained by distributing external diffusive chemical cues (such as growth factors, glucose, and oxygen) throughout the computational domain. Throughout a simulation run, the concentrations of the chemical cues are continuously diffused and updated at a fixed rate with partial differential equations (PDEs). Each cell has a self-maintained EGFR signaling network. As a simulation progresses, cells in distinct locations are likely to experience different external microenvironmental conditions. Thus, even though their internal states (including cell phenotype and concentrations of pathway components) are set to be identical initially, they will exhibit different phenotypes after a certain lapse of time due to their respective molecular changes.

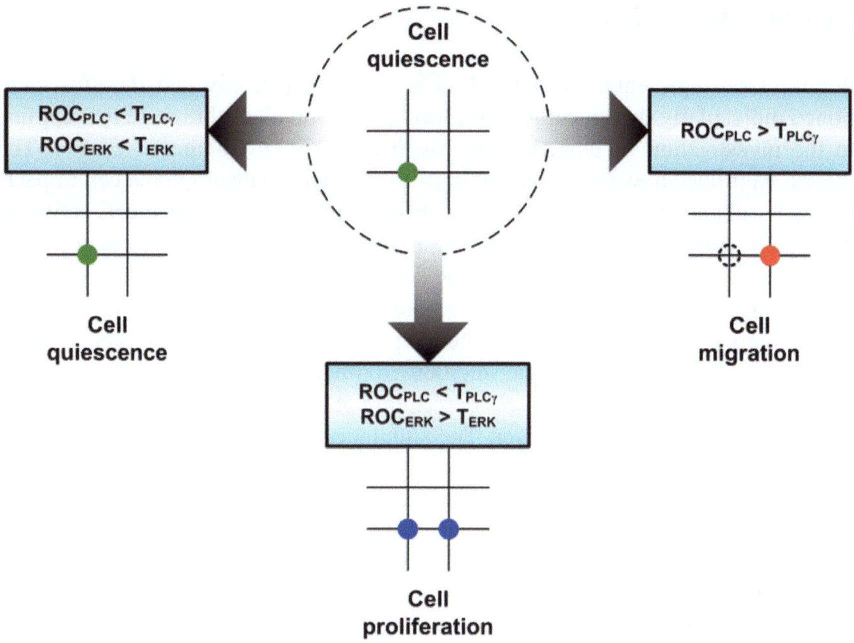

Fig. 1 Assume a cell's current phenotype is quiescence and the on-site nutrition is sufficient so the cell will not die. The cell's phenotype at the next step is determined as follows. The cell will remain quiescent if both $ROC_{PLC\gamma}$ and ROC_{ERK} remain below their corresponding thresholds (T_{PLC} and T_{ERK}, respectively); the cell will proliferate (and a new cell will then occupy an adjacent free location) if only ROC_{ERK} exceeds T_{ERK}; and the cell will migrate to an adjacent free location if $ROC_{PLC\gamma}$ (regardless of ERK) exceeds T_{PLC}

2.1.3 Molecularly Driven Cellular Phenotype Decision

PLCγ is known to be involved in directional cell movement in response to EGF [30] and is activated transiently in cancer cells, to a greater extent during migration and more gradually in the proliferation mode [31]. Derived from this finding, PLCγ is modeled as the decision molecule for determining a cell's migratory fate by comparing the current rate of change ($ROC_{PLC\gamma}$) to a pre-specified threshold T_{PLC}. That is, if ROC_{PLC} exceeds T_{PLC}, the cell then has the "potential" to migrate. However, a cell additionally has to meet other microenvironmental requirements, such as sufficient local nutrient conditions and available adjacent space, in order to process any phenotype transitions. If any of these conditions are not met, then the cell will have to remain in its current location, waiting for the next iteration in the simulation when conditions will be re-evaluated. Figure 1 schematically illustrates the cell phenotype decision algorithm. In addition to using PLCγ to determine the cell migration fate, ERK has been employed in making cell proliferation decision, also based on experimental evidence [32].

2.1.4 Modeling Examples

A simplified representation of the EGFR signaling pathway and the aforementioned molecularly-driven cellular phenotype decision algorithm were first applied to the investigation of brain tumor growth in a 2D environment [33, 34]. The models examined how the molecular profile of each individual glioma cell impacts the cell's phenotypic switch, and how such context-specific single-cell activities potentially affect the dynamics of the entire tumor system. In particular, the models found that increasing the EGFR density per cell results in an acceleration of the entire tumor system's spatiotemporal expansion dynamics [34], a finding that is in good agreement with published experimental data [35]. To simulate brain tumor growth in a more realistic microenvironment, a 3D model was developed [36], where a simplified cell-cycle description at the sub-cellular scale based on [37] was added to molecular layer. The simulation results not only confirmed the impact of regulation of EGFR signaling on tumor behavior (both on the single cell and multi-cellular level), but also indicated that over time, proliferative and migratory cell populations oscillate and have a direct effect on the entire spatiotemporal tumor expansion pattern. A recent extension study [38] further studied the emergence of heterogeneous tumor cell clones through introducing an element of genetic instability and analyzed how heterogeneity impacts brain tumor progression patterns. Simulation results showed that cell clones with higher EGFR density were comprised of a larger migratory fraction and smaller proliferative and quiescent fractions, which corresponds well with reported experimental data [39].

EGFR also plays an important role in progression and metastasis of NSCLC. Can the same modeling method be applied to NSCLC as well? A 2D model with a revised EGF-induced, EGFR-mediated pathway specific to NSCLC was developed to quantitatively understand the relationship between extrinsic chemotactic stimuli, the underlying properties of signaling networks, and the cellular biological responses they trigger in NSCLC from a systemic view [40]. In addition to confirming the experimentally known fact that increasing the amount of available growth factors leads to a spatially more aggressive cancer system [41, 42], the model found that in the cancer cell closest to the nutrient source, a minimal increase in EGF concentration can temporarily abolish its proliferative phenotype. More recently, the model was extended to a 3D case in which both EGF and transforming growth factor β (TGFβ) and their interplay were taken into account [43]. This physiologically and clinically motivated extension of the NSCLC modeling platform allowed for investigating how the effects of individual and combinatorial change in EGF and TGFβ concentrations at the molecular level alter tumor growth dynamics (including tumor volume and expansion rate) on the multi-cellular level. A particular region of tumor system stability, generated by unique pairs of EGF and TGFβ concentration variations, was discovered. Figure 2 shows the simulation results from changing EGF and TGFβ concentrations both simultaneously and asynchronously. As can be seen, the common stable phenotypic region is generated by [2-7]-fold variation of EGF and [0.3–3]-fold variation of TGFβ. This result indicates that when the variation-pair of EGF and TGFβ concentrations occurred

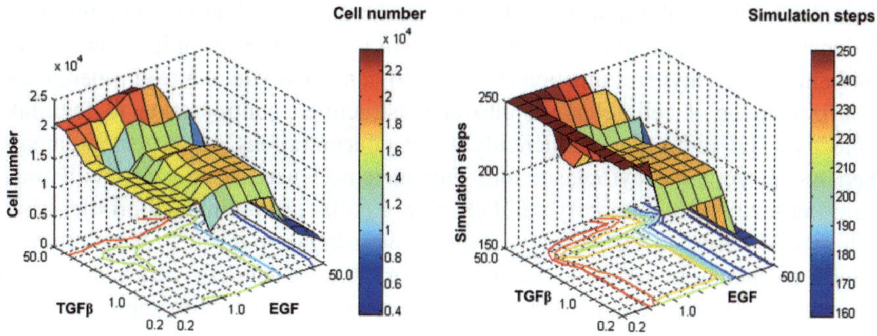

Fig. 2 The effects of asynchronous combinatorial change in EGF and TGFβ concentrations on tumor volume represented by cell number (*left* panel) and tumor expansion rate represented by inverse simulation step (*right* panel). In tumor volume evaluation, the largest tumor volume is reached under conditions of high TGFβ and low or standard (with a variation of 1.0-fold) EGF concentrations. However, in tumor expansion rate evaluation, the most aggressive tumor expansion rate (fewest simulation steps) occurs under conditions of high EGF, regardless of TGFβ concentrations. Reproduced with permission from [43]

within this region, changes caused by the two growth factors did not effectively transmit to the downstream activation cascade, potentially explaining the resulting robustness of the tumor system at the multi-cellular level. However, the tumor system becomes sensitive to external variations in EGF and/or TGFβ when they occur outside this region, processing a phenotypic switch once the microenvironment becomes more permissive.

2.2 *E-cadherin Signaling and Cell–Cell Adhesion*

E-cadherin mediates cell–cell adhesion and plays a critical role in the formation and maintenance of cell contact. E-cadherin mutation has been correlated with malignant transformation and invasive behavior [45], whereas increasing the E-cadherin expression level has been shown to reduce the growth of malignant cell lines [46]. Only recently, cancer modelers began to explicitly investigate cell–cell adhesion related intracellular signaling in a multicellular context to understand the possible effects of changes in E-cadherin on the growth characteristics of cancer cell populations [44, 47, 48].

2.2.1 E-cadherin Signaling

When a cell adheres to adjacent neighbors, E-cadherin binds to β-catenin to form a complex which can interact with neighboring cells to form bonds. When cells detach from one another, β-catenin is released into the cytoplasm, targeted for degradation, and at the same time, E-cadherin is sequestered into the cytoplasm by endocytosis. Upregulation of soluble β-catenin is related to cell migration and the epithelial-mesenchymal transition (EMT) [49], a process where a well ordered and polarized layer of cells changes into an unstructured configuration to facilitate collective cell migration. Sufficiently large concentrations of soluble β-catenin then move from the cytoplasm into the nucleus, where it interacts with transcription factors which modify cell behavior, e.g., by promoting cell proliferation. It has been observed that invasive cells show a higher nuclear accumulation of soluble β-catenin. In particular, as proposed in [50], β-catenin may translocate into the nucleus above a certain concentration threshold, leading to downregulation of E-cadherin-mediated adhesion.

2.2.2 Microenvironment

A lattice-free environment is constructed to explore linking the intracellular dynamics of E-cadherin and β-catenin interaction pathways, physical forces on the cells, and the extracellular microenvironment. Each cell is considered as an individual entity in which intracellular dynamics are governed by the a set of mass conservation chemical equations, using the same form of Eq. (1). A simplified β-catenin pathway which captures the key features of the cell adhesion process is implemented. Since cells in isolation tend to aggregate, it is assumed that an invasive cell can change into a noninvasive state again if it comes into contact with other cells. Cell movement is modeled by a stochastic equation of motion:

$$\gamma \bar{v}_i = \sum_{j nn i} \bar{F}_{ij} + \sum_{j nn i} \bar{F}^a_{ij} + \bar{f}_i(t), \tag{2}$$

where $\bar{v}_i$ is the velocity of the cell i at time t; $\bar{F}_{ij}$ and $\bar{F}^a_{ij}$ are the repulsive force and adhesive force of cell j on cell i, respectively, and the sums are over the nearest neighbors in contact with cell i; $\bar{f}_i(t)$ represents a noise term for cell i at time t; and γ is the cell-substrate friction constant. Interested readers are encouraged to refer to the original article [48] for more details on each term. Briefly, this governing equation accounts for the influence of forces and a random contribution to the locomotion which results from the local exploration of space. The adhesion forces between cells are controlled by the density of E-cadherin in the cell membrane within the cell–cell contact zone.

2.2.3 Cell Migration Determination upon Expression Levels of β-Catenin

As explained above, upregulation of soluble β-catenin is assumed to interact with transcription factors in the nucleus, and intend to induce cell migration [51]. In fact, both attachment and detachment of cells lead to an exchange of E-cadherin between the membranes in the contact zone of the interacting cells. The expression levels of β-catenin (denoted by [β]) can be used to determine a cell's migration potential by comparing with a threshold. Specifically, if [β] exceeds the threshold, the soluble β-catenin in cytoplasm is considered to be large enough, so it is free to enter the nucleus and interact with transcription factors, resulting in cell migration. Note that, the decision for a cell to migrate can be triggered in a number of different ways, all of them involving an upregulation of the soluble β-catenin which needs to exceed the pre-specified threshold. For example, the cytoplasmic concentration of β-catenin may be upregulated due to a failure in the proteasome system or detachment of local neighbors; in both cases, free β-catenin enters the nucleus and triggers cell migration.

2.2.4 Modeling Examples

A number of computational studies have been carried out to probe the relative importance of E-cadherin and β-catenin intracellular signaling properties in determining tumor tissue characteristics. A multiscale lattice-free ABM was first developed to study how cell adhesion may be regulated by the interactions between E-cadherin and β-catenin [48]. Simulation results showed that down-regulation of β-catenin can be mainly driven by cell–cell contacts, and EMT can be achieved depending on the regulation of soluble β-catenin by local contacts. The intra- and intercellular protein interactions that govern cell–cell adhesion combined with cellular physical properties are also the driving forces of an essential mechanism that a cancer cell uses to attach to the endothelial wall, i.e., transendothelial migration (TEM) [52]. In a subsequent study [47], the influence of different protein pathways in the achievement of TEM was investigated by adding the Src pathway to the molecular layer. Four cancer cell genotypes that differ in the adhesion protein pathways were considered. The genotypes were characterized by their capacity of creating N-cadherin-mediated bonds with the tunica intima and by their capacity of inducing a detachment of the endothelial–endothelial bonds by Src activity. Simulation results indicate that the slowest migration was found in the case when both N-cadherin and Scr were knocked out, while the fastest case occurred when both N-cadherin and Scr remained active.

In a more recent study from the same group [44], three cell–cell adhesion sub-pathways were proposed, and the influence of these pathways on tumor profiles was studied. Sub-pathway 1 considers bond formation as a cause of the interaction between β-catenin and E-cadherin; sub-pathway 2 considers in addition the degradation of the adhesion complex after β-catenin is phosphorylated by Src; and sub-pathway 3 considers intracellular interactions between Src, β-catenin, E-cadherin, and PI3. Model analysis finds that cells with sub-pathway 2 and slow synthesis rate of Src associated sub-pathway 3 generated the largest subpopulation due to an advantageous position close to the tumor border that permits them to more easily form clones of large size. Figure 3 shows the transversal section of

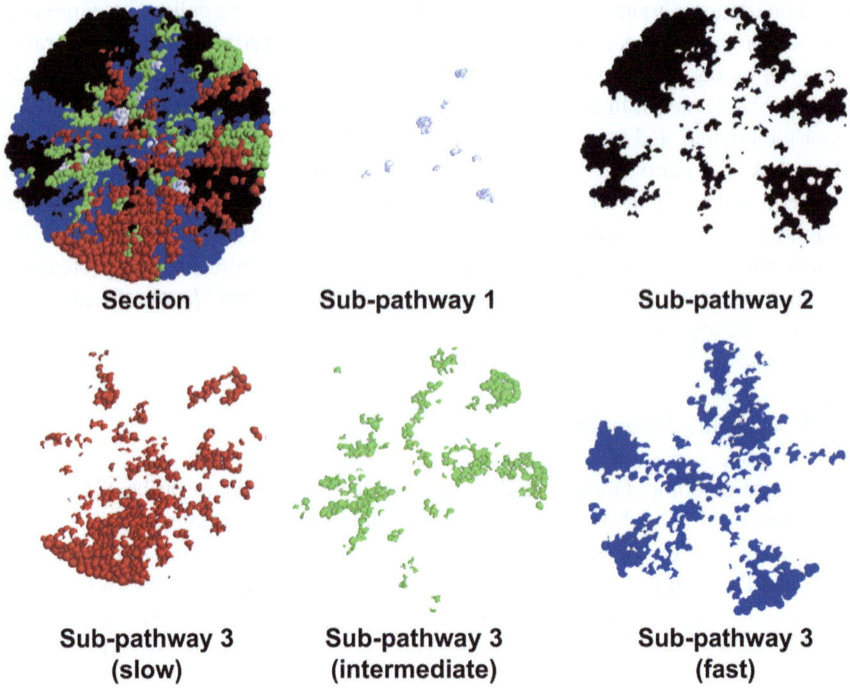

Fig. 3 A transversal section of a simulated tumor formed by different cell types. The external rim of the tumor is mostly formed by weak adhesive cell types (*black*, *red* and *blue*), while sub-pathway 1 cell and the sub-pathway 3 cell phenotype associated with intermediate Src synthesis rate (*white* and *green*) remain mostly at the interior. Slow and fast Src synthesis cells and sub-pathway 2 cells dominate the positions at the tumor border, outcompeting cell types with strong cell–cell adhesion (sub-pathway 1 and sub-pathway 3 with intermediate Src synthesis rate). Reproduced with permission from [44]

one simulation example with a tumor of 100,000 cells. This finding agrees well with an experimental study [53] where cell–cell adhesion behavior was found to be only possible at intermediate concentrations of Src, and when Src was absent or when it was at high concentrations, the cells were not able to express E-cadherin-mediated bonds. This model suggested that therapies eradicating Src may not be effective to prevent cancer invasion because sub-pathway 3 phenotypes associated with low Src expression levels were one of the most aggressive.

3 Discussion and Future Directions

We have reviewed the methods and achievements of a specific type of multiscale agent-based modeling which encompass molecular and multicellular scales. Although still at an early stage, this approach has demonstrated its ability to help understand the relationship between extracellular stimuli, intracellular signaling dynamics, and multicellular tumor growth. Simulation results can be validated with in vitro or in vivo experiments, or verified with other theoretical studies. For using this approach, an algorithm for linking molecular and multicellular scales is indispensable and sometimes needs experimentally supported creative thoughts. On the basis of the modeling works presented here, we argue that this type of ABM is highly suited to modeling complex emergent behaviors of cancerous systems, which are generated as an outcome of direct and indirect interactions between large numbers of individual cells.

The molecular-multicellular ABM not only enables the monitoring of multicellular dynamics in response to molecular changes, but also facilitates the tracking of the fate of molecular components per cell and cell cluster as the entire tumor system evolves. It is now possible to ascertain the cause of a specific tumor growth pattern at the multicellular level by exploring the time-course history of intracellular signaling profiles within individual cells. For example, the EMT process has been studied by using the ABMs presented in [48]; it is very difficult for an averaged population-based continuum model to study this process. It is also noteworthy that the molecular-multicellular ABM has been employed in other biomedical fields as well other than cancer, e.g., in epithelial cell study [54, 55] and in acute inflammation study [56–58], highlighting the promise of this type of ABM in translating mechanistic knowledge into an integrated experimental and computational framework.

There are a number of technical challenges in transitioning these ABMs to biomedical/clinical practice. These include the more common issues such as obtaining access to relevant data to validate simulation results and defining standards for model definitions. The most severe issue, however, is the compute intensity associated with these discrete-based hybrid models. In modeling cancer, it is generally accepted that the higher a model's spatial and temporal resolution, the higher its compute power demand [4]. ABMs are generally too detailed to simulate over a long period of time, particularly in a large, 3D domain. Parallelizing the code and then running the model on a cluster of supercomputers is a possible but not always practical solution that still may not resolve all the difficulties in handling the enormous amount of experimental and clinical data. We and others have begun to turn to *hybrid*, *multiscale* and *multi-resolution* modeling [6, 12, 59], where multi-resolution means that cells at distinct topographic regions are treated differently in terms of the modeling approach applied. This approach has the potential to achieve discretely high resolution wherever and whenever necessary to improve the model's predictive power, while at the same time reducing compute intensity as much as possible to support scalability of the approach to clinically relevant levels. By drawing on the strengths of the multi-resolution approach, and integrating it into the next generation ABM models with a hierarchy of processes at varying time and space scales, we can produce computationally efficient models to simulate tumor progression, predict treatment impact, and ultimately, be applicable in clinical practice.

Acknowledgments This work has been supported in part by National Institutes of Health (NIH) grant CA 113004. VC acknowledges the NIH for support through 1U54CA143837, 1U54CA143907, and 1U54CA149196, and the National Science Foundation (NSF) for support under grant DMS-0818104.

References

1. Al-Hajj, M., Clarke, M.F.: Self-renewal and solid tumor stem cells. Oncogene **23**(43), 7274–7282 (2004)
2. Balmain, A., Gray, J., Ponder, B.: The genetics and genomics of cancer. Nat. Genet. **33**(Suppl), 238–244 (2003)
3. Kitano, H.: Computational systems biology. Nature **420**(6912), 206–210 (2002)
4. Deisboeck, T.S., Zhang, L., Yoon, J., Costa, J.: In silico cancer modeling: is it ready for prime time? Nat. Clin. Pract. **6**(1), 34–42 (2009)
5. Schnell, S., Grima, R., Maini, P.K.: Multiscale modeling in biology—new insights into cancer illustrate how mathematical tools are enhancing the understanding of life from the smallest scale to the grandest. Am. Sci. **95**(2), 134–142 (2007)
6. Anderson, A.R., Quaranta, V.: Integrative mathematical oncology. Nat. Rev. Cancer **8**(3), 227–234 (2008). doi:10.1038/nrc2329, nrc2329 [pii]
7. Tracqui, P.: Biophysical models of tumour growth. Rep. Prog. Phys. **72**(5) (2009). doi:056701 10.1088/0034-4885/72/5/056701
8. Cristini, V., Lowengrub, J.S.: Multiscale Modeling of Cancer: An Integrated Experimental and Mathematical Modeling Approach. Cambridge University Press, Cambridge (2010)
9. Lowengrub, J.S., Frieboes, H.B., Jin, F., Chuang, Y.L., Li, X., Macklin, P., Wise, S.M., Cristini, V.: Nonlinear modelling of cancer: bridging the gap between cells and tumours. Nonlinearity **23**(1), R1–R91 (2010). doi:10.1088/0951-7715/23/1/r01
10. Schaller, G., Meyer-Hermann, M.: Continuum versus discrete model: a comparison for multicellular tumour spheroids. Philos. Trans. **364**(1843), 1443–1464 (2006)
11. Gatenby, R.A., Silva, A.S., Gillies, R.J., Frieden, B.R.: Adaptive therapy. Cancer Res. **69**(11), 4894–4903 (2009)
12. Wang, Z., Deisboeck, T.S.: Computational modeling of brain tumors: discrete, continuum or hybrid? Sci. Model. Simul. **15**, 381–393 (2008)
13. Bonabeau, E.: Agent-based modeling: methods and techniques for simulating human systems. Proc. Nat. Acad. Sci. USA. **99**(Suppl 3), 7280–7287 (2002)
14. Thorne, B.C., Bailey, A.M., Peirce, S.M.: Combining experiments with multi-cell agent-based modeling to study biological tissue patterning. Br. Bioinform. **8**(4), 245–257 (2007). 10.1093/bib/bbm024, bbm024[pii]
15. Abbott, R.G., Forrest, S., Pienta, K.J.: Simulating the hallmarks of cancer. Artif. Life **12**(4), 617–634 (2006)
16. Spencer, S.L., Gerety, R.A., Pienta, K.J., Forrest, S.: Modeling somatic evolution in tumorigenesis. PLoS Comput. Biol. **2**(8), e108 (2006)
17. Pepper, J.W., Sprouffske, K., Maley, C.C.: Animal cell differentiation patterns suppress somatic evolution. PLoS Comput. Biol. **3**(12), e250 (2007)
18. Schaller, G., Meyer-Hermann, M.: Multicellular tumor spheroid in an off-lattice voronoi-delaunay cell model. Phys. Rev. **71**(5 Pt 1), 051910 (2005)
19. Pearce, I.G., Chaplain, M.A., Schofield, P.G., Anderson, A.R., Hubbard, S.F.: Chemotaxis-induced spatio-temporal heterogeneity in multi-species host-parasitoid systems. J. Math. Biol. **55**(3), 365–388 (2007)
20. Ramis-Conde, I., Chaplain, M.A., Anderson, A.R.: Mathematical modelling of cancer cell invasion of tissue. Math. Comp. Model. **47**, 533–545 (2006)

21. Hlavacek, W.S., Faeder, J.R., Blinov, M.L., Posner, R.G., Hucka, M., Fontana, W.: Rules for modeling signal-transduction systems. Sci. STKE. **2006**(344), re6 (2006)
22. Postovit, L.M., Seftor, E.A., Seftor, R.E., Hendrix, M.J.: Influence of the microenvironment on melanoma cell fate determination and phenotype. Cancer Res. **66**(16), 7833–7836 (2006)
23. Hendrix, M.J., Seftor, E.A., Seftor, R.E., Kasemeier-Kulesa, J., Kulesa, P.M., Postovit, L.M.: Reprogramming metastatic tumour cells with embryonic microenvironments. Nat. Rev. Cancer **7**(4), 246–255 (2007). 10.1038/nrc2108, nrc2108 [pii]
24. Zhang, L., Wang, Z., Sagotsky, J.A., Deisboeck, T.S.: Multiscale agent-based cancer modeling. J. Math. Biol. **58**(4–5), 545–559 (2009)
25. Oda, K., Matsuoka, Y., Funahashi, A., Kitano, H.: A comprehensive pathway map of epidermal growth factor receptor signaling. Mol. Syst. Biol. **1**(2005), 0010 (2005)
26. Friedl, P., Wolf, K.: Tumour-cell invasion and migration: diversity and escape mechanisms. Nat. Rev. Cancer **3**(5), 362–374 (2003)
27. Hatakeyama, M., Kimura, S., Naka, T., Kawasaki, T., Yumoto, N., Ichikawa, M., Kim, J.H., Saito, K., Saeki, M., Shirouzu, M., Yokoyama, S., Konagaya, A.: A computational model on the modulation of mitogen-activated protein kinase (MAPK) and Akt pathways in heregulin-induced ErbB signalling. Biochem. J. **373**(Pt 2), 451–463 (2003)
28. Kholodenko, B.N., Demin, O.V., Moehren, G., Hoek, J.B.: Quantification of short term signaling by the epidermal growth factor receptor. J. Biol. Chem. **274**(42), 30169–30181 (1999)
29. Schoeberl, B., Eichler-Jonsson, C., Gilles, E.D., Muller, G.: Computational modeling of the dynamics of the MAP kinase cascade activated by surface and internalized EGF receptors. Nat. Biotechnol. **20**(4), 370–375 (2002)
30. Mouneimne, G., Soon, L., DesMarais, V., Sidani, M., Song, X., Yip, S.C., Ghosh, M., Eddy, R., Backer, J.M., Condeelis, J.: Phospholipase C and cofilin are required for carcinoma cell directionality in response to EGF stimulation. J. Cell Biol. **166**(5), 697–708 (2004)
31. Dittmar, T., Husemann, A., Schewe, Y., Nofer, J.R., Niggemann, B., Zanker, K.S., Brandt, B.H.: Induction of cancer cell migration by epidermal growth factor is initiated by specific phosphorylation of tyrosine 1248 of c-erbB-2 receptor via EGFR. Faseb. J. **16**(13), 1823–1825 (2002)
32. Santos, S.D., Verveer, P.J., Bastiaens, P.I.: Growth factor-induced MAPK network topology shapes Erk response determining PC-12 cell fate. Nat. Cell Biol. **9**(3), 324–330 (2007)
33. Athale, C., Mansury, Y., Deisboeck, T.S.: Simulating the impact of a molecular decision-process on cellular phenotype and multicellular patterns in brain tumors. J. Theor. Biol. **233**(4), 469–481 (2005)
34. Athale, C.A., Deisboeck, T.S.: The effects of EGF-receptor density on multiscale tumor growth patterns. J. Theor. Biol. **238**(4), 771–779 (2006)
35. Lund-Johansen, M., Bjerkvig, R., Humphrey, P.A., Bigner, S.H., Bigner, D.D., Laerum, O.D.: Effect of epidermal growth factor on glioma cell growth, migration, and invasion in vitro. Cancer Res. **50**(18), 6039–6044 (1990)
36. Zhang, L., Athale, C.A., Deisboeck, T.S.: Development of a three-dimensional multiscale agent-based tumor model: simulating gene-protein interaction profiles, cell phenotypes and multicellular patterns in brain cancer. J. Theor. Biol. **244**(1), 96–107 (2007)
37. Alarcon, T., Byrne, H.M., Maini, P.K.: A mathematical model of the effects of hypoxia on the cell-cycle of normal and cancer cells. J. Theor. Biol. **229**(3), 395–411 (2004)
38. Zhang, L., Strouthos, C.G., Wang, Z., Deisboeck, T.S.: Simulating brain tumor heterogeneity with a multiscale agent-based model: linking molecular signatures, phenotypes and expansion rate. Math. Comput. Model. **49**(1–2), 307–319 (2009)
39. Steinbach, J.P., Klumpp, A., Wolburg, H., Weller, M.: Inhibition of epidermal growth factor receptor signaling protects human malignant glioma cells from hypoxia-induced cell death. Cancer Res. **64**(5), 1575–1578 (2004)
40. Wang, Z., Zhang, L., Sagotsky, J., Deisboeck, T.S.: Simulating non-small cell lung cancer with a multiscale agent-based model. Theor. Biol. Med. Model. **4**(1), 50 (2007)
41. Price, J.T., Wilson, H.M., Haites, N.E.: Epidermal growth factor (EGF) increases the in vitro invasion, motility and adhesion interactions of the primary renal carcinoma cell line, A704. Eur. J. Cancer **32A**(11), 1977–1982 (1996)

42. Xue, C., Wyckoff, J., Liang, F., Sidani, M., Violini, S., Tsai, K.L., Zhang, Z.Y., Sahai, E., Condeelis, J., Segall, J.E.: Epidermal growth factor receptor overexpression results in increased tumor cell motility in vivo coordinately with enhanced intravasation and metastasis. Cancer Res. **66**(1), 192–197 (2006)
43. Wang, Z., Birch, C.M., Sagotsky, J., Deisboeck, T.S.: Cross-scale, cross-pathway evaluation using an agent-based non-small cell lung cancer model. Bioinformatics **25**, 2389–2396 (Oxford, England) (2009)
44. Ramis-Conde, I., Drasdo, D.: From genotypes to phenotypes: classification of the tumour profiles for different variants of the cadherin adhesion pathway. Phys. Biol. **9**(3), 036008 (2012). doi:10.1088/1478-3975/9/3/036008
45. Conacci-Sorrell, M., Zhurinsky, J., Ben-Ze'ev, A.: The cadherin-catenin adhesion system in signaling and cancer. J. Clin. Investig. **109**(8), 987–991 (2002). doi:10.1172/JCI15429
46. Foty, R.A., Steinberg, M.S.: Cadherin-mediated cell–cell adhesion and tissue segregation in relation to malignancy. Int. J. Dev. Biol. **48**(5–6), 397–409 (2004). doi:10.1387/ijdb.041810rf
47. Ramis-Conde, I., Chaplain, M.A., Anderson, A.R., Drasdo, D.: Multi-scale modelling of cancer cell intravasation: the role of cadherins in metastasis. Phys. Biol. **6**(1), 16008 (2009)
48. Ramis-Conde, I., Drasdo, D., Anderson, A.R., Chaplain, M.A.: Modeling the influence of the E-cadherin-beta-catenin pathway in cancer cell invasion: a multiscale approach. Biophys. J. **95**(1), 155–165 (2008)
49. Christiansen, J.J., Rajasekaran, A.K.: Reassessing epithelial to mesenchymal transition as a prerequisite for carcinoma invasion and metastasis. Cancer Res. **66**(17), 8319–8326 (2006). 10.1158/0008-5472.CAN-06-0410, 66/17/8319 [pii]
50. Kemler, R., Hierholzer, A., Kanzler, B., Kuppig, S., Hansen, K., Taketo, M.M., de Vries, W.N., Knowles, B.B., Solter, D.: Stabilization of beta-catenin in the mouse zygote leads to premature epithelial-mesenchymal transition in the epiblast. Development **131**(23), 5817–5824 (2004). doi:10.1242/dev.01458
51. Jankowski, J.A., Bruton, R., Shepherd, N., Sanders, D.S.: Cadherin and catenin biology represent a global mechanism for epithelial cancer progression. Mol. Pathol. **50**(6), 289–290 (1997)
52. Qi, J., Wang, J., Romanyuk, O., Siu, C.H.: Involvement of Src family kinases in N-cadherin phosphorylation and beta-catenin dissociation during transendothelial migration of melanoma cells. Mol. Biol. Cell **17**(3), 1261–1272 (2006). 10.1091/mbc.E05-10-0927, E05-10-0927 [pii]
53. McLachlan, R.W., Kraemer, A., Helwani, F.M., Kovacs, E.M., Yap, A.S.: E-cadherin adhesion activates c-Src signaling at cell–cell contacts. Mol. Biol. Cell **18**(8), 3214–3223 (2007). doi:10.1091/mbc.E06-12-1154
54. Walker, D., Wood, S., Southgate, J., Holcombe, M., Smallwood, R.: An integrated agent-mathematical model of the effect of intercellular signalling via the epidermal growth factor receptor on cell proliferation. J. Theor. Biol. **242**(3), 774–789 (2006)
55. Walker, D.C., Georgopoulos, N.T., Southgate, J.: From pathway to population–a multiscale model of juxtacrine EGFR-MAPK signalling. BMC Syst. Biol. **2**, 102 (2008)
56. An, G.: Introduction of an agent-based multi-scale modular architecture for dynamic knowledge representation of acute inflammation. Theor. Biol. Med. Model. **5**, 11 (2008)
57. An, G.: A model of TLR4 signaling and tolerance using a qualitative, particle-event-based method: introduction of spatially configured stochastic reaction chambers (SCSRC). Math. Biosci. **217**(1), 43–52 (2009)
58. An, G.C., Faeder, J.R.: Detailed qualitative dynamic knowledge representation using a BioNetGen model of TLR-4 signaling and preconditioning. Math. Biosci. **217**(1), 53–63 (2009)
59. Sanga, S., Frieboes, H.B., Zheng, X., Gatenby, R., Bearer, E.L., Cristini, V.: Predictive oncology: a review of multidisciplinary, multiscale in silico modeling linking phenotype, morphology and growth. NeuroImage **37**(Suppl 1), S120–S134 (2007)

Author Index

Stud Mechanobiol Tissue Eng Biomater (2013) 14: 395–396
DOI: 10.1007/978-3-642-36482-2

MIX
Papier aus verantwortungsvollen Quellen
Paper from responsible sources
FSC® C105338

If you have any concerns about our products,
you can contact us on
ProductSafety@springernature.com

In case Publisher is established outside the EU,
the EU authorized representative is:
Springer Nature Customer Service Center GmbH
Europaplatz 3, 69115 Heidelberg, Germany

Printed by Libri Plureos GmbH
in Hamburg, Germany